# OPTICAL MULTI-BOUND SOLITONS

# Optics and Photonics

*Series Editor*

**Le Nguyen Binh**

*Huawei Technologies, European Research Center, Munich, Germany*

# OPTICAL MULTI-BOUND SOLITONS

## Le Nguyen Binh

European Research Center
Huawei Technologies
Munich, Germany

CRC Press
Taylor & Francis Group
Boca Raton London New York

CRC Press is an imprint of the
Taylor & Francis Group, an **informa** business

CRC Press
Taylor & Francis Group
6000 Broken Sound Parkway NW, Suite 300
Boca Raton, FL 33487-2742

First issued in paperback 2017

© 2016 by Taylor & Francis Group, LLC
CRC Press is an imprint of Taylor & Francis Group, an Informa business

No claim to original U.S. Government works

ISBN-13: 978-1-4822-3763-4 (hbk)
ISBN-13: 978-1-138-74962-7 (pbk)

### Library of Congress Cataloging-in-Publication Data

Binh, Le Nguyen.
    Optical multi-bound solitons / Le Nguyen Binh.
        pages cm. -- (Optics and photonics ; 13)
    Includes bibliographical references and index.
    ISBN 978-1-4822-3763-4
    1. Solitons. 2. Wave packets. 3. Photonics. 4. Energy transfer. I. Title.

QC174.26.W28B55 2016
530.12'4--dc23                                                          2015007179

**Visit the Taylor & Francis Web site at**
**http://www.taylorandfrancis.com**

**and the CRC Press Web site at**
**http://www.crcpress.com**

*To my parents*
*To Phuong and Lam*

# Contents

# Preface

The rates of competition of limited resources from different subsystems of a generic system would lead to chaotic states. This could happen in natural systems as well as in engineering systems, in particular nonlinear photonic ones. In energy storage systems, if there is more than one storage element or subsystem, then competition for energy would occur, and when they try to extract from or deposit their energy into each other, chaotic, bifurcation, or both phenomena would occur. The system can thus behave in a nonlinear manner, and hence the term "nonlinear systems." However, controlled phases of lightwave carriers, which transport pulse envelopes around a fiber ring or the nonlinear phase due to high pulse energy, can balance linear dispersive effects. These are called solitons. Solitons can split into several subsolitons in the time domain, which can then be bound together when their phases satisfy a difference of $\pi$ or a multiple of $\pi$ rads. The overlapping tails of these solitons do not suffer any dispersive effects and thus keep these solitons bounded. This binding phenomenon can repeat to generate higher-order bound solitons if the energy of the system is high enough to supply optical power into these higher-order solitons for them to keep splitting and binding with each other. The highest order of multi-bound solitons that we have achieved experimentally is sixth as we could not supply more optical power into the system or were limited by the nonlinear breakdown of optical guided media.

The most commonly known energy storage systems are those containing capacitors and inductors as charging and discharging elements, and their counterparts in photonics are ring resonators incorporating optical amplification to compensate for the optical loss. Thus, the loss represents the resistance in optical circuits, the resonance represents the inductance and capacitance, and the amplification represents the gain. When the gain equals the loss, we do observe complex conjugations in optical circuits, and hence lasing. An optical modulator manipulates the phases of the lightwaves and hence the formation of resonating pulse sequence, solitons, and splitting into multiple solutions and self phase binding to form multiple-bound soliton states. Unlike in electrical circuits, the feedback in the optical domain is positive if both the feedback and the optical signals received have the same phases; if the phase is opposite then the feedback is negative. Hence a negative feedback mechanism is necessary for the stable generation of optical pulse sequences.

Chaotic and frequency doubling states can be created during these transition states in such nonlinear optical systems.

Higher-order differential equations can be employed to represent the dynamics of the evolution of the amplitude of current or voltage or field amplitudes of the lightwaves in such systems. This book deals with nonlinear systems in terms of fundamental principles and associated phenomena as well as their applications in signal processing in contemporary optical systems for communications and laser systems with a touch of mathematical representation of nonlinear equations, which provide some insight into the nonlinear dynamics at different phases.

The chapters in this book are classified into three parts: (1) Fundamental understanding of solitons and the formation of multi-bound solitons (Chapters 2 through 5), where sequences of single soliton as well as multiple numbers of solitons assembled in a group, the multi-bound solitons, are demonstrated. (2) Nonlinear chaotic and frequency higher-order multiplication states in fiber ring lasers (Chapters 6 and 7); bispectrum nonlinear processing technique is described for the representation of the amplitude and nonlinear phases of multi-bound solitons (Chapter 5). (3) Transmission of multi-bound solitons over guided optical medium (Chapter 8). In addition, five appendices are given to supplement a number of essential definitions, mathematical representations, and derivations.

The motivation and material for this book were provided mainly by advanced research works over the years in various Australian universities, at Nanayang Technological University of Singapore, at the Advanced Technology Research Laboratories of Siemens, at Huawei's European Research Center, and at Nortel Networks, where the author spent several fruitful years. The research endeavors of many people have contributed significantly to some sections of this book, especially the author's research doctoral graduates over the years, in particular the following scholars: Dr. Lam Quoc Huy, Dr. Wenn Jing Lai, Dr. Nam Quoc Ngo, and Dr. Nguyen Duc Nhan, and Professor D.Y. Tang and his former research fellow Dr. L.M. Zhao of the Singapore Nanyang Technological University. The author thanks Ashley Gasque of CRC Press for her encouragement toward the formation of this book.

**Le Nguyen Binh, Dr Eng**
*Muenchen, Deutschland*

MATLAB® is a registered trademark of The MathWorks, Inc. For product information, please contact:

The Math Works, Inc.
3 Apple Hill Drive
Natick, MA 01760-2098 USA
Tel: 508-647-7000
Fax: 508-647-7001
E-mail: info@mathworks.com
Web: www.mathworks.com

# Author

**Le Nguyen Binh** received his BE (Hons) and PhD in electronic engineering and integrated photonics in 1975 and 1978, respectively, both from the University of Western Australia. He is currently with Huawei's European Research Center as technical director in Munich, Germany.

He has authored and coauthored more than 300 papers in learned journals and refereed conferences and 8 books in the field of photonic signal processing, digital optical communications, and integrated optics, all published by CRC Press. His current research interests are in advanced modulation formats for long-haul optical transmission, electronic equalization techniques for optical transmission systems, ultrashort pulse lasers and photonic signal processing, optical transmission systems and network engineering, and Si-on-SiO$_2$ integrated photonics, especially the 5G optical switched flexible grid wireless-optical networks. He was chair of Commission D (Electronics and Photonics) of the National Committee for Radio Sciences of the Australian Academy of Sciences (1995–2005). Dr. Binh was professorial fellow at Nanyang Technological University of Singapore and the Christian-Albrechts-Universität zu Kiel of Germany and numerous Australian universities.

Dr. Binh has been awarded three Huawei Technologies gold medals for his work on advanced optical communication technologies.

He is an alumnus of the Phan Chu Trinh and Phan Boi Chau high schools of Phan Thiet Vietnam.

Currently, he is the series editor of "Optics and Photonics" for CRC Press.

# 1 Introduction

Optical solitons and multiple or multi-bound solitons have been the subject of research over the last three decades with a view that their pulse shapes are nondistorted over long transmission distances. The formation of multi-bound solitons relies on the phase differences of the carriers between adjacent solitons. Hence it requires control of the phase modulation of the carrier.

The generations of these solitons are thus important, in the first instance, for such potential applications and are the main focus of this book. The nonlinear phase distortion compensates for the linear part in an optical resonant cavity. This chapter introduces the main conceptual understanding of these solitonic pulses, especially their formations by passive or active mode-locking in recirculating resonant fiber lasers.

## 1.1 ULTRASHORT PULSE AND MULTI-BOUND SOLITONS

The science and technology of ultrashort pulse lasers have reached the stage of rapid evolution. They are the subjects of ultrafast optics research over a few decades since the invention of laser in the early 1960s. Optical pulses generated from ultrashort pulse lasers can be in the range from femtoseconds to picoseconds, which have been used in several applications. Due to their availability in the market and their improved performance, applications of these ultrashort pulse sources can range from testing and measurement applications [1] to high technology applications such as electronics and biomedicine [2–4]. Ultrashort pulse lasers have increasingly expanded their applications because of their useful characteristics such as high peak power, broad bandwidth, and high temporal and spatial resolution. Because various laser systems generate ultrashort pulses of different characteristics, each system can be suitably used in one or more applications. For example, the ultrashort pulse laser with high peak power is preferably used in nonlinear microscopy [5] and micromachining and marking [6,7], which are useful techniques in many industrial areas such as electronics, medicine, automotive industry, and many others. Frequency metrology and optical spectroscopy require laser systems with ultrashort pulses and long-term coherence [8,9]. While strict requirements such as high repetition rate, low noise, and low jitter are set by telecommunication systems, most of the research activities on ultrashort pulse lasers have been accomplished to target toward telecommunication applications. On the other hand, optical fiber communication has driven the development of ultrashort pulse sources.

In the telecommunication field, ultrashort pulse lasers have become a key component in various subsystems employed in optical fiber communication systems. The progress in telecommunication systems has led to the increasing use of ultrashort pulse lasers. First, ultrashort pulse lasers with the ability to transform limited hyperbolic secant pulses at rates from 10 to 40 Gbps can be employed in ultralong-haul transmission systems based on soliton transmission technology. Soliton transmission distances up to 10,000 km at 80 Gbps and 70,000 km at 40 Gbps have been proven in the laboratory [10,11]. With a pulse width of 200 fs at 10 GHz, an ultrahigh speed transmission at 1.28 and 5.1 Tbps has been demonstrated in optical time-division multiplexing (OTDM) systems [12,13]. Moreover, ultrashort pulse lasers also prove their importance in wavelength-division multiplexing (WDM) systems in which a broad bandwidth of ultranarrow pulses are sliced into many bands or wavelength channels. For those systems, a super-continuum source can be built by a chirped short pulse laser, which is passed through a special fiber such as a dispersion-decreasing fiber or a photonic crystal fiber. A flat optical spectrum with a bandwidth exceeding 150 nm can be obtained by using this technique [14]; thus it is possible to produce thousands of wavelength channels using spectral slicing techniques. The broadband coherence of the ultrashort pulse laser that is a specific and well-defined frequency-dependent phase relation across the spectrum can be used in encoding various types of optical data in optical code-division multiple accesses (O-CDMA) [15]. Besides the employment in optical data generators, the ultrashort pulse laser has become an indispensable component for optical signal processing. Many processing functions such as OTDM, demultiplexing, optical sampling, and optical analogue-to-digital converter (ADC) require high-performance ultrashort pulse lasers.

The classification of ultrashort pulse laser usually depends on the gain medium that includes solid-state, semiconductor, and fiber lasers. Each category has its own advantages and disadvantages. On the other hand, they have different performance parameters to be able to adapt to a specific application. In optical telecommunications, semiconductor and fiber lasers are the most commonly used for the generation of ultrashort pulses using some techniques such as mode-locking, gain switching, or externally modulating. Although semiconductor lasers offer a good engineering solution, mode-locked fiber lasers have undergone drastic development for optical communication systems. The reason these lasers have remained a research interest is that they are based on the fiber that offers a lot of advantages such as simple doping procedures, low loss, and the possibility of pumping with compact, efficient diodes. The availability of various fiber components minimizes the need for bulk optics and mechanical alignment that is very important to reduce the complexity as well as the size and the cost of the laser system for practical use. Moreover, research interests focused on mode-locked fiber lasers come from the fact that they can directly produce hyperbolic secant pulses or optical solitons. Soliton pulses are desirable in ultralong-haul transmission systems because they remain stable against perturbations of transmitting medium and keep the pulse shape undistorted during propagation. There has been a significant advance in developing mode-locked fiber lasers to reduce the size and cost as well as to improve their performance. Although the techniques in mode-locked fiber lasers have matured practical applications today, they have still been a rich research field. Current research efforts focus on exploring new properties in various operation schemes, especially in highly nonlinear regime, of generated short pulses for future potential applications.

## 1.2  MODE-LOCKED FIBER LASERS AS SOLITON AND MULTI-BOUND SOLITON GENERATORS

Mode-locking is the most popular technique to generate ultrafast pulses in fiber systems. When hundreds or thousands of longitudinal oscillation modes in a fiber cavity are forced in phase, short pulses are formed. The pulse width depends on the number of phase-locked longitudinal modes or is inversely proportional to the spectral bandwidth of the generated pulses. Like any mode-locked laser, mode-locked fiber lasers can be classified into two basic categories: active and passive sources.

Passively mode-locked fiber lasers perform mode-locking based on the exploitation of some optical effects in an active fiber cavity without using any external driving signal. On the other hand, the short pulses are generated by the loss modulation of the passive elements, which also determine the physical mechanism of passive mode-locking in the fiber cavity. There are several typical configurations of passively mode-locked fiber lasers that are based on mechanisms such as saturable absorption, nonlinear amplification loop mirror (NALM), and nonlinear polarization evolution [16–18]. Pulse shaping in all passive mode-locking techniques is commonly based on the intensity-dependent discrimination where the peak of optical pulse acquires the lowest loss while the wings of the pulse experience a higher loss per roundtrip in the fiber cavity. After many roundtrips, the pulse is shortened until its bandwidth is comparable to the gain bandwidth of the medium. The reduced gain in spectral wings then provides a broadening mechanism that stabilizes the pulse width to a specific value. Because the nonlinear amplification loop mirror (NALM) and nonlinear polarization rotation (NPR) are considered fast saturable absorption processes, the width of generated pulses based on these mechanisms is often very narrow that can be of the order of sub-100 fs. Furthermore, it has been theoretically and experimentally shown that a ring configuration of mode-locked fiber laser initiates more easily at lower threshold [19,20].

Although the passively mode-locked fiber lasers can easily generate very short pulses with high peak power, the repetition rate of generated pulse train is limited only at scale of the order of 100 MHz. The reason for this limited repetition rate is the length of the fiber cavity that determines the fundamental frequency is of the order of meters. This limitation prevents passively mode-locked fiber lasers to be used in high-speed transmission applications. To increase the repetition rate in these fiber lasers, a harmonic mode-locking, where multiple pulses coexist inside the fiber cavity, can be employed. However, the generated pulses in this scheme experience amplitude and timing fluctuations that are unable to meet strict requirements in communication systems. Therefore, active mode-locking is an alternative technique to generate mode-locked pulses for communication-oriented applications.

Unlike passive mode-locking, active mode-locking requires an active element driven by an external electronic signal to modulate the loss inside the fiber cavity. Although a bulk modulator can be used for mode-locking, most actively mode-locked fiber lasers use the integrated electro-optic (EO) LiNbO$_3$ modulators for mode-locking. There are two types of modulators, amplitude modulator and phase modulator, which correspond to two active mode-locking mechanisms: amplitude modulation (AM) and frequency modulation (FM). A typical configuration of

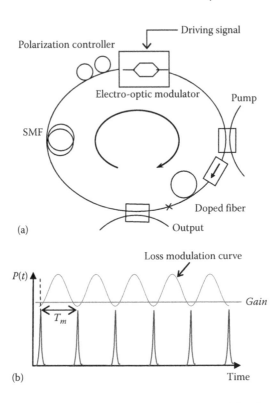

**FIGURE 1.1** Typical configurations of an actively mode-locked fiber laser: (a) A fiber ring configuration and (b) pulse formation in active mode-locking. $P(t)$ instant power.

an active mode-locked fiber laser using an amplitude modulator or a phase modulator for shaping a pulse in the cavity is shown in Figure 1.1a. To provide a high-speed pulse train, the active modulator is often driven at harmonics of the fundamental mode spacing. In AM mode-locking, the loss of cavity is periodically modulated and the pulse is built up at the positions having minimum loss in the fiber cavity as described in Figure 1.1b. FM mode-locking is carried out with repeated up or down frequency chirping and after many roundtrips the pulse is built up and shortened because the chirped light is swept out of the gain bandwidth of the laser. There are differences in characteristics between these two types of actively mode-locked fiber lasers. Recent studies have shown the superiority of active FM mode-locked fiber lasers compared to AM mode-locked fiber lasers to become a valuable candidate for ultrashort pulse generation at very high speeds. First, a phase modulator is more attractive than an amplitude modulator because it has no DC bias that can avoid the DC bias drift of the modulator. Second, the pulses obtained from the FM mode-locked fiber laser are generally shorter than those obtained from the AM mode-locked fiber laser due to the pulse compression induced by group velocity dispersion (GVD) and chirping from phase modulation [21]. The performance of mode-locked pulses from the FM mode-locked fiber laser can be improved further in optimal combination between GVD, phase modulation, and nonlinear effects in the design of the fiber cavity [22]. Furthermore, when FM mode-locked fiber lasers use the rational harmonic mode-locking technique to increase repetition rate, the output pulses do not experience the problem of unequal amplitudes due to the unique property of phase modulation [23]. Some other advantages such as the reduction of timing jitter or smaller quantum limited jitter also make the FM mode-locked fiber laser more attractive in very high speed optical communication systems [24].

However, the generation of stable short pulse train in the long term to achieve ultrahigh-speed optical communication is still a challenge. Because an active mode-locked fiber laser often operates in harmonic mode-locking regime, stability of pulse train is always a concern in the design of these fiber lasers. Hence if the modulation frequency is not equal to a harmonic of the fundamental frequency, the phases of the cavity modes are not locked and consequentially unstable pulsing occurs. The polarization states, the cavity length or the fundamental frequency change is sensitive to the temperature variations and/or mechanical vibrations and this causes a mismatch between the modulation frequency and the cavity mode. On the other hand, it is difficult to maintain an optimum operation condition in a long term. A solution for this problem is to utilize a regenerative mode-locking technique that uses a feedback loop to adjust the phase between the pulse and the clock signal driving the modulator [25,26]. To stabilize the repetition rate at a fixed frequency, the phase locking can be carried out by controlling the fiber cavity length with a fiber stretcher [27–29]. These stabilization techniques really have made the actively FM mode-locked fiber laser more attractive and more feasible for practical optical fiber transmission applications.

Although the actively mode-locked fiber laser can produce short pulses at speeds of 10–40 GHz, the pulse width is only of the order of picoseconds. The generation of femtosecond pulses in the GHz region is especially important in ultrahigh-speed transmission systems using OTDM technique. Therefore, it would be desirable to produce a femtosecond pulse train at a repetition rate of 10–40 GHz with a conventional high-speed modulator. Hence there are some techniques to compress the mode-locked pulses from mode-locked fiber lasers. A promising technique is to use hybrid mode-locking, which combines passive and active mode-locking in a fiber laser using phase modulator [30]. In this hybrid scheme, the mechanism of passive mode-locking such as NPR is used to further shorten the pulse width. The asynchronously driven phase modulator is treated as a polarizing element, which is responsible for both timing of pulse and shortening the pulse through NPR. This configuration has been used to produce pulse train with 816 fs pulse width at a repetition rate of 10 GHz repetition rate [31], and a shorter pulse width of 400 fs with a repetition rate from 2.5 to 12 GHz has been recently reported when additionally using a dispersion-shifted fiber (DSF) to compensate the chirp of the output pulses [32].

Another promising technique is the adiabatic soliton pulse compression through dispersion-decreasing fibers. The mechanism of the compression is based on the principle of keeping the soliton energy constant. Thus when a soliton propagates through a fiber with decreasing the GVD, the pulse width of the soliton automatically shortens corresponding to the ratio between the decrease in the dispersion and the original dispersion. By using this technique, a generation of 54 fs pulse train at

10 GHz has been successfully demonstrated recently [33] and it has been used to realize the first 1.28 Tbps OTDM transmission experiment [12]. However, when the pulse width is shortened to femtosecond scale, more effects such as higher-order dispersion, polarization mode dispersion, and the finite nonlinear response time of the fiber are impossibly negligible. They limit the compression ratio as well as the performance of shortened pulses.

## 1.3 NONLINEAR EFFECTS AND HIGHER-ORDER SPECTRAL ANALYSES

### 1.3.1 NONLINEAR EFFECTS

Nonlinearity is a fundamental property of optical waveguides, including both circular and planar waveguides such as optical fibers and, for example, Si photonic integrated rib or $LiNbO_3$-diffused waveguides. The origin of the nonlinear response is related to the high-order susceptibility of material under the influence of an applied field. For silica glass, which is the material of optical fibers and some nonlinear waveguides, only third-order susceptibility is responsible for nonlinear effects through the contribution to the total polarization $\vec{P}$, which is given by [34]

$$\vec{P} = \varepsilon_0(\chi^{(1)} \cdot \vec{E} + \chi^{(3)} \vdots \vec{E}\vec{E}\vec{E} + \cdots) \tag{1.1}$$

where
   $\varepsilon_0$ is the vacuum permittivity
   $\chi^{(j)}$ is $j$th order susceptibility.

In general, the nonlinear effects in optical fiber can be classified into two classes. One class refers to the energy transfer from the optical field to the propagation medium that is the result of stimulated inelastic scattering such as stimulated Raman scattering (SRS) and stimulated Brillouin scattering (SBS).

   The other class is governed by the third-order susceptibility $\chi^{(3)}$ and responsible for the most important nonlinear effect, called the Kerr effect. The Kerr effect is also known as the nonlinear refractive index because of the intensity dependence of the refractive index. In a high-intensity regime, the refractive index can be given by

$$n = n_0 + n_2 |E|^2 \tag{1.2}$$

where
   $n_0$ is the linear part of the refractive index that determines the material dispersion of a propagation medium
   $n_2$ is the nonlinear refractive index that relates to $\chi^{(3)}$
   $|E|^2$ is the intensity of optical field.

Important effects that originate from the nonlinear refractive index consist of third-harmonic generation (THG), four-wave mixing (FWM), self-phase modulation (SPM), and cross-phase modulation (XPM). First, SPM refers to the self-induced phase shift experienced by an optical field during its propagation. The phase of an optical field can be represented by

$$\phi_{NL} = n_2 k_0 L |E|^2 \tag{1.3}$$

where
$k_0 = 2\pi/\lambda$
$L$ is the length of fiber.

Because the phase change is proportional to the intensity, an optical pulse modulates its own phase with its intensity. The alteration of optical phase leads to frequency chirping according to the pulse intensity profile. Therefore, SPM is responsible for spectral broadening and formation of optical soliton in the balance with anomalous dispersion effect. Contrast to SPM, XPM and FWM effects require the participation of more than one optical field having different wavelengths, directions, or polarization states.

The main difference between XPM and FWM is that the occurrence of FWM requires a specific phase-matching condition. For XPM, an optical field with high intensity imposes nonlinear phase shift (NPS) on another field and this results in a spectral broadening similar to SPM. The interaction between optical fields in FWM creates new optical waves according to the conservation of energy-momentum. SPM can severely affect a single transmission channel, while XPM and FWM effects are the main concern in WDM systems, especially at the zero-dispersion spectral region. However, these Kerr effects are very significant in photonic signal processing systems due to the instantaneous response of the third-order nonlinearity.

Indeed, nonlinear effects play an important role in the design of fiber systems including mode-locked fiber lasers. With a strong pump power and hence a high optical gain, mode-locked pulses in the fiber cavity attain high intensity causing sufficient NPS, which affects the pulse formation in the cavity. In the nonlinear effects, the Kerr effects influence the formation of the soliton sequences of the output. When the optical power level in the fiber laser is sufficiently high, the NPS produced by SPM and XPM can change the power-dependent transmission of the nonlinear fiber loop mirror such as in Figure of Eight laser type or the polarization states as in non-linear polarization rotator fiber laser type [17,18]. Owing to ultrafast response of fiber nonlinearity, passive mode-locked fiber lasers can easily produce ultrashort soliton pulses at the femtosecond scale. Together with dispersion, the NPS plays an important role in the evolution of mode-locked pulses as well as in shaping output pulses. Particularly, in normal dispersion schemes, parabolic pulses with high energy are normally formed rather than Gaussian pulses [35,36]; otherwise soliton pulses are generated in the anomalous dispersion regime.

Mode-locking for actively mode-locked fiber lasers is based on harmonic modulation of active elements as mentioned in Section 1.2. Although nonlinear effects do not totally determine mode-locking, they also influence significantly to the shapes and the performance of output pulses. Some reports that analyze the actively mode-locked fiber laser taking the nonlinear effects into account indicated the influence of SPM on ultrashort pulse generation. Depending on the optical power levels in the cavity, the actively mode-locked fiber laser operates in various regimes that correspond to a certain performance of output pulses [37–40]. In general, the important influence of the SPM in the active mode-locked fiber can influence in two aspects. The first is pulse compression. When the average power in fiber cavity increases, the SPM effect becomes important and shortens the pulse based on soliton effect in the anomalous dispersion regime, but a large amount of dropouts may occur. The stable operating regime is only attained when the power exceeds a certain value [38]. The other impact of SPM in active mode-locking is the suppression of supermode noises superimposed by ASE noises that cause amplitude fluctuations. The peak intensity of a high power pulse with a short pulse width is eventually limited to within the filter bandwidth, so the energy at the pulse center shifts to either the right- or the left-hand side of the principal carrier frequency, due to SPM. Hence, the pulse amplitude is clipped to a certain intensity level at which the pulse energy is stabilized when the pulse spectral width reaches a certain ratio of the bandwidth of optical filter incorporated in the ring. When the SPM is strong enough, stabilization occurs through a filtering effect to effectively suppress the supermode noises [41].

Furthermore, for actively FM mode-locked fiber lasers in linear operating regimes, it is well known that the pulses of FM mode-locked fiber lasers tend to switch back and forth between the two phase states: up-chirp and down-chirp [21]. In fact such unstable operation does not occur due to the presence of dispersion in the cavity and the stable pulse state with the pulses locked at up-chirped or down-chirped positions depends on the sign of average cavity dispersion [22,42]. When the optical power increases, the presence of SPM in the cavity greatly modifies the up- and down-chirp caused by the FM. It should be noted that the FM directly interacts with the SPM. There are two ways of applying an imbalanced FM. One is where the up-chirped modulation produced by the phase modulator is in phase with the chirp caused by the SPM. In this case, the total chirp is increased and the pulse in the cavity is further compressed by anomalous GVD. Thus the soliton effect helps to generate a shorter pulse. Another case is where the phase modulator produces a down-chirp, and here the chirping is out of phase with the chirp caused by the SPM. In this case, the total down chirp is reduced and weakens both the pulse formation via phase modulation and filtering, and the soliton effect. Furthermore, in asynchronous mode-locking schemes, both the SPM effect and the filtering effectively suppress the relaxation oscillation that causes the intensity fluctuations of the output pulse sequence [43].

Appendix E gives further details on nonlinear effects in optical waveguides.

### 1.3.2 Nonlinear Processing and Higher-Order Spectral Analyses

Optical fiber communication systems can now support Tbps information capacity over thousands of kilometers [12,44,45]. They form the core technology for high capacity telecommunication networks. Recently, broadband applications such as high-speed Internet access and multibroadcast systems have been dramatically growing, which require huge transport capacity in core communication networks, especially at 100, 160, or even 320 Gbps operating speed. The intense demand for increased system reach and transport capacity as well as reduced cost of systems is a driving factor for research to develop advanced photonic technologies, especially high-speed signal processing. Besides the matured technologies such as WDM and optical fiber amplification, new transmission techniques such as OTDM and advanced modulation formats have set new challenges to signal processing in future optical networks, which is expected to operate at tens of Tbps.

The OTDM technology seems to be a good option to implement an Ethernet transmission channel at the rate of Tbps. However, the realization of this technology requires all signal processing functions in the system, such as the pulse generator, switch, and buffer, operating at ultrahigh speed. At the rate higher than 40 GHz and toward hundreds of GHz, the electronic domain is technologically and economically limited to perform signal processing functions. Thus the limited operating bandwidth of functional blocks in optical networks is a major challenge. Furthermore, several technological advancements such as advanced modulation formats have recently drawn a lot of attention in order to extend the reach and capacity [46]. By multilevel-coding data bits into the phase of optical carrier in advanced modulation formats such as Mary-phase shifted keying (M-PSK) or quadrature amplitude modulation (M-QAM), the capacity can be increased several times while the bandwidth of the optical signal remains unchanged, which is very important in high-capacity communication networks. A high spectral efficiency 1.6 bps/Hz transmission has been demonstrated by using differential quadrature phase shift keying (DQPSK) modulation and polarization multiplexing where 40 Gbps channels can be allocated into a WDM system with a wavelength spacing of only 25 GHz [47]. For a single channel system, a transmission rate of 5 Tbps has recently been demonstrated using the OTDM technique and 8-PSK and 16-QAM formats [48]. The introduction of advanced modulation formats such as DPSK, DQPSK, and QAM in optical networks results in new challenges that require signal processing techniques transparent to different modulation formats and transmission rates.

In order to overcome challenges in future optical networks, ultrafast photonic processing or all-optical signal processing is expected. By shifting signal processing from electronic domain to photonic domain, the limitation of operation bandwidth is easily eliminated by a reduction of optical-to-electrical (O/E) conversions. Ultrafast photonic signal processing can be implemented by exploiting nonlinearity in optical guided systems such as optical fibers and nonlinear waveguides. Kerr effects with ultrafast response can be applied in photonic signal processing systems to overcome the bandwidth limitation. Besides the formation of solitons and pulse compression as well as pulse shaping mechanism in the mode-locked fiber system, many signal processing functions such as signal regeneration and supercontinuum generation can be implemented by exploiting SPM in high-speed transmission systems [49,50]. Similarly, wavelength conversion, signal regeneration, and high-speed optical switching can be also implemented by a NALM based on XPM effect. These functions are very essential to high-capacity communication networks using OTDM/WDM technologies. However, FWM have recently attracted a lot of attention in photonic signal processing. A broad range of signal processing functions from wavelength conversion, ultrahigh speed switching to distortion compensation can be implemented by exploiting high-performance FWM processes. Another attractive feature of FWM-based signal processing is that it is format independent, because FWM can preserve both amplitude and phase information. Therefore, it is advantageous for future high-speed transmission systems using advanced modulation formats.

Rapid progress of nonlinear signal processing is due to the development of nonlinear propagation devices. Initially, the DSF was used as a highly nonlinear fiber (HNLF) because of the possibility to modify the fiber geometry [51]. The value of nonlinear coefficient of this fiber is normally ten times higher than standard single mode fiber (SSMF) that is mainly achieved by reducing the fiber core. Consequently, the DS-HNLF length of hundreds of meters is required in some nonlinear signal processing applications. Recent advances in the development of HNLFs have targeted to reduce the propagation length and operation power in practical applications. By modifying the fiber structures as well as the glass composition, many novel HNLFs, such as photonic crystal fibers [52], microstructure fibers [53], bismuth-based or chalcogenide glass fibers [54,55] or Si-integrated photonic waveguides, have been developed with a nonlinear coefficient more than thousands times that of SSMF. Another option, based on nonlinear waveguides, is to enable photonic signal processing in compact devices, which has recently attracted a lot research attention. The benefit of using nonlinear waveguides is the feasibility of the development of photonic integrated circuits (PICs) for all-optical signal processing. It recently has been shown that a dispersion-engineered planar chalcogenide waveguide is a potential candidate for photonic signal processing devices in future optical networks. With the nonlinear coefficient of about $10,000$ $W^{-1}$ $km^{-1}$, which is several times greater than that of SSMF, multifunction processing of optical signal at ultrahigh speed has been demonstrated in compact devices [56,57].

Properties of nonlinear processes, especially the multi-bound solitons can be identified using higher-order spectral analyses. This tool is very useful to identify the linear and nonlinear effects as both the amplitude and phase planes can be produced via such high-order spectrum techniques. Appendix C gives the calculation procedures for the bispectrum algorithms.

## 1.4  MOTIVATION AND OBJECTIVES OF THE BOOK CHAPTERS

Thus an actively FM mode-locked fiber laser can offer a lot of advantages such as generation of a stable ultrashort pulse sequence at a common transmission rate, easy synchronization with other electronic components in communication systems, and low timing jitter. Moreover, active phase modulation plays an important role in the shortening and the stability of mode-locked pulses, especially in the presence of dispersion and nonlinear effects. Although actively mode-locked fiber lasers have been available in commercial markets, they are still the subject for fundamental research of pulse dynamics in various operating conditions. On the other hand, exploration of new states and their characteristics in the actively FM mode-locked fiber laser is of importance in explicit interpretation of the mode-locking and generation mechanisms inside the fiber cavity.

In general, mode-locked fiber ring lasers are ideal systems in nonlinear fiber optics area. The attention of research has been not only to generate stable solitons with ultrashort width but also to investigate the dynamic characteristics of solitons in these systems. In periodic systems like mode-locked fiber lasers, solitons that can be in fact considered dissipative solitons experience periodical loss and gain effects and they behave in various interesting manners when circulating inside the fiber cavity. When the fiber ring cavity operates in strongly nonlinear regime, soliton pulses can exhibit several exciting characteristics such as bifurcation, period-doubling, chaos, and specially bound states, observed recently [58–62]. These complicated nonlinear dynamics have attracted a lot of researchers because the knowledge of soliton is very significant for an insight into the mode-locked fiber system as well as to explore new application areas of soliton. However, most research efforts have focused on the passively mode-locked fiber laser system [63]; it is only recently that attention has been paid to the actively mode-locked fiber laser system. On the other hand, the dynamic aspects of the actively soliton mode-locked fiber lasers remain very challenging. Moreover, the dynamic behaviors of soliton in the active mode-locked fiber laser are different from that in the passive mode-locked fiber laser due to the presence of an active photonic component, for example, an EO phase modulator.

The presence of bound solitons in passively mode-locked fiber lasers has shown complicated behavior of the fiber cavity through various interactions. The competition with the CW component and interaction between bound soliton pairs can cause the position of bound solitons not fixed and varying from time to time [59]. To generate periodic bound solitons with a fixed time separation between pulses, the requirements in design and tuning parameters of passively mode-locked fiber lasers are very strict. The stability of the bound soliton train is very sensitive to the change of operating conditions in the fiber cavity. Furthermore, the limitation of repetition rate may prevent passively mode-locked fiber lasers from potential applications in the telecommunication area. New behaviors of solitons such as bound states can be also exhibited in the actively mode-locked fiber laser. However, not much attention has been paid in bound solitons in actively mode-locked fiber lasers. When bound solitons are generated by active mode-locking, a high repetition rate of bound soliton train is easily achieved and synchronized with other electronic components that can create a new soliton source for potential applications in optical communication systems.

The relative phase difference between adjacent pulses is a very important parameter that contributes to the stability of multi-bound solitons in the actively mode-locked fiber ring cavity. Similar to phase-modulated signals such as DQPSK or QAM in advanced optical communication networks, the phase information becomes significant not only for monitoring but also for signal processing, especially signal recovery. For a conventional transmission system where information bits modulate the intensity of an optical carrier only, a power spectrum or a Fourier transform of autocorrelation is a popular technique to characterize the modulated signal. However, this is not sufficient for phase-modulated signals because the phase information is blind in common power spectrum analysis. Instead of power spectrum characterization, multidimensional spectra, which are known as Fourier transforms of high order correlation functions, can provide us not only the magnitude information but also the phase information. One of multidimensional spectrum analysis is bispectrum, which can be estimated by Fourier transform of the triple correlation. Especially it is useful in characterizing the non-Gaussian or nonlinear processes and applicable in many various fields such as signal processing, biomedicine, and image reconstruction [64]. However, not much attention has been paid to apply this technique in the fiber systems. Therefore, this technique is very useful to characterize the short pulse generated from mode-locked fiber lasers, especially from those operating at high power and exhibiting remarkably nonlinear effects such as doubling, multipulsing. Due to the phase information obtained from the bispectrum, it is a useful tool to analyze the behaviors of signals generated from these systems such as multi-bound solitons.

In this book, we focus on two issues that investigate the nonlinearity in the guided wave systems including the actively mode-locked fiber ring and nonlinear optical waveguides for signal processing. First issue is the generation of multi-bound solitons where the actively FM mode-locked fiber ring laser is the subject of this book. Mechanism of multi-bound soliton formation in an actively fiber ring under high nonlinear scheme is investigated. In addition, the phase relationship between bound solitons and the influence of active phase modulation on the formation and the stability of multi-bound solitons are also discussed. Dynamic behaviors of multi-bound solitons in the fiber ring cavity as well as transmission in fibers are investigated. Second, we also focus on aspects of nonlinearity in photonic signal processing, especially exploiting parametric process such as FWM to implement some signal processing functions such as parametric amplification, high-speed de-multiplexing, and bispectrum estimation. The bispectrum technique is proposed to characterize phase-modulated signals, which include multi-bound solitons and other phase-modulated signals.

## 1.5  ORGANIZATION OF THE CHAPTERS

The chapters in the book are classified under two main parts: first part is devoted to experimental and numerical investigation of multi-bound solitons in the actively FM mode-locked fiber ring laser. The second part focuses on numerical investigations on the third-order nonlinearity parametric process as well as bispectrum technique for photonic signal processing; in addition, nonlinear fiber lasers are presented. The chapter organization is as follows.

Chapter 2 gives an introduction of soliton generation from fiber lasers. First, fundamentals of pulse propagation in nonlinear dispersive medium are introduced through important nonlinear Schrodinger equations (NLSE) and their analytical solutions. These equations also provide the theoretical basis for building the numerical model of the FM mode-locked fiber laser, the main subject of this book. It is more important to give a foundation of soliton and multi-bound soliton generation in the nonlinear fiber ring resonators based on both passive and active mode-locking techniques. Experimental works on the actively FM mode-locked fiber lasers are conducted to investigate soliton generation in single pulse scheme. Numerical simulations are also used to investigate the pulse formation inside the fiber cavity as well as to verify the experimental results. Besides, detuning effects in the FM mode-locked fiber laser are experimentally and numerically investigated. All results obtained in this chapter are the foundation for next chapters.

Chapter 3 describes the process of multi-bound solitons in nonlinear fiber ring lasers. First, the review of bound solitons generated by passively mode-locked fiber lasers is given and the mechanisms of multi-bound soliton formation and dynamics in the fiber ring cavity are introduced. Then the conditions of multi-bound soliton in an actively mode-locked fiber ring laser is determined and discussed with the participation of the active phase modulation in the balance of soliton interactions inside the fiber ring cavity. Experimental and simulation works on the generation of multi-bound solitons in the actively FM mode-locked fiber ring are conducted and described in this chapter. For the first time, the existence of multi-bound solitons up to the fourth and sixth orders has been demonstrated in papers [1,5,8]. This chapter describes this formation, especially the spectral properties of such multi-mutual binding effects. States of multi-bound solitons at higher order up to the sixth order have been published in papers [3,10].

In an actively FM mode-locked fiber ring laser, the EO phase modulator is a key component for not only mode-locking but also multi-bound soliton generation. Therefore, Chapter 4 is devoted to the influence of EO phase modulator on pulse shaping and multi-bounding solitons. Important characteristic measurements of the phase modulator such as dynamic response and half-wave voltage are implemented to show the difference between two typical types of the EO phase modulators: lumped type and traveling-wave type. The influence of the EO phase modulators on the operating states of actively FM mode-locked fiber ring laser exhibits on two aspects: the artificial comb-filtering generated from the modal birefringence of the phase modulator and the chirping caused by the radio frequency (RF) driving signal. Obtained results have shown these effects on wavelength tunability of the fiber laser as well as stability and formation of multi-bound solitons in our system. The main part of this chapter has been published in References 3,5,9,12.

One of the main functionalities of the actively mode-locked fiber ring is to generate a stable periodic train of multi-bound solitons. Therefore, we experimentally demonstrate the generations and propagation of multi-bound solitons in optical fibers in Chapters 3 through 5. A theoretical basis of soliton transmission in optical fibers is also reviewed. Numerical simulations are also implemented to investigate dynamic behaviors of the different multi-bound solitons during propagation along the fibers. Some results of this chapter are presented in References 3,11.

In Chapter 5, another approach to characterize multi-bound solitons by multi-dimensional transformation, the bispectrum analysis to produce high-order spectra, is proposed. Therefore, the basis for bispectrum analysis is introduced prior to applying it in the analysis of multi-bound solitons. Furthermore, we also consider a possibility of bispectrum estimation based, particularly, on the FWM process. The limitation of this estimation as well as the application in optical receivers of the transmission systems are also discussed and presented in References 13,15.

Chapter 6 gives details of generations of multiple solitons by nonlinear dual polarization passive mode-locked fiber lasers. The two polarization modes of the lightwaves form two parallel nonlinear fiber lasers in which the nonlinearity comes from the saturation of the medium, hence the chaotic and frequency doubling of the soliton sequences. The competition of the solitons generated in each polarized ring leads to the formation of multi-bound solitons and their dynamics are illustrated.

In Chapter 7, we then describe materials related to nonlinear fiber ring lasers to illustrate the behavior of the formation of solitons and their dynamics as a general summation of all nonlinear dynamical mechanisms on the circulating lightwaves in such resonant ring.

Finally, Chapter 8 presents the transmission of multi-bound solitons over long lengths of optical fibers.

Appendices A through E are given at the end of the book to provide some basic analyses of the dynamic behavior of the lightwave governed by NLSE or systems of NLSE and MATLAB® m-file for the simulation of the generation of bound solitons. The basics of optical waveguides are also introduced so that the dispersion properties and linear and nonlinear guiding and propagation can be summarized for readers who are not familiar with lightwaves guided in circular or planar structures.

## REFERENCES

1. K. Minoshima and H. Matsumoto, High-accuracy measurement of 240-m distance in an optical tunnel by use of a compact femtosecond laser, *Appl. Opt.*, 39, 5512–5517, 2000.
2. Y. Ding, R. M. Brubaker, D. D. Nolte, M. R. Melloch, and A. M. Weiner, Femtosecond pulse shaping by dynamic holograms in photorefractive multiple quantum wells, *Opt. Lett.*, 22, 718–720, 1997.
3. M. E. Zevallos, S. K. Gayen, B. B. Das, M. Alrubaiee, and R. R. Alfano, Picosecond electronic time-gated imaging of bones in tissues, *IEEE J. Select. Top. Quant. Electron.*, 5, 916–922, 1999.
4. R. Cubeddu, A. Pifferi, P. Taroni, A. Torricelli, G. Valentini, and E. Sorbellini, Fluorescence lifetime imaging: An application to the detection of skin tumors, *IEEE J. Select. Top. Quant. Electron.*, 5, 923–929, 1999.
5. H. N. Paulsen, K. M. Hilligse, J. Thøgersen, S. R. Keiding, and J. J. Larsen, Coherent anti-Stokes Raman scattering microscopy with a photonic crystal fiber based light source, *Opt. Lett.*, 28, 1123–1125, 2003.
6. A. Tünnermann, J. Limpert, and S. Nolte, Ultrashort pulse fiber lasers and amplifiers, in F. Dausinger et al. (Eds.), *Femtosecond Technology for Technical and Medical Applications*, vol. 96, Springer, Berlin, Germany, pp. 35–54, 2004.
7. D. Breitling, H. Lubatschowski, and F. Lichtner, Drilling of metals, in F. Dausinger et al. (Eds.), *Femtosecond Technology for Technical and Medical Applications*, vol. 96, Springer, Berlin, Germany, pp. 131–156, 2004.
8. T. Udem, R. Holzwarth, and T. W. Hänsch, Optical frequency metrology, *Nature*, 416, 233–237, 2002.
9. I. Hartl, X. D. Li, C. Chudoba, R. K. Ghanta, T. H. Ko, J. G. Fujimoto, J. K. Ranka, and R. S. Windeler, Ultrahigh-resolution optical coherence tomography using continuum generation in an air-silica microstructure optical fiber, *Opt. Lett.*, 26, 608–610, 2001.
10. K. Suzuki, H. Kubota, A. Sahara, and M. Nakazawa, 40 Gbit/s single channel optical soliton transmission over 70000 km using in-line synchronous modulation and optical filtering, *Electron. Lett.*, 34, 98–100, 1998.
11. M. Nakazawa, K. Suzuki, and H. Kubota, Single-channel 80 Gbit/s soliton transmission over 10000 km using in-line synchronous modulation, *Electron. Lett.*, 35, 162–164, 1999.

12. M. Nakazawa, T. Yamamoto, and K. R. Tamura, 1.28 Tbit/s-70 km OTDM transmission using third- and fourth-order simultaneous dispersion compensation with a phase modulator, *Electron. Lett.*, 36, 2027–2029, 2000.

13. H. G. Weber, R. Ludwig, S. Ferber, C. Schmidt-Langhorst, M. Kroh, V. Marembert, C. Boerner, and C. Schubert, Ultrahigh-speed OTDM-transmission technology, *J. Lightwave Technol.*, 24, 4616–4627, 2006.

14. K. Mori, H. Takara, S. Kawanishi, M. Saruwatari, and T. Morioka, Flatly broadened supercontinuum spectrum generated in a dispersion decreasing fibre with convex dispersion profile, *Electron. Lett.*, 33, 1806–1808, 1997.

15. C. C. Chang, H. P. Sardesai, and A. M. Weiner, Code-division multiple-access encoding and decoding of femtosecond optical pulses over a 2.5-km fiber link, *IEEE Photon. Technol. Lett.*, 10, 171–173, 1998.

16. M. Zirngibl, L. W. Stulz, J. Stone, J. Hugi, D. DiGiovanni, and P. B. Hansen, 1.2 ps pulses from passively mode-locked laser diode pumped Er-doped fibre ring laser, *Electron. Lett.*, 27, 1734–1735, 1991.

17. I. N. Duling, III, Subpicosecond all-fibre erbium laser, *Electron. Lett.*, 27, 544–545, 1991.

18. K. Tamura, H. A. Haus, and E. P. Ippen, Self-starting additive pulse mode-locked erbium fibre ring laser, *Electron. Lett.*, 28, 2226–2228, 1992.

19. K. Tamura, J. Jacobson, E. P. Ippen, H. A. Haus, and J. G. Fujimoto, Unidirectional ring resonators for self-starting passively mode-locked lasers, *Opt. Lett.*, 18, 220–222, 1993.

20. F. Krausz and T. Brabec, Passive mode locking in standing-wave laser resonators, *Opt. Lett.*, 18, 888–890, 1993.

21. D. Kuizenga and A. Siegman, FM and AM mode locking of the homogeneous laser—Part I: Theory, *IEEE J. Quant. Electron.*, 6, 694–708, 1970.

22. K. Tamura and M. Nakazawa, Pulse energy equalization in harmonically FM mode-locked lasers with slow gain, *Opt. Lett.*, 21, 1930–1932, 1996.

23. Y. Shiquan and B. Xiaoyi, Rational harmonic mode-locking in a phase-modulated fiber laser, *IEEE Photon. Technol. Lett.*, 18, 1332–1334, 2006.

24. M. E. Grein, H. A. Haus, Y. Chen, and E. P. Ippen, Quantum-limited timing jitter in actively modelocked lasers, *IEEE J. Quant. Electron.*, 40, 1458–1470, 2004.

25. H. Takara, S. Kawanishi, and M. Saruwatari, Stabilisation of a modelocked Er-doped fibre laser by suppressing the relaxation oscillation frequency component, *Electron. Lett.*, 31, 292–293, 1995.

26. M. Nakazawa and E. Yoshida, A 40-GHz 850-fs regeneratively FM mode-locked polarization-maintaining erbium fiber ring laser, *IEEE Photon. Technol. Lett.*, 12, 1613–1615, 2000.

27. M. Nakazawa, E. Yoshida, and K. Tamura, Ideal phase-locked-loop (PLL) operation of a 10 GHz erbium-doped fibre laser using regenerative modelocking as an optical voltage controlled oscillator, *Electron. Lett.*, 33, 1318–1320, 1997.

28. W.-W. Hsiang, C. Y. Lin, N. Sooi, and Y. Lai, Long-term stabilization of a 10 GHz 0.8 ps asynchronously mode-locked Er-fiber soliton laser by deviation-frequency locking, *Opt. Expr.* 14, 1822–1828, 2006.

29. Y. Shiquan, J. Cameron, and X. Bao, Stabilized phase-modulated rational harmonic mode-locking soliton fiber laser, *IEEE Photon. Technol. Lett.*, 19, 393–395, 2007.

30. C. R. Doerr, H. A. Haus, and E. P. Ippen, Asynchronous soliton mode locking, *Opt. Lett.*, 19, 1958–1960, 1994.

31. W.-W. Hsiang, C. Y. Lin, M. F. Tien, and Y. Lai, Direct generation of a 10 GHz 816 fs pulse train from an erbium-fiber soliton laser with asynchronous phase modulation, *Opt. Lett.*, 30, 2493–2495, 2005.

32. S. B. Eduardo, L. A. M. Saito, and E. A. DeSouza, 396 fs, 2.5–12 GHz asynchronous mode-locking erbium fiber laser, *Conference on Lasers and Electro-Optics, CLEO 2007*, May 6–11, 2007, Baltimore, MD, p. CMC2, 2007.

33. K. R. Tamura and M. Nakazawa, 54-fs, 10-GHz soliton generation from a polarization-maintaining dispersion-flattened dispersion-decreasing fiber pulse compressor, *Opt. Lett.*, 26, 762–764, 2001.

34. G. P. Agrawal, *Nonlinear Fiber Optics*, 3rd edn., Academic Press, San Diego, CA, 2001.

35. F. Ilday, A. Chong, and F. W. Wise, Self-similar evolution of parabolic pulses in a laser, *Phys. Rev. Lett.*, 92, 213902, 2004.

36. A. Chong, W. H. Renninger, and F. W. Wise, Properties of normal-dispersion femtosecond fiber lasers, *J. Opt. Soc. Am. B*, 25, 140–148, 2008.

37. F. X. Kärtner, D. Kopf, and U. Keller, Solitary-pulse stabilization and shortening in actively mode-locked lasers, *J. Opt. Soc. Am. B*, 12, 486–496, 1995.

38. T. F. Carruthers and I. N. Duling III, 10-GHz, 1.3-ps erbium fiber laser employing soliton pulse shortening, *Opt. Lett.*, 21, 1927–1929, 1996.

39. M. Horowitz, C. R. Menyuk, T. F. Carruthers, and I. N. Duling, III, Theoretical and experimental study of harmonically modelocked fiber lasers for optical communication systems, *J. Lightwave Technol.*, 18, 1565–1574, 2000.

40. A. Zeitouny, Y. N. Parkhomenko, and M. Horowitz, Stable operating region in a harmonically actively mode-locked fiber laser, *IEEE J. Quant. Electron.*, 41, 1380–1387, 2005.

41. M. Nakazawa, K. Tamura, and E. Yoshida, Supermode noise suppression in a harmonically modelocked fibre laser by selfphase modulation and spectral filtering, *Electron. Lett.*, 32, 461, 1996.

42. N. G. Usechak, G. P. Agrawal, and J. D. Zuegel, FM mode-locked fiber lasers operating in the autosoliton regime, *IEEE J. Quant. Electron.*, 41, 753–761, 2005.

43. S. Yang, E. A. Ponomarev, and X. Bao, Experimental study on relaxation oscillation in a detuned FM harmonic mode-locked Er-doped fiber laser, *Optics Commun.*, 245, 371–376, 2005.

44. Y. Frignac et al., Transmission of 256 wavelength-division and polarization-division-multiplexed channels at 42.7 Gb/s (10.2 Tb/s capacity) over 3 × 100 km of TeraLight™ fiber, in *Optical Fiber Communication Conference and Exhibit, 2002 (OFC 2002)*, Los Angeles, CA, pp. FC5-1–FC5-3, 2002.

45. A. H. Gnauck, P. Tran, P. J. Winzer, C. R. Doerr, J. C. Centanni, E. C. Burrows, T. Kawanishi, T. Sakamoto, and K. Higuma, 25.6-Tb/s WDM transmission of polarization-multiplexed RZ-DQPSK signals, *J. Lightwave Technol.*, 26, 79–84, 2008.

46. P. J. Winzer and R.-J. Essiambre, Advanced modulation formats for high-capacity optical transport networks, *J. Lightwave Technol.*, 24, 4711–4728, 2006.

47. C. Wree, N. Hecker-Denschlag, E. Gottwald, P. Krummrich, J. Leibrich, E.-D. Schmidt, and B. Lankl, High spectral efficiency 1.6-b/s/Hz transmission (8 × 40 Gb/s with a 25-GHz grid) over 200-km SSMF using RZ-DQPSK and polarization multiplexing, *IEEE Photon. Technol. Lett.*, 15, 1303–1305, 2003.

48. C. Schmidt-Langhorst, R. Ludwig, D.-D. Gross, L. Molle, M. Seimetz, R. Freund, and C. Schubert, Generation and coherent time-division demultiplexing of up to 5.1 Tb/s single-channel 8-PSK and 16-QAM signals, in *Conference on Optical Fiber Communication—Includes Post Deadline Papers, 2009 (OFC 2009)*, San Diego, CA, pp. 1–3, 2009.

49. A. G. Striegler and B. Schmauss, Analysis and optimization of SPM-based 2R signal regeneration at 40 Gb/s, *J. Lightwave Technol.*, 24, 2835, 2006.

50. H. Murai, Y. Kanda, M. Kagawa, and S. Arahira, Regenerative SPM-based wavelength conversion and field demonstration of 160-Gb/s all-optical 3R operation, *J. Lightwave Technol.*, 28, 910–921, 2010.

51. M. Takahashi, R. Sugizaki, J. Hiroishi, M. Tadakuma, Y. Taniguchi, and T. Yagi, Low-loss and low-dispersion-slope highly nonlinear fibers, *J. Lightwave Technol.*, 23, 3615–3624, 2005.

52. J. C. Knight, T. A. Birks, A. Ortigosa-Blanch, W. J. Wadsworth, and P. S. J. Russell, Anomalous dispersion in photonic crystal fiber, *IEEE Photon. Technol. Lett.*, 12, 807–809, 2000.

53. J. K. Ranka, R. S. Windeler, and A. J. Stentz, Optical properties of high-delta air silica microstructure optical fibers, *Opt. Lett.*, 25, 796–798, 2000.

54. S. Naoki, T. Nagashima, T. Hasegawa, S. Ohara, K. Taira, and K. Kikuchi, Bismuth-based optical fiber with nonlinear coefficient of 1360 $W^{-1}$ $km^{-1}$, *Optical Fiber Communication Conference, OFC 2004*, Vol. 2, Los Angeles, CA, p. PD26, 2004.

55. R. E. Slusher, G. Lenz, J. Hodelin, J. Sanghera, L. B. Shaw, and I. D. Aggarwal, Large Raman gain and nonlinear phase shifts in high-purity As2Se3 chalcogenide fibers, *J. Opt. Soc. Am. B*, 21, 1146–1155, 2004.

56. M. D. Pelusi, F. Luan, S. Madden, D. Y. Choi, D. A. Bulla, B. Luther-Davies, and B. J. Eggleton, Wavelength conversion of high-speed phase and intensity modulated signals using a highly nonlinear chalcogenide glass chip, *IEEE Photon. Technol. Lett.*, 22, 3–5, 2010.

57. D. V. Trung et al., Photonic chip based 1.28 Tbaud transmitter optimization and receiver OTDM demultiplexing, *Optical Fiber Communication (OFC), Collocated National Fiber Optic Engineers Conference, 2010 Conference on (OFC/NFOEC)*, San Diego, CA, p. PDPC5, 2010.

58. D. Y. Tang, W. S. Man, H. Y. Tam, and P. D. Drummond, Observation of bound states of solitons in a passively mode-locked fiber laser, *Phys. Rev. A*, 64, 033814, 2001.

59. D. Y. Tang, B. Zhao, D. Y. Shen, C. Lu, W. S. Man, and H. Y. Tam, Bound-soliton fiber laser, *Phys. Rev. A*, 66, 033806, 2002.

60. Ph. Grelu, F. Belhache, F. Gutty, and J.-M. Soto-Crespo, Phase-locked soliton pairs in a stretched-pulse fiber laser, *Opt. Lett.*, 27, 966–968, 2002.

61. L. M. Zhao, D. Y. Tang, and B. Zhao, Period-doubling and quadrupling of bound solitons in a passively mode-locked fiber laser, *Opt. Commun.*, 252, 167–172, 2005.

62. A. Haboucha, H. Leblond, M. Salhi, A. Komarov, and F. Sanchez, Coherent soliton pattern formation in a fiber laser, *Opt. Lett.*, 33, 524–526, 2008.

63. S. T. Cundiff, Soliton dynamics in mode-locked lasers, in N. Akhmediev and A. Ankiewicz (Eds.), *Dissipative Solitons*, vol. 661, Springer, Berlin, Germany, pp. 183–206, 2005.

64. C. L. Nikias and J. M. Mendel, Signal processing with higher-order spectra, *IEEE Signal Process. Mag.*, 10, 10–37, 1993.

# 2 Generations of Solitons in Optical Fiber Ring Resonators

This chapter gives an introduction on the analysis of the generation of solitons in a mode-locked fiber ring structure in which optical modulator and optical gain device are incorporated. Conditions for passive and active mode-locking are given in association with the nonlinear Schrodinger equation (NLSE). An experimental setup and simulation system are described to demonstrate the analytical, simulation, and experimental agreement in the generations of solitons.

## 2.1 NONLINEAR SCHRODINGER EQUATIONS

### 2.1.1 NONLINEAR RESPONSE

To understand the propagation of optical pulses in nonlinear medium, Maxwell's equations are employed to obtain the nonlinear wave equation (see further details given in Appendix B) that describes light waves confinement and propagation in optical waveguides including optical fibers as follows [1]:

$$\nabla \times \nabla \times \vec{E} + \frac{1}{c^2}\frac{\partial^2 \vec{E}}{\partial t^2} = -\mu_0 \frac{\partial^2 \vec{P}}{\partial t^2} \tag{2.1}$$

Using the vector identity $\nabla \times \nabla \times \vec{E} = \nabla(\nabla \cdot \vec{E}) - \nabla^2 \vec{E}$ and $\nabla \cdot \vec{E} = 0$ for dielectric materials, the nonlinear wave propagation in nonlinear waveguide in the time–spatial domain in vector form can be expressed as follows:

$$\nabla^2 \vec{E} - \frac{1}{c^2}\frac{\partial^2 \vec{E}}{\partial t^2} = \mu_0 \left( \frac{\partial^2 \vec{P}_L}{\partial t^2} + \frac{\partial^2 \vec{P}_{NL}}{\partial t^2} \right) \tag{2.2}$$

where
  $\vec{E}$ is the electric field vector of the optical wave
  $\mu_0$ is the vacuum permeability assuming a nonmagnetic wave-guiding medium
  $c$ is the speed of light in vacuum
  $\vec{P}_L, \vec{P}_{NL}$ are, respectively, the linear and nonlinear parts of polarization vector $\vec{P}$, which are formed as follows:

$$\vec{P}_L(\vec{r},t) = \varepsilon_0 \int_{-\infty}^{\infty} \chi^{(1)}(t-t') \cdot \vec{E}(\vec{r},t')\,dt' \qquad (2.3)$$

$$\vec{P}_{NL}(\vec{r},t) = \varepsilon_0 \int_{-\infty}^{\infty}\int\int \chi^{(3)}(t-t_1,t-t_2,t-t_3) \times \vec{E}(\vec{r},t_1)\vec{E}(\vec{r},t_2)\vec{E}(\vec{r},t_3) \qquad (2.4)$$

where $\chi^{(1)}$ and $\chi^{(3)}$ are the first- and third-order susceptibility tensors. Thus the linear and nonlinear coupling effects in optical waveguides can be described by Equation 2.2. In optical waveguides such as optical fibers, only the third-order nonlinearity is of special importance because it is responsible for all nonlinear effects. The second term on the right-hand side (RHS) of Equation 2.2 is responsible for nonlinear processes, including interaction between optical waves through the third-order susceptibility. If the nonlinear response is assumed to be instantaneous, the time dependence of $\chi^{(3)}$ is given by the product of three delta functions of the form $\delta(t-t_1)$ and then Equation 2.4 reduces to

$$\vec{P}_{NL}(\vec{r},t) = \varepsilon_0 \chi^{(3)} \vec{E}(\vec{r},t)\vec{E}(\vec{r},t)\vec{E}(\vec{r},t) \qquad (2.5)$$

where $\chi^{(3)}$ is the fourth-rank tensor.

To simplify the analysis of Equation 2.2, the following assumptions are made: (1) Only one mode is present in the waveguide, or a single-mode fiber is considered. (2) The field is linearly polarized (LP) in the same direction and the polarization state remains unchanged during the propagation. (3) The nonlinearity can be seen as a small perturbation because nonlinear change in the refractive index is $\Delta n/n < 10^{-6}$ in practice. (4) The variation of the carrier wave is much faster than that of the envelope of the optical pulse. On the other hand, the bandwidth of the optical pulse $\Delta \omega$ is much smaller than the carrier frequency $\omega_0$.

In this approximation of the slowly varying envelope, the electric field can be written in the form

$$\vec{E}(\vec{r},t) = \frac{1}{2}\hat{x}\,[E(\vec{r},t)\exp(-j\omega_0 t)+c.c.] \qquad (2.6)$$

where
$\omega_0$ is the carrier frequency
$\hat{x}$ is the polarization unit vector
$E(\vec{r},t)$ is a slowly varying function of time
$\vec{P}_L, \vec{P}_{NL}$ can also be expressed in a similar manner.

Introducing Equation 2.6 into Equations 2.3 and 2.5 yields

$$P_L = \varepsilon_0 \chi_{xx}^{(1)} E \qquad (2.7)$$

$$P_{NL} = \frac{1}{8} \varepsilon_0 \chi_{xxxx}^{(3)} (E e^{-j\omega_0 t} + E^* e^{j\omega_0 t})^3$$

$$= \frac{1}{8} \varepsilon_0 \chi_{xxxx}^{(3)} (E^3 e^{-j3\omega_0 t} + 3E^2 E^* e^{-j\omega_0 t} + 3E E^{2*} e^{j\omega_0 t} + E^{3*} e^{j3\omega_0 t})$$

$$= \frac{1}{8} \varepsilon_0 \chi_{xxxx}^{(3)} ((E^3 e^{-j3\omega_0 t} + c.c.) + 3 |E|^2 (E e^{-j\omega_0 t} + c.c.)) \qquad (2.8)$$

The first term in Equation 2.8 describes the nonlinear part of the polarization at three times the original carrier frequency, which is responsible for the third harmonic generation and requires phase matching. The second term describes the nonlinear part of the polarization at the carrier frequency and is responsible for most of the important nonlinear effects relating to the nonlinear refractive index.

The linear and nonlinear parts of polarization (see also Section E.5) are related to the dielectric constant as

$$\varepsilon(\omega) = \varepsilon_L + \varepsilon_{NL} \qquad (2.9)$$

where $\varepsilon_L$, $\varepsilon_{NL}$ are the linear and nonlinear contributions to the dielectric constant and obtained from Equation 2.7 and Equation 2.8.

$$\varepsilon_L = 1 + \chi_{xx}^{(1)} \quad \text{and} \quad \varepsilon_{NL} = \frac{3}{4} \chi_{xxxx}^{(3)} |E|^2 \qquad (2.10)$$

This dielectric constant can be used to define the refractive index $n(\omega)$ and the absorption coefficient $\alpha(\omega)$ of the nonlinear medium as follows:

$$\varepsilon(\omega) = \left[ n(\omega) + \frac{j\alpha(\omega)c}{2\omega} \right]^2 \approx n^2(\omega) + j \frac{n(\omega)\alpha(\omega)c}{\omega} \qquad (2.11)$$

Both $n(\omega)$ and $\alpha(\omega)$ relate to the linear and nonlinear parts of $\varepsilon(\omega)$ by introducing

$$n(\omega) = n_0 + n_2 |E|^2 \quad \text{and} \quad \alpha(\omega) = \alpha_0 + \alpha_2 |E|^2 \qquad (2.12)$$

where the linear index $n_0$ and the absorption coefficient $\alpha_0$ are related to the real and imaginary parts of $\chi_{xx}^{(1)}$, while the nonlinear index $n_2$ and the two-photon absorption coefficient $\alpha_2$ are related to the real and imaginary parts of $\varepsilon_{NL}$ by

$$n_2 = \frac{3}{8n} \mathrm{Re}\left(\chi_{xxxx}^{(3)}\right) \quad \text{and} \quad \alpha_2 = \frac{3\omega_0}{4nc} \mathrm{Im}\left(\chi_{xxxx}^{(3)}\right) \qquad (2.13)$$

For some optical waveguides such as optical fibers, the coefficient $\alpha_2$ is negligible and the intensity-dependent nonlinear refractive index is responsible for the nonlinear response of the propagation medium.

### 2.1.2 Nonlinear Schrodinger Equation

The NLSE plays an important role in the description of nonlinear effects in optical pulse propagation. It can be derived from the wave equation Equation 2.2 after some algebra by using the method of separating variables (see Appendix B).

$$\frac{\partial A(z,t)}{\partial z} + \frac{\alpha}{2} A(z,t) - j \sum_{n=1}^{\infty} \frac{j^n \beta_n}{n!} \frac{\partial^n A(z,t)}{\partial t^n} = j\gamma \left( 1 + \frac{j}{\omega_0} \frac{\partial}{\partial t} \right)$$

$$\times A(z,t) \int_{-\infty}^{\infty} R(t') |A(z, t - t')|^2 dt' \tag{2.14}$$

where $A(z,t)$ is the slowly varying complex envelope propagating along $z$ in the propagation medium, the effect of the propagation constant and its higher derivatives $\beta$ and $\beta_n$ (see Appendix E on optical fibers) in the neighborhood of the optical carrier radial frequency $\omega_0$ in which a Taylor-series expansion is valid, $R(t)$ is the nonlinear response function, and $\gamma = \omega_0 n_2 / c A_{eff}$ is the nonlinear coefficient of the guided medium. In most cases, related to optical communication transmission, optical pulses of a width wider than 100 fs are employed. Equation 2.14 can be further simplified as follows:

$$\frac{\partial A}{\partial z} + \frac{\alpha}{2} A + \frac{j\beta_2}{2} \frac{\partial^2 A}{\partial \tau^2} - \frac{\beta_3}{6} \frac{\partial^3 A}{\partial \tau^3} = j\gamma \left[ |A|^2 A + \frac{j}{\omega_0} \frac{\partial(|A|^2 A)}{\partial \tau} - T_R A \frac{\partial(|A|^2)}{\partial \tau} \right] \tag{2.15}$$

where a frame of reference moving with the pulse at the group velocity $v_g$ is used by making the transformation $\tau = t - z/v_g \equiv t - \beta_1 z$, and the propagation constant is expanded up to the third order term that includes the group velocity dispersion ($\beta_2$) and the third-order dispersion ($\beta_3$). In Equation 2.15, the first moment of the nonlinear response function is defined as follows:

$$T_R \equiv \int_0^{\infty} t R(t') dt' \tag{2.16}$$

which is responsible for the Raman scattering effect and the second term on the RHS of Equation 2.15 is responsible for the self-steepening effect. However, if the width of optical pulses is of the order of picoseconds, the high order effects such as self-steepening and Raman scattering can be ignored. Hence, Equation 2.15 becomes

$$\frac{\partial A}{\partial z} + \frac{\alpha}{2} A + \frac{j\beta_2}{2} \frac{\partial^2 A}{\partial \tau^2} - \frac{\beta_3}{6} \frac{\partial^3 A}{\partial \tau^3} = j\gamma |A|^2 A \tag{2.17}$$

This equation can describe the most important linear and nonlinear propagation effects of the optical pulse sequence propagating in optical fibers.

If we can further simplify Equation 2.17 by setting the attenuation factor and the third-order dispersion coefficient to zero, to obtain the well-known NLSE as

$$\frac{\partial A}{\partial z} + \frac{j\beta_2}{2}\frac{\partial^2 A}{\partial \tau^2} = i\gamma \mid A \mid^2 A \tag{2.18}$$

Equation 2.18 is normally used to examine the dynamics of the propagation of optical solitary waves in nonlinear dispersive media.

### 2.1.3  GINZBURG–LANDAU EQUATION: A MODIFIED NLSE

Although the NLSEs described can be used to explain most nonlinear effects including higher-order effects, they only describe the pulse propagation in passive nonlinear media without gain. In a propagation medium with gain as fiber amplifiers, the gain effect is required to be included into the NLSE. The general equation that governs the pulse propagation in active fibers is given by ignoring other effects for simplification as follows:

$$\frac{\partial A(z,\omega)}{\partial z} = \frac{1}{2}g(\omega)A(z,\omega) \tag{2.19}$$

where $g(\omega)$ is the gain coefficient of the active fiber. For the approximation of a homogeneously broadened system, the gain spectral shape takes a Lorentzian profile [2]:

$$g(\omega) = \frac{g_0}{1 + (\omega - \omega_g)^2 / \Delta\omega_g^2} \tag{2.20}$$

where
$g_0$ is the maximum small signal gain
$\omega_g$ is the atomic transition frequency
$\Delta\omega_g$ is the gain bandwidth that relates to the dipole relaxation time.

The gain spectrum can be approximated by a Taylor expand in around $\omega_g$:

$$g(\omega) \approx g_0\left(1 - \frac{(\omega - \omega_g)^2}{\Delta\omega_g^2}\right) \tag{2.21}$$

By substituting Equation 2.21 into Equation 2.19, and taking the inverse Fourier transform with the assumption of the carrier frequency $\omega_0$ close to $\omega_g$, the propagation equation with amplification is obtained as

$$\frac{\partial A(z,t)}{\partial z} = \frac{g_0}{2}\left[A(z,t) + \frac{1}{\Delta\omega_g^2}\frac{\partial^2 A(z,t)}{\partial t^2}\right] \tag{2.22}$$

However, in many cases of pulse propagation, especially in mode-locked fiber laser systems, the gain saturation plays an important role in pulse amplification. Therefore, the saturation power is to be included in Equation 2.22 by replacing $g_0$ with $g_{sat}$ [2] given as

$$g_{sat}(z) = \frac{g_0(z)}{1 + P_{av}(z)/P_{sat}} \tag{2.23}$$

Thence, Equation 2.22 can be modified by using Equation 2.23 as

$$\frac{\partial A(z,t)}{\partial z} = \frac{g_{sat}}{2}\left[ A(z,t) + \frac{1}{\Delta\omega_g^2}\frac{\partial^2 A(z,t)}{\partial t^2} \right] \approx \frac{g_0}{2(1 + P_{av}(z)/P_{sat})}$$

$$\times \left[ A(z,t) + \frac{1}{\Delta\omega_g^2}\frac{\partial^2 A(z,t)}{\partial t^2} \right] \tag{2.24}$$

where
  $g_0$ is assumed to be constant along the length of the active fiber
  $P_{sat}$ is the saturated power of the gain medium
  $P_{av}$ is the average power of the signal at position $z$ in the active fiber as follows:

$$P_{av}(z) = \frac{1}{T_m}\int_{-T_m/2}^{T_m/2} |A(z,t)|^2 \, dt \tag{2.25}$$

For a full model of the pulse evolution in gain medium, other effects such as dispersion and nonlinear effects are also required to be included in Equation 2.24. Hence, by replacing the gain term in Equation 2.17 with that in Equation 2.24, a modified NLSE can be obtained as follows:

$$\frac{\partial A}{\partial z} + \frac{j}{2}\left( \beta_2 + \frac{jg_{sat}}{\Delta\omega_g^2} \right)\frac{\partial^2 A}{\partial \tau^2} - \frac{\beta_3}{6}\frac{\partial^3 A}{\partial \tau^3} = i\gamma\,|A|^2\,A + \frac{1}{2}(g_{sat} - \alpha)A \tag{2.26}$$

This equation is also called the Ginzburg–Landau equation (GLE), which can also be derived from the wave equation [3]. Besides the cubic GLE, an extended version is the quintic cubic G-L equation (QCGLE), which has also attracted considerable attention [4]. The GLE plays an important role in the description of nonlinear systems [2,4,5] as well as fiber lasers [6].

## 2.1.4 COUPLED NONLINEAR SCHRODINGER EQUATIONS

In a birefringent fiber, pulse sequences of more than one polarization state can be copropagating, therefore, Equation 2.6 can be replaced by

$$\vec{E}(\vec{r},t) = \frac{1}{2}[(\hat{x}E_x + \hat{y}E_y)\exp(-j\omega_0 t) + c.c.]  \tag{2.27}$$

where $E_x$, $E_y$ are the orthogonal polarization components of the optical field. The polarized components of the nonlinear induced medium can be expressed as

$$P_{NL,x}(\vec{r},t) = -\varepsilon_0 2n_2 n(\omega_0)\left[\left(|E_x|^2 + \frac{2}{3}|E_y|^2\right)E_x + \frac{1}{3}\left(E_x^* E_y\right)E_y\right]  \tag{2.28}$$

$$P_{NL,y}(\vec{r},t) = -\varepsilon_0 2n_2 n(\omega_0)\left[\left(|E_y|^2 + \frac{2}{3}|E_x|^2\right)E_y + \frac{1}{3}\left(E_y^* E_x\right)E_x\right]  \tag{2.29}$$

In a similar manner, two coupled equations for the slowly varying components of $E_x$ and $E_y$ can be derived as [7]

$$\frac{\partial A_x}{\partial z} + \frac{\Delta\beta_1}{2}\frac{\partial A_x}{\partial \tau} + \frac{j}{2}\left(\beta_2 + \frac{jg_{sat}}{\Delta\omega_g^2}\right)\frac{\partial^2 A_x}{\partial \tau^2} - \frac{\beta_3}{6}\frac{\partial^3 A_x}{\partial \tau^3}$$
$$= \frac{1}{2}(g_{sat} - \alpha)A_x + j\gamma\left[\left(|A_x|^2 + \frac{2}{3}|A_y|^2\right)A_x + \frac{j}{3}A_x^* A_y^2 \exp(-2j\Delta\beta_0 z)\right]  \tag{2.30}$$

$$\frac{\partial A_y}{\partial z} - \frac{\Delta\beta_1}{2}\frac{\partial A_y}{\partial \tau} + \frac{j}{2}\left(\beta_2 + \frac{jg_{sat}}{\Delta\omega_g^2}\right)\frac{\partial^2 A_y}{\partial \tau^2} - \frac{\beta_3}{6}\frac{\partial^3 A_y}{\partial \tau^3}$$
$$= \frac{1}{2}(g_{sat} - \alpha)A_y + j\gamma\left[\left(|A_y|^2 + \frac{2}{3}|A_x|^2\right)A_y + \frac{j}{3}A_y^* A_x^2 \exp(+2j\Delta\beta_0 z)\right]  \tag{2.31}$$

where
   $A_x$, $A_y$ are the slowly varying envelopes of orthogonal polarized components
   $\Delta\beta_1 = \beta_{1,x} - \beta_{1,y} = \Delta n_g/c$ is the difference in the propagation phase velocity of the polarized components
   $\Delta n_g$ is the group birefringence

Under the case of highly birefringent fibers, the terms $\exp(-2j\Delta\beta_0 z)$ and $\exp(+2j\Delta\beta_0 z)$ can be neglected due to their rapid oscillations and Equations 2.30 and 2.31 become

$$\frac{\partial A_x}{\partial z} + \frac{\Delta\beta_1}{2}\frac{\partial A_x}{\partial \tau} + \frac{j}{2}\left(\beta_2 + \frac{jg_{sat}}{\Delta\omega_g^2}\right)\frac{\partial^2 A_x}{\partial \tau^2} - \frac{\beta_3}{6}\frac{\partial^3 A_x}{\partial \tau^3}$$
$$= j\gamma\left(|A_x|^2 + \frac{2}{3}|A_y|^2\right)A_x + \frac{1}{2}(g_{sat} - \alpha)A_x  \tag{2.32}$$

$$\frac{\partial A_y}{\partial z} + \frac{\Delta\beta_1}{2}\frac{\partial A_y}{\partial \tau} + \frac{j}{2}\left(\beta_2 + \frac{jg_{sat}}{\Delta\omega_g^2}\right)\frac{\partial^2 A_y}{\partial \tau^2} - \frac{\beta_3}{6}\frac{\partial^3 A_y}{\partial \tau^3}$$

$$= j\gamma\left(\left|A_y\right|^2 + \frac{2}{3}\left|A_x\right|^2\right)A_y + \frac{1}{2}(g_{sat} - \alpha)A_y \tag{2.33}$$

These equations are significant in the representation of the propagation of guided light waves relating to the polarization states such as polarization mode dispersion in fiber transmission and nonlinear polarization rotation (NPR) in passive mode-locking systems [7–9]. For an optical fiber with a length $L$, the phase variation of polarization components due to nonlinearity can be derived by considering the non-linear term only in Equations 2.32 and 2.33 for simplicity.

$$\phi_{NL}^x = \gamma L\left(\left|A_x\right|^2 + \frac{2}{3}\left|A_y\right|^2\right) \tag{2.34}$$

$$\phi_{NL}^y = \gamma L\left(\left|A_y\right|^2 + \frac{2}{3}\left|A_x\right|^2\right) \tag{2.35}$$

and hence, the angle of polarization rotation is given by

$$\phi_{NPR} = \phi_{NL}^y - \phi_{NL}^x = \frac{\gamma L}{3}\left(\left|A_y\right|^2 - \left|A_x\right|^2\right) \tag{2.36}$$

Note that this angle is zero when the guided light waves are LP because of the equality $|A_x|^2 = |A_y|^2$. On the other hand, the polarization ellipse rotates with waves propagating in the fiber.

## 2.2 OPTICAL SOLITONS

### 2.2.1 TEMPORAL SOLITONS

The NLSE Equation 2.18 obtained in Section 2.1.2 governs the propagation of optical pulse in nonlinear dispersive media such as optical waveguides or fibers. By using the following transformation of variables of

$$T = \frac{\tau}{\tau_0}, \quad \xi = \frac{z}{L_D}, \quad u = \sqrt{\gamma L_D}\,A \tag{2.37}$$

where
  $\tau_0$ is a temporal scaling parameter often taken to be the input pulse width
  $L_D = \tau_0^2 / |\beta_2|$ is the dispersion length.

Equation 2.18 can be normalized to the (1 + 1)-dimensional NLSE as follows:

$$j\frac{\partial u}{\partial Z} - \frac{s}{2}\frac{\partial^2 u}{\partial \tau^2} \pm |u|^2 u = 0 \qquad (2.38)$$

where $s = \text{sgn}(\beta_2) = \pm 1$ stands for the sign of the group velocity dispersion (GVD) parameter,* which can be positive or negative, depending on the wavelength. The nonlinear term is positive (+1) for optical fibers but may become negative (−1) for waveguides made of semiconductor materials such as Si photonic waveguides. Thus, there are two cases of basic propagation in optical fibers relating to the dispersion that is normally measured by the parameter

$$D(\lambda) = \frac{d}{d\lambda}\left(\frac{1}{v_g}\right) = -\frac{2\pi c}{\lambda^2}\beta_2 \qquad (2.39)$$

where the dispersion parameter $D(\lambda)$ is expressed in units of ps/(nm km). For standard single mode fiber (SSMF), $D(\lambda)$ is positive or GVD is anomalous at wavelengths >1.3 μm at which the dispersion becomes zero for a standard single mode optical fiber (SSMF). Hence, $D(\lambda)$ is negative or GVD is normal at wavelengths shorter than 1.3 μm (see Appendix E).

Because of two different signs of the GVD parameter, optical fibers can support two different types of solitons which are the eigensolutions of Equation 2.38. In particular, Equation 2.38 has solutions in the form of dark temporal solitons in case of normal GVD ($s = +1$) and bright temporal solitons in case of anomalous GVD ($s = -1$). These solutions can be found by the inverse scattering method [10] (see also Chapter 10 for more details). In case of anomalous GVD, Equation 2.38 takes the form

$$j\frac{\partial u}{\partial Z} + \frac{1}{2}\frac{\partial^2 u}{\partial \tau^2} + |u|^2 u = 0 \qquad (2.40)$$

and the most interesting solution of Equation 2.40 is the fundamental soliton with general form given by

$$u(\xi, \tau) = \text{sech}(\tau)\exp\left(\frac{j\xi}{2}\right) \qquad (2.41)$$

Thus at the input of the fiber, the soliton is given as $u(0,\tau) = \text{sech}(\tau)$ and it can be converted into real units as

$$A(0,T) = \sqrt{P_0}\,\text{sech}\left(\frac{T}{T_0}\right) = \left(\frac{|\beta_2|}{\gamma T_0^2}\right)^{1/2}\text{sech}\left(\frac{T}{T_0}\right) \qquad (2.42)$$

---

* See Appendix E for detailed definitions of GVD.

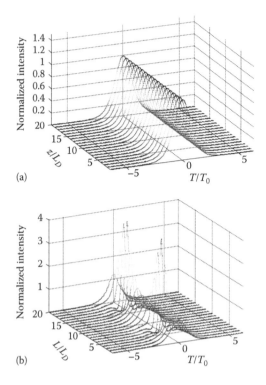

**FIGURE 2.1**  Evolution of the (a) first-order and (b) second-order solitons.

The expression Equation 2.42 indicates that if a hyperbolic-secant pulse has peak power $P_0$ that satisfies Equation 2.42, then it can propagate nondistorted with no change in its temporal and spectral shapes in a lossless optical fiber at arbitrary distances, as shown in Figure 2.1. This important feature results from the balancing between GVD and self-phase modulation (SPM) effects.

Now assuming that a pulse sequence whose peak power launched into the fiber is sufficiently high, a higher-order soliton can be excited if the shape of the pulse takes the form

$$u(0, \tau) = N \operatorname{sech}(\tau) \tag{2.43}$$

where $N$ is an integer representing the soliton order and determined by

$$N^2 = \frac{L_D}{L_{NL}} = \frac{\gamma P_0 T_0^2}{|\beta_2|} \tag{2.44}$$

Different from the fundamental soliton, the temporal and spectral shapes of higher-order solitons vary periodically during propagation with the period $\xi_0 = \pi/2$ or in real units $z_0 = \pi L_D/2$.

With $s=+1$ in case of normal GVD, Equation 2.38 takes the form

$$j\frac{\partial u}{\partial Z}-\frac{1}{2}\frac{\partial^2 u}{\partial \tau^2}+|u|^2 u=0 \tag{2.45}$$

and the solutions of Equation 2.45 are dark solitons found by the inverse scattering method similar to the case of bright solitons (see also Chapter 9). The solution of a fundamental dark soliton can be written as [1,11]

$$u(\xi,\tau)=\cos\phi\tanh[\cos\phi(\tau-\xi\sin\phi)-j\sin\phi]\exp(j\xi) \tag{2.46}$$

The features of the dark soliton are a high constant power level with an intensity dip and an abrupt phase change at the center depending on parameter $\phi$. For $\phi=0$, the dark soliton is a black soliton with the zero dip and a phase jump of $\pi$ at the center. When $\phi\neq 0$, the dip intensity is nonzero and such solitons are called gray solitons as shown in Figure 2.2.

In general, solitons in fiber have attracted a lot of interest in research from fundamentals to practical applications due to their unique features. In communication

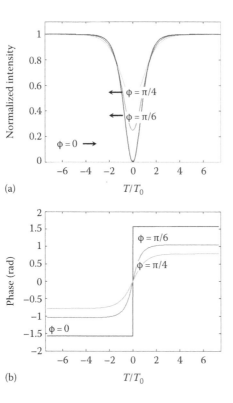

(a)

(b)

**FIGURE 2.2**   (a) Intensity and (b) phase profiles of dark solitons for various values of the internal phase $\phi$.

systems, the understanding of solitons is important for both transmission and signal processing applications [2,12].

## 2.2.2 DISSIPATIVE SOLITONS

Optical solitons mentioned discussed only exist in an ideal propagation medium that has no perturbations such as loss and gain. In practical systems such as mode-locked fiber lasers, the optical signal is periodically amplified to compensate the loss that it experiences in the fiber cavity. Thus there is a periodic variation of the pulse power that can form an index grating and induce modulation stability. It has been demonstrated that solitons are able to exist in these systems [2]. Therefore, propagation of optical pulse in this case should be described by complex cubic GLE Equation 2.26, which includes the loss and gain effects. Similar to NLSE, it is useful to introduce the dimensionless variables and Equation 2.26 becomes

$$j\frac{\partial u}{\partial \xi} - \frac{1}{2}(s + jd)\frac{\partial^2 u}{\partial \tau^2} + |u|^2\, u = \frac{j}{2}\mu u \tag{2.47}$$

where

$$d = \frac{g_{sat}L_D T_2^2}{T_0^2}, \quad \mu = (g_{sat} - \alpha)L_D \tag{2.48}$$

Because Equation 2.47 is not integrable, a solitary-wave solution can be initially guessed and the soliton form can be [2]

$$u(\xi,\tau) = N_s \left[ \operatorname{sech}\left(\frac{\tau}{\tau_0}\right) \right]^{(1+jq)} \exp(jk\xi) \tag{2.49}$$

The parameters of the solution can be determined by substituting Equation 2.49 into Equation 2.47 as

$$N_s^2 = \frac{1}{2\tau_0^2}[s(q^2 - 2) + 3qd] \tag{2.50}$$

$$\tau_0^2 = -\frac{[d(1 - q^2) + 2sq]}{\mu} \tag{2.51}$$

$$\kappa = -\frac{1}{2\tau_0^2}[s(1 - q^2) - 2qd] \tag{2.52}$$

$$q = \frac{3s \pm \sqrt{9 + 8d^2}}{2d} \tag{2.53}$$

Thus the width and the peak power of the soliton are determined by the system parameters such as loss, gain, and its finite bandwidth. Different from solitons supported by NLSE, the influence of these parameters plays an important role in the existence of solitons in these fiber systems. Due to periodic perturbations during propagation, solitons dissipate their energy that plays an essential role in their formation and stabilization. Such a soliton is often called a dissipative soliton or an autosoliton due to its mechanism of self-organization [5]. In order to preserve the shape and energy, a balance between gain and loss mechanisms is required besides the balance between GVD and SPM. A frequency chirping can help to maintain the balance between gain and loss during propagation in a bandwidth-limited system. This explains why dissipative solitons are normally chirped pulses.

## 2.3 GENERATION OF SOLITONS USING NONLINEAR OPTICAL FIBER RING RESONATORS

### 2.3.1 MASTER EQUATION FOR MODE-LOCKING

Mode-locking is a technique to force axial modes of a laser in phase to generate short pulses. For mode-locked fiber ring lasers, an optical pulse circulating in every round-trip experiences the same effects, such as optical loss and gain. Thus the circulation of the pulse inside the ring resonators is approximately equivalent to a propagation of the pulse in a fiber transmission link with infinite length. Therefore, Equation 2.26 can be used to describe mode-locking in fiber lasers. However, the variation of the pulse in round-trip time scale is considered rather than that in distance scale. In addition, a modulation function $M$, which is responsible for various mode-locking mechanisms, can be superimposed into Equation 2.26. Thus Equation 2.26 can be modified by introducing a new variable $T=\xi/v_g$, in which $v_g$ is the group velocity of the optical pulse, to obtain

$$\frac{1}{v_g}\frac{\partial A(T,\tau)}{\partial T} + \frac{j}{2}\left(\beta_2 + \frac{jg_{sat}}{\Delta\omega_g^2}\right)\frac{\partial^2 A(T,\tau)}{\partial \tau^2} - \frac{\beta_3}{6}\frac{\partial^3 A(T,\tau)}{\partial \tau^3}$$

$$= j\gamma|A(T,\tau)|^2 A(T,\tau) + \frac{1}{2}(g_{sat}-\alpha)A(T,\tau) + M(A,T,\tau) \quad (2.54)$$

A further modification can be implemented by multiplying both sides of Equation 2.54 with $L_c$, the ring cavity length, and setting $T_c=L_c/v_g$ as the superimposed cavity temporal period. Then Equation 2.54 becomes

$$T_c\frac{\partial A(T,\tau)}{\partial T} + \frac{j}{2}\left(\beta_2 + \frac{jg_{sat}}{\Delta\omega_g^2}\right)L_c\frac{\partial^2 A(T,\tau)}{\partial \tau^2} - \frac{\beta_3 L_c}{6}\frac{\partial^3 A(T,\tau)}{\partial \tau^3}$$

$$= j\gamma L_c|A(T,\tau)|^2 A(T,\tau) + \frac{L_c}{2}(g_{sat}-\alpha)A(T,\tau) + M(A,T,\tau) \quad (2.55)$$

Equation 2.55 is the well-known master equation that was first derived by Haus to describe the mode-locking phenomenon in the time domain [13]. Thus there are two time scales in this equation: the time $T$ is measured in terms of the cavity period or the round-trip time $T_c$, while the time $\tau$ is measured in terms of the pulse temporal window. The term $M$ is a function of the amplitude and time in terms of temporal scale unit $\tau$ in every round-trip. This term depends on whether the mode-locking mechanism is passive or active. In several theoretical studies on mode-locking, Equation 2.55 is employed for investigating the pulse shape evolution around the cavity. We note that the parameters in the master equation (Equation 2.55) are averaged over the cavity length for the ease of analysis.

### 2.3.2 PASSIVE MODE-LOCKING

There are three popular structures of the passively mode-locked fiber laser as shown in Figure 2.3a through c. In the first structure, a saturable absorber acts as a passive mode locker to attenuate the lower intensity parts of a pulse, whereas higher intensity parts of a pulse are minimally attenuated since they quickly saturate the absorber and pass through without loss (see Figure 2.3a). A saturable absorber is normally a semiconductor device that can be a bulk InGaAsP saturable absorber or a saturable Bragg reflector (SBR) based on InGaAs/InP multiple quantum wells. In some cases, a mirror attached to the saturable absorber is also made using a periodic arrangement of thin layers that forms a grating and reflects light through Bragg diffraction. Such a device is also called saturable Bragg reflector (SBR). In practice, most SBRs are slow saturable absorbers because their response is much longer than the time scale of the pulse width. For the passively mode-locked fiber laser using SBR, the width of the mod-locked pulses is usually at the picoseconds time scale [14,15]; however, shorter pulses of less than 500 femtoseconds (fs) can be generated by carefully managing the dispersion property of the fiber cavity [16].

The second configuration is based on a nonlinear fiber loop mirror (NFLM), known as an all-optical switch. The NFLM is a Sagnac interferometer as described in Figure 2.4a whose intensity-dependent transmission can shorten optical pulses propagating inside the cavity. In passively mode-locked fiber lasers based on this configuration, the NFLM connects to a main fiber ring through a 3 dB coupler that splits the entering pulses into two equal counter-propagation parts, as shown in Figure 2.3b. Because the optical amplifier is unequally located in the Sagnac ring, the counter-propagation pulses acquire different nonlinear phase shifts after a roundtrip inside such a Sagnac loop. Moreover, the phase difference also depends on temporal profile of the optical pulse, thus the peak of the pulse is passed without loss while the pulse wings are reflected due to their lower intensity and smaller phase shift. It can be shown that the transmittance of the NFLM varies as a function of pulse power $P$ as given in References 17 and 18:

$$T = 1 - 0.5\{1 + \cos[0.5\gamma(G-1)PL]\} \qquad (2.56)$$

where
    $G$ is the amplification factor
    $L$ is the length of the Sagnac loop

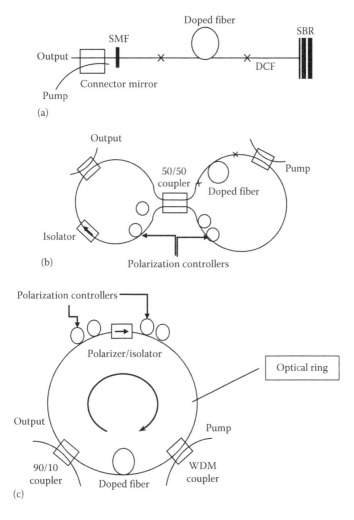

**FIGURE 2.3** Typical configurations of passively mode-locked fiber laser: (a) a linear cavity configuration, (b) a configuration based on nonlinear fiber loop mirror, and (c) a ring configuration based on nonlinear polarization rotation.

Thus a complete transmission is implemented when the peak power $P_p$ satisfies the condition

$$P_p = \frac{2\pi}{\gamma L(G-1)} \tag{2.57}$$

In other words, the peak of the pulse experiences a higher net gain per roundtrip than its side wings to shorten the pulse. This mechanism is sometimes known as

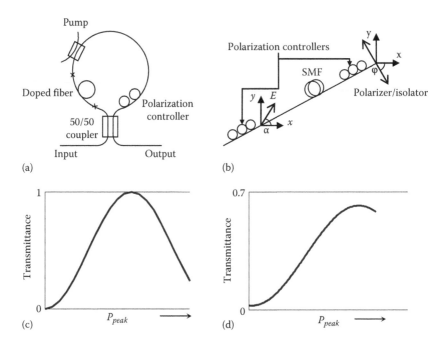

**FIGURE 2.4** A description of operation and transmittance of artificial fast saturable absorption based on (a and c) NFLM and (b and d) NPR, respectively.

additive pulse mode-locking (APM) and the NFLM can be considered as a fast or artificial saturable absorber as shown in Figure 2.4c. The fiber lasers using NFLM were first proposed for mode-locking in 1991 to obtain generated pulse-width of 0.4 ps [19] and a much shorter width of 290 fs was obtained from this laser by optimizing the dispersion and nonlinearity in the fiber cavity [20].

In the third configuration using nonlinear polarization evolution, the intensity-dependent change in polarization state is explored for mode-locking through a polarizing element. Figure 2.3c shows a setup of the passively mode-locked fiber laser based on this principle. The physical mechanism behind the mode-locking makes use of the nonlinear birefringence. A polarizer that can also be an isolator combined with two polarization controllers (PCs) acts as a mode-locker, as described in Figure 2.4b. The polarizer makes the optical wave LP and then the following PC changes the polarization state of the wave to elliptical. The polarization state evolves nonlinearly during propagation of the pulse due to the nonlinear phase shift of two orthogonal polarization components in the birefringent fiber ring. The transmittance of this structure is given by

$$T = \cos^2 \alpha \cos^2 \varphi + sin^2\alpha \sin^2 \varphi + \frac{1}{2}\sin 2\varphi \cos(\Delta\phi_L + \Delta\phi_{NL}) \qquad (2.58)$$

where

α is the angle between the polarization directions of the input light and the fast axis of optical fiber

φ is the angle between the fast axis of optical fiber and the polarization direction of the polarizer

$\Delta\phi_L$ and $\Delta\phi_{NL}$ are linear and the nonlinear phase difference between the two orthogonal polarization components, respectively.

They are given as follows:

$$\Delta\phi_L = \phi_L^y - \phi_L^x = \frac{2\pi L}{\lambda}(n_y - n_x) \qquad (2.59)$$

$$\Delta\phi_{NL} = \phi_{NL}^x - \phi_{NL}^y = \frac{\gamma L P \cos 2\alpha}{3} \qquad (2.60)$$

where

$n_x$ and $n_y$ are the refractive indices of the fast and slow axes of the optical fiber respectively

$L$ is the length of the fiber in the cavity.

Because of the intensity dependence of the nonlinear phase shift in expression (Equation 2.60), as shown in Figure 2.4d, the state of polarization varies across the pulse profile. The PC before the polarizer is adjusted to force the polarization to be linear in the peak of the pulse; hence, the high intensity part passes the polarizer without loss while the lower intensity wings are blocked. Thus the pulse is shortened after every roundtrip inside the fiber ring. This configuration can easily generate very narrow pulses of sub-100 fs by careful dispersion optimization [21–23]. The shortest pulse of 47 fs has been generated from an erbium-doped fiber laser using this technique [24].

In these techniques, the ring configurations based on NFLM and nonlinear polarization rotation (NPR) are normally applicable to soliton fiber lasers due to their fast response of the saturable absorption process as well as a possibility of self-initialization. A fast saturable absorption mode-locking can be theoretically described by introducing the saturable loss modulation into the master equation (Equation 2.55). The modulation of a fast saturable absorber $M_{sa}(t)$ can be modeled by [25]

$$M_{sa}(t) = \frac{s_0}{1 + |A|^2 / (A_{eff} I_{sat})} \qquad (2.61)$$

where

$s_0$ is the unsaturated loss

$I_{sat}$ is the saturation intensity of the absorber.

In the case of weak saturation, that is, $|A|^2 \cong I_{sat}$, Equation 2.61 can be expanded by using Taylor series as follows:

$$M_{sa}(t) \approx s_0 - \frac{s_0}{A_{eff}I_{sat}}|A|^2 = s_0 - s_{SAM}|A|^2 \tag{2.62}$$

where $s_{SAM}$ is termed as the self amplitude modulation (SAM) coefficient. The master equation of passive mode-locking thus can be derived by using Equation 2.55 and Equation 2.62 as follows:

$$T_c \frac{\partial A}{\partial T} + \frac{j}{2}\left[\beta_2 + \frac{jg_{sat}}{\Delta\omega_g^2}\right]L_c \frac{\partial^2 A}{\partial\tau^2} - \frac{\beta_3 L_c}{6}\frac{\partial^3 A}{\partial\tau^3} = j\gamma L_c|A|^2 A + \frac{L_c}{2}(g_{sat} - \alpha)$$

$$\times A - \frac{s_0}{2}A + s_{SAM}|A|^2 A \tag{2.63}$$

We can simplify Equation 2.63 by incorporating the unsaturated loss $s_0$ into the loss coefficient and ignoring the effects of dispersion and nonlinearity.

$$T_c \frac{\partial A}{\partial T} - \frac{g_{sat}}{2\Delta\omega_g^2}L_c \frac{\partial^2 A}{\partial\tau^2} = \frac{L_c}{2}(g_{sat} - \alpha')A + s_{SAM}|A|^2 A \tag{2.64}$$

This is the simplest case of passive mode-locking where the pulse shaping is based on purely saturable absorption. The solution of Equation 2.64 is a simple soliton pulse given by

$$A(\tau) = A_0 \operatorname{sech}\left(\frac{\tau}{\tau_0}\right) \tag{2.65}$$

By substituting Equation 2.65 into Equation 2.64, the pulse width and relations between parameters of the system can be obtained [25]:

$$\tau_0^2 = \frac{g_{sat}L_c}{s_{SAM}\Delta\omega_g^2|A_0|^2} \tag{2.66}$$

and

$$g_{sat} - \alpha' = -\frac{g_{sat}}{\Delta\omega_g^2\tau_0^2} \tag{2.67}$$

Equation 2.66 explains that the pulse width in passive mode-locking can be much shorter than that in active mode-locking due to the fact that the loss modulation and the curvature are proportional to $s_{SAM}|A_0|^2/\tau_0^2$. Therefore, the curvature of loss modulation under passive mode-locking increases faster when the pulse is shorter, while in active mode-locking it remains unchanged.

However, the effects of GVD and SPM are always significant to pulse shaping in practical passively mode-locked fiber lasers. Hence, the master equation needs to include these effects and is given by

$$T_c \frac{\partial A}{\partial T} + \frac{j}{2}\left[\beta_2 + \frac{jg_{sat}}{\Delta\omega_g^2}\right]L_c\frac{\partial^2 A}{\partial\tau^2} = j\gamma L_c|A|^2 A + \frac{L_c}{2}(g_{sat} - \alpha')A + s_{SAM}|A|^2 A \quad (2.68)$$

This equation has a simple steady-state solution given by [26]:

$$A(T,\tau) = A_0\left[\mathrm{sech}\left(\frac{\tau}{\tau_0}\right)\right]^{1+jq} e^{j\kappa T} \quad (2.69)$$

By using Equation 2.69 as an ansatz and balancing terms, the following pulse parameters and relations can be obtained:

$$\tau_0^2 = \frac{L_c[g_{sat}(2-q^2)/\Delta\omega_g^2 + 3\beta_2 q]}{2s_{SAM}|A_0|^2} \quad (2.70)$$

$$q = -\frac{3g_{sat}}{2\beta_2\Delta\omega_g^2} \pm \sqrt{\left(\frac{g_{sat}}{2\beta_2\Delta\omega_g^2}\right)^2 + \frac{2(\beta_2 + \gamma\tau_0|A_0|^2)}{\beta_2}} \quad (2.71)$$

$$\kappa = \frac{L_c}{2\tau_0^2 T_c}\left(\beta_2(q^2-1) + \frac{g_{sat}q}{\Delta\omega_g^2}\right) \quad (2.72)$$

$$g_{sat} - \alpha' = \frac{1}{\tau_0^2}\left(\frac{g_{sat}(q^2-1)}{\Delta\omega_g^2} - 2\beta_2 q\right) \quad (2.73)$$

Equation 2.71 indicates that a combination of the anomalous GVD ($\beta_2 < 0$) and SPM effect can give a zero chirp solution and the shortest pulses can be obtained in this case. For a small SAM coefficient and weak filtering, a soliton can be formed via the balance of anomalous GVD and SPM. The SAM and filtering effects can be considered as weak perturbations exerted on the fiber cavity. However, these perturbations play an important role in the stabilization of the pulse against the accumulated noises in the periods between the pulse circulating rounds [25].

When the solitons are periodically perturbed by the gain, loss, filtering, and SAM effects inside the fiber ring cavity, they generate continuum or dispersive waves. If the continuum components shed by the soliton is phase-matched from pulse to pulse, its energy can build up and the sidebands appear in the spectrum. These spectral components have their relative phase changes by an integer multiple ($n$) of $2\pi$ per round trip. These parasitic sidebands were first described and explained by Kelly [27]. This phenomenon is observed in most passively mode-locked soliton fiber lasers. The locations of the sidebands in the spectrum depend strongly on the dispersion of the fiber cavity via [28]

$$\Delta\lambda_n = \pm n\lambda_0 \sqrt{\frac{2n}{cDL_c} - 0.0787 \frac{\lambda_0^2}{(c\tau_{FWHM})^2}} \qquad (2.74)$$

where

$n$ is the order of sideband

$D$ is the fiber dispersion parameter in the cavity

$\tau_{FWHM}$ is the full-width at half-maximum of the pulse

$\lambda_0$ is the center wavelength.

Thus the dispersion of the cavity can be estimated from the locations of the generated sidebands.

### 2.3.3 ACTIVE MODE-LOCKING

#### 2.3.3.1 AM Mode-Locking

In this type of mode-locking, the amplitude modulation (AM) provides a time-dependent loss. The pulse would form at the time slots where the loss dip levels are below that of the gain. The modulation of an amplitude modulator can be mathematically described as

$$M_A(T,\tau) = -m_{AM}[1 - \cos(\omega_m\tau)] \qquad (2.75)$$

and the pulse evolution equation can be derived from general master equation (Equation 2.55) to describe AM mode-locking as follows:

$$T_c\frac{\partial A}{\partial T} + \frac{j}{2}\left(\beta_2 + \frac{jg_{sat}}{\Delta\omega_g^2}\right)L_c\frac{\partial^2 A}{\partial\tau^2} - \frac{\beta_3 L_c}{6}\frac{\partial^3 A}{\partial\tau^3}$$

$$= j\gamma L_c|A|^2 A + \frac{L_c}{2}(g_{sat} - \alpha)A - m_{AM}[1 - \cos(\omega_m\tau)]A \qquad (2.76)$$

where

$m_{AM}$ is the modulation index

$\omega_m = 2\pi f_m$ is the angular modulation frequency.

For active AM mode-locking, the modulation frequency is normally much higher than that of the cavity. The harmonics of the cavity's fundamental frequency ($f_c$) are $f_m = Nf_c$, with $N$ as the order of harmonic.

In the case of purely AM mode-locking, we can ignore GVD and SPM effects. Additionally, we can expand the modulation function via the use of the Taylor series expansion, to the second order in the time domain. This is due to the fact that the pulse can only be positioned at the minimum of the modulation curve (Figure 2.4). Then Equation 2.76 becomes

$$T_c \frac{\partial A}{\partial T} - \frac{g_{sat} L_c}{\Delta\omega_g^2} \frac{\partial^2 A}{\partial\tau^2} = \frac{L_c}{2}(g_{sat} - \alpha)A - \frac{m_{AM}}{2}\omega_m^2 \tau^2 A \tag{2.77}$$

The solution of this equation can be a Gaussian pulse given by

$$A = A_0 \exp\left[-\frac{\tau^2}{2\tau_0^2}\right] \tag{2.78}$$

where

$$\tau_0 = \sqrt[4]{\frac{g_{sat} L_c}{\Delta\omega_g^2 m_{AM}\omega_m^2}} \tag{2.79}$$

This result, predicted by Kuizenga–Siegman [29] shows that the pulse width is proportional to the inverse of the gain bandwidth and the modulation frequency. The eigenvalue of the equation can give the important condition of the mode-locked laser through the expression

$$g - l = \sqrt{\frac{m_{AM}\omega_m^2 g}{2\Delta\omega_g^2}} \tag{2.80}$$

where
$g = (g_{sat} L_c)/2$
$l = (\alpha L_c)/2$

$g$ and $l$ parameters are considered as the gain and the loss within one round trip of the fiber cavity.

Thus the expression (Equation 2.80) also indicates that the gain must be fixed at a certain value higher than the loss to compensate for the excess loss caused by the modulator and the filtering from the limited gain bandwidth. This condition requires that there be sufficient gain for compensating the loss of the ring cavity. Optical fiber amplifiers are thus preferred to operate in the saturation region in the cases when the ring is either under modulation or no modulation, so as to achieve stability of the total energy distributed in the ring.

With the presence of the GVD effect in the fiber lasers, the evolution of the pulse inside the fiber ring cavity satisfies the following equation:

$$T_r \frac{\partial A}{\partial T} + \frac{j}{2}\left(\beta_2 + \frac{j g_{sat}}{\Delta\omega_g^2}\right) L_c \frac{\partial^2 A}{\partial\tau^2} = \frac{L_c}{2}(g_{sat} - \alpha)A - \frac{m_{AM}}{2}(\omega_m\tau)^2 A \tag{2.81}$$

This is Hermite's differential equation and a stable solution of this equation takes the form [25]

$$A = A_0 \exp\left[-\frac{\tau^2(1+jq)}{2\tau_0^2}\right]\exp(j\kappa T) \qquad (2.82)$$

Which represents a chirped Gaussian pulse with the pulse parameters obtainable by balancing terms in Equation 2.81 as

$$\tau_0 = \sqrt[4]{\frac{L_c\left[g_{sat}(1-q^2)+2q\beta_2\Delta\omega_g^2\right]}{m_{AM}\omega_m^2\Delta\omega_g^2}} \qquad (2.83)$$

$$q = -\frac{g_{sat}}{\Delta\omega_g^2\beta_2} \pm \sqrt{\left(\frac{g_{sat}}{\Delta\omega_g^2\beta_2}\right)^2+1} \qquad (2.84)$$

$$\kappa = -\frac{L_c}{\tau_0^2 T_c}\left(\frac{g_{sat}q}{\Delta\omega_g^2}-\beta_2\right) \qquad (2.85)$$

These results indicate that if $\beta_2 = 0$ (i.e., ignoring the GVD effect or nondispersive medium), then the chirp factor $q = 0$ and $\kappa = 0$, and subsequently the pulse width in Equation 2.83 returns to Equation 2.79. Thus the presence of GVD can cause the generated pulse to chirp.

When the fiber cavity is pumped with sufficiently high power, the SPM effect is not negligible and included into the master equation as fully described in Equation 2.76. With the addition of sufficient negative GVD and SPM, the solitary pulse formation can be obtained and the solution of Equation 2.76 is assumed to be a chirped secant hyperbole pulse [30].

$$A = A_0\left[\text{sech}\left(\frac{\tau}{\tau_0}\right)\right]^{(1+jq)} \qquad (2.86)$$

In the nonlinear regime, the pulse is shortened by a combination of SPM and negative GVD similar to passive mode-locking. However, in order to shorten the pulse by soliton compression, the following two conditions must be satisfied [31]:

1. The synchronization between the modulator and the pulse train must be maintained or the solitons must be exactly re-timed on each round trip. This is a common condition for the stable operation of mode-locked fiber lasers against the thermal drift of the cavity length. This condition also ensures the phase matching is satisfied, that is, the total phase of the light waves circulating in the ring cavity, is a multiple of $2\pi$.

2. The excess loss of the continuum, which is determined by the eigenvalue in Equation 2.76, must be higher than that of the soliton. The resulting condition is [32]

$$\text{Re}\sqrt{\frac{m_{AM}\omega_m^2}{2L_c}\left[\frac{g_{sat}}{\Delta\omega_g^2}-j\beta_2\right]} > \frac{\pi^2}{24}m_{AM}\omega_m^2\tau_0^2+\frac{g_{sat}}{3\Delta\omega_g^2\tau_0^2} \qquad (2.87)$$

In Equation 2.87, on the RHS is the loss experienced by the soliton and on the left-hand side (LHS) is the loss of the continuum. In addition, the modulation must not drive the soliton unstable. The condition for suppression of energy fluctuations of the soliton can be obtained from soliton perturbation theory [31–33]:

$$\frac{g_{sat}}{3\Delta\omega_g^2\tau_0^2} > \frac{\pi^2}{24}m_{AM}\omega_m^2\tau_0^2 \qquad (2.88)$$

From these conditions, the mode-locked pulse can be compressed with the width much shorter than that predicted by Kuizenga–Siegman. The factor of pulse width shortening $R$ can be found as follows [34]:

$$R \leq 1.37\left(\frac{\beta_2 L}{g_{sat}/(\Delta\omega_g)^2}\right)^{1/4} \qquad (2.89)$$

The condition (Equation 2.89) determines the lower limit of the pulse width with the help of SPM and negative GVD. The possible pulse width reduction is proportional to the fourth root of dispersion that indicates the need for an excessive amount of dispersion to maintain a stable soliton while suppressing the continuum spectra parts.

### 2.3.3.2 FM Mode-Locking

Contrast to AM mode-locking, frequency modulation (FM) mode-locking is based on a periodic chirping caused by phase modulation. When the modulation frequency is a multiple harmonic of the fundamental frequency, the phase matching condition can be satisfied to obtain the resonance of optical waves in the cavity. In the frequency domain, the sidebands generated by phase modulation are matched to the axial modes of the cavity. While in the time domain, the pulses are built up at the extremes of the modulation cycles. At these temporal positions, the optical waves are not chirped and thus are constructively summed when they are in phase, while they are destructively interfered at other temporal positions due to repeatedly linear chirping in the cavity.

A phase modulation of the optical field can be represented by the function

$$M_F(T,\tau) = jm_{FM}\cos(\omega_m\tau) \qquad (2.90)$$

where $m_{FM}$ is the phase modulation index. Under the phase modulation exerted in a mode-locking laser, FM mode-locking can be described by the master equation as follows:

$$T_c \frac{\partial A}{\partial T} + \frac{j}{2}\left(\beta_2 + \frac{jg_{sat}}{\Delta\omega_g^2}\right)L_c \frac{\partial^2 A}{\partial\tau^2} - \frac{\beta_3 L_c}{6}\frac{\partial^3 A}{\partial\tau^3} = j\gamma L_c |A|^2 A + \frac{L_c}{2}$$

$$\times (g_{sat} - \alpha)A + jm_{FM}\cos(\omega_m\tau)A \qquad (2.91)$$

Because the pulse is formed within a narrow temporal window of the modulation period, the modulation function can be approximated to the second order of the Taylor expansion, thus Equation 2.91 can be simplified by ignoring the contribution of the nonlinear effect as

$$T_c \frac{\partial A}{\partial T} + \frac{j}{2}\left(\beta_2 + \frac{jg_{sat}}{\Delta\omega_g^2}\right)L_c \frac{\partial^2 A}{\partial\tau^2} = \frac{L_c}{2}(g_{sat} - \alpha)A + jm_{FM}\left(1 - \frac{\omega_m^2\tau^2}{2}\right)A \qquad (2.92)$$

Therefore, Equation 2.92 describes FM mode-locking in the linear regime. Similar to AM mode-locking, the solution of this equation for a chirped Gaussian pulse is given in the form

$$A = A_0 \exp\left[-\frac{\tau^2(1+jq)}{2\tau_0^2}\right]\exp(j\kappa T) \qquad (2.93)$$

Similar to the techniques employed in Section 2.3.3.1, the pulse parameters can be obtained as

$$\tau_0 = \sqrt[4]{\frac{L_c\left[\beta_2\Delta\omega_g^2(1-q^2)+2qg_{sat}\right]}{m_{FM}\omega_m^2\Delta\omega_g^2}} \qquad (2.94)$$

$$q = \frac{\beta_2\Delta\omega_g^2}{g_{sat}} \pm \sqrt{\left(\frac{\beta_2\Delta\omega_g^2}{g_{sat}}\right)^2 + 1} \qquad (2.95)$$

$$\kappa = \frac{m_{FM}}{T_c} - \frac{L_c}{2\tau_0^2 T_c}\left(\frac{g_{sat}q}{\Delta\omega_g^2} - \beta_2\right) \qquad (2.96)$$

$$g_{sat} - \alpha = \frac{1}{\tau_0^2}\left(\beta_2 q + \frac{g_{sat}}{\Delta\omega_g^2}\right) \qquad (2.97)$$

In the simplest case, that is, pure FM mode-locking under nondispersive propagation, $\beta_2 = 0$, the generated pulse is always chirped with $q = \pm 1$ due to the applied phase modulation. On the other hand, the pulses generated from an FM mode-locked laser can be located at either extreme limits of up-chirp ($q > 0$) or down-chirp ($q < 0$) of the modulation cycle, in the absence of dispersion and nonlinearity that can create a switching between these two states in a random manner [29].

However, in the presence of the dispersion effect this switching can possibly be suppressed as indicated in Equation 2.97, which includes the effects on the formation of soliton pulse due to the excess loss of the cavity. If $\beta_2 < 0$ (anomalous dispersion), the excess loss at the up-chirp half-cycle is lower than that at the down-chirp cycle ($q > 0$) and vice versa. Thus the pulses located at positive half-cycles are preferred in the *anomalous* dispersive fiber cavity while those located at negative half-cycles are preferred in the case of *normal* dispersion. On the other hand, the pulse in the up-chirp cycle is compressed by the dispersion while the down-chirp pulse is broadened in the anomalous GVD cavity. This shortened up-chirp pulse experiences less chirp after passing the phase modulator, which reduces the loss due to the filtering and gain bandwidth limitation effects. The up-chirp pulse is finally dominant in the anomalous GVD cavity. This stability of the mode-locked pulse in the FM fiber ring laser has been theoretically demonstrated in Reference 35.

In nonlinear regime, the SPM is also significant in pulse shaping. Similar to AM mode-locking, the soliton is formed inside the fiber cavity with the balance between GVD and SPM effects. To generate stable solitons, the required gain for noise must be higher than that for the soliton and the condition for stability can be obtained by the perturbation soliton theory [36].

$$\Re e \left\{ \sqrt{\frac{jm_{FM}\omega_m^2}{2L_c} \left[ \frac{g_{sat}}{\Delta\omega_g^2} + j\beta_2 \right]} \right\} > \frac{2g_{sat}}{3\Delta\omega_g^2\tau_0^2} \tag{2.98}$$

In this equation, the LHS represents the loss of the ASE noise and the RHS the attenuation of the soliton. Tamura and Nakazawa [35,37] also indicated that pulse energy equalization occurs in support of the SPM and filtering effects when the dispersion of the cavity is anomalous. This stability of the soliton in the presence of the third-order dispersion in a FM mode-locked fiber laser has been numerically investigated in Reference 38.

Besides synchronous mode-locking in FM mode-locked fiber lasers, another mechanism for mode-locking based on asynchronous phase modulation has been proposed [39]. In this scheme, asynchronous modulation is obtained by proper detuning of the modulation frequency. However, in order to generate a stable soliton train, the detuning is required to remain within a limit that satisfies the following condition [36]:

$$\left| \Delta f_{\lim} \right| \ll \frac{1}{T_c} \Re e \left\{ \sqrt{\frac{jm_{FM}\omega_m^2}{2L_c} \left[ \frac{g_{sat}}{\Delta\omega_g^2} + j\beta_2 \right]} \right\} \tag{2.99}$$

With a small detuning within this limit, the soliton can overcome the frequency shift to remain in the mode-locked state. When the detuning exceeds the above limit, the noise can build up and destroy the solitons. The fiber laser will operate in FM oscillation state if the modulation frequency is moderately detuned. In this regime, the output has a constant intensity in the time domain but with periodical chirp and its optical spectrum is broadened [40]. The transition from FM oscillation state to phase mode-locking exhibits complex behavior at a smaller detuning rate, where relaxation oscillations (RO) with different properties can occur [41]. In this state, the noise can build up faster than the soliton to become the new pulse and replace the old one. The cause of RO is the change of the cavity loss when the modulation frequency is detuned. The pulse passes through the modulator with a small time shift from the extremes of the modulation cycles, which increase the loss of the pulse, while the ASE noise gets more gain at the extremes of modulation cycles where the loss in the cavity is the lowest. When the detuning becomes larger, a new pulse can build up from the ASE noise while the old pulse decays and ultimately disappears. This process is periodically repeated and this repetition can satisfy the phase matching condition for resonance in the fiber ring cavity, which leads to RO. In this state, the output displays periodically strong spikes, or high peak power, whose average power can be lower because more energy is now stored in the cavity due to the ring resonance. The RO can occur several times due to the central mode hopping in detuning process. The supermode noise can be dramatically enhanced between these transitions by accumulating the excess energy from the cavity through matching between the RO frequency and the frequency of the beatings between the modulation sidebands and lasing modes [42]. All these interesting phenomena can only exist in the FM mode-locked lasers. They have been theoretically and experimentally investigated as reported in References 40–42.

### 2.3.3.3 Rational Harmonic Mode-Locking

In active mode-locking, there is one way to increase the repetition rate that is to use the rational harmonic mode-locking by detuning the modulation frequency to a rational number of the cavity fundamental frequency [43] given as

$$f_m = Nf_c \pm \frac{f_c}{M} \tag{2.100}$$

where
  $N, M$ are integers
  $N$ is the harmonic order
  $M$ can be considered a multiplication factor

To understand the rate multiplication in rational harmonic mode-locking, a simple description in the frequency domain can be depicted as shown in Figure 2.5a. In the frequency domain, the harmonics of modulation frequency would be only matched to the different multiples of the cavity modes when the modulation frequency is detuned with the amount of $f_c/M$. Therefore, the repetition rate of the output pulse can be multiplied by a factor of $M$. The mechanism can be understood in the time domain as described in Figure 2.5b. When the fiber laser is detuned by a ratio $f_c/M$, the difference between the cavity round-trip time and $N$ times the modulation frequency is equal to the time delay experienced by a pulse after one round trip. On the other hand, the phase delay of a pulse between consecutive round trips is proportional to $2\pi N/M$ and the pulse returns to its original positions after $M$ round trips. As a result, there are $M$ sets of pulses in one round trip window resulting in a multiplied repetition rate.

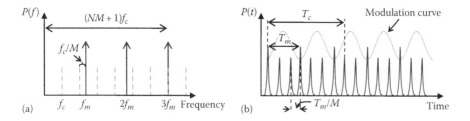

**FIGURE 2.5**   A description of rational harmonic mode-locking in (a) the frequency domain and (b) the time domain with $N=2$ and $M=3$ as an example.

Because of the pulse distribution over a nonuniform modulation window, the output pulses suffer from large amplitude fluctuations that severely limit the applications of rational harmonic AM mode-locking techniques in practical systems. Several methods such as nonlinear optics methods [44,45] and modulator transmittance adjustment [44,46,47] have been proposed for pulse amplitude equalization.

The situation of FM mode-locking is different from AM mode-locking in that it is shown that the rational harmonic mode-locking is due to the contributions of the harmonics of the modulation frequency in the amplified electrical driving signal [48]. Higher-order harmonics can be generated from the power amplifier operating in the saturation regime. Therefore, the phase of the optical field at the output of the EO phase modulator is modulated and given as

$$E_{out} = E_{in}e^{j\varphi(t)} \tag{2.101}$$

where $E_{in}$, $E_{out}$ are the optical fields at the input and output of the phase modulator, respectively, and $\varphi(t)$ varies correspondingly to the electrical driving signal. In case of the amplified signal, this variation can be represented by a summation of a series of trigonometric cosine functions as follows:

$$\varphi(t) = \sum_{k=1}^{\infty} m_k \cos(k2\pi f_m t + \theta_k) \tag{2.102}$$

where
   $f_m$ is the modulation frequency
   $m_k$ and $\theta_k$ are the modulation index and the phase delay bias for each frequency component, respectively.

Following the analysis given in Reference 48, the field of the optical signal experiences an average phase modulation after $M$ round trips by

$$\overline{\varphi}_M(t) = \sum_{k=1}^{\infty} M m_{kM} \cos(kM2\pi f_m t + \theta_{kM}) \tag{2.103}$$

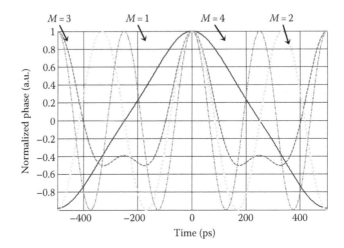

**FIGURE 2.6**   Average phase modulation profile at different detuning $f_c/M$ with $M = 1$–4 to achieve rational harmonic mode-locking in the FM mode-locked fiber laser. The driving signal of the phase modulator is modeled with magnitudes of higher harmonic components as follows: $m_2 = 0.008\ m_1$, $m_3 = 0.06\ m_1$, $m_4 = 0.001\ m_1$, and $m_5 = 0.0008\ m_1$.

Thus when the modulation frequency $f_m$ can be detuned as specified in Equation 2.100, the modulation effect of lower-order harmonics of the $f_m$ are cancelled and only the frequency component of $M$ times the modulation frequency is enhanced and becomes dominant after $M$ round trips. Otherwise, the effective phase modulation variation with respect to time can be multiplied by a multiple number $M$ within the same temporal window as  shown in Figure 2.6. Because mode-locking is based on frequency chirping, the generated pulse train in FM rational mode-locking does not suffer from the unequal amplitude problem.

## 2.4   ACTIVELY FM MODE-LOCKED FIBER RINGS: AN EXPERIMENT

### 2.4.1   Experimental Setup

By using an electro-optic (EO) phase modulator as a mode locker, the actively mode-locked fiber ring laser can generate a high-speed pulse train with low jitter. Moreover, it is simple in synchronization between the fiber cavity and other electronic devices. In this section, an actively FM mode-locked fiber ring laser will be constructed for demonstrating soliton generation at a high repetition rate in which an erbium gain medium is used for amplification inside the cavity.

Figure 2.7 shows the experimental setup of the FM mode-locked fiber ring laser. In this setup, an erbium-doped fiber amplifier (EDFA) pumped at 980 nm is used in the fiber ring to moderate the optical power in the loop for mode-locking. This amplifier operates in the saturation region and its output power can be adjusted by varying the pump laser power via the current of the pump laser diode. A 50 m Corning SMF-28 fiber is inserted after the EDFA with the aim of

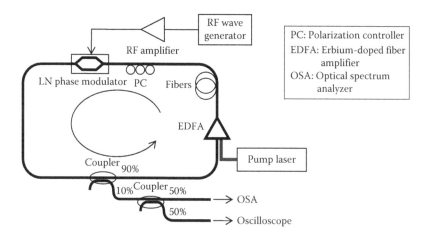

**FIGURE 2.7**   Experimental setup of an actively FM mode-locked fiber laser.

enhancing the nonlinear phase shift through the SPM effect as well as ensuring that the average dispersion in the fiber ring is *anomalous*, which is important to obtain stability of the soliton fiber laser. By measuring the input and output powers of the EDFA, the total loss of the cavity can be estimated. This loss consisting of the insertion loss of the phase modulator and connections is approximately 10.5–11 dB in the proposed setup. We note that the loss can vary due to a change in the polarization state.

An EO integrated phase modulator, model PM-315P of Crystal Technology, Inc., assumes the role as a mode locker and controls the states of locking in the fiber ring. At the input of the phase modulator, a PC consisting of two quarter-wave plates and one half-wave plate is used to control the polarization of the light wave that is required to minimize the loss cavity and influence stable formation of the solitons. The phase modulator with a half-wave voltage of 9 V is driven by a sinusoidal signal generated from a synthesizer HP-8647A in the frequency region of 1.0 GHz. The RF sinusoidal wave is amplified by a broadband RF power amplifier DC7000H with 18 dB gain that can provide a maximum saturated power of approximately 19 dBm at the output. Thus the phase modulation index of ~1 rad can be achieved for mode-locking. The fundamental frequency of the fiber laser cavity is determined by tuning the modulation frequency to lock the fiber laser at different harmonics. In our setup, the fundamental frequency of the fiber cavity is 2.2862 MHz, which is equivalent to the 90 m length of the fiber ring.

The outputs of the mode-locked laser extracted from the coupler 90:10 are monitored by an optical spectrum analyzer (OSA) HP-70952B and an oscilloscope, Agilent DCA-J 86100C with an optical bandwidth of 65 GHz. Because of the fiber laser is operating only at 1.0 GHz, the bandwidth of the oscilloscope is wide enough for measuring pulse widths larger than 10 ps. In case of pulse widths less than 10 ps, the rise time of the oscilloscope of 7.4 ps should be taken into account in the estimation of the pulse width by the relationship $\tau_p = \sqrt{\tau_{meas}^2 - \tau_{equi}^2}$, where $\tau_p$, $\tau_{meas}$, and $\tau_{equi}$ are the estimated and measured pulse widths and the rise time of the oscilloscope, respectively. A RF spectrum analyzer FS315 is also used to determine the stability of the generated pulse train.

Beside the conventional ring structure described, another setup of the actively mode-locked fiber ring laser using EO phase modulator based on the Sagnac loop interferometer has also been implemented in our experiment. When the phase modulator is placed at the middle of a fiber Sagnac loop as shown in Figure 2.8, phase modulation is converted into AM by interference between clockwise light and counterclockwise light, which have a phase difference between them. This effect comes from the fact that the phase modulator is optimized for only one transmission direction. The transmission of the phase modulated Sagnac loop (PMSL) output is given by [49]

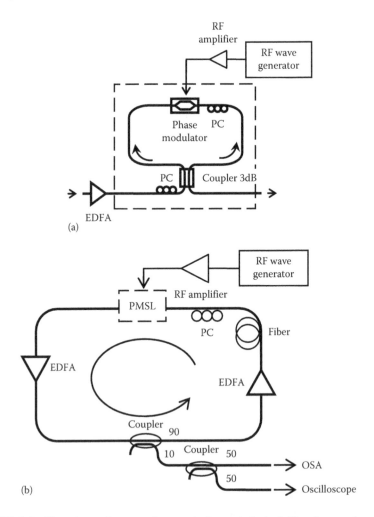

**FIGURE 2.8** Experimental setup of an actively mode-locked fiber laser using phase-modulated Sagnac loop (PMSL): (a) schematic diagram of a whole fiber ring laser and (b) detailed diagram of the PMSL.

$$T(t) = \sin^2 \left[ \frac{\Delta\varphi(t) + \phi}{2} \right] \qquad (2.104)$$

where

$\Delta\varphi(t)$ is the differential phase caused by the driving signal between two passage directions

$\phi$ is the bias differential phase.

Thus in this configuration of the actively mode-locked fiber laser, the PMSL can be considered as an amplitude modulator without the bias drift problem and possibly polarization dependence [49,50]. Equation 2.104 shows that the intensity modulation of the PMSL can operate at double modulation frequency. However, the mode-locked pulses at peaks of the transmission acquire residual chirp from phase modulation [51]. This chirp will affect the pulse characteristics as well as stability in the same way as in the actively FM mode-locked fiber ring laser. Similar to rational mode-locking, the repetition rate multiplication can be archived by detuning with different rational number of harmonics. The detuning allows the pulses to experience the opposite chirp in consecutive round trips, and consequently a uniform intensity transmission window can be obtained. In fact, the intensity modulation curve of the PMSL depends strongly on the phase modulation profile. If the phase modulation curve is distorted, the higher-order harmonics in the resulted intensity modulation signal are also strongly enhanced to facilitate the rational harmonic mode-locking.

### 2.4.2 RESULTS AND DISCUSSION

#### 2.4.2.1 Soliton Generation

By setting the saturated power of the EDFA of 5 dBm, a stable pulse train is generated by tuning the modulation frequency to the 438th order harmonics of fundamental frequency. Figure 2.9a through c shows the time trace and spectrum of the generated mode-locked pulse. The temporal and spectral widths of the pulse are 12.5 ps and 0.23 nm, respectively. Thus this result indicates that the pulse is transform-limited with the product of time-bandwidth of 0.36. Because no optical band-pass filter is used in the setup, the emission wavelength of the fiber laser is around 1560 nm where the gain of the EDFA is maximized and flat after optimizing the polarization state of the cavity. Stability of the mode-locked pulse train is demonstrated by the RF spectrum analysis, as shown in Figure 2.9d. The sideband suppression ratio (SSR) of higher than 50 dB was achieved without any feedback circuit for stabilization.

With the aim of increase in the phase modulation index, we have replaced the phase modulator PM-315P by the phase modulator Covega's Mach-40-27 which has a half-wave voltage of only 4 V at 1 GHz modulation frequency. It is surprising that a larger width of generated pulses at the higher modulation index of 2.3 rad was obtained in this setup with the same intracavity optical power as in the previous setup. Figure 2.10 shows the time trace and the corresponding optical spectrum of

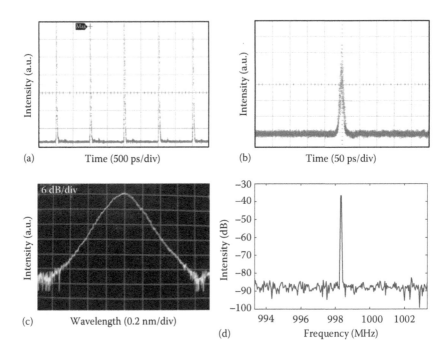

**FIGURE 2.9** Time traces of (a) the pulse train, (b) the single pulse, (c) the corresponding optical spectrum, and (d) the RF spectrum of the mode-locked pulse generated from the FM mode-locked fiber ring.

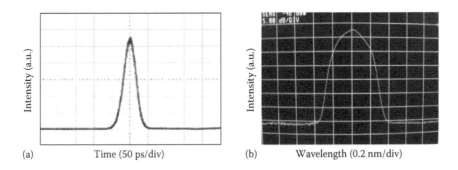

**FIGURE 2.10** (a) Time trace and (b) spectrum of the mode-locked pulse generated from the FM mode-locked fiber ring laser using the phase modulator Mach-40-27.

the mode-locked pulse generated from this setup. The pulse broadening in this case indicates a limitation of pulse spectrum or gain bandwidth in the cavity that can relate to the property of the phase modulators. This is examined in more detail in Chapter 4.

In the configuration of mode-locked fiber laser using PMSL, the generated pulse train has a high stability but wide pulse width. Figure 2.11 shows the time trace of the pulse train at 1 GHz generated from this configuration. Because of the insertion

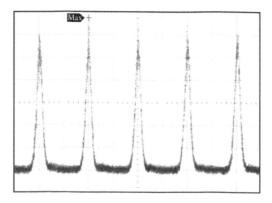

**FIGURE 2.11** Time trace of the pulse train generated from the mode-locked fiber laser using the PMSL configuration.

of the 3 dB coupler in the PMSL, the total loss of the ring cavity is about 16 dB, which is much higher than that of the conventional FM mode-locked fiber ring laser. High cavity loss reduces the efficiency of the soliton compression effect in the fiber laser. Additionally, the PMSL is equivalent to an amplitude modulator so that the modulation index depends on the intensity modulation curve converted from the phase modulation. With small phase modulation index, the modulation index of the PMSL is also small. Therefore, the width of the generated pulse is about 100 ps and the pulse shape is a Gaussian pulse rather than a soliton.

### 2.4.2.2 Detuning Effect and Relaxation Oscillation

When the modulation frequency $f_m$ is detuned, the fiber laser can experience various regimes. Especially, the transition state shows complex behaviors such as RO and excess noise enhancement, as mentioned in Section 2.3.3.2. In our setup, the fiber laser experiences three main regimes—mode-locked regimes, FM oscillation, and transition regimes—depending on the amount of detuning $|\Delta f_m|$. Because the range of $|\Delta f_m|$ in each regime depends on the cavity length or the cavity dispersion, we have inserted 100 m SMF-28 fiber into the fiber ring (Scenario A) beside 50 m SMF-28 fiber in the original setup (Scenario B). By detuning the FM, the important regimes in both cases are identified as follows:

- When $|\Delta f_m| < 2$ kHz for the Scenario A and $|\Delta f_m| < 4$ kHz for Scenario B, the fiber ring laser operates in FM oscillation regime, which can be identified by its optical spectrum. At large $|\Delta f_m|$, the optical spectrum is similar to the continuous wave (CW) signal due to the limitation of resolution in the OSA, while the RF spectrum cannot identify the first harmonic component of $f_m$ as shown in Figure 2.14a. When the effective modulation index is sufficient by decreasing $|\Delta f_m|$, the optical spectrum is broadened with a double-peak shape due to the energy going to the optical frequencies far from the center carrier mode, as shown in Figure 2.12a. The spectrum keeps broadening when the $|\Delta f_m|$ decreases close to 2 and 4 kHz for the cases of $\bar{\beta}_2 > 0$ and

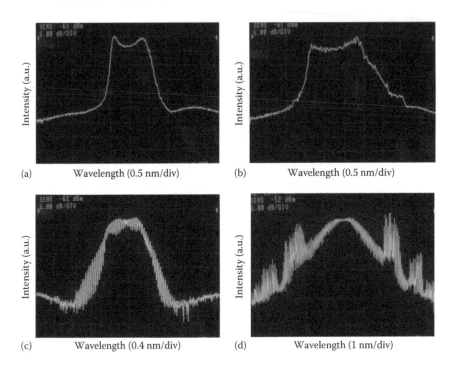

**FIGURE 2.12** Optical spectra at different frequency detuning regimes: (a) FM oscillation, (b) entering the transition regime, (c) enhanced relaxation oscillation in the transition regime, and (d) relaxation oscillation at a higher optical power level.

$\bar{\beta}_2 < 0$, respectively. Figure 2.14b also shows the typical RF spectrum in this regime, which indicates a strong supermode noise and a broad linewidth of each side mode due to the beating between the cavity modes and the modulation frequency. Moreover, the strength of the first harmonic is increased in proportion with decreasing $|\Delta f_m|$.

- When 0.3 kHz < $|\Delta f_m|$ < 2 kHz for Scenario A and 0.8 kHz < $|\Delta f_m|$ < 4 kHz for Scenario B, the fiber laser enters a transition regime where many complex behaviors such as the RO as well as the enhanced supermode noise status can occur [42]. The double-peak spectrum broadening reaches its maximum level before the energy at the main carrier grows up according to the reduction of detuning as shown in Figure 2.12b. Between the FM oscillation and mode-locking regimes, the RO as well as the enhanced supermode noise status have also been observed. In the first stage, the time trace show a high constant intensity that varies continuously and rapidly in the time domain. When the detuning decreases further, the envelope of the high intensity is more deeply modulated, as shown in Figure 2.13a. Because the supermode noise is still the dominant noise source, the RF spectrum exhibits a behavior similar to the one shown in Figure 2.14b. In the narrower resolution bandwidth, the beat noise between the modulation sidebands and the cavity modes of about 10 kHz can be observed,

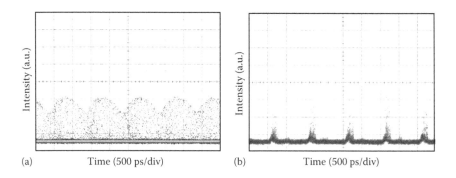

**FIGURE 2.13** Time traces in the transition regime: (a) at the initial stage and (b) at a latter stage when the relaxation oscillation is enhanced.

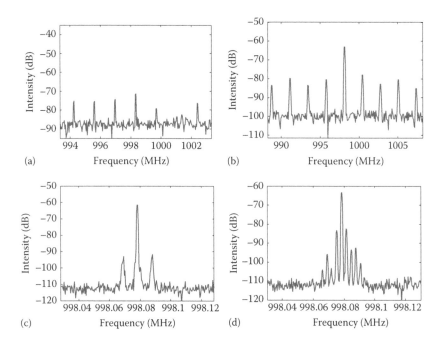

**FIGURE 2.14** RF spectra at different regimes: (a) CW-like regime, (b) FM oscillation regime and transition regime in a span of 20 MHz, (c) initial stage of transition regime in a span of 100 kHz, and (d) enhanced relaxation oscillation in transition regime in a span of 100 kHz.

as seen in Figure 2.14c. In the latter stage of the transition regime, when the detuning is decreased to around 500 Hz, the RO becomes stronger as an enhanced excess noise. In this state, the building up of new pulses and the decay of old pulses can occur at the same time that exhibits a rapid variation of both amplitude and time position. Therefore, in this state the time trace of the signal is observed as noisy pulses as shown in

Figure 2.13b and the corresponding optical spectrum shows ripples in the envelope as shown in Figure 2.12c. If the optical power in the cavity is further increased, the optical spectrum can exhibit sidebands as seen in Figure 2.12d, which indicates the existence of ultrashort pulses with high peak power in this stage. Figure 2.14d shows the RF spectrum with strong sidebands of 2 kHz formed by beating between the modulation frequency and the RO frequency.

- When $|\Delta f_m| < 0.3$ kHz for Scenario A and $|\Delta f_m| < 0.8$ kHz for Scenario B, the mode-locked state can be achieved. When the detuning is decreased to a small amount that is within a specific limitation as specified in Equation 2.99 or when there is no detuning, mode-locking can be achieved to generate the stable pulse train as shown in Figure 2.9. By providing a sufficient gain in the cavity, an RO noise suppression greater than 40 dB and a super-mode noise suppression greater than 45 dB can be achieved in our setup without using any stabilization technique.

In the transition regime, it is really interesting that the existence of ultrashort pulses with very high peak power is observed in this regime when the detuning is about 500 Hz incorporating with the adjustment of PC. Figure 2.15 shows the spectrum and the time trace of this state in Scenario A. In this figure, the time trace shows the generated pulses with very narrow width and fixed high peak power but strong timing jitter, while the optical spectrum shows a broad spectral width and Kelly sidebands generated by the resonance of dispersive waves and the generated pulses. These results indicate clearly the existence of solitons in this state. Based on the spectral width of 1.58 nm, the pulse width is approximately about 1.8 ps, which is clearly impossible to be resolved by the oscilloscope. From the sideband locations, the total cavity dispersion estimated by Equation 2.74 is of about $-0.0172$ ps$^2$/m. The high stability of the optical spectrum also demonstrates that solitons are stably formed inside the cavity. We believe that the passive mode-locking based on NPR plays a key role in this state. This process can be explained as follows: when the cavity is slightly detuned, the relaxation oscillation occurs in

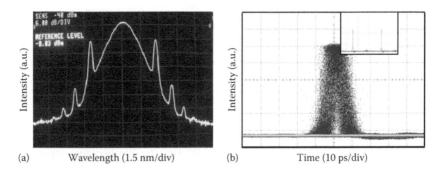

(a)         Wavelength (1.5 nm/div)          (b)          Time (10 ps/div)

**FIGURE 2.15**    (a) Spectrum and (b) time trace of the hybrid mode-locking state in Scenario A with an insertion of 100 m SMF-28 fiber into the ring cavity.

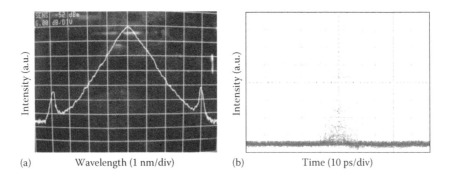

(a)          Wavelength (1 nm/div)          (b)          Time (10 ps/div)

**FIGURE 2.16**    (a) Spectrum and (b) time trace of the hybrid mode-locking state in Scenario B with an insertion of 50 m SMF-28 fiber into the ring cavity.

which the pulses in the form of spikes acquire a peak power so high that the NPR becomes significant in the weak birefringence cavity. After creating a favorable condition by changing the settings of the PC, passive mode-locking based on NPR can be achieved to shape the pulse circulating in the cavity. Thus the fiber laser in this state operates similar to a hybrid passive–active mode-locked laser [52–54]. In order to verify this passive mechanism, the RF modulation signal was turned off; however, the optical spectrum with sidebands remained for at least 5 min before disappearing. Owing to the detuning, solitons experience a frequency shift that results in temporal variation of the pulses or timing jitter. Moreover, this state operates in APM regime; it is easily prone to dropout as demonstrated in Figure 2.15b by the base line at the bottom of the pulse trace. This state is also observed in Scenario B, although it is more difficult for adjustment due to insufficient NPR effect in a shorter fiber cavity. By carefully adjusting the PC, solitons generated by this mechanism in Scenario B can be obtained as shown in Figure 2.16. With the first sideband spacing of 3.95 nm, the estimated average GVD of the cavity is about −0.0144 ps²/m. The estimation of the average GVD is valid due to a reduction of the dispersion in Scenario B of −1.1 ps² which is exactly equivalent to 50 m SMF-28 fiber.

### 2.4.2.3    Rational Harmonic Mode-Locking

As described in Section 2.3.3.3, a rational harmonic mode-locking can be implemented in the actively FM mode-locked fiber laser by an excitation of the higher-order harmonics of the amplified driving signal. By using a radio frequency (RF) spectrum analyzer, the magnitude of the harmonics at the output of the power amplifier is measured as a function of the RF input power. Figure 2.17 shows the measured results, which indicate a strong increase of the second harmonics while the magnitude of the first-order harmonic remains unchanged at the saturated value of 19 dBm at the RF input power higher than 2 dBm. Higher-order harmonics, such as third- and fourth-order harmonics, are slightly enhanced at an input power higher than 5 dBm. The magnitude of high-order harmonics

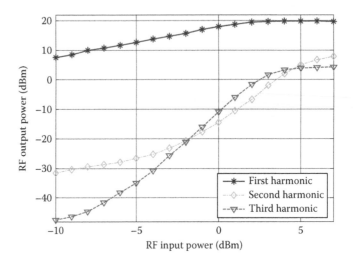

**FIGURE 2.17** Output power of the first, second, and third harmonics as a function of the RF input power.

determines the modulation index of rational harmonic mode-locking at corresponding orders.

When the modulation frequency is detuned by an amount of $\pm f_c/2$ and $\pm f_c/3$ from the 438th harmonics of the cavity fundamental frequency, the pulse trains at the repetition rate of double and triple modulation frequency are generated as shown in Figure 2.18 at the RF input power of 7 dBm. The pulse train at the second-order rational harmonic mode-locking shows a better performance than that at the third-order rational harmonic mode-locking due to higher modulation index of the second harmonic component. Higher-order rational harmonic mode-locking, such as the fourth and fifth orders, also can be obtained by an appropriate detuning, but gives very poor performance due to the weakness of the corresponding harmonic components in the driving signal.

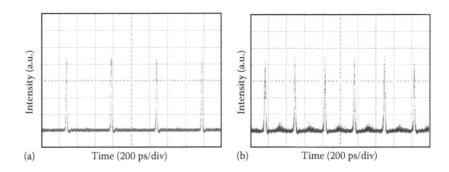

**FIGURE 2.18** Time traces of the pulse trains in the (a) second- and (b) third-order rational harmonic mode-locking.

Similarly, rational harmonic mode-locking can be achieved in the mode-locked fiber laser using PMSL. Although the phase modulator operates in small signal modulation region, higher-order harmonic components in AM of the PMSL can be easily enhanced by the distortion of phase modulation through phase-amplitude conversion. By detuning the modulation frequency of $\pm f_c/M$ with $M$ from 2 to 6, the repetition rate of the pulse train is multiplied by a factor $M$, as shown in Figure 2.19. However, the amplitude of pulses is unequal because of the nonuniformity of the intensity modulation profile.

Thus the pulse trains at the output of the mode-locked fiber ring laser using PMSL show the problem of nonuniform amplitude, which is a disadvantage of rational harmonic mode-locking using AM, while the uniform pulse trains are always generated

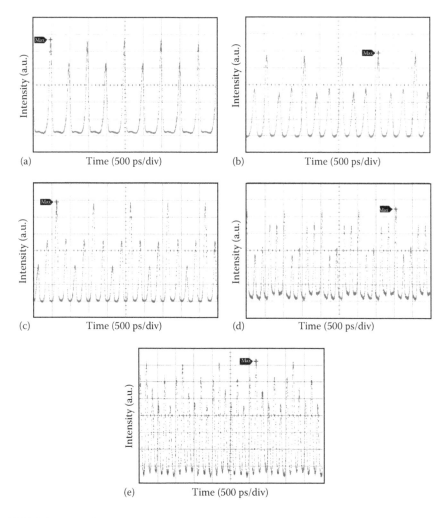

FIGURE 2.19 Time traces of (a) second-, (b) third-, (c) fourth-, (d) fifth-, and (e) sixth-order rational harmonic mode-locking in the fiber ring laser using PMSL.

by phase modulation in rational harmonic mode-locking if the strength of the higher-order harmonic component is sufficiently high in the driving signal.

## 2.5 SIMULATION OF ACTIVELY FM MODE-LOCKED FIBER LASER

### 2.5.1 NUMERICAL SIMULATION MODEL

Although FM mode-locking can be theoretically described by the master equation with averaged parameters, it is difficult in solving this equation to find an analytical solution with the involvement of all important effects. Therefore, in order to understand the physical processes occurring inside the fiber ring cavity, a numerical model is developed in this chapter. The model of mode-locked fiber laser with a ring configuration consists of basic components similar to that in the experimental setup.

Figure 2.20 shows the block diagram of the numerical model for an actively FM mode-locked fiber ring laser. In this model, a slowly varying envelope of the optical pulse passes through each component incorporated in the ring, once in each round trip. This process is repeated until a desired solution is obtained. Thus the envelope function of the pulse at the $n$th round-trip can be given by

$$A^n(t) = \hat{L}\,\hat{M}\,\hat{F}\,\hat{F}_g \cdot A^{n-1}(t) \qquad (2.105)$$

where

$A^n$, $A^{n-1}$ are the complex amplitude of the pulse at the $n$th and the $(n-1)$th round-trips respectively

$\hat{L}$, $\hat{M}$, $\hat{F}$, and $\hat{F}_g$ are the operators representing for the loss of the cavity, the modulation mechanism, and the passive and active fibers, respectively.

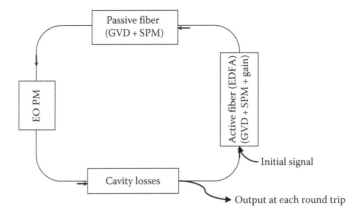

**FIGURE 2.20** Schematic of the numerical simulation model for generation of solitons in an actively FM mode-locked fiber ring laser.

Based on well-understood physical mechanisms, the operators or the models of the components inside the ring cavity can be exactly described and the effect of each component individually considered in the model described next.

First, the operators $\hat{F}$ and $\hat{F}_g$, which describe the propagation of the optical field in optical fibers, can be modeled by the generalized NLSE (Equation 2.26), which can be rewritten for convenience as follows:

$$\frac{\partial A}{\partial z} + \frac{j}{2}\left(\beta_2 + \frac{jg_{sat}}{\Delta\omega_g^2}\right)\frac{\partial^2 A}{\partial\tau^2} - \frac{\beta_3}{6}\frac{\partial^3 A}{\partial\tau^3} = i\gamma|A|^2 A + \frac{1}{2}(g_{sat} - \alpha)A \qquad (2.106)$$

In case of passive fiber, the gain factor is set to null, but the gain factor with saturation is finite in the active fiber for amplification in the EDFA and is modeled by using Equations 2.23 through 2.25. The ASE noise generated from the EDFA is also included at the end of the active fiber. The ASE noise is modeled as an additive complex Gaussian-distributed noise with a variance given by

$$\sigma_{ASE}^2 = hvn_{sp}(G-1)B_{ASE} \qquad (2.107)$$

where
  $h$ is the Planck constant
  $v$ is the optical carrier frequency
  $G$ is the gain coefficient
  $B_{ASE}$ is the optical noise bandwidth
  $n_{sp}$ is the spontaneous factor that relates to the noise figure $NF$ of the EDFA as
    follows:

$$n_{sp} = \frac{NF \times G - 1}{2(G-1)} \qquad (2.108)$$

Equation 2.106, applicable for both active and passive fibers, can be solved by using the well-known split-step Fourier method in which the fiber is split into small sections of length and the linear and nonlinear effects are alternatively evaluated between two Fourier-transform domains, respectively [1].

Second, the operator $\hat{M}$ for an EO phase modulator is given by

$$\hat{M} = \exp[jm\cos(\omega_m(\tau + \Delta\tau_s) + \phi_0] \qquad (2.109)$$

where
  $m$ is the phase modulation index
  $\phi_0$ is the initial phase
  $\omega_m = 2\pi f_m$ is the angular modulation frequency, assumed to be a harmonic of the
    fundamental frequency of the fiber ring
  $\Delta\tau_s$ is the time shift caused by detuning and is given by

$$\Delta\tau_s = \frac{T_m - T_h}{T_c} T \tag{2.110}$$

where

$T_m = 1/f_m$ is the modulation period

$T_h = T_c/N$, where $T_c$ is the cavity period and $N$ is the harmonic order

$T$ is the time scale in terms of the round trip scale

Besides the attenuation of optical fibers, there are some losses inside the cavity such as coupling loss and insertion losses of the modulator and connectors. These losses need to be included into the simulation and combined into the total cavity loss factor. The influence of these losses is given as follows:

$$\hat{L} = 10^{l_{dB}/20} \tag{2.111}$$

where $l_{dB}$ is the total loss of the cavity in dB. Thus this effect in turn determines the required gain coefficient of the EDFA to ensure that the gain is sufficient to compensate the total loss in a single round trip for stable lasing operation.

Equations 2.106 through 2.111 provide a full set of equations for numerical simulation of an actively FM mode-locked fiber ring laser. Due to recursive nature of pulse propagation in a ring cavity, the operation of the operators is repeatedly applied to the complex envelope of the pulse to find a stable solution. The complex amplitude of the output is used as the input of the next round-trip and stored for display and analysis. In the simulation of pulse formation, a complex Gaussian-distributed noise of −10 dBm is used as a seeding signal. Depending on the strength of the effects in the model, a stable pulse can be found in 500 round trips or even up to 10,000 round trips. The number of samples in the simulation window as well as the step size in the spatial domain is properly chosen to minimize numerical errors in the calculation.

## 2.5.2 SIMULATION RESULTS AND DISCUSSION

### 2.5.2.1 Mode-Locked Pulse Formation

By using the numerical model described, the pulse formation in the FM mode-locked fiber ring laser can be investigated. Table 2.1 summarizes all parameters used in the simulations. Figure 2.21a shows an evolution of the mode-locked pulse built up from the noise at $m \sim 1$ radian in the ring cavity with an anomalous $\overline{\beta}_2$ of −0.0171 ps²/m. Figure 2.21b plots the peak power as a function of round trip number. The steady state is only reached after 5000 round trips and a damped oscillation occurs in the initial stage of pulse formation process. Figure 2.22 shows the time trace and the spectrum of mode-locked pulse at steady state. We note that the pulse with the width of 11.6 ps is well fitted to a secant hyperbolic pulse rather than a Gaussian pulse; however, its spectrum exhibits no sideband due to weak dispersive waves in the cavity.

The effect of phase modulation index $m$ on the mode-locked pulse is also numerically investigated by varying the index in a range from 0.175 to $\pi$ rad and the results

## TABLE 2.1

## Parameter Values Used in the Simulations of FM Mode-Locked Fiber Lasers

| | | |
|---|---|---|
| $\beta_2^{SMF} = -21.7$ ps$^2$/km for $\bar{\beta}_2 < 0$ | $\beta_2^{EDF} = 19$ ps$^2$/km | $\Delta\omega_g = 16$ nm |
| $\beta_2^{SMF} = +21.7$ ps$^2$/km for $\beta_2 > 0$ | $\gamma^{EDF} = 0.0023$ W$^{-1}$ m$^{-1}$ | $f_m \approx 1$ GHz |
| $\gamma^{SMF} = 0.0014$ W$^{-1}$ m$^{-1}$ | $\alpha^{EDF} = 0.5$ dB/km | $m = 0.05\pi - 1\pi$ rad |
| $\alpha^{SMF} = 0.2$ dB/km | $P_{sat} = 5-8$ dBm | $\lambda = 1559$ nm |
| $L_{SMF} = 80$ m | $g_0 = 0.315$ m$^{-1}$ | |
| $L_{EDF} = 10$ m | $NF = 5$ dB | |

SMF, SSMF; EDF, erbium-doped fiber; NF, noise figure of the EDFA.

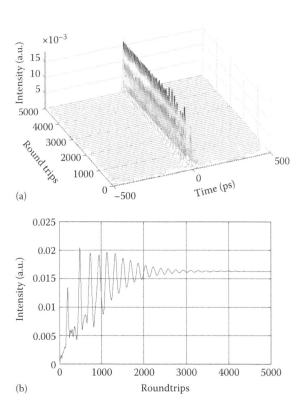

(a)

(b)

**FIGURE 2.21** (a) Numerically simulated evolution of the mode-locked pulse formation from noise and (b) variation of the peak power during 5000 round trips in the anomalous average dispersion cavity.

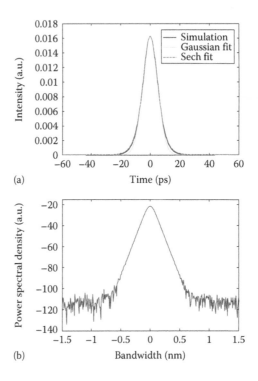

(a)

(b)

**FIGURE 2.22** (a) Numerically simulated time trace and (b) the corresponding spectrum of mode-locked pulse at steady state in the anomalous average dispersion cavity.

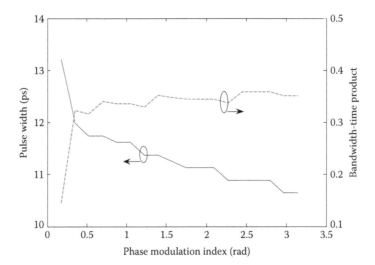

**FIGURE 2.23** Variation of mode-locked pulse width, hence, bandwidth–time product, as a function of the phase modulation index in the anomalous average dispersion cavity.

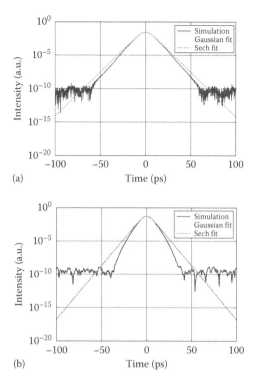

(a)

(b)

**FIGURE 2.24** Numerical simulated waveforms in steady state of mode-locked pulses at two different modulation indices: (a) $m = 0.87$ radian and (b) $m = \pi$ radian.

are shown in Figure 2.23. The increase in the modulation index shortens the pulse width. At the index lower than 0.5 rad, the rate of pulse shortening is higher due to the soliton compression effect, which results from the dominant SPM effect in pulse shaping. However, at the modulation index higher than 0.5 rad where the active phase modulation becomes stronger in pulse shaping, the reduction of the pulse width is slow and almost linear. The chirp of pulse is increased with the increase in the phase modulation index and the pulse diverges from the secant hyperbolic profile and approaches a Gaussian pulse shape of higher exponential index. Figure 2.24 shows the waveforms of the generated pulse at two different modulation indices with the Gaussian fit and secant hyperbolic fit curves.

Instead of an anomalous dispersion cavity, the sign of dispersion in the fibers is reversed to provide a normal dispersion cavity with $\bar{\beta}_2$ of +0.0213 ps²/nm. With the same conditions of mode-locking, the mode-locked pulse in the normal dispersion cavity is wider than that in the anomalous dispersion cavity. The parameters of the mode-locked pulse as a function of the modulation index in the normal dispersion cavity are depicted in Figure 2.25. The pulse is also shortened with the increase in the phase modulation index. The evolution of a mode-locked pulse in the normal average dispersion cavity at $m \sim 1$ rad is shown in Figure 2.26a. Figure 2.27 shows

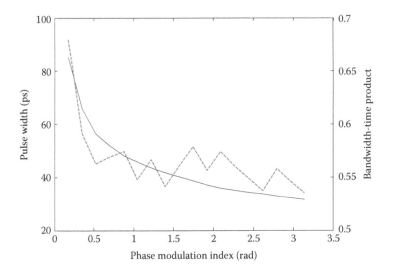

**FIGURE 2.25**   Variation of mode-locked pulse width, hence, bandwidth–time product, as a function of the phase modulation index in the normal average dispersion cavity.

the pulse profile in the time domain and its spectrum at steady state, which indicates a parabolic pulse rather than a soliton or Gaussian pulse. Furthermore, in the initial stage of pulse formation, dark solitons now exist and are embedded into the background pulses. The amplitude of the dark soliton can increase, as shown in Figure 2.26b. It is understood that in this stage the accumulated phase modulation is relatively weak, making the SPM effect dominant due to the high gain from the EDFA. The dark soliton formation occurs from the balance of normal GVD and SPM effects, yet the dark soliton is unstable. It experiences a periodic variation of time position (dotted line in Figure 2.26b) and decays due to the chirping caused by active phase modulation. Figure 2.28 shows the waveform with a dip at near center of the background pulse and its phase profile with a phase change of about $\pi/2$ at the dip at the 2900th round trip. Because the intensity of the dip that also varies along the evolution is nonzero, the formed dark soliton is referred to as the gray soliton rather than a black soliton. The dark soliton exists only until the accumulated phase shift is sufficient to lock a pulse at the minimum of the modulation cycle.

In FM mode-locked fiber lasers, the pulse formation at up-chirping or down-chirping cycles depends on the dispersion of the cavity which is anomalous or normal. On the other hand, the pulse switching between positive and negative modulation cycles is suppressed due to the presence of GVD effect. Figure 2.29 shows evolutions of the mode-locked pulses built up from noise in normal and anomalous dispersive fiber rings in one modulation period. The modulation curve is phase-shifted by $\pi/2$ to display both up-chirp and down-chirp cycles in the same simulation window. The dashed lines in the graphs indicate the phase modulation curve applied in every round-trip. Thus the pulse is built up only at the extreme of the up-chirp cycle in

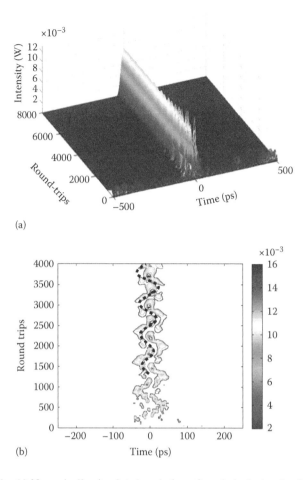

(a)

(b)

**FIGURE 2.26** (a) Numerically simulated evolution of mode-locked pulse formation from noise in the normal average dispersion cavity over 8000 round trips and (b) the evolution in the first 4000 round trips showing the dark soliton formation (black dotted line) embedded in the background pulse.

the anomalous dispersion cavity or only at the extreme of the down-chirp cycle in the normal dispersion cavity.

### 2.5.2.2 Detuning Operation

In the detuning operation, the modulation frequency is moved away from the harmonic of the cavity frequency. On the other hand, the pulse passes through the modulator at different positions during each round trip, that is, not only at the extreme of the modulation cycle, and experiences a frequency shifting. During propagation through the dispersive fiber, the frequency shift is converted into a temporal position variation in the simulation window. However, a large detuning can destroy a stable

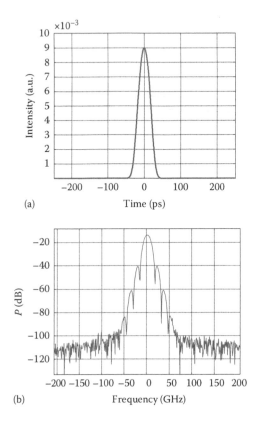

**FIGURE 2.27** (a) Simulated waveform and (b) spectrum of mode-locked pulse at steady state in the normal average dispersion cavity.

mode-locking state due to fast variations in the modulation cycle between successive round trips and the fiber laser falls into the FM oscillation regime, which generates a highly chirped signal. Figure 2.30 shows the numerically simulated results in the time and frequency domains when the modulation frequency is detuned by 6 kHz. The evolution presented in Figure 2.30a indicates patterns like noise changing from one round trip to the next one. An example of unstable waveform in the time domain at the 5000th round trip is shown in Figure 2.30b, and its spectrum in Figure 2.30c is broadened with two peaks as observed in the experiment. By taking the average of the waveforms over the last 500 round trips, a waveform with the envelope modulated at $f_m$ can be observed in Figure 2.30d.

When the detuning is slightly moderate, the phase variation of the modulation cycle between successive round trips is sufficiently slow to enable the pulse to build up in the cavity with adequately high gain. However, the built-up pulses experience the frequency shift induced by detuning that leads to the variation of temporal position and higher loss to be decayed. RO behavior occurs in this state as shown in Figure 2.31a at the detuning of 1 kHz. The repetition of the process consisting

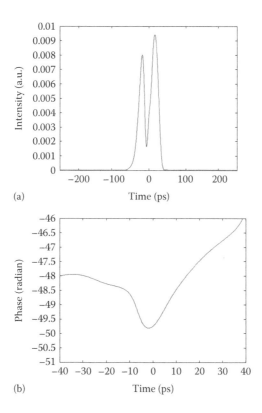

(a)

(b)

**FIGURE 2.28**    (a) Simulated waveform and (b) its phase profile at the 2900th round trip showing a gray soliton embedded in the building-up pulse.

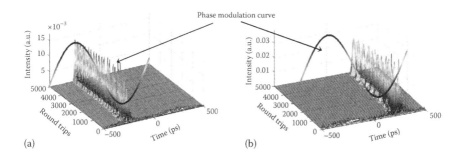

(a)                                                          (b)

**FIGURE 2.29**    Numerically simulated evolution of mode-locked pulse built up from noise in (a) anomalous dispersion cavity and (b) normal dispersion cavity at the same phase of modulation curve.

of the pulse decay and the pulse building up exhibits a turbulence-like behavior as seen in the contour plot view in Figure 2.31b. Figure 2.31c shows a typical time trace of this state at the 5000th round trip in which there are three pulses existing simultaneously in the same modulation cycle. In this figure, the lowest pulse close to the extreme of the modulation cycle is the new pulse building up from noise, the middle pulse with the highest peak power and narrow width is experiencing the time shift, while the last pulse with the lower peak, which is far from the extreme, is decaying due to higher loss. The built-up pulses can survive in around 500–1000

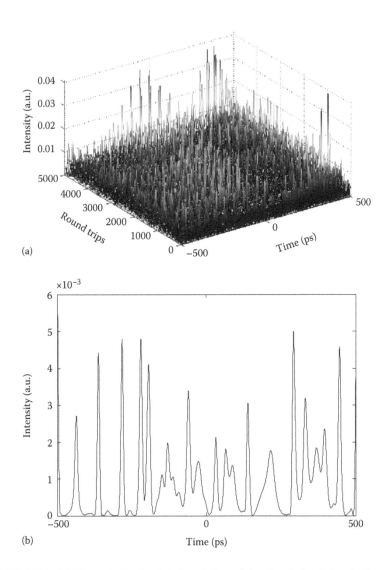

**FIGURE 2.30** (a) Numerically simulated evolution of the signal circulating in the cavity, (b) the time trace of the output over 5000 round trips. *(Continued)*

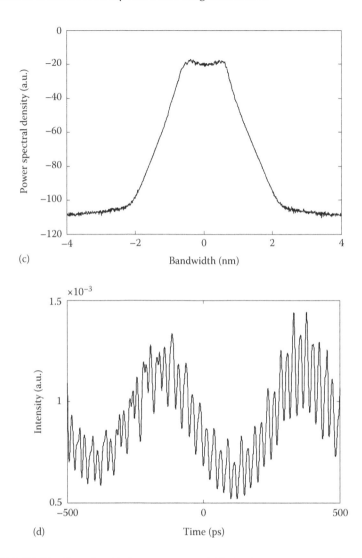

**FIGURE 2.30 (*Continued*)** (c) the spectrum, and (d) the time trace averaged over the last 500 round trips when the modulation frequency is detuned by 6 kHz.

round trips. Because very narrow pulses of 2.5 ps are generated in this case, the corresponding spectrum exhibits sidebands as shown in Figure 2.31d.

When the detuning is sufficiently small, the cavity still remains in a mode-locked state as in the case of asynchronous mode-locking. However, the mode-locked pulse can experience a variation of temporal position induced by the frequency shift and the dispersion of the cavity. Figure 2.32 shows the behavior of the mode-locked pulse when the modulation frequency is detuned by 0.25 kHz. The contour plot in Figure 2.32b indicates that the pulse still can overcome the detuning problem to stabilize in the cavity.

## 2.6   CONCLUDING REMARKS

In this chapter, the fundamentals of optical pulse propagation and mode-locking mechanisms were reviewed. In order to obtain a stable pulse train from active mode-locking, some conditions have been summarized and explained. In an active mode-locked fiber laser with sufficiently high gain, the SPM effect becomes significant for pulse compression inside the anomalous dispersion average cavity to generate solitons.

Actively mode-locked fiber ring lasers using the EO phase modulator have been experimentally demonstrated for the generation of a sequence of short pulses. The obtained results also indicate that the FM mode-locked fiber laser offers better performance in terms of pulse width and stability than that using PMSL acting as an amplitude modulator. Rational harmonic mode-locking has been achieved by the enhancement of higher-order harmonics in the electrical driving signal applied to

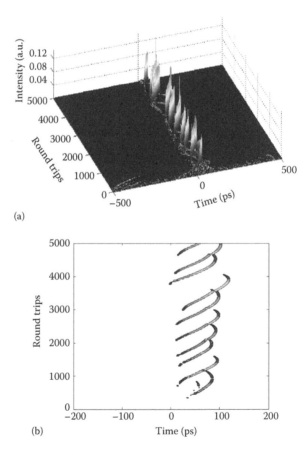

(a)

(b)

**FIGURE 2.31**   (a) Numerically simulated evolution of the signal circulating in the cavity, (b) contour plot view of the evolution.                                                                 (*Continued*)

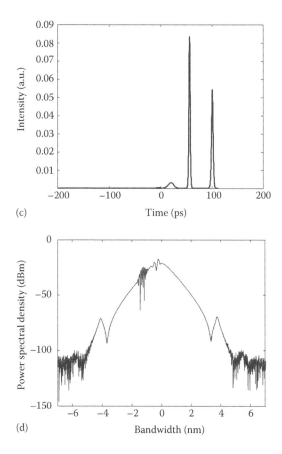

(c)

(d)

**FIGURE 2.31 (Continued)** (c) the time trace of the output over 5000 round trips, and (d) the spectrum averaged over the last 500 round trips when the modulation frequency is detuned by 1 kHz with a higher gain factor $g_0=0.315$ m$^{-1}$ and $P_{sat}=8$ dBm.

the optical modulator. Moreover, the detuning effect in the fiber laser has been characterized through the time traces, optical spectrum, and RF spectrum analysis. All obtained results indicated complex behaviors in the transition regime in which the hybrid passive–active mode-locking can be achieved to generate solitons of narrow pulse width.

Furthermore, a numerical model of the FM mode-locked fiber ring laser was developed to investigate the physical processes occurring inside the fiber cavity. The formation of mode-locked pulses in various conditions of the cavity has been numerically studied to demonstrate that short pulses generated from the FM mode-locked fiber laser system are, indeed, chirped solitons. The model has also been used to reproduce all possible operations of the FM mode-locked fiber laser, including the detuning effect in the experiment, to demonstrate the validity of the model that will be used in the other chapters.

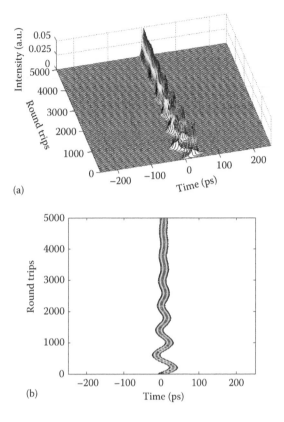

(a)

(b)

**FIGURE 2.32** (a) Numerically simulated evolution of the mode-locked pulse circulating in the cavity and (b) contour plot view of the evolution when detuning the modulation frequency by only 250 Hz.

## REFERENCES

1. G. P. Agrawal, *Nonlinear Fiber Optics*, 3rd edn., Academic Press, San Diego, CA, 2001.
2. G. P. Agrawal, *Applications of Nonlinear Fiber Optics*, Academic Press, San Diego, CA, 2001.
3. G. P. Agrawal, Optical pulse propagation in doped fiber amplifiers, *Phys. Rev. A*, 44, 7493, 1991.
4. S. C. Mancas and S. R. Choudhury, The complex cubic-quintic Ginzburg-Landau equation: Hopf bifurcations yielding traveling waves, *Math. Comput. Simul.*, 74, 281–291, 2007.
5. N. Akhmediev and A. Ankiewicz (Eds.), *Dissipative Solitons* (Lecture Notes in Physics 661). Springer, Berlin, Germany, 2005.
6. H. A. Haus, J. G. Fujimoto, and E. P. Ippen, Structures for additive pulse mode locking, *J. Opt. Soc. Am. B*, 8, 2068–2076, 1991.
7. C. Menyuk, Nonlinear pulse propagation in birefringent optical fibers, *IEEE J. Quant. Electron.*, 23, 174–176, 1987.

8. D. Y. Tang, S. Fleming, W. S. Man, H. Y. Tam, and M. S. Demokan, Subsideband generation and modulational instability lasing in a fiber soliton laser, *J. Opt. Soc. Am. B*, 18, 1443–1450, 2001.

9. K. M. Spaulding, D. H. Yong, A. D. Kim, and J. Nathan Kutz, Nonlinear dynamics of mode-locking optical fiber ring lasers, *J. Opt. Soc. Am. B*, 19, 1045–1054, 2002.

10. V. E. Zakharov and A. B. Shabat, Exact theory of two-dimensional self-focusing and one-dimensional self-modulation of waves in nonlinear media, *Soviet Phys.—JETP*, 34, 62–69, 1972.

11. Y. S. Kivshar and B. Luther-Davies, Dark optical solitons: Physics and applications, *Phys. Rep.*, 298, 81–197, 1998.

12. L. N. Binh, *Photonic Signal Processing: Techniques and Applications*, CRC Press, Boca Raton, FL, 2008.

13. H. Haus, A theory of forced mode locking, *IEEE J. Quant. Electron.*, 11, 323–330, 1975.

14. M. Zirngibl, L. W. Stulz, J. Stone, and J. Hugi, 1.2 ps pulses from passively mode-locked laser diode pumped Er-doped fiber ring laser, *Electron. Lett.*, 27, 1734–1735, 1991.

15. D. Abraham, R. Nagar, V. Mikhelashvili, and G. Eisenstein, Transient dynamics in a self-starting passively mode-locked fiber-based soliton laser, *Appl. Phys. Lett.*, 63, 2857–2859, 1993.

16. B. C. Collings, K. Bergman, and W. H. Knox, Stable multigigahertz pulse-train formation in a short-cavity passively harmonic mode-locked erbium/ytterbium fiber laser, *Opt. Lett.*, 23, 123–125, 1998.

17. N. J. Doran and D. Wood, Nonlinear-optical loop mirror, *Opt. Lett.*, 13, 56–58, 1988.

18. M. E. Fermann, F. Haberl, M. Hofer, and H. Hochreiter, Nonlinear amplifying loop mirror, *Opt. Lett.*, 15, 752–754, 1990.

19. I. N. Duling, III, Subpicosecond all-fiber erbium laser, *Electron. Lett.*, 27, 544–545, 1991.

20. M. Nakazawa et al., Low threshold, 290 fs Erbium-doped fiber laser with a nonlinear amplifying loop mirror pumped by InGaAsP laser diodes, *Appl. Phys. Lett.*, 59, 2073–2075, 1991.

21. K. Tamura, E. P. Ippen, H. A. Haus, and L. E. Nelson, 77-fs pulse generation from a stretched-pulse mode-locked all-fiber ring laser, *Opt. Lett.*, 18, 1080–1082, 1993.

22. K. Tamura, C. R. Doerr, L. E. Nelson, H. A. Haus, and E. P. Ippen, Technique for obtaining high-energy ultrashort pulses from an additive-pulse mode-locked erbium-doped fiber ring laser, *Opt. Lett.*, 19, 46–48, 1994.

23. G. Lenz et al., All-solid-state femtosecond source at 1.55 μm, *Opt. Lett.*, 20, 1289–1291, 1995.

24. D. Y. Tang and L. M. Zhao, Generation of 47-fs pulses directly from an erbium-doped fiber laser, *Opt. Lett.*, 32, 41–43, 2007.

25. H. A. Haus, Mode-locking of lasers, *IEEE J. Select. Top. Quant. Electron.*, 6, 1173–1185, 2000.

26. O. E. Martinez, R. L. Fork, and J. P. Gordon, Theory of passively mode-locked lasers including self-phase modulation and group-velocity dispersion, *Opt. Lett.*, 9, 156–158, 1984.

27. S. M. J. Kelly, Characteristic sideband instability of periodically amplified average soliton, *Electron. Lett.*, 28, 806–807, 1992.

28. M. L. Dennis and I. N. Duling, III, Experimental study of sideband generation in femtosecond fiber lasers, *IEEE J. Quant. Electron.*, 30, 1469–1477, 1994.

29. D. Kuizenga and A. Siegman, FM and AM mode locking of the homogeneous laser—Part I: Theory, *IEEE J. Quant. Electron.*, 6, 694–708, 1970.

30. H. Sotobayashi and K. Kikuchi, Design theory of ultra-short pulse generation from actively mode-locked fiber lasers, *IEICE Trans. Electron.*, E81-C, 201–207, 1998.

31. D. J. Jones, H. A. Haus, and E. P. Ippen, Subpicosecond solitons in an actively mode-locked fiber laser, *Opt. Lett.*, 21, 1818–1820, 1996.

32. H. A. Haus and A. Mecozzi, Long-term storage of a bit stream of solitons, *Opt. Lett.*, 17, 1500–1502, 1992.

33. H. A. Haus and A. Mecozzi, Noise of mode-locked lasers, *IEEE J. Quant. Electron.*, 29, 983–996, 1993.

34. F. X. Kärtner, D. Kopf, and U. Keller, Solitary-pulse stabilization and shortening in actively mode-locked lasers, *J. Opt. Soc. Am. B*, 12, 486–496, 1995.

35. K. Tamura and M. Nakazawa, Pulse energy equalization in harmonically FM mode-locked lasers with slow gain, *Opt. Lett.*, 21, 1930–1932, 1996.

36. H. A. Haus, D. J. Jones, E. P. Ippen, and W. S. Wong, Theory of soliton stability in asynchronous modelocking, *J. Lightwave Technol.*, 14, 622–627, 1996.

37. O. Pottiez, O. Deparis, R. Kiyan, M. Haelterman, P. Emplit, P. Mégret, and M. Blondel, Supermode noise of harmonically mode-locked erbium fiber lasers with composite cavity, *IEEE J. Quant. Electron.*, 38(3), 252–259, March 2002.

38. N. G. Usechak, G. P. Agrawal, and J. D. Zuegel, FM mode-locked fiber lasers operating in the autosoliton regime, *IEEE J. Quant. Electron.*, 41, 753–761, 2005.

39. C. R. Doerr, H. A. Haus, and E. P. Ippen, Asynchronous soliton mode locking, *Opt. Lett.*, 19, 1958–1960, 1994.

40. S. Longhi and P. Laporta, Floquet theory of intracavity laser frequency modulation, *Phys. Rev. A*, 60, 4016, 1999.

41. S. Yang, E. A. Ponomarev, and X. Bao, Experimental study on relaxation oscillation in a detuned FM harmonic mode-locked Er-doped fiber laser, *Opt. Commun.*, 245, 371–376, 2005.

42. S. Yang and X. Bao, Experimental observation of excess noise in a detuned phase-modulation harmonic mode-locking laser, *Phys. Rev. A*, 74, 033805, 2006.

43. Z. Ahmed and N. Onodera, High repetition rate optical pulse generation by frequency multiplication in actively modelocked fib, *Electron. Lett.*, 32, 455, 1996.

44. Y. J. Kim, C. G. Lee, Y. Y. Chun, and C.-S. Park, Pulse-amplitude equalization in a rational harmonic mode-locked semiconductor fiber ring laser using a dual-drive Mach-Zehnder modulator, *Opt. Express*, 12(5), 907–915, 2004.

45. C. G. Lee, Y. J. Kim, H. K. Choi, and C.-S. Park, Pulse-amplitude equalization in a rational harmonic mode-locked semiconductor ring laser using optical feedback, *Opt. Commun.*, 209, 417–425, 2002.

46. G. Zhu and N. K. Dutta, Eighth-order rational harmonic mode-locked fiber laser with amplitude-equalized output operating at 80 Gbits/s, *Opt. Lett.*, 30, 2212–2214, 2005.

47. F. Xinhuan et al., Pulse-amplitude equalization in a rational harmonic mode-locked fiber laser using nonlinear modulation, *IEEE Photon. Technol. Lett.*, 16, 1813–1815, 2004.

48. Y. Shiquan and B. Xiaoyi, "Rational harmonic mode-locking" in a phase-modulated fiber laser, *IEEE Photon. Technol. Lett.*, 18, 1332–1334, 2006.

49. M. L. Dennis, I. N. Duling, III, and W. K. Burns, Inherently bias drift free amplitude modulator, *Electron. Lett.*, 32, 547–548, 1996.

50. M. L. Dennis and I. N. Duling, III, Polarisation-independent intensity modulator based on lithium niobate, *Electron. Lett.*, 36, 1857–1858, 2000.

51. Y. Shiquan and B. Xiaoyi, Repetition-rate-multiplication in actively mode-locking fiber laser by using phase modulated fiber loop mirror, *IEEE J. Quant. Electron.*, 41, 1285–1292, 2005.

52. H. A. Haus, E. P. Ippen, and K. Tamura, Additive-pulse modelocking in fiber lasers, *IEEE J. Quant. Electron.*, 30, 200–208, 1994.
53. T. F. Carruthers, I. N. Duling, III, and M. L. Dennis, Active–passive modelocking in a single-polarisation erbium fiber laser, *Electron. Lett.*, 30, 1051–1053, 1994.
54. C. R. Doerr, H. A. Haus, E. P. Ippen, M. Shirasaki, and K. Tamura, Additive-pulse limiting, *Opt. Lett.*, 19, 31–33, 1994.

# 3 Multi-Bound Solitons
## *Fundamentals and Generations*

## 3.1 INTRODUCTORY REMARKS

Optical solitons have attracted considerable attention in research and practical applications on not only their generation techniques but also their remarkable dynamics. Understanding the dynamics of solotonic waves are of much significance for soliton applications in communications and signal processing systems. Therefore, extensive studies on soliton interactions have been conducted. For practical dissipative systems such as mode-locked fiber lasers, the interaction between solitons can be theoretically described by the complex Ginzburg–Landau equation (CGLE) instead of the nonlinear Schrodinger equation, which is valid only in conservative systems. Hence, the interaction states of solitons based on CGLE have attracted considerable attention in theoretical study. Malomed [1] analyzes the interaction of slightly overlapping CGLE solitons and was the first to predict the formation of effectively stable dual- and multipulse bound states of solitons [1,2]. Then Akhmediev and Ankiewicz numerically investigated the interaction of CGLE solitons by using the two-dimensional phase space approach [3]. They also found a stable solution of dual and multipulse bound solitons with a $\pi/2$ phase difference between them. These theoretical results opened a new research frontier on the states and dynamics of short pulses in mode-locked fiber lasers.

It took more than one decade since Malomed's prediction to experimentally demonstrate the existence of bound solitons in a passively mode-locked fiber laser using the nonlinear phase rotation (NLPR) technique [4] (see also descriptions given in Chapter 6). Then, bound solitons were also reported in the passively mode-locked fiber laser using nonlinear fiber loop mirror (NFLM) [5]. Most numerical and experimental investigations on various operation modes and dynamics of bound solitons have been implemented in passively mode-locked fiber lasers [4–18]. However, there was an interesting question about the existence of bound solitons in actively mode-locked fiber lasers. Bound soliton pairs, but not multi-bound solitons, were experimentally observed for the first time in an actively frequency modulation (FM) mode-locked fiber laser [19]. Then, the existence of multi-bound solitons with more than two solitons in bound states was experimentally demonstrated [20]. In this chapter, we describe the generation and the formation mechanism of multi-bound solitons from the actively FM mode-locked fiber laser. The characteristics of multi-bound solitons are also explained and analyzed.

This chapter is organized as follows. Section 3.2 gives a brief description of mode-locking in passive ring laser for generating multi-bound solitons. Theoretical simulations and experiments on multi-bound solitons are then described in Section 3.3. The relative phases of the carriers of multi-bound solitons are examined in Section 3.4. Section 3.5 presents a new type of bound-solitons, the saddle bound solitons. Then some concluding remarks are given in Section 3.6.

## 3.2   BOUND SOLITONS BY PASSIVE MODE-LOCKING

### 3.2.1   MULTIPULSING OPERATION

Bound soliton states can be considered as one of the multiple soliton operation modes of the passively mode-locked fiber laser. On the other hand, the formation of bound solitons relates to multipulse operation and effective interactions between solitons inside the fiber cavity. In the strongly nonlinear regime of operation, passively mode-locked fiber lasers exhibit a multipulsing behavior. The existence of multiple solitons in the cavity also refers to the soliton energy quantization effect, which limits the peak power of solitons in specific conditions. Efforts in finding mechanisms of multipulse formation that is important for optimization and design of soliton fiber lasers have been achieved [21–23].

There are three main mechanisms of multipulse formation that limit the peak power of the pulse circulating in the cavity. The first mechanism relates to the pulse splitting due to the limited gain bandwidth [22]. Under highly nonlinearity due to strong pumping, the mode-locked pulse is compressed by the self-phase modulation (SPM) effect so that its spectrum is broadened so wide that some frequency components can exceed limited bandwidth of the gain medium. Consequently, the pulse experiences higher extra loss of the cavity and splits into two pulses with wider pulse width or smaller spectral width to avoid this loss. Accordingly, a pulse would experience a compression process until its width is so narrow to split into two pulses. In the case of the fiber laser, however, solitons are considerably chirped by the SPM and the bandwidth limitation effect of the optical gain factor. A generation of subpulses from an initial pulse can occur during circulation in the high power cavity because of strong chirping after a number of round trips [21]. These subpulses can keep growing until they reach the same width and amplitude.

According to dissipative soliton theory, this second mechanism of the multipulse generation can be also explained as follows: Due to the dissipative nature of the fiber lasers, the dispersive waves are generated by shedding the energy of the soliton when circulating in the cavity. In a strongly pumping gain medium, a part of dispersive waves acquires sufficient gain to grow and evolve into a new dissipative soliton with the width and amplitude specified by the fiber cavity parameters [24].

The third and final mechanism of multipulsing is based on the cavity effect that relates to the transmittance of the cavity formed by nonlinear polarization rotation (NPR) or nonlinear fiber loop mirror (NFLM). As described in Section 2.3.2, the transmittance of the fiber cavity in both techniques using NPR and NFLM is a sinusoidal function of the nonlinear phase delay (modulation). Thus, there are two distinct regimes of operation in a period of the transmittance curve (see Figure 2.1). In the first half, the transmittance increases with increasing intensity, while in the second half it decreases with increasing intensity. The peak of the transmittance curve is also the transition point between two regimes. For mode-locking, the cavity operates in the first regime due to the characteristic of saturable absorption. When the nonlinear phase delay is sufficient in soliton operation of the fiber laser, the cavity can be dynamically switched from the saturable absorption mode to the saturable amplification mode that limits the peak of solitons generated inside the cavity. For the NPR technique, the switching point between two modes can be adjusted by changing the setting of polarization controllers to change the linear phase delay. As a result of the cavity feedback, the peak power of soliton is limited, which is responsible for the multipulse operation in the cavity [25].

### 3.2.2 BOUND STATES IN A PASSIVELY MODE-LOCKED FIBER RING

Although the multipulsing operation was experimentally observed sometime ago [26], only in 2001 were the bound states of solitons confirmed for the first time by Tang et al. in an experiment on soliton formation in a passively mode-locked fiber laser using the nonlinear polarization rotation technique [4]. This observation accelerated stimulating research interest in new states of solitons in the mode-locked fibers. Grelu et al. also reported experimental observation of two, three, and multi-bound solitons with the separation between adjacent pulses varying under different conditions [6,9]. By using a strictly dispersion-managed fiber cavity, multipulse solitons can be formed with a fixed pulse separation [15]. In the bound states, the multipulse bound solitons function as a unit and they also behave in similar dynamics to a single soliton in the cavity such as forming states of bound multipulse solitons or period-doubling, period-quadrupling, and chaotic states [16]. The investigation of bound solitons was extended in fiber lasers operating in large normal cavity dispersion regimes where the pulse shape is parabolic rather than hyperbolic secant [27]. Besides, bound soliton pairs were also observed in nonlinear optical loop mirror (NOLM) Figure-8 fiber lasers under the condition of nonbalancing in the total dispersion [8]. Research efforts on bound solitons have been in progress and the latest report has shown the observation of the bound state of 350 pulses or a "soliton crystal" in an NPR mode-locked fiber laser, a record in the number of pulses in the bound state [28].

The formation of bound states following the pulse splitting process in passive mode-locking can be affected by various interaction mechanisms among solitons inside the cavity. Depending on the setup of the fiber laser, one or more than one interaction mechanisms become stronger than the others and determine the formation and the dynamics of bound solitons. In passively mode-locked fiber lasers, there are a variety of possible interaction mechanisms that relate to different modes of bound solitons as follows:

- *Gain depletion and recovery*: The interaction between pulses with the transient depletion and recovery dynamics of the gain medium. The pulses effectively repel each other due to a group velocity drift caused by the time-dependent gain depletion acting in conjunction with the gain recovery. This mechanism is significant in the stabilization of pulse spacing in harmonic mode-locking [29].

- *Acoustic effect and electrostriction*: Due to the intense electric field, the optical pulses can interact with materials forming the fiber guided medium to generate acoustic waves in the propagation of the consecutive pulses. This electrostriction induces a small frequency shift between the light-waves under the pulses leading to effective pulse-to-pulse attraction [30].
- *CW component and soliton interaction*: In some given settings of the passively mode-locked fiber laser using NPR, a CW component can coexist with the solitons in the cavity. It has been experimentally shown that the CW component causes the central frequency shift in each soliton [12]. When the CW component becomes unstable, the solitons acquire different frequency shifts that result in their various relative velocities in the cavity. This interaction is also responsible for motion mode and harmonic mode-locking in passively mode-locked lasers [31].
- *Soliton–soliton interaction*: This direct interaction between solitons is from the nature of fundamental solitons that can attract or repel each other depending on their relative phases. A repulsive force appears between quadrature solitons, that is, when their phase difference is in multiples of $\pi/2$, while an attraction occurs between in-phase solitons. In general, this interaction is considerably effective only when the solitons are sufficiently close together [32].
- *Soliton–dispersive wave interaction*: During the circulating in the cavity, solitons radiate dispersive waves or continuum due to periodically varying perturbations of the cavity such as losses and limited bandwidth gain [32]. The dispersive waves create a local interaction that can generate an attractive or repulsive force between the solitons [33].

When individual solitons interact with each other through these mechanisms, different bound states can be formed inside the fiber cavity. Depending on the binding strength between solitons, bound solitons can be classified into two basic types: weakly and tightly bound solitons.

In the former type, solitons are often bound together into a bunch with the temporal separation between the pulses much larger than their pulse width. Therefore, the weakly bound solitons exhibit weak binding energy and would be easily destructed by random environmental perturbations. Due to the wide separation between solitons, long-range interactions such as the first two mechanisms play a dominant role in the formation of weakly bound solitons. Self-stabilization of this state can only be achieved by a balance of the interplay between these mechanisms [34].

For tightly bound solitons, the last two mechanisms play a major role in the formation of the bound states. Because solitons in this type are closely separated, local interactions are attributed to the binding energy between individual solitons, which determine the characteristics and the dynamics of bound solitons. It has been shown that the direct soliton interaction mainly contributes to the formation of tightly bound solitons with discrete fixed separations [15]. We note that pulses formed in the fiber lasers are often chirped solitons, the direct interaction among them is not like that of nonlinear Schrodinger equation (NLSE) solitons. Therefore, the overlap of oscillating tails of solitons results in an effective binding between them to obtain a stable bound state. However, it was believed that the dispersive wave–soliton interaction also affects the state to generate various dynamics of bound solitons [35]. Due to the strong binding energy, the tightly bound soliton behaves as a unit and exhibits all features of single soliton pulse such as collision, harmonic mode-locking, and an evenly bound state.

In summary, all the various interactions among solitons always coexist in a passive mode-locked fiber laser. However, only one or two of them play a key role in the formation and the characteristics of bound solitons. Furthermore, the existence of other interactions also contributes to the dynamics of their bound states. In other words, the bound soliton falls into only one of multipulse operation under some specific conditions of the cavity settings.

## 3.3   BOUND SOLITONS BY ACTIVE MODE-LOCKING

### 3.3.1   MULTI-BOUND SOLITONS CONDITIONS

In active mode-locking, although the multipulsing operation is more difficult to achieve than in passive mode-locking due to higher loss of the cavity, it still occurs under strong pumping mode of the gain medium. Mechanisms of pulse splitting in active mode-locking are also similar to those for passive mode-locking. However, the width of the pulse in active mode-locking is much wider than that in passive mode-locking, the contribution of the band-limited gain medium to the peak power is negligible, unless when either a band-pass filter is inserted into the ring or an intrinsically filtering effect from the ring cavity sufficiently shorten the gain-bandwidth product. Therefore, splitting of a single pulse into multipulses can only occur in an active fiber cavity when the power in the fiber ring increases above a certain mode-locking threshold. At higher power, higher-order solitons can be excited. In addition, the accumulated nonlinear phase shift in the loop must be sufficiently high that a single pulse can split into many pulses [36]. The number of split pulses depends on the optical power preserved in the ring, so there is a specific range of power for each splitting level. The fluctuation of pulses may occur at the region of power where there is a transition from the lower to the higher splitting level. Moreover, the chirping caused by phase modulator in the loop also makes the process of pulse conversion from a chirped single pulse into multipulses occur more easily [37,38]. In hybrid mode-locked fiber lasers using NPR for pulse shaping, the peak power limiting effect is caused by the additive pulse limiting (APL) effect [39]. By a proper setting of the cavity parameters, the cavity can be switched from the additive-pulse mode-locking (APM) regime to the APL regime that can clamp the peak of solitons at a certain level.

Because pulse splitting can only occur at some specific positions located around the loss minima of the modulation curve in the cavity, the split pulses can be closely separated. Thus, the tightly bound states can be considered the only mode formed in the actively mode-locked fiber laser. Hence, effective interactions exist among these pulses to stabilize the multi-bound states in the cavity. In other words, multi-bound solitons can be formed and stabilized by a balanced interplay between these interactions. The direct soliton interaction, which is influenced by the relative phase difference of the optical carrier under the envelope of adjacent pulses, is believed to contribute significantly to the formation of bound solitons. However, for an actively FM mode-locked fiber laser, another interaction among these pulses need to be considered due to the linear chirping effect caused by active phase modulation. Particularly, the multi-bound soliton sequence is only stably formed at the up-chirping half-cycle for the case of an anomalous path-averaged dispersion fiber ring.

On the other hand, the bound soliton pulses should be symmetrically distributed around the extreme of the positive phase modulation half-cycle where the pulses acquire up-chirping as described in Figure 3.1. Thus, the group velocity of the light-wave, induced by anomalous dispersion, contained within the first pulse of the soliton bunch is decreased after passing through the phase modulator. While the last pulse of the bunch experiences an opposite effect, that is, enhancement of group velocity.

The variation of the group velocity between pulses creates an attractive force that pulls them to the extreme of modulation half-cycle similar to the jittering control in soliton transmission systems [40]. Thus, another interaction between the solitons is required to balance the effective attraction by linear chirping and it is the direct soliton interaction. A repulsive force that is sufficient enough to balance with the effective attractive force appears only in the case of the $\pi$ phase difference between adjacent pulses [41–43]. For other phase differences, there is an energy exchange or an attraction between pulses that are possibly not conforming to the stable existence of multi-bound pulse sequence in a FM mode-locked fiber laser. Thus, we can generate multisoliton bound states using an appropriate phase modulation profile that can also be the high-order harmonics of the modulation frequency. When stable pulse sequence is formed, the pulses would have traveled at least 500 rounds of the fiber ring and thus, significant delays occur between the lightwave carriers under the bound soliton pulses. If a longer distance is traveled by the pulse sequence, there would be breaking up of the sequence and any nonlinear phase shifts can push the sequence into chaotic states. Currently, it is difficult to compensate for the different delays between the lightwaves contained within the pulses. Therefore, in a specific mode-locked fiber laser setup, beside the optical power level and net dispersion of the fiber cavity, the modulator-induced chirp or the phase modulation index determine not only the pulse width but also the temporal separation of bound-soliton pulses at which the interactive effects cancel each other.

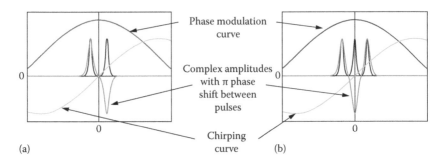

**FIGURE 3.1** A description of effective interactions in multi-bound solitons: (a) dual-soliton bound state and (b) triple-soliton bound state.

The presence of a phase modulator in the cavity is to balance the effective inter-actions among bound-soliton pulses that are similar to the use of this device in a long-haul soliton transmission system to reduce the timing jitter [40,44]. Due to this reason, a simple perturbation technique can be applied to determine the role of phase modulation on the formation mechanism of multi-bound solitons. The optical field of a multisoliton bound state can be described as follows:

$$u_{bs} = \sum_{i=1}^{N} u_i(z,t) \tag{3.1}$$

and

$$u_i = A_i \operatorname{sech}\left\{ A_i \left[ \frac{t - T_i}{T_0} \right] \right\} \exp(j\theta_i - j\omega_i t) \tag{3.2}$$

where
  $N$ is the number of solitons in the bound state
  $T_0$ is pulse width of the soliton
  $A_i$, $T_i$, $\theta_i$, $\omega_i$ represent the amplitude, position, phase, and frequency of soliton, respectively

In the simplest case of multisoliton bound state, $N=2$ or we consider the dual-soliton bound state with the identical amplitude of pulse and the phase difference of $\pi$ value ($\Delta\theta = \theta_{i+1} - \theta_i = \pi$), the ordinary differential equations for the frequency difference and the pulse separation can be derived by using the perturbation method [40,44]

$$\frac{d\omega}{dz} = -\frac{4\beta_2}{T_0^3} \exp\left[ -\frac{\Delta T}{T_0} \right] - 2\alpha_m \Delta T \tag{3.3}$$

$$\frac{d\Delta T}{dz} \beta_2 \omega \tag{3.4}$$

where
  $\beta_2$ is the averaged group-velocity dispersion of the fiber loop
  $\Delta T$ is the temporal separation between two adjacent solitons ($T_{i+1} - T_i = \Delta T$)
  $\alpha_m = m\omega_m^2/(2L_{cav})$, $L_{cav}$ is the total length of the ring, $m$ is the phase modulation index, $\omega_m$ is the angular modulation frequency

Equations 3.3 and 3.4 show the evolution of frequency difference and position of bound solitons in the fiber ring in which the first term on the right-hand side repre-sents the accumulated frequency difference of two adjacent pulses during a round trip of the fiber ring and the second one represents the relative frequency difference of these pulses when passing through the phase modulator. At steady state, pulse separation is constant and the induced frequency differences cancel each other. On the other hand, if Equation 3.3 is set to zero, we have

$$-\frac{4\beta_2}{T_0^3}\exp\left[-\frac{\Delta T}{T_0}\right] - 2\alpha_m \Delta T = 0 \quad \text{or} \quad \Delta T \exp\left[\frac{\Delta T}{T_0}\right] = -\frac{4\beta_2}{T_0^3}\frac{L_{cav}}{m\omega_m^2} \quad (3.5)$$

We can see from Equation 3.5 the effect of phase modulation to the pulse separation and that $\beta_2$ and $\alpha_m$ must have opposite signs, which means that in an anomalous dispersion fiber ring with a negative value of $\beta_2$, the pulses should be up-chirped. With a specific setup of the actively FM mode-locked fiber laser, when the magnitude of chirping increases, the bound pulse separation can decrease subsequently. The pulse width is also reduced according to the increase in the phase modulation index and modulation frequency, so that the ratio $\Delta T/T_0$ cannot change substantially. Thus, the binding of solitons in the FM mode-locked fiber laser is assisted by phase modulation. Bound solitons in the ring experience periodically the frequency shift and hence, their velocity in response to changes in their temporal positions by the interactive forces in equilibrium state.

Thus, in principle, multi-bound solitons can be also generated in any actively mode-locked fiber laser including amplitude mode-locking with an appropriate frequency chirping in the cavity. Controllability of linear chirping in the amplitude modulator is hence, required to maintain a balance in the steady state. Moreover, a frequency shifting facilitates a pulse splitting due to broadening the pulse spectrum in a limited gain bandwidth [38].

### 3.3.2 Experimental Generation

#### 3.3.2.1 Experimental Setup

An actively FM mode-locked fiber ring laser has been set up to generate multi-bound solitons. Initially, we used the experimental setup of the FM mode-locked fiber laser as shown in Figure 3.2a. Because the erbium-doped fiber amplifier (EDFA) in this setup has low gain and saturated power that was insufficient for multi-pulse operation, two EDFAs with 12 dBm maximum output were used for the experiment of multi-bound soliton generation. However, this setup was limited in number of bound solitons due to the limitation of the saturated power of the EDFA. Moreover, the amplification stimulated emission (ASE) noise was enhanced by the presence of two EDFAs in the ring that degraded the performance of the bound solitons. Therefore, we have replaced this setup with the one similar to that in the generation of the single-soliton train described in Chapter 2. We show again the experimental setup of the actively FM mode-locked fiber laser for the generation of multi-bound solitons in Figure 3.2b. In this setup, only one EDFA with higher gain and a maximum saturated power of 17 dBm is used. The modulation frequency of about 1 GHz is selected to generate the bound solitons. The output is characterized by the instruments described in Chapter 2 to estimate the performance of the pulse train.

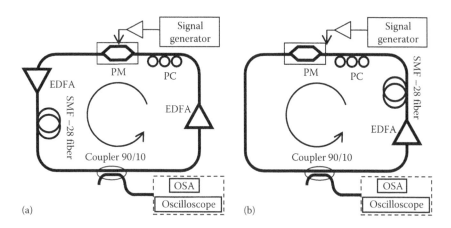

**FIGURE 3.2** The experimental setups for multi-bound soliton generation (a) using two EDFA with low gain and saturated power and (b) using one EDFA with high gain and saturated power.

### 3.3.3 RESULTS AND DISCUSSION

In our initial setup, the generation of multi-bound solitons in the active mode-locked-fiber laser (MLFL) has been achieved [45], yet the number of bound solitons has been limited up to the quadruple state due to the limitation of the EDFA's saturated power. We have then extended the experimental investigation with the bound states of up to the sixth order by adjusting the polarization states in association with the total circulating optical power in the cavity. The multisoliton bound states are depicted in Figure 3.3. When the average optical power is sufficiently increased to a certain level, the dual bound states are correspondingly switched to higher-order states. A slight adjustment of the polarization controller is necessary to stabilize the bound states. Figure 3.3a through e show the traces and spectra of the lowest to highest order bound states, the sextuple-soliton bound state, as observed in our experiment. The significant advantage of the active fiber laser is the ease of the generation of multisoliton bound sequence at a moderately high modulation frequency, as shown in the inset of Figure 3.3c1.

In the tightly bound states, solitons are closely separated, and then the overlap between solitons causes their spectra modulated with the shape and the symmetry depending on their relative phase difference. Hence, the existence of the bound states is also confirmed through the shapes of modulated optical spectra as compared to the conventional spectrum of the single soliton state.

When the multi-bound states appear after the optical power in the cavity is increased to an appropriate level, there is a sudden change of the optical spectrum. The symmetry of spectra in Figure 3.3a2 through e2 indicate a relative phase difference of $\pi$ between two adjacent bound solitons. The dashed-dot lines show the envelope of modulated spectra that correspond to the spectrum of a single soliton pulse. The suppression of the carrier at the center of the pass-band spectrum further confirms the $\pi$ phase difference between adjacent pulses, especially the pair of the dual-bound solitons. The distance between the two spectral main lobes is exactly correlated to the temporal separation between two adjacent pulses in time domain. The specific shape of the spectrum depends on the number of solitons in the bound states, which can be odd or even. In the case of even-soliton bound states such as dual-soliton and quadruple-soliton bound states, there is always a dip at the center of the spectrum, while there is a shallow hump in the case of odd-soliton bound state such as triple- or quintuple-soliton bound states. The shallow hump at the center of the spectrum is formed by far interaction between next neighboring pulses that are in-phase in the bound states. This is similar to the case for quadrature phase shift keying modulation format, which is well-known in the field of digital communications [46]. When the phase difference moves away from the $\pi$ value, the modulation and the symmetry of the spectrum vary accordingly. The spectrum in the sextuple-soliton bound state is weakly modulated due to the variation of phase difference induced by enhanced ASE noise. The change in the relative phase locking influences the interaction of adjacent pulses or the performance of bound soliton output, as shown in Figure 3.3e1.

All critical experimentally measured parameters of multi-bound solitons are summarized in Figure 3.4, plotting the average values of the pulsewidth ($\bar{\tau}$) and the time separation between pulses ($\Delta\tau$), the maximum peak power of a soliton pulse ($P_0$),

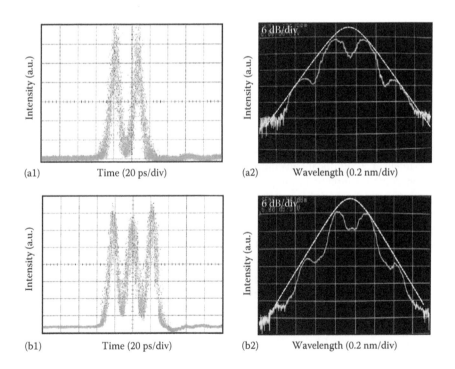

**FIGURE 3.3** Time-domain oscilloscope traces (1) and optical spectra (2) of (a1 and a2) dual-soliton, (b1 and b2) triple-soliton. *(Continued)*

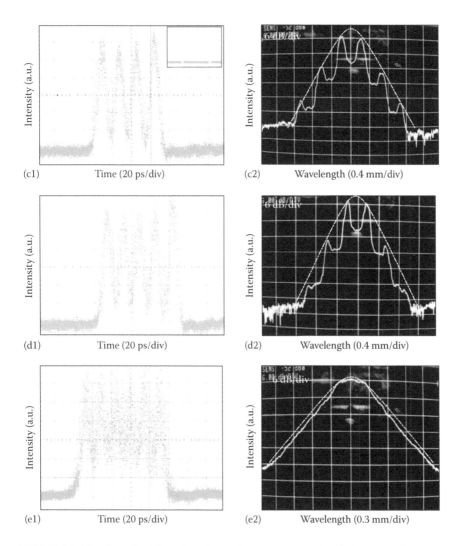

**FIGURE 3.3 (*Continued*)** Time-domain oscilloscope traces (c1 and c2) quadruple-soliton, (d1 and d2) quintuple-soliton, and (e1 and e2) sextuple-soliton bound states.

and the average optical power ($\bar{P}$) inside the fiber ring against the number of bound solitons.

Figure 3.4a shows the variation of the pulsewidth decreasing with respect to the increase in the number of pulses in bound states. This is due to the pulse compression effect, the SPM, enhanced by higher optical power level in the cavity at higher-order bound states. While the temporal separation between pulses in the multi-bound state also decreases correspondingly, the ratio between the pulse separation and the pulses' width remains nearly unchanged at an approximate value of 3, as shown in Figure 3.4b. The onset power level of the energy stored in the ring is 7.5, 12,

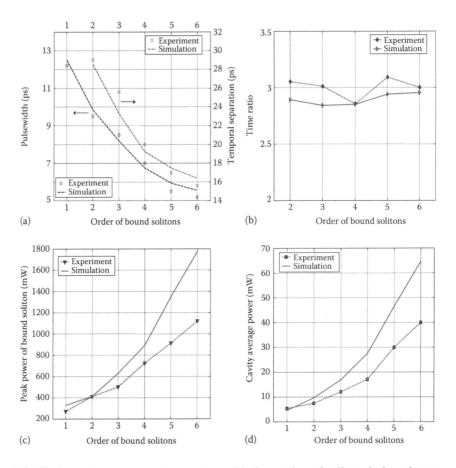

**FIGURE 3.4** The variation of parameters with the number of solitons in bound states (a) pulse width and temporal separation, (b) ratio between pulse width and separation, (c) peak power, and (d) corresponding average power of the cavity.

17, 30, and 40 mW for the generation of dual, triple, quadruple, quintuple, and sextuple states, respectively.

Similarly Figure 3.4c and d show the variation of the peak power and corresponding average power of the cavity with respect to the order of solitons.

Similar to the case of single-soliton state, the multi-bound solitons can be also generated using the rational harmonic mode-locking phenomena. Figure 3.5 shows the time traces of dual- and triple-bound solitons at the second and third rational harmonic mode-locking after adequately increasing the intra-cavity saturated power as well as the radio frequency (RF) input power to enhance the second harmonic of modulation frequency. With an appropriate phase modulation profile which is formed by a detuning amount of $\pm f_c/2$, the multi-bound solitons in rational harmonic mode-locking are generated in the same interaction mechanism with the same characteristic as those in the conventional harmonic mode-locking. However, the temporal separation between solitons in this case is smaller than that in the conventional

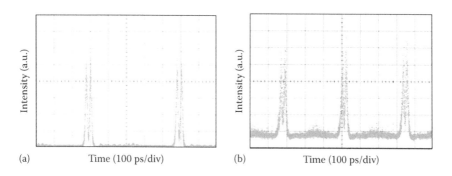

(a)            Time (100 ps/div)                (b)            Time (100 ps/div)

**FIGURE 3.5** Time traces of dual-bound solitons in rational harmonic mode-locking schemes: (a) the second-order rational harmonic and (b) the third-order rational harmonic.

harmonic mode-locking due to higher chirp rate of higher-order harmonics of modulation frequency. Thus, multi-bound solitons can exist in various regimes of operation similar to the conventional single pulse mode.

When higher-order bound solitons operate at a high power level, they are more sensitive to changes in the polarization states of the lightwaves under the soliton envelopes. This effect is disadvantageous to the multi-bound soliton operation due to the reduction of energy of the operating wavelength from other excited wavelengths. On the other hand, higher-order bound solitons are more sensitive to fluctuations of the environmental condition.

In experimental conditions, these polarization effects are usually controlled by the polarization controller, especially at the input of integrated optical modulator so that only one polarized state can be normally preferred through the modulator and hence, the forcing of the matching condition of this polarized state. Moreover, the increase in saturated power in the fiber cavity allows generation of higher-order bound soliton states, yet the ASE noise is also enhanced under strong pumping schemes in the EDFAs. If the phase noise induced by the ASE noise is sufficient, it affects not only the phase-matching conditions of the fiber ring but also the phase-locking between adjacent pulses, as discussed in Section 3.2.

Figure 3.6a1 and b1 show the RF spectra of the dual- and quadruple-bound soliton trains, respectively, to estimate the stability of the multi-bound soliton train. From the results of the RF spectrum analysis, the signal to carrier ratio (SCR) is 45 dB higher for the dual-bound soliton, but it is reduced to 40 dB for the quadruple-bound soliton. Obviously, there are more fluctuations in higher-order bound states or they are more sensitive to the environmental conditions. However, there are small fluctuations in amplitude and temporal position of solitons in bound states as indicated by a broadening of the spectral line shown in Figure 3.6a2 and b2. In the experiment, it has been found that the tuning of the polarization controller becomes harder, making it difficult to obtain a stable bound state when the number of solitons in the bound state is larger.

Although the electro-optic (EO) phase modulator plays a key role as a mode locker in our active mode-locked fiber laser, we should note that the polarization effect shows an important influence to multi-bound solitons. At high power levels of the cavity, polarization-dependent mechanisms, such as polarization-dependent loss (PDL) and

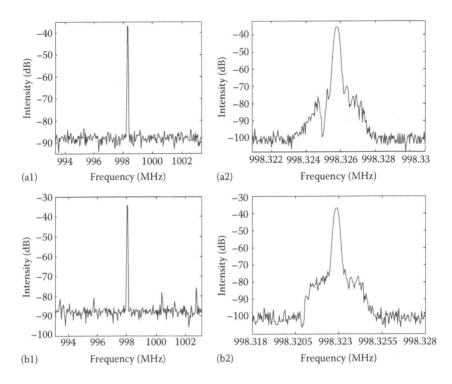

**FIGURE 3.6**   The RF spectra in coarse (10 MHz full span) and high resolution (10 kHz full span), respectively, of (a1 and a2) the dual-bound soliton and (b1 and b2) the quadruple-bound soliton.

polarization-dependent gain (PDG), are enhanced as the phase modulator is also a polarizing element [47–50]. Hence, the variation of the polarization state changes the total loss or the gain of the cavity that affects the shaping, the formation, as well as the parameters of multi-bound solitons.

We have found that a stable multi-bound state is only obtained by correct polarization settings. If a setting of the polarization controller is altered, besides the stable multi-bound states observed earlier, a noise-like pulse regime results. Figure 3.7 shows the time traces of the noise-like "square" pulses at the power levels of the quadruple- and quintuple-bound solitons, respectively. Although the pulses cannot be resolved in the traces, the widths of these pulses are exactly the same width of the bunch of corresponding bound solitons. This regime seems to be a multipulsing operation but unstable. The optical spectra of these states are slightly modulated and varied similar to Figure 3.3e2. The pulses in the bunch, which might be in phase, move and even collide with each other or oscillate very fast around the extreme of the modulation cycle. As a result, the pulses cannot be clearly isolated in the bunch. Therefore, the traces of these states on oscilloscope are seen like noisy waveforms.

Different from passively mode-locked fiber lasers, active phase modulation contributes significantly to the formation and the stability of multi-bound solitons states.

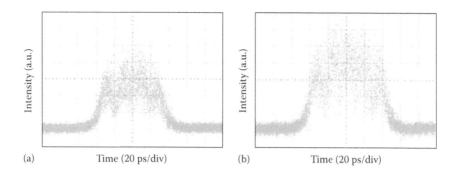

(a)                 Time (20 ps/div)          (b)                 Time (20 ps/div)

**FIGURE 3.7**   Time traces of the noise-like pulse at the optical power levels of (a) quadruple-bound and (b) quintuple-bound solitons respectively.

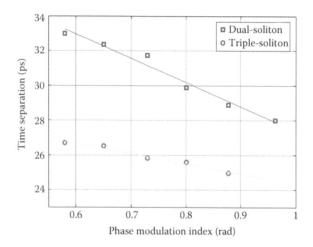

**FIGURE 3.8**   Variation of time-separation of dual and triple-bound soliton states with respect to the phase modulation index.

Figure 3.8 shows experimental results of the influence of chirping, which is directly proportional to the phase modulation index, on temporal separation between adjacent pulses at different bound states. When the chirp rate increases, the relative variation of group velocity between adjacent pulses is enhanced.

Thus, the increase in frequency chirping reduces the temporal separation that is necessary to keep the frequency shift between these pulses unchanged in the cavity with a specific average dispersion. The decrease in time separation in experimental results shows a nearly linear function of the phase modulation index that should be an exponential function as theoretically analyzed in Reference 45. This can be due to the small range of the phase modulation index. In addition, because the pulse width of the triple-bound soliton is narrower that of the dual-bound soliton, the pulses acquire a smaller chirp that results in a low rate of decrease in time separation in the triple-bound soliton, as seen in Figure 3.8.

### 3.3.4 SIMULATION GENERATION

#### 3.3.4.1 Formation of Multisoliton Bound States

To understand the operation of multi-bound solitons in the actively FM mode-locked fiber laser, we have numerically investigated the process using the numerical simulation model described in Chapter 2. First, we have simulated the formation process of bound states in the FM mode-locked fiber laser whose main parameters are shown in Table 3.1. The lengths of the active fiber and passive fiber are chosen to get the cavity's average dispersion $\bar{\beta}_2 = -10.7$ ps$^2$/km. Figure 3.9 shows a simulated dual-soliton

---

**TABLE 3.1**

**Simulation Parameters of Multi-Bound Soliton Formation**

| | | |
|---|---|---|
| $\beta_2^{SMF} = -21$ ps$^2$/km | $\beta_2^{ErF} = 6.43$ ps$^2$/km | $\Delta\omega_g = 16$ nm |
| $\gamma_{SMF} = 0.0019$ W$^{-1}$ m$^{-1}$ | $\gamma_{EDF} = 0.003$ W$^{-1}$ m$^{-1}$ | $f_m \approx 1$ GHz |
| $\alpha_{SMF} = 0.2$ dB/km | $\alpha_{EDF} = 0.5$ dB/km | $m = \pi/3$ |
| $L_{SMF} = 80$ m | $P_{sat} = 8-14$ dBm | $\lambda = 1559$ nm |
| $L_{EDF} = 10$ m | $g_0 = 0.35-0.45$ m$^{-1}$, $NF = 5$ dB | $l_{cav} = 11$ dB |

---

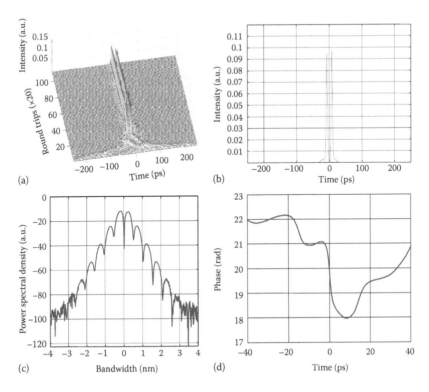

**FIGURE 3.9** (a) Numerically simulated evolution of the dual-soliton bound state formation from noise, (b) the waveform, (c) the spectrum, and (d) the phase at the 2000th round trip.

bound state building up from initial Gaussian-distributed noise as an input seed over the first 2000 round trips with a $P_{sat}$ value of 8 dBm and a $g_0$ of 0.36 m$^{-1}$. The built-up pulse experiences transitions with large fluctuations of intensity, position, and pulse width during the first 1000 round trips before the formation of the bound soliton state. Figure 3.9b and c show the time waveform and the spectrum of the output signal at the 2000th round trip.

Bound states with a higher number of pulses can be formed at a higher gain of the cavity, hence, when $P_{sat}$ and $g_0$ are increased to 11 dBm and 0.38 m$^{-1}$, respectively, which enhances the average optical power in the ring, the triple-soliton bound steady-state is formed from the noise seeded via simulation, as shown in Figure 3.10. In the case of higher optical power, the fluctuation of signal at initial transitions is stronger and it needs more round trips to reach a more stable triple-bound state. The waveform and spectrum of the output signal from the FM mode-locked fiber laser at the 5000th round trip are shown in Figure 3.10b and c, respectively.

Although the amplitude of pulses is not equal, indicating the bound state requires a larger number of round trips before the effects in the ring balance, the phase difference of pulses accumulated during circulating in the fiber loop is approximately of $\pi$ value that is indicated by strongly modulated spectra. In particular, from the

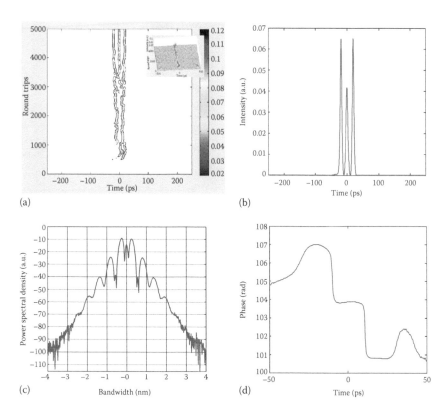

**FIGURE 3.10** (a) Numerically simulated evolution of the triple-soliton bound state formation from noise, (b) the waveform, (c) the spectrum, and (d) the phase at the 5000th round trip.

simulation result, the phase difference between adjacent pulses is $0.98\pi$ in case of the dual-pulse bound state and $0.89\pi$ in case of the triple-pulse bound state. These simulation results reproduce the experimental results (shown in Figures 3.9d and 3.10d) discussed earlier to confirm the existence of multisoliton bound states in an FM mode-locked fiber laser.

It is found that the cavity requires higher gain to form the multi-bound states than that in stable or steady states. By adjusting the parameters of gain medium after a multipulse state is formed in the cavity, stable multi-bound soliton states can be generated after at least 10,000 round trips. Figure 3.10a shows the contour of the dual-bound soliton formation in 10,000 round trips of the ring cavity. In this simulation, the gain factor is reduced from 0.37 to 0.35 m$^{-1}$ at the 2000th round trip (the dash line). In the first 2000 round trips, the dual-bound soliton is formed, yet unstable with strong fluctuation of the peak power. Then the fluctuation is reduced by the reduction of the gain factor and the generated dual-bound soliton converges to stable state after 10,000 round trips as shown in Figure 3.10b. Figure 3.10c and d show the time-dependent amplitude and phase of the output at the 10,000th round trip.

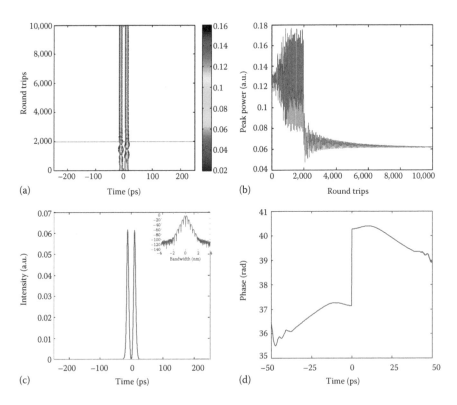

**FIGURE 3.11** (a) Contour plot of simulated evolution of the dual-soliton bound state formation from noise, (b) variation of the peak power with the gain switching at the 2,000th round trip, (c) the waveform (inset: the corresponding spectrum), and (d) the phase at the 10,000th round trip.

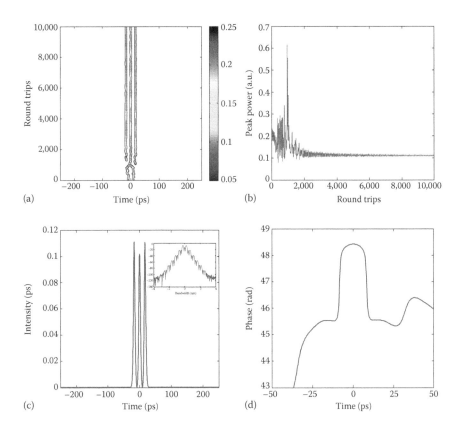

**FIGURE 3.12**  (a) Contour plot of simulated evolution of the triple-soliton bound state formation from noise, (b) variation of the peak power with the gain switching at the 2,000th round trip, (c) the waveform (inset: the corresponding spectrum), and (d) the phase at the 10,000th round trip.

Similarly, the parameters of gain medium are also adjusted to obtain stable higher-order multi-bound solitons. Figures 3.11 and 3.12 show the simulated results of the triple-bound soliton and the quadruple-bound soliton formation processes, respectively. For the triple-bound soliton, the gain factors are kept at the same values as those in case of the dual-bound soliton, but the saturated power $P_{sat}$ is increased to 11 dBm instead of 8 dBm. Variation of peak power and evolution of the triple-bound soliton formation with the gain switching at the 2000th round trip are shown in Figure 3.12a and b. For the quadruple-bound soliton, both the initial gain and adjusted gain factors are increased to higher values, which are 0.4 and 0.377 m$^{-1}$, respectively, at a $P_{sat}$ of 11 dBm. Figure 3.13b shows a damping of the peak power variation after gain adjustment to reach a stable triple-bound soliton state; however, the peak power of the quadruple-bound soliton still oscillates evenly after 10,000 round trips as seen in Figure 3.13b. Thus, it indicates higher sensitivity of the higher-order multi-bound soliton to the operating condition. The sensitivity also exhibits through the uniform of pulse intensity in the bound state

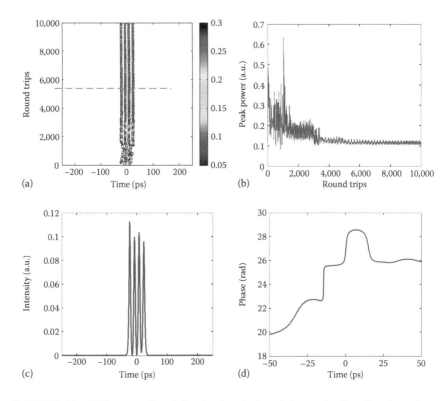

**FIGURE 3.13** (a) Contour plot of simulated evolution of the quadruple-soliton bound state formation from noise, (b) variation of the peak power with the gain switching at the 3,000th round trip, (c) the waveform (Inset: the corresponding spectrum), and (d) the phase at the 10,000th round trip.

and the relative phase difference, which decreases from $\pi$ in the dual-bound soliton to $0.87\pi$ in the quadruple-bound soliton, as shown in Figures 3.11c and d, 3.12c and d, and 3.13c and d.

We note that a stable multi-bound soliton is difficult to be formed in the cavity without adjusting the gain parameters. With the high gain, a multi-bound soliton can be generated, but it will be unstable or will become quasi-stable with a periodic variation of soliton parameters in the bound state. With the low gain, the nonlinear phase shift is not sufficient to generate the desirable higher-order multipulse state, while it is too strong for stable, lower-order multi-bound state. Different from passive mode-locking, pulse splitting in our system is caused by the excitation of higher-order soliton and nonlinear chirping rather than an energy quantization mechanism.

### 3.3.4.2 Evolution of the Bound Soliton States in an FM Fiber Loop

Obviously, a multi-bound soliton can be stably generated in the phase-modulated fiber cavity. On the other hand, the bound soliton with the relative difference of $\pi$ given by Equations 3.1 and 3.2 is considered as a stable solution of the actively FM mode-locked fiber laser. By using the multisoliton waveform in Equation 3.1 and

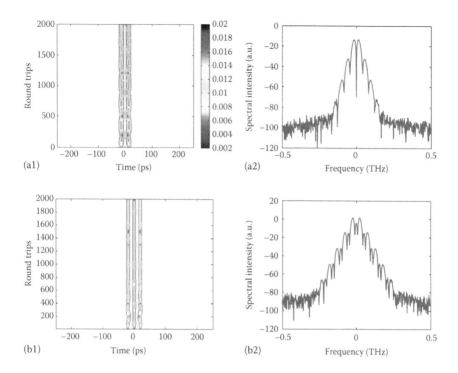

**FIGURE 3.14** Evolution of (a1) dual-bound soliton in time domain and (a2) spectral density distribution and (b1) triple-bound soliton with a relative phase difference of $\pi$ in the FM fiber ring cavity with (b2) corresponding spectral density distribution.

Equation 3.2 as input, we have simulated the stability of multi-bound solitons in the active FM mode-locked fiber laser. Figure 3.14 shows the evolutions in 2000 round trips of dual- and triple-bound solitons in the cavity. Because bound solitons are really chirped pulses, while the input in simulation is unchirped, there is a damping oscillation of bound solitons in the initial stage that is considered as a transition of solitons to adjust their own parameters to match the parameters of the cavity. However, multi-bound solitons easily reach stable states after only a few hundred round trips.

Simulation results on parameters of multi-bound solitons after 5000 round trips are also shown in Figure 3.14 for comparison by using the experimental parameters of multi-bound solitons as the initial parameters. Generally, the simulated results agree with the experimental results. However, as observed in Figure 3.14c and d, there are discrepancies of the peak and average power levels between the simulated and experimental pulses, especially at higher-order soliton bound states. The level of discrepancy varies from 0.1 to 3 dB for peak power of 1000–1800 mW, respectively. While the experimental results show a nearly linear dependence of the peak power on the order of bound states, the simulation results show an exponential variation. Hence, there is also a difference in the average power of the cavity between them. However, both sets of results indicate an exponential dependence of the cavity average power on the order of bound soliton states. Taking into account the discrepancy, some reasons could be as follows: First, the parameters of the fiber cavity used in the simulation is not totally matched to those in our experiment; second, when the real pulse width at the higher-order bound soliton state is narrower, the accuracy of the pulse width measurement on the oscilloscope is reduced, although the influence of the rise-time of the oscilloscope was considered in the estimation. Furthermore, the variation of the polarization state becomes stronger at a higher power level of the cavity, which also increases the error in power measurement. Hence, the error between simulation and experimental results increases at a higher order of the soliton bound state.

In addition, the higher-order multi-bound soliton is more sensitive to the cavity parameter settings. Figure 3.15a shows an unstable state evolving in the FM fiber ring cavity. Multiple pulses are generated in the cavity, yet it is difficult to acquire the phase difference of $\pi$ and uniformity between the pulses. In other words, the balance in the effective interaction between pulses is difficult to be achieved, the pulses can therefore, collide and vary rapidly in both amplitude and time position. However, this rapid variation only occurs in a limited time window around the extreme of the modulation cycle as indicated by the dashed lines in Figure 3.15b. By taking the average over the last 2000 round trips, the waveform and its spectrum, which can be represented for a dynamical state, are shown in

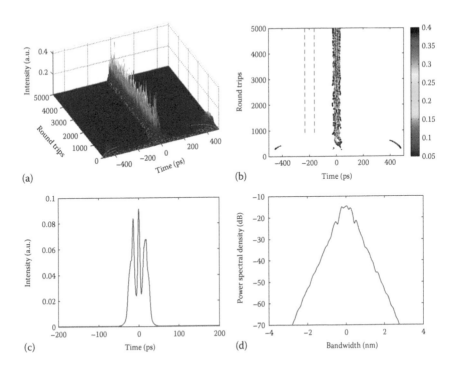

(a)

(b)

(c)

(d)

**FIGURE 3.15** (a) Simulated evolution of multi-bound soliton in unstable condition over 5000 round trips, (b) contour plot view of the evolution, (c) the waveform, and (d) corresponding spectrum averaged over the last 2000 round trips.

Figure 3.15c and d, respectively. The wave form, which is like a noisy pulse, and the spectrum, which is slightly modulated, are similar to what is observed in the experiment (see Figure 3.7).

## 3.4   RELATIVE PHASE DIFFERENCE OF MULTI-BOUND SOLITONS

### 3.4.1   INTERFEROMETER MEASUREMENT AND EXPERIMENTAL SETUP

The relative phase difference plays a key factor in stability as well as determination of various modes of the bound soliton states. For passive mode-locking, although the relative phase between the bound solitons of $\pi$ has been confirmed in some experimental works [8,11], other relative phases have also been demonstrated [6,8,10]. The relative phase difference is of importance in the determination of dynamics of the bound solitons in the fiber laser system. In particular, the bound solitons with $\pi/2$ phase difference can collide either elastically or nonelastically with a single soliton depending on its initial phase [9]. On the other hand, the bound solitons can change their relative phase when the settings in the cavity change.

Different from passive mode-locking, only the bound solitons with $\pi$ phase difference are stably generated in the actively mode-locked fiber laser system, which has been confirmed by their symmetrically modulated spectra. In each bunch of bound solitons, adjacent pulses are out of phase to form a stable bound state through the balanced interplay of effective interactions. However, if the multi-bound soliton is considered as a unit like the single pulse state, there might be two possibilities of multi-bound soliton trains: one in which solitons between neighboring bunches are in phase and another in which solitons between neighboring bunches are out of phase, as described in Figure 3.16. On the other hand, there may be a phase inversion between bunches of bound solitons. To check the dynamics of the relative phase difference in multi-bound soliton train, which is impossible to identify through the optical spectrum measurement, an interferometer measurement

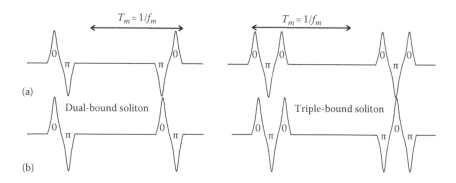

**FIGURE 3.16**   A description of two possibilities of phase difference between neighboring pulses: (a) solitons between neighboring bunches are in phase and (b) solitons between neighboring bunches are out of phase.

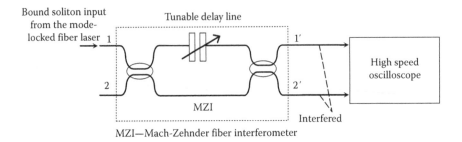

**FIGURE 3.17**   Experimental setup and principle of the interferometer measurement.

has been proposed and implemented. Figure 3.17 shows the schematic of the measurement based on an asymmetrical Mach–Zehnder fiber interferometer (MZI). The intensities of the interference patterns at two output ports of the asymmetrical MZI can be given by

$$I^{1'} = \left|E_{out}^{1'}\right|^2 = \left|\frac{1}{2} j^2 E_{in}(t) + \frac{1}{2} E_{in}(t - T_d)\right|^2 \tag{3.6}$$

$$I^{2'} = \left|E_{out}^{2'}\right|^2 = \left|\frac{1}{2} j E_{in}(t) + \frac{1}{2} j E_{in}(t - T_d)\right|^2 \tag{3.7}$$

where

$E_{in}$ is the input field of the MZI, which is the field of multi-bound soliton train

$T_d$ is the variable time delay between two arms

Depending on the adjustment of the time delay, the pulses of two multi-bound solitons on two arms would be interfered constructively or destructively over the overlapped positions at the output coupler.

Thus, if there is no dynamic phase inversion between bunches of bound solitons or the phase difference of $\pm\pi$ between multi-bound solitons remains unchanged, the interference patterns of two outputs would be fixed and contrary to each other when the multi-bound solitons between two arms are overlapped over one or two pulses. In the case of triple-bound solitons, for example, the calculated patterns of constructive and destructive interferences with two overlapped pulses at the outputs are shown in Figure 3.18. On the contrary, the interference patterns are dynamically varied between two output ports depending on the phase states of multi-bound solitons between two arms.

In the experimental setup as shown in Figure 3.17, the asymmetric fiber interferometer is built by two 3 dB couplers that are connected together by optical fibers to form two arms of MZI. A tunable delay line of 80 ps delay time is inserted into an arm of MZI to sufficiently provide an overlapping of multi-bound solitons between two arms at the output coupler. The amplitude of overlapped pulses at the output

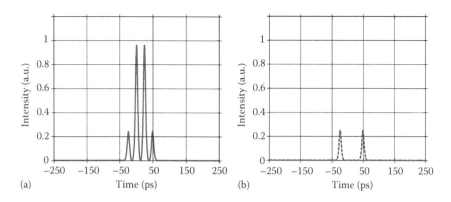

**FIGURE 3.18** Calculated patterns of triple-bound solitons overlapped over two pulses at two outputs of MZI: (a) constructive interference and (b) destructive interference.

ports (1′ and 2′) of MZI depends on the phase difference of solitons in bound states. This determines either constructive and/or destructive interferences in the time domain at overlapped positions. The interference patterns at two output ports are simultaneously monitored by two ports on the high-speed oscilloscope.

### 3.4.2 Results and Discussion

The initial phase delay between two arms of the MZI is adjusted by the fiber length difference between two arms; on the other hand, the time difference between multi-bound solitons in two arms is initially about 75 ps as shown in Figure 3.19a. By adjusting the tunable delay line, a specific overlapping between two triple-bound solitons can be achieved. Figure 3.19b and c show, as an example, the interference patterns of triple-bound soliton state over two overlapped pulses at ports 1′ and 2′, respectively. The results also indicate that the practical value of phase difference is not totally equal to π due to the peak power of overlapped pulses at port 1′ and port 2′

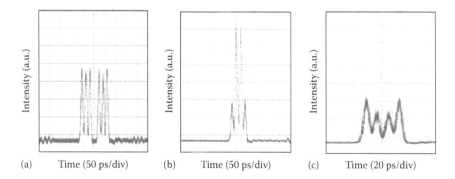

**FIGURE 3.19** The time traces of triple-bound solitons: (a) before overlap, (b) overlapped at port 1′, and (c) overlapped at port 2′ of MZI.

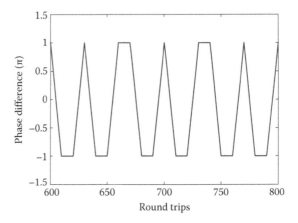

**FIGURE 3.20** Periodic variation of the phase difference of triple-bound soliton after 200 round trips.

which is only three times higher and lower than that of input pulses, respectively. However, the interference patterns are not steady but alternatively changed between two ports.

The alternating change of the patterns between two ports indicates that a phase inversion periodically occurs. For the single pulse active-mode-locking scheme, O'Neil et al. [51] has theoretically demonstrated that only that pulse train in which neighboring pulses are out of phase is stable in the cavity. Therefore, in multi-bound states, it is understandable when solitons are out of phase not only within bunches but also between bunches. Because there is a phase shift in each round trip that is accumulated during circulating in the cavity, the phase inversion occurs after the accumulated phase shift is an integer multiple of $2\pi$. This was also confirmed by the simulation result, as shown in Figure 3.20, to indicate a periodic variation of the phase difference between $-\pi$ and $\pi$.

## 3.5 MULTI-BOUND AND SADDLE SOLITONS: EXPERIMENTAL OBSERVATIONS

Once again, multi-bound solitons can be formed by a group of individual solitons whose optical carriers are different in phase at some specific values, thus they can be attracted and remain bounded to each other. These multi-bound solitons can be generated using mode-locked fiber ring lasers (MLFRL). They would find applications as the coded header for optical packet switching in such ultrahigh-speed systems. MLFRL can be formed by active or passive mode-locking. In an active mode of operation, an optical modulator is normally integrated in the optically amplified fiber ring. As shown in Figure 3.2a, the experimental setup of the MLFRL consists of a ring of fiber, an integrated optical phase modulator (OPM), an optical fiber amplifier (OFA) to compensate for the propagation loss and providing sufficient optical power circulating in the ring. The OFA operates in the saturation region and the output power can be adjusted by varying the pump power. A 50 m standard single mode fiber (SSMF) is inserted after the OFA to enhance the nonlinear phase shift as well as ensure that the average dispersion in the fiber ring is anomalous.

We have observed that the resonance condition of the optical circular cavity depends on the total phase exerted on the optical waves circulating in the ring. Thus, the embedded optical phase modulator plays a major role in the control of the formation of bound solitons. Multi-bound solitons occur when the phase difference between adjacent solitons equals π radians. This phase difference leads to the cancellation of the dispersive effects that are due to the differential group delay and interference between the spectral components of the solitons propagating through the SSMF. However, the characteristics of the optical phase modulator depend on the interactions between the electric fields created by the signals applied to the electrodes and that of the lightwaves. The RF waves traveling along the electrodes can be of traveling wave or lumped types, hence, its influence on the behavior of the bounding stability of the multi-bound solitons. Various bound soliton states with the number of individual solitons can be generated with the maximum number of six as shown in Figure 3.21a through d. Higher-order solitons can only be obtained with much higher optical power at which the fiber ring is under intense nonlinear effects leading to a chaotic state.

This chapter examines the stability of solitons in multi-bound states under the influences of the phase and polarization states of the optical modulators incorporated within the MLFRL as well as reporting a new type of soliton—saddle solitons that result from strongly multi-bound states and soliton resonance.

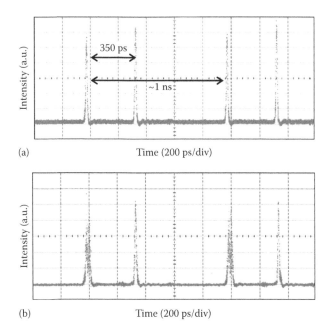

**FIGURE 3.21** Observation of generated multi-bound solitons: (a) dual-bound, (b) triple-bound. *(Continued)*

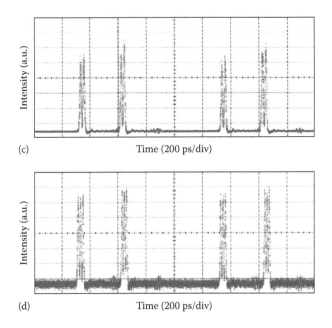

(c)                                           Time (200 ps/div)

(d)                                           Time (200 ps/div)

**FIGURE 3.21 (*Continued*)**   Observation of generated multi-bound solitons: (c) quadruple-bound, and (d) sextuple-bound solitons.

The existence of a comb-like filtering effect in the cavity using an EO phase modulator remarkably influences the characteristics of the generated pulses. This artificial filtering based on the birefringence of the cavity limits the gain bandwidth and contributes to the mechanism of pulse broadening due to the limitation of pulse spectrum. This effect has been experimentally demonstrated by measuring the pulse width from two setups of the MLFRL incorporating either a linear phase modulator (LPM) or a traveling-wave phase modulator (TWPM). Although the lumped type is driven by much higher modulation index as compared to the traveling type of lower $V_\pi$, the pulse generated from the setup using TWPM is more than twice as wide as that generated from the setup using a lumped-type modulator. Under the use of the TWPM, solitons of much narrower width can be achieved due to TWPM's broader gain bandwidth. However, in the soliton pulses generated in the case of LPM-type optical modulator, a high peak power can be achieved, creating sufficiently high nonlinear phase shift that may lead to the pulse splitting in the cavity. Moreover, the effective interactions between pulses in bound states also depend on the pulse width through the overlap of long pulse tails and the chirping caused by phase modulation. In the case of LPM, the chirp imposed on the generated pulses with a sufficiently small width allows an effective attraction to balance the repulsion to generate a stable bound state. In contrast, a distinct bound state cannot be observed in the case of TWPM with pulses generated with a wider width. In this case, wider pulses are obtained from higher chirp by the TWPM to achieve stronger attraction. At once the repulsion force from direct interaction becomes weaker. Therefore, a balanced interaction cannot be achieved in this case to generate a bound state with obviously resolved pulses although the cavity is

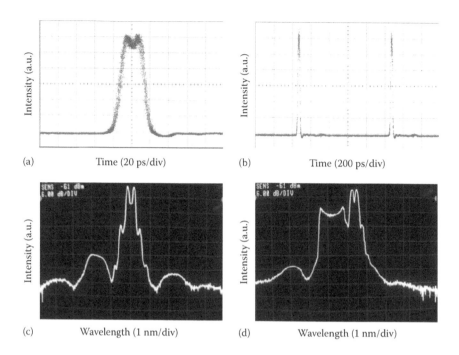

**FIGURE 3.22** Observation of a *twin-hump saddle pulse* in the cavity using a TWPM: (a) single pulse trace of the saddle soliton and (b) its pulse train sequence trace; and switching from the triple-bound soliton into the dual-bound soliton by the polarization manipulation (c) spectrum of saddle solitons and (d) its sequence.

strongly pumped. A state of mode-locked pulses with a mode-locked sequence of solitons having a two-hump structure generated from the setup using TWPM is shown in Figure 3.22a and b. The corresponding spectra of the saddle solitons and sequence are shown in Figure 3.22c and d, respectively. The spacing between the two highest peaks of the spectrum corresponds to a quarter of the bit rate, and the depletion of the carrier at the center of the spectrum indicates that the phase difference between the overlapping pulse (the saddle soliton) is $\pi$ and the phase is continuous, similar to the case of the minimum shift keying modulation scheme [52,53].

This is a new type of soliton of the fiber laser system, termed as saddle solitons. The time separation between two humps less than 20 ps indicates that there is a strong possibility that a strong attraction between two adjacent pulses occurs at the instant of the chirping at which both the binding of the two individual solitons and the resonance condition for locking of ultrashort pulses, the train of solitons in such MLFRL can be satisfied.

Figure 3.23 shows the threshold power variation and that of the average power circulating in the ring cavity with respect to the variation in the modulation index. The phase chirping rate of the lightwaves determines the separation between bound solitons. The chirp effect due to the modulation index exerted on the OPM, and the number of soliton in the bound states are correlated. Figure 3.23a and b, respectively, depict and measure the experimental threshold power for the creation of dual- and

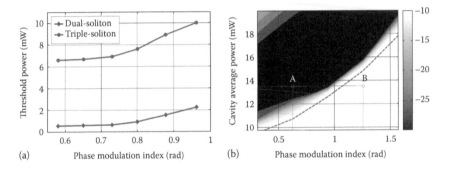

**FIGURE 3.23** (a) Experimentally measured variation of threshold splitting power with the phase modulation index of the bound soliton states and (b) simulated variation of the peak power in dB of the triple-bound state with the phase modulation index.

triple-soliton bound states against the phase modulation index, which indicates an increase in the required splitting optical power when the chirping is increased. An increase in the chirp rate requires a higher threshold power to maintain a specific number of pulses in the bound state. When the phase modulation index increases, the expansion of the signal spectrum is enhanced.

However, the increase in the chirping rate also causes an enhanced energy transfer to the sidebands of the spectrum of the multi-bound soliton that can degrade a certain number of solitons in the sequence. Although the up-chirped pulse is further compressed by anomalous dispersion of the fiber ring resonator at a higher modulation index, larger frequency chirp at pulse edges may also require higher energy and the momentum preserved in the pulses. This tendency agrees with the theoretical analysis of soliton formation from chirped pulses. At a low phase modulation index, the chirping is almost unaffected by the threshold power due to the weak binding of the bound solitons. Consequently, the waveforms of the bound states are noisier and more sensitive to phase fluctuations caused by the ASE noise or the random polarization variation of the guided medium. A simulation result of the power limit of the cavity to maintain the bound state agrees with the observed spectra shown in Figure 3.22c and d. As shown in Figure 3.23b, the rates of increasing and decreasing peak power of bound solitons are indicating two distinct regions of operations, the triple- and the double-bound separated by the dashed bold line. This limit line (dashed bold line) is obtained when a minimum power level is required during circulation in the cavity to maintain the bound state that divides the graph into the triple- and double-bound regions. At a specific modulation index, when the average power of the cavity decreases, the variation of the peak power is stronger. When the power is lower than the limit line, the higher-order soliton bound state switches to the lower-order soliton bound state. This tendency also indicates that the increase in chirping diminishes the number of solitons in the bound state at a specific optical power level. These figures show the evolution of the triple-soliton bound state in the cavity at the same average power level, but different phase modulation indices that correspond to points A and B, respectively, in Figure 3.23b.

## 3.6 CONCLUDING REMARKS

We have reviewed the important mechanisms as well as interactions in bound soliton formation of mode-locked fiber lasers. Formation of multi-bound solitons in active mode-locking has also been explained to show the role of phase modulation in the balanced interplay between interactions of multi-bound solitons.

The generation of stable multisoliton bound states in an FM mode-locked fiber laser has been experimentally and numerically demonstrated. We have demonstrated that it is possible to generate bound states from dual to sextuple states provided that there is sufficient optical energy circulating in the fiber ring. Multibound soliton states in a phase-modulated fiber ring are rigorously explained based on the phase matching and chirping effects of the lightwave and the velocity variation of the optical pulses in a dispersive fiber ring. Experimental and numerical results have confirmed the stable existence of multi-bound solitons with a phase difference of $\pm\pi$ between neighboring solitons. However, it is more sensitive to the cavity settings as well as the external perturbations at higher-order multi-bound soliton states.

## REFERENCES

1. B. A. Malomed, Bound solitons in the nonlinear Schrodinger and Ginzburg–Landau equation, *Phys. Rev. A*, 44, 6954, 1991.
2. B. A. Malomed, Bound solitons in coupled nonlinear Schrodinger equations, *Phys. Rev. A*, 45, R8321, 1992.
3. N. N. Akhmediev and A. Ankiewicz, Multisoliton solutions of the complex Ginzburg–Landau equation, *Phys. Rev. Lett.*, 79, 4047, 1997.
4. D. Y. Tang et al., Observation of bound states of solitons in a passively mode-locked fiber laser, *Phys. Rev. A*, 64, 033814, 2001.
5. N. H. Seong and D. Y. Kim, Experimental observation of stable bound solitons in a figure-eight fiber laser, *Opt. Lett.*, 27, 1321–1323, 2002.
6. P. Grelu, F. Belhache, F. Gutty, and J.-M. Soto-Crespo, Phase-locked soliton pairs in a stretched-pulse fiber laser, *Opt. Lett.*, 27, 966–968, 2002.
7. D. Y. Tang, B. Zhao, D. Y. Shen, and C. Lu, Bound-soliton fiber laser, *Phys. Rev. A*, 66, 033806, 2002.
8. Y. D. Gong, P. Shum, D. Y. Tang, C. Lu, Z. W. Qi, W. J. Lai, W. S. Man, and H. Y. Tam, Close spaced ultra-short bound solitons from DI-NOLM Figure-8 fiber laser, *Opt. Commun.*, 220, 297–302, 2003.
9. P. Grelu, F. Belhache, F. Gutty, and J. M. Soto-Crespo, Relative phase locking of pulses in a passively mode-locked fiber laser, *J. Opt. Soc. Am. B*, 20, 863–870, 2003.
10. D. Y. Tang, B. Zhao, D. Y. Shen, C. Lu, W. S. Man, and H.-Y. Tam, Compound pulse solitons in a fiber ring laser, *Phys. Rev. A*, 68, 013816, 2003.
11. B. Zhao, D. Tang, Y. Gong, and W. S. Man, Energy quantization of twin-pulse solitons in a passively mode-locked fiber ring laser, *Appl. Phys. B: Lasers Opt.*, 77, 585–588, 2003.
12. Y. D. Gong, P. Shum, D. Y. Tang, C. Lu, T. H. Cheng, Z. W. Qi, Y. L. Guan, W. J. Lai, W. S. Man, and H. Y. Tam, Bound solitons with 103-fs pulse width and 585.5-fs separation from DI-NOLM figure-8 fiber laser, *Microw. Opt. Technol. Lett.*, 39(2), 163–164, October 20, 2003.

13. B. Ortac, A. Hideur, T. Chartier, M. Brunel, P. Grelu, H. Leblond, and F. Sanchez, Generation of bound states of three ultrashort pulses with a passively mode-locked high-power Yb-doped double-clad fiber laser, *IEEE Photon. Technol. Lett.*, 16, 1274–1276, 2004.

14. B. Zhao, D. Y. Tang, P. Shum, W. S. Man, H. Y. Tam, Y. D. Gong, and C. Lu, Passive harmonic mode locking of twin-pulse solitons in an erbium-doped fiber ring laser, *Opt. Commun.*, 229, 363–370, 2004.

15. D. Y. Tang, L. M. Zhao, and B. Zhao, Multipulse bound solitons with fixed pulse separations formed by direct soliton interaction, *Appl. Phys. B: Lasers Opt.*, 80, 239–242, 2005.

16. L. M. Zhao, D. Y. Tang, and B. Zhao, Period-doubling and quadrupling of bound solitons in a passively mode-locked fiber laser, *Opt. Commun.*, 252, 167–172, 2005.

17. M. Grapinet and P. Grelu, Vibrating soliton pairs in a mode-locked laser cavity, *Opt. Lett.*, 31, 2115–2117, 2006.

18. L. M. Zhao, D. Y. Tang, T. H. Cheng, H. Y. Tam, and C. Lu, Bound states of dispersion-managed solitons in a fiber laser at near zero dispersion, *Appl. Opt.*, 46, 4768–4773, 2007.

19. W.-W. Hsiang, C.-Y. Lin, and Y. Lai, Stable new bound soliton pairs in a 10 GHz hybrid frequency modulation mode-locked Er-fiber laser, *Opt. Lett.*, 31, 1627–1629, 2006.

20. L. N. Binh, N. D. Nguyen, and T. L. Huynh, Multi-bound solitons in a FM mode-locked fiber laser, in *Conference on Optical Fiber Communication/National Fiber Optic Engineers Conference, 2008 (OFC/NFOEC 2008)*, Anaheim, CA, pp. 1–3, 2008.

21. G. P. Agrawal, Optical pulse propagation in doped fiber amplifiers, *Phys. Rev. A*, 44, 7493, 1991.

22. M. J. Lederer, B. Luther-Davies, H. H. Tan, C. Jagadish, N. N. Akhmediev, and J. M. Soto-Crespo, Multipulse operation of a Ti:sapphire laser mode locked by an ion-implanted semiconductor saturable-absorber mirror, *J. Opt. Soc. Am. B*, 16, 895–904, 1999.

23. F. X. Kurtner, J. A. derAu, and U. Keller, Mode-locking with slow and fast saturable absorbers—What's the difference? *IEEE J. Select. Top. Quant. Electron.*, 4, 159–168, 1998.

24. D. Y. Tang, W. S. Man, and H. Y. Tam, Stimulated soliton pulse formation and its mechanism in a passively mode-locked fiber soliton laser, *Opt. Commun.*, 165, 189–194, 1999.

25. D. Y. Tang, L. M. Zhao, B. Zhao, and A. Q. Liu, Mechanism of multisoliton formation and soliton energy quantization in passively mode-locked fiber lasers, *Phys. Rev. A*, 72, 043816, 2005.

26. A. B. Grudinin, D. J. Richardson, and D. N. Payne, Energy quantisation in figure eight fiber laser, *Electron. Lett.*, 28, 67–68, 1992.

27. G. Martel, C. Chédot, V. Réglier, A. Hideur, B. Ortaç, and Ph. Grelu, On the possibility of observing bound soliton pairs in a wave-breaking-free mode-locked fiber laser, *Opt. Lett.*, 32, 343–345, 2007.

28. A. Komarov, A. Haboucha, and F. Sanchez, Ultrahigh-repetition-rate bound-soliton harmonic passive mode-locked fiber lasers, *Opt. Lett.*, 33, 2254–2256, 2008.

29. J. N. Kutz, B. C. Collings, K. Bergman, and W. H. Knox, Stabilized pulse spacing in soliton lasers due to gain depletion and recovery, *IEEE J. Quant. Electron.*, 34, 1749–1757, 1998.

30. A. N. Pilipetskii, E. A. Golovchenko, and C. R. Menyuk, Acoustic effect in passively mode-locked fiber ring lasers, *Opt. Lett.*, 20, 907–909, 1995.

31. W. H. Loh, A. B. Grudinin, V. V. Afanasjev, and D. N. Payne, Soliton interaction in the presence of a weak nonsoliton component, *Opt. Lett.*, 19, 698–700, 1994.

32. G. P. Agrawal, *Nonlinear Fiber Optics*, 3rd edn., Academic Press, San Diego, CA, 2001.
33. L. Socci and M. Romagnoli, Long-range soliton interactions in periodically amplified fiber links, *J. Opt. Soc. Am. B*, 16, 12–17, 1999.
34. F. Gutty, Ph. Grelu, N. Huot, G. Vienne, and G. Millot, Stabilisation of modelocking in fiber ring laser through pulse bunching, *Electron. Lett.*, 37, 745–746, 2001.
35. D. Y. Tang, B. Zhao, L. M. Zhao, and H. Y. Tam, Soliton interaction in a fiber ring laser, *Phys. Rev. E*, 72, 016616, 2005.
36. R. P. Davey, N. Langford, and A. I. Ferguson, Interacting solutions in erbium fiber laser, *Electron. Lett.*, 27, 1257–1259, 1991.
37. D. Krylov, L. Leng, K. Bergman, J. C. Bronski, and J. NathanKutz, Observation of the breakup of a prechirped N-soliton in an optical fiber, *Opt. Lett.*, 24, 1191–1193, 1999.
38. S. Longhi, Pulse dynamics in actively mode-locked lasers with frequency shifting, *Phys. Rev. E*, 66, 056607, 2002.
39. C. R. Doerr, H. A. Haus, E. P. Ippen, M. Shirasaki, and K. Tamura, Additive-pulse limiting, *Opt. Lett.*, 19, 31–33, 1994.
40. S. Wabnitz, Suppression of soliton interactions by phase modulation, *Elect. Lett.*, 29, 1711–1713, 1993.
41. J. P. Gordon, Interaction forces among solitons in optical fibers, *Opt. Lett.*, 8, 596–598, 1983.
42. Y. Kodama and K. Nozaki, Soliton interaction in optical fibers, *Opt. Lett.*, 12, 1038–1040, 1987.
43. F. M. Mitschke and L. F. Mollenauer, Experimental observation of interaction forces between solitons in optical fibers, *Opt. Lett.*, 12, 355–357, 1987.
44. T. Georges and F. Favre, Modulation, filtering, and initial phase control of interacting solitons, *J. Opt. Soc. Am. B*, 10, 1880–1889, 1993.
45. N. D Nguyen and L. N. Binh, Generation of bound-solitons in actively phase modulation mode-locked fiber ring resonators, *Opt. Commun.*, 281, 2012–2022, 2008.
46. J. G. Proakis, *Digital Communications*, McGraw-Hill, New York, 2001.
47. L. J. Wang, J. T. Lin, and P. Ye, Analysis of polarization-dependent gain in fiber amplifiers, *IEEE J. Quant. Electron.*, 34, 413–418, 1998.
48. H. A. Haus, E. P. Ippen, and K. Tamura, Additive-pulse modelocking in fiber lasers, *IEEE J. Quantum Electron.*, 30, 200–208, 1994.
49. T. F. Carruthers, I. N. Duling III, and M. L. Dennis, Active–passive modelocking in a single polarization erbium fiber laser, *Electron. Lett.*, 30, 1051–1053, 1994.
50. C. R. Doerr, H. A. Haus, and E. P. Ippen, Asynchronous soliton mode locking, *Opt. Lett.*, 19, 1958–1960, 1994.
51. J. J. O'Neil, J. N. Kutz, and B. Sandstede, Theory and simulation of the dynamics and stability of actively modelocked lasers, *IEEE J. Quant. Electron.*, 38, 1412–1419, 2002.
52. J. G. Proakis, *Digital Communications*, 4th edn., McGraw-Hill, New York, 1983.
53. L. N. Binh, *Optical Digital Communications*, CRC Press, Boca Raton, FL, 2008,

# 4 Multi-Bound Solitons under Carrier Phase Modulation

As shown in Chapters 2 and 3, an electro-optic (EO) phase or amplitude modulator is commonly integrated in a fiber ring laser in order to tune the generated soliton train into multi-bound solitons. The influence of the phase modulation is reflected through two aspects of the phase modulator: the inherent birefringence in the Ti:LiNbO$_3$ waveguide and the phase modulation profile or chirp rate caused by the modulation signal. Thus, there are two polarized ring laser competing for resonant energy of the system. This phenomenon is very interesting to observe and may lead to potential applications. In this chapter, first, the characteristics of optical phase modulators are described and, second, their phase modulation properties corroborated with the dynamic behavior of the generated multi-bound solitons are described.

## 4.1 ELECTRO-OPTIC PHASE MODULATORS

As described in Chapter 3, the modulation of the phases of the guided lightwaves by integrated EO modulators incorporated in the mode-locked fiber lasers is essential for the generations and stability control of the solitons and multi-bound solitons. These EO phase modulators (EOPMs) are also employed as the principal components in optical transmission and photonic signal processing systems due to their small size and inline compatibility with single-mode optical fiber [1].

In an actively mode-locked fiber laser, an EOPM acts as a mode-locker, hence its characteristics significantly influence the performance as well as the operation modes of the fiber cavity. In our thesis we have used two models of integrated EOPMs: one is the model PM-315P of Crystal Technology and the other is the model Mach-40-27 of Covega. Depending on the geometry of the electrodes, an integrated LiNbO$_3$ modulator can belong to one of two types: lumped-type modulator or traveling-wave type modulator.

### 4.1.1 LUMPED-TYPE MODULATOR

A LiNbO$_3$ phase modulator consists of a titanium-diffused monomode waveguide and a pair of modulation electrodes. Titanium waveguides can support both transverse electric (TE) and transverse magnetic (TM) optical polarizations. Due to the crystal symmetry in LiNbO$_3$, there are two useful crystal orientations Z-cut and

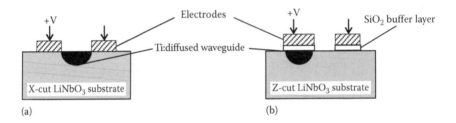

**FIGURE 4.1** Geometry of LiNbO$_3$ phase modulators with the electrodes for field distribution across the optical waveguide cross section for maximizing the electro-optic interaction effect depending on (a) X- or (b) Z-cut LiNbO$_3$ crystal substrate.

X-cut, which take advantage of the strongest EO coefficient. For an X-cut device, the waveguide is symmetrically located between two electrodes as shown in Figure 4.1. In a Z-cut device with the optical axis perpendicular to the surface, one of the electrodes is directly placed on the waveguide and an optical isolation layer is inserted between the waveguide and the electrode to avoid increased optical losses. Application of the electrical driving voltage to the electrodes causes a small change in the refractive index of the waveguide, and the phase of the optical signal passing through the modulator can be consequently changed as follows:

$$\Delta\varphi = \pi\frac{V}{V_\pi} = \frac{\pi L}{\lambda} r_{eff} n_{eff}^3 \frac{V}{d}\Gamma \tag{4.1}$$

where
   $V$ is the applied voltage
   $L$ is the length of the electrode
   $\lambda$ is the wavelength
   $r_{eff}$ is the appropriate electro-optic coefficient
   $n_{eff}$ is the unperturbed refractive index
   $d$ is the separation between electrodes
   $\Gamma$ is the overlap factor between the electric field and the optical mode field, which
      is an important parameter to optimize the modulator design.

Depending on the type of electrode, the EOPMs can be classified into two types: the lumped-type modulator and the traveling-wave modulator. A lumped-type EOPM is illustrated in Figure 4.2. In this configuration, the radio frequency (RF) driving voltage is directly fed to the electrodes whose lengths are small compared to the drive-signal wavelength, and the modulation bandwidth is limited by the $RC$ time constant of the electrode capacitance ($C$) and the parallel matching resistance ($R$) as follows:

$$f_{3dB} = \frac{1}{\pi RC} \tag{4.2}$$

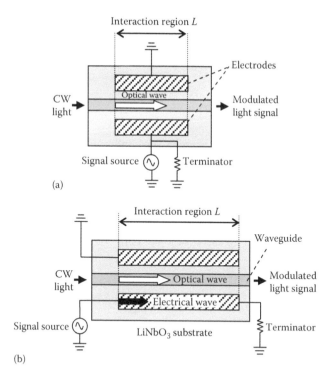

**FIGURE 4.2** Two types of EO phase modulators: (a) lumped and (b) traveling-wave.

The matching resistance is normally set to 50 Ω to allow broadband matching to a 50 Ω driving source. For low operating voltage, the modulator of longer interactive lengths is required, but the capacitance of the electrode increases, which limits the maximum frequency of operation. It is difficult to fabricate a lumped-type EOPM with a broad bandwidth and low operating voltage. In other words, lumped-type EOPMs can operate at a frequency of a few gigahertz with the expense of low $V_\pi$.

### 4.1.2 TRAVELING-WAVE MODULATOR

In order to increase the modulation bandwidth, traveling-wave electrodes are preferably used in the modulator design. Traveling-wave electrodes are designed as transmission lines, fed at one end and terminated with a resistive load at the other end, as shown in Figure 4.2b. The optical signal in the waveguide and the electrical signal in the electrode propagate in the same direction. Because the effective refractive index of the electrical wave $n_m$ is about twice that of the optical wave $n_o$, there is a velocity mismatch between the electrical wave and the optical wave that limits the modulation bandwidth as follows:

$$f_{3dB} = \frac{1.4c}{\left[ L(n_o - n_m) \right]} \qquad (4.3)$$

where
  $c$ is the velocity of light in vacuum
  $L$ is the length of the electrode.

Much effort has been put into designing traveling-wave modulators to match the velocities of the electrical and optical waves, and thus to improve the frequency response characteristics. To obtain the velocity matching, the effective index of the electric wave $n_m$ is lowered by using effectively a dielectric buffer layer between the electrode and the waveguide. Based on this concept, many structures of the traveling-wave electrode have been proposed to produce the EO modulators with broad bandwidth of tens of gigahertz or lower $V_\pi$ [2–4].

## 4.2  CHARACTERIZATION MEASUREMENTS

### 4.2.1  HALF-WAVE VOLTAGE

In some options for half-wave voltage ($V_\pi$) measurement, the direct optical spectrum analysis offers an accurate and simple solution in case the modulation sidebands can be resolved by the optical spectrum analyzer (OSA). When a continuous wave (CW) light passes through a phase modulator, its spectrum is broadened by the generation of modulation sidebands. The strength of sidebands is proportional to the modulation index and the sideband positions. The optical field in the frequency domain of a phase-modulated CW signal can be mathematically expressed by Fourier expansion as follows:

$$E_o(\omega) = E_i \exp(j\omega_0 t) \sum_{k=-\infty}^{\infty} (j)^k J_k(m) \exp(jk\omega_m t) \qquad (4.4)$$

where
  $\omega_0$, $\omega_m$ are the carrier frequency and RF modulation frequency, respectively,
    $\omega = \omega_0 + k\omega_m$
  $k$ is the sidemode index, which is an integer
  $J_k$ is the $k$th-order Bessel function of the first kind
  $E_o$, $E_i$ are the output and input fields of the phase modulator, respectively.

Thus, the intensity of each sidemode measured on an OSA is

$$I_o(\omega_0 + k\omega_m) = I_i J_k^2(m) \qquad (4.5)$$

where $m$ is the modulation index and is related to the $V_\pi$ by Equation 4.1 and $I = |E|^2$. Expression (Equation 4.5) shows a relationship between the modulation index and the intensity at a specific sideband that varies as a Bessel function, as shown in Figure 4.3. Based on optical spectrum analysis, there are some approaches for the

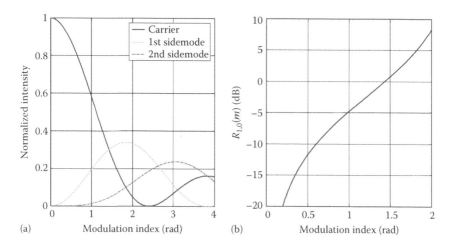

**FIGURE 4.3**    Variation of (a) normalized optical intensity for carrier and sidemodes and (b) ratio $R_{1,0}$ as a function of the phase modulation index $m$.

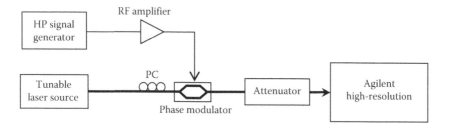

**FIGURE 4.4**    The OSA-based $V_\pi$ measurement setup.

$V_\pi$ measurement such as carrier nulling method [5] and the relative sideband/carrier intensity ratio method [6]. In power limitation of RF amplifiers, the relative first sideband/carrier intensity ratio is the most suitable method for small signal modulation that is used for measurement.

The measurement setup and the components are depicted in Figure 4.4. The phase modulators consisting of the model PM-315P (phase modulator) and the model (Mach–Zehnder interferometer) MZIM-40-27 (intensity modulator biased to phase modulation) Covega were characterized by OSA at a modulation frequency of 1 GHz. For a conventional OSA, the resolution is limited to resolve the modulation sidemodes at such low frequency. Fortunately, we borrowed a high-resolution spectrum analyzer Agilent 83453B for this measurement. By the OSA with a resolution of <0.008 pm (~1 MHz), the intensity of sidemodes with a spacing of 1 GHz can be clearly displayed, as shown in Figure 4.5. At a specific RF driving power, the modulation index $m$ is calculated from the relative intensity ratio between the carrier and the first sidemode $R_{1,0}$:

$$R_{1,0}(m) = \frac{I_o(\omega_0 \pm \omega_m)}{I_o(\omega_0)} = \frac{J_1^2(m)}{J_0^2(m)} \qquad (4.6)$$

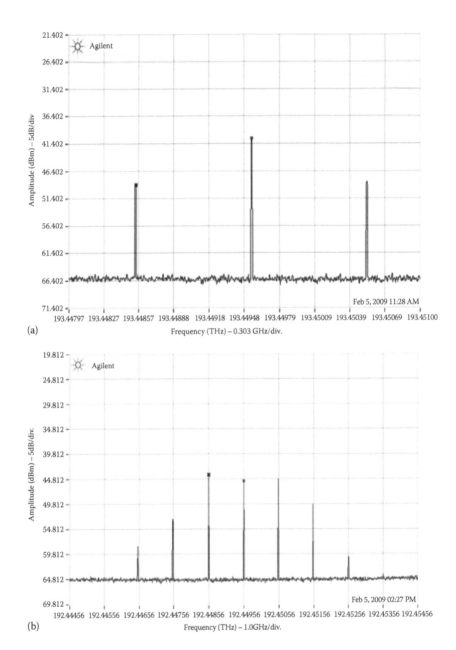

**FIGURE 4.5** Optical spectra of the signal modulated by phase modulators: (a) PM-315P and (b) Mach-40-27 at the same RF driving level of 19 dBm.

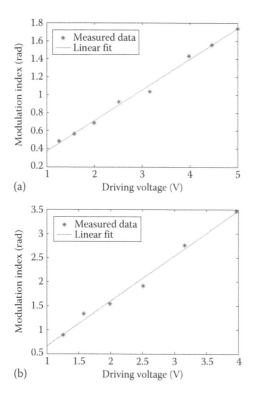

**FIGURE 4.6** Phase modulation index calculated from measured relative intensity ratio $R_{1,0}$ at 1 GHz as a function of RF driving voltage for two models: (a) PM-315P and (b) MZIM-40-27.

which is measured by the OSA. By varying the RF driving voltages, the $V_\pi$ of two phase modulators can be then determined from the slope of the linear fit line as shown in Figure 4.6. The estimated $V_\pi$ of PM-315 and MZIM-40-27 are 3.93 and 8.87 V, respectively. These results agree with the specifications provided by the manufacturers.

## 4.2.2 DYNAMIC RESPONSE

The dynamic response of the EOPM is an important characteristic, which indicates the variation of modulation efficiency over a range of frequencies. Because optical phase modulators only modulate the phase of optical carrier, it is impossible to measure directly the modulator response. A conversion of phase modulation into amplitude modulation needs to be done for this measurement. There are several techniques that have been proposed to implement this operation consisting of using a MZIM [5], using a Fabry–Perot interferometer as an optical discriminator [7], and using a Sagnac loop configuration [8]. However, the two former techniques are limited in use due to their complexity and reliability, the last technique provides an efficient and simple way of measuring the dynamic response for all types of EOPMs

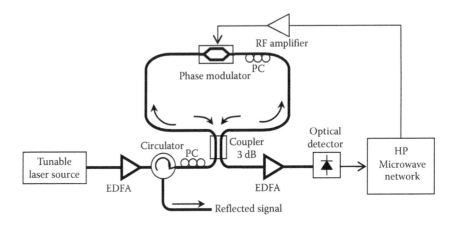

**FIGURE 4.7** Schematic setup for measuring the dynamic response of the EO phase modulator.

with high resolution. Especially, various types of phase modulators can be identified through this measurement.

The setup of the dynamic response measurement based on a Sagnac loop is shown in Figure 4.7. This configuration is similar to the configuration of phase modulated Sagnac loop (PMSL) described in Chapter 2. The only main difference is the position of the phase modulator in the Sagnac loop, which is off center. Two counter-propagating optical waves are phase-modulated and then coherently summed at the output 3 dB coupler. This process converts phase modulation into intensity modulation, so that the response of a phase modulator can be measured by a network analyzer HP8753D and an S-parameter test set HP85046A. The polarization is optimized by the polarization controller to maximize the optical power passing through the lithium niobate optical waveguide. Thus, the transfer function for this structure, which can be detected by the network analyzer, is given by [8]

$$H(f) = \frac{1}{2} l_{PM}^2 K_0(f) \Re \left[ \left( 1 + \eta^2(f) \right) - 2\eta(f) \cos \phi(f) \right] R_0 \qquad (4.7)$$

where
   $l_{PM}$ is the insertion loss of phase modulator
   $K_0(f)$ is the modulation response parameter
   $R_0$ is the load resistance
   $\Re$ is the detector responsivity
   $\phi(f)$ is the phase difference
   $\eta(f)$ is the ratio of backward to forward phase modulation index, which relates to
      the signal transit time $\tau_L$ [9] as follows:

$$\eta(f) = \frac{\sin(2\pi f \tau_L)}{2\pi f \tau_L} \qquad (4.8)$$

This structure is also a notch filter applied in photonic signal processing [10], so that the dynamic response of the phase modulator should be the envelope of the periodic notch filter response. The free spectral range (FSR) of the notch response relates to the phase difference by

$$\phi(f) = 2\pi f \tau \quad \text{and} \quad \tau = \frac{1}{\text{FSR}_{\text{notch}}} = \frac{\Delta L}{c/n} \tag{4.9}$$

where $\tau$, $\Delta L$ are the time delay and the length difference between two sides of the Sagnac loop, respectively. In order to measure accurately the dynamic response of the modulator, FSR of the notch response must be as small as possible to give a high resolution, which is the reason the phase modulator must be located off-center of the loop. In our setup, the length difference $\Delta L$ is about 3.5 m that gives an FSR of 58 MHz.

With a frequency range up to 3 GHz, the network analyzer may not cover the whole bandwidth of the phase modulators. However, this measurement allows an identification of different types of the phase modulator. Figure 4.8 shows the dynamic responses for two phase modulators, and indicates the difference between them. For a lumped-type modulator such as PM-315P, the ratio $\eta$ keeps unchanged over the bandwidth of the phase modulator, so the notch response shows deep notches and a flat passband within the bandwidth as shown in Figure 4.8a. The envelope of this response gives exactly the dynamics response of the phase modulator with the measured 3 dB bandwidth of 2.7 GHz. For a traveling-wave modulator, the velocity mismatch effect causes the notch depth to disappear at certain frequencies $f_k$ that are related to the transit time $\tau_L$ as follows [9]:

$$f_k = \frac{k}{2\tau_L} \quad \text{with} \quad k = 1, 2, \dots \tag{4.10}$$

The measured notch response of the phase modulator Mach-40-27 with the null frequency $f_1$ of 2.27 GHz is shown in Figure 4.8, which indicates a response of the traveling-wave modulator. And the net dynamic response of the modulator can be also obtained by a correction of the envelope of the measured notch response from the known function $\eta(f)$. More importantly, the interactive length of the electrode in the traveling-wave modulator can be estimated from the transit time $\tau_L$. For the Mach-40-27 modulator, the interactive length about 32 mm is calculated from the transit time of 0.455 ns. We note that the length of the waveguide in this modulator would be much longer than that of the electrode due to a polarizer integrated in the phase modulator [11].

Thus, from the characterization measurements of the two phase modulators, the difference between lumped-type modulator and traveling-wave modulator has been indicated. For the lumped-type modulator PM-315P, the short length of electrodes has been verified by its high $V_\pi$ and flat frequency response over the measurement bandwidth. Similarly, for the traveling-wave-type modulator Mach-40-27, long

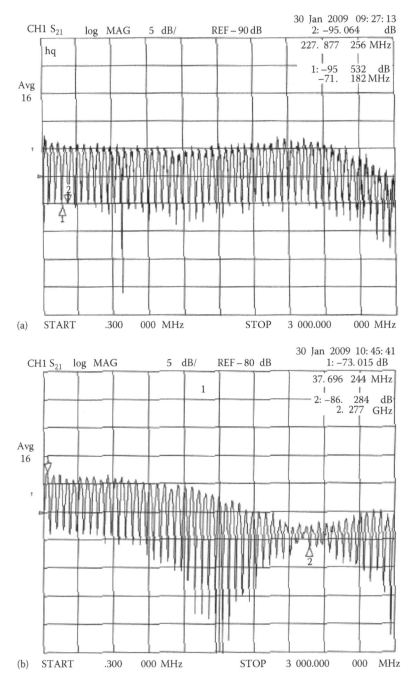

**FIGURE 4.8** Measured frequency notch response of the phase modulators (a) lumped-type PM-315P and (b) traveling-wave type Mach-40-27 within a measured frequency range up to 3 GHz. Vertical axis: amplitude (arbitrary unit); horizontal scale: frequency (full scale as noted).

electrodes and waveguide have been verified by its low $V_\pi$ and the velocity mismatch in the frequency response measurement.

## 4.3   COMB SPECTRUM IN ACTIVELY MODE-LOCKED FIBER RING RESONATOR INCORPORATING PHASE MODULATOR

### 4.3.1   BIREFRINGENCE AND COMB SPECTRUM IN THE FIBER RING USING PHASE MODULATOR

Because the integrated EO modulators are normally polarization-dependent elements, a birefringence effect always exists in an actively mode-locked fiber ring laser, even if the birefringence of other components such as optical fibers can be ignored. In fact, a Ti-diffused electro-optic LiNbO$_3$ phase modulator (PM) can support both TE and TM modes propagating at different ordinary and extraordinary effective indices. Although the polarization state is usually controlled by the polarization controller, especially at the input of integrated EO modulator so as only one polarized state can be preferred through the modulator and hence, forcing the matching condition of this polarized state, there is still an asymmetrically simultaneous existence of two polarization modes in the ring, especially in the Ti-diffused waveguide. Thus, these modes with different phase delays can couple and interfere at the output of the modulator to form an artificial birefringence filter known as Lyot filter in the ring cavity. Hence, the output spectral response of the ring cavity is similar to that of a MZIM, the all-pass filter with nulls and maxima [12,13]. The output transmittance of the ring cavity using an EOPM is simply given by [12]

$$T = \Gamma_g \cos^2\left(\frac{\Delta\phi}{2}\right) \tag{4.11}$$

where
  $\Gamma_g$ is the insertion loss of the waveguide depending on the input polarization states and the gain bandwidth of the EDFA
  $\Delta\phi$ is the effective phase difference between TE and TM modes.

Because no polarization maintaining (PM) fiber is used, the effective phase difference in the cavity is dominated by birefringence of the Ti-diffused waveguide. Therefore, the phase difference is given as follows:

$$\Delta\phi = \frac{2\pi l \Delta n_{eff}}{\lambda} \tag{4.12}$$

where
  $l$ is the waveguide length
  $\Delta n_{eff}$ is the effective index difference between TE and TM modes

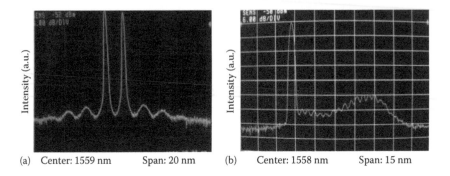

(a)    Center: 1559 nm          Span: 20 nm      (b)    Center: 1558 nm          Span: 15 nm

**FIGURE 4.9**  The comb-like optical spectra in two setups of the fiber ring laser using (a) model PM-315P and (b) model Mach-40-27, respectively.

The interference between two polarization modes generates a comb-like spectrum with spacing or FSR

$$\delta\lambda = \frac{\lambda^2}{l\Delta n_{eff}} \tag{4.13}$$

On the other hand, the gain spectrum in the fiber cavity is modulated when the waves propagate through the phase modulator.

With two models of phase modulator used in the experiment, the comb-like optical spectra were measured by appropriate setting to display in OSA. Figure 4.9 shows the optical spectra in CW operation mode of the ring cavities using models PM-315P and Mach-40-27 respectively. The emission wavelength is located on one of the maxima of the comb-like response where the gain is maximized. For the ring using model PM-315P, the average FSR is 2.15 nm in a bandwidth of about 12 nm, while the average FSR of the ring using model Mach-40-027 is only 0.45 nm in a bandwidth of about 10 nm as observed in Figure 4.9. Based on the characterization and the physical dimensions of these two models, the waveguide lengths of PM-315P and Mach-40-27 are assumed to be 16 and 64 mm, respectively. Thus, the FSRs of comb-like spectral response in cases of two models, calculated by Equation 4.13, are 2.0254 and 0.4668 nm, respectively, when the effective index difference of TE and TM modes of 0.08 is used [14]. We have examined birefringence of the ring cavity in both cases by changing the optical fibers of different lengths. The FSRs remain unchanged in all cases that has proven that the birefringence is mainly determined by the Ti-diffused waveguide. These results are reasonable and agree with the results obtained from OSA. The existence of parasitic birefringence in the integrated EOPMs forms naturally a comb-like filter that affects mode-locking schemes and the characteristics of mode-locked pulse trains.

### 4.3.2  Discrete Wavelength Tuning

One of the influences of the comb-like filtering effect is a discrete wavelength tuning in the FM mode-locked fiber ring laser. In our mode-locked fiber lasers, the mode-locking happens only at the peaks of the comb-like optical spectrum where the lightwaves acquire sufficient energy gain to satisfy the mode-locking condition. The wavelength of the sequence of the mode-locked pulses can be tuned by only changing the modulation frequency of the phase modulator. Thus, the wavelength tuning is based on the matching of the dispersed spectrum of the pulse sequence, hence tuning of the pulse central passband [15–17]. Hence, the tuning range of the laser depends on the profile of gain spectrum and the total dispersion of the cavity. Because the gain spectrum is modulated by the interference between two polarization modes, the modulation frequency tuning is required at an amount of [15]

$$\delta f_m = -\frac{f_m^2 D L_{cav} \delta \lambda}{m} \tag{4.14}$$

where

$f_m$ is the modulation frequency

$m$ is the harmonic order

$D$ and $L_{cav}$ are the average dispersion and the total length of the ring cavity, respectively

We note that the tuning is only achieved in a specific range of wavelengths to prevent the competition of the CW modes that suppress the mode-locking at other matched wavelengths although the comb-like spectrum can be observed in most phase modulators.

In the setup using the phase modulator Mach-40-027 with an integrated polarizer at its output, mode-locking is achieved at a modulation frequency of about 1 GHz. A 100 m long Corning SMF-28 fiber is inserted after the PM to ensure that the average dispersion in the loop is anomalous. The total loop length is 190 m corresponding to a mode spacing of 1.075 MHz. By tuning the modulation frequency, the light-waves in the cavity are mode-locked to generate a short pulse sequence at the wavelength corresponding to one of the peaks of the comb spectrum, as shown in Figure 4.9. A 0.46 nm average spacing between two adjacent mode-locked wavelengths is achieved. By adjusting the PC to optimize the gain spectrum, we can tune over 18 different wavelengths (1554–1562 nm) by simply changing the modulation frequency of 1.35 kHz as shown in Figure 4.10a. These measured parameters agree with the theoretically predicted values, which are of 0.4578 nm and 1.27 kHz, respectively. Figure 4.10b shows the pulse width and bandwidth-time (BT) product (of mode-locked pulses at 18 wavelengths. The average BT of 0.8 indicates that the output pulses are highly chirped and they can be compressed using a suitable dispersive fiber at the output of the laser. In wavelength-tunable harmonically mode-locked fiber laser, the supermode noise is normally an important issue that requires some methods of suppression to improve performance of the generated pulse train [17]. It is worth to note that the mode-locked pulses at all 18 wavelengths are very clean, which can be seen from the time trace and its corresponding RF spectrum shown in Figure 4.11. When the $f_m$ is tuned by an amount of $\delta f_m/2$ a strong mode competition occurs, which degrades the waveform of the output, as shown

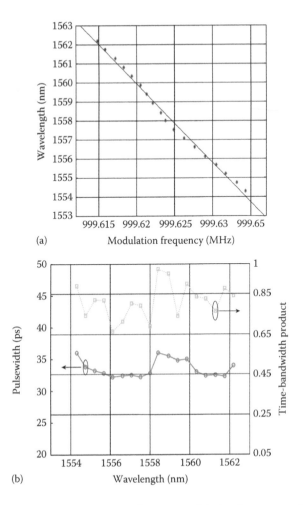

**FIGURE 4.10** (a) Mode-locked wavelength versus modulation frequency and (b) measured characteristics of a mode-locked pulse over the tuning range.

in Figure 4.12. On the other hand, this is the transition state for switching to adjacent spectral peak; total mode-locking cannot be achieved. When the wavelength is tuned to the edge of gain spectrum, a multiwavelength operation of the laser can be easily excited and it also affects the performance of mode-locked pulses, as shown in Figure 4.12.

Similarly, a wavelength tuning operation is also observed in the setup using the phase modulator PM-315P, although it is more difficult to adjust the polarization controller to obtain a sufficient gain bandwidth covering the tuning range of wavelength because of the wide spacing in between (Figure 4.13). Only three wavelengths at the peaks of the comb-like spectrum are tuned with changing the modulation frequency of 3.1 kHz in this setup to obtain a high-quality pulse train, whose characteristics are summarized in Table 4.1.

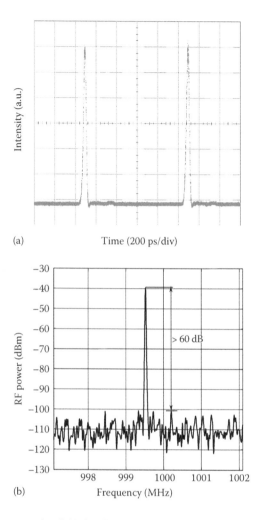

(a)                                    Time (200 ps/div)

(b)                                    Frequency (MHz)

**FIGURE 4.11**    An example of (a) the time trace and (b) RF spectrum of the mode-locked pulse sequence at the tuned wavelength.

## 4.4    INFLUENCE OF PHASE MODULATOR ON MULTI-BOUND SOLITONS

### 4.4.1    FORMATION OF MULTI-BOUND SOLITONS

The existence of a comb-like filtering effect in the cavity using EOPM remarkably influences the characteristics of the generated pulses. The artificial filter based on the birefringence of the cavity limits the gain bandwidth and contributes to the mechanism of pulse broadening due to the limitation of pulse spectrum. This effect has been experimentally demonstrated by the results of pulse width obtained from two setups of the mode-locked fiber lasers using the modulators PM-315P and Mach-40-27.

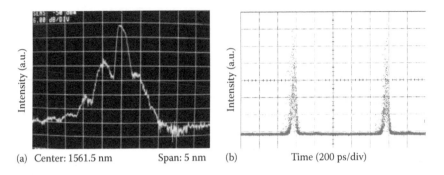

(a)   Center: 1561.5 nm              Span: 5 nm      (b)           Time (200 ps/div)

**FIGURE 4.12**   (a) Optical spectrum and (b) time trace of the output when modulation frequency $f_m$ is tuned by $\pm\delta f_m/2$.

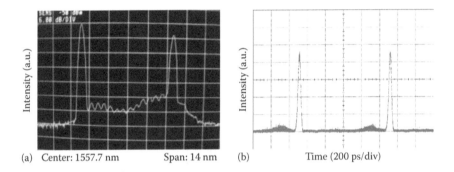

(a)   Center: 1557.7 nm              Span: 14 nm     (b)           Time (200 ps/div)

**FIGURE 4.13**   (a) Optical spectrum and (b) time trace of mode-locked pulse when the wavelength is tuned at the edge of gain spectrum.

---

**TABLE 4.1**

**Pulse Characteristics at Tuned Wavelengths of the FM Mode-Locked Laser Using the Modulator PM-315P**

| Wavelength (nm) | 1555.98 | 1558.03 | 1560.14 |
|---|---|---|---|
| Pulse width (ps) | 11.9 | 13 | 11.3 |
| Spectral width (nm) | 0.237 | 0.24 | 0.26 |
| BT product | 0.3525 | 0.405 | 0.3672 |

---

Although the modulator Mach-40-27 is driven at much higher modulation index compared to the modulator PM-315P because of lower $V_\pi$, the pulse generated from the setup using Mach-40-27 modulator is more than twice as wide as that generated from the setup using PM-315P modulator. With a much smaller width due to broader gain bandwidth, the pulse generated in the case of PM-315P has a high peak power to create a sufficiently nonlinear phase shift that is necessary for pulse splitting in the cavity. Moreover, the effective interactions between pulses in bound states also depend on the pulse width caused by the overlap of long pulse tails and the chirping caused by phase modulation. In the case of PM-315P, the chirp imposed on the generated pulses with sufficient small width allows an effective attraction to balance the repulsion to generate a stable bound state. In contrast, a distinct bound state cannot be observed in the case of Mach-40-27 with wider generated pulses, as mentioned in Chapter 3. In this case, wider pulses obtain higher chirp from the phase modulator to result in a stronger attraction. At once the repulsion force becomes weaker due to direct interaction. Therefore, a balanced interaction cannot be achieved in this case to generate a bound state with obviously resolved pulses although the cavity is strongly pumped. A state of mode-locked pulse with two humps generated from the setup using Mach-40-27 has been observed, as shown in Figure 4.14. This result may be a new solution to the fiber laser system, yet the time separation between two humps less than 20 ps may also indicate a strong attraction between two adjacent pulses.

Furthermore, the polarization sensitivity of the cavity also increases because of the inherent birefringence of the waveguide. Under a strong pumping scheme, some effects such as spectral hole burning (SHB) and polarization hole burning (PHB) are enhanced in the gain medium, and these make polarization-dependent loss and gain more complex [18–21]. Inhomogeneous broadening of the gain spectrum associated with SHB and PHB effects excites multiwavelength emission in the fiber ring laser, which results in a gain competition between different emission wavelengths. Normally only one emission wavelength satisfies the phase-locking condition in the cavity to generate a pulse train, while other wavelengths may operate in the CW or FM mode depending on the polarization setting in the cavity. Thus, if the polarization setting is not optimized, the gain of the mode-locked wavelength can be reduced, which results in a switching from higher-order bound solitons to lower-order bound solitons. Figure 4.15 shows a typical example of switching from the triple-bound soliton to the dual-bound soliton after the polarization controller is slightly adjusted. The gain reduction of the mode-locked wavelength always accompanies the gain enhancement of adjacent wavelengths, as shown in Figure 4.15b. Therefore, it is understandable that the higher-order bound states are more sensitive

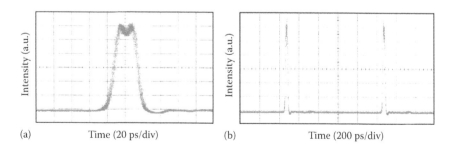

(a)          Time (20 ps/div)          (b)          Time (200 ps/div)

**FIGURE 4.14** A mode-locked state generating a two-hump pulse in the cavity using the phase modulator Mach-40-27: (a) single pulse trace, the saddle soliton, and (b) pulse train trace.

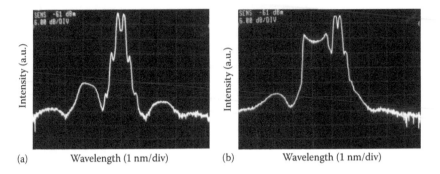

**FIGURE 4.15** Switching phenomenon from the triple-bound soliton to the dual-bound soliton after an adjustment in the polarization controller: (a) spectrum of triple-bound soliton and (b) spectrum of dual-bound soliton with an adjacent wavelength in the FM mode.

to polarization settings. However, the polarization-dependent gain effect also provides a flexible mechanism for the stabilization of bound state by adjusting the input polarization state of the phase modulator. If the gain is too high, multi-bound solitons can fall into an unstable state because of the phase fluctuation, which leads to a breakdown of the interactive balance between pulses. Therefore, a reduction of gain from tuning the polarization state can pull the ring laser back to the stable operation region.

### 4.4.2 LIMITATION OF MULTI-BOUND SOLITON STATES

According to the interaction mechanism of bound solitons in the actively mode-locked fiber ring laser, an effective repulsion is induced by linear chirping caused by periodic phase modulation, or a balanced interaction between solitons is only achieved around the extremes of the modulation cycle, where the chirp can be approximated by a parabolic function. However, this approximation is no longer valid if the width of the bunch is too wide. Deviation from linear chirping increases when the number of solitons in bound states increases. Thus, the pulses at the edge of a bunch would experience a nonlinear chirping rather than a linear chirping, which influences the balanced interaction of bound solitons. This explains why the trace of sextuple-bound solitons looks noisy because of oscillation of solitons in the bound state and why the spectrum is slightly modulated due to change in the relative phase difference. It has also been shown that higher-order multi-bound solitons are more sensitive to environmental perturbations such as temperature variations. When the fluctuation in the cavity length causes a slight detuning, the outer pulses in multi-bound states would be most affected by the frequency shifting, which breaks the balance in soliton interaction.

Besides, the chirping rate determines the separation between bound solitons as described in Chapter 3, the chirp effect and the number of soliton in the bound states are also correlated. Figure 4.16a depicts the experimental threshold power for the creation of dual- and triple-soliton bound states against the phase modulation index, which indicates the requirement for an increase in the splitting optical

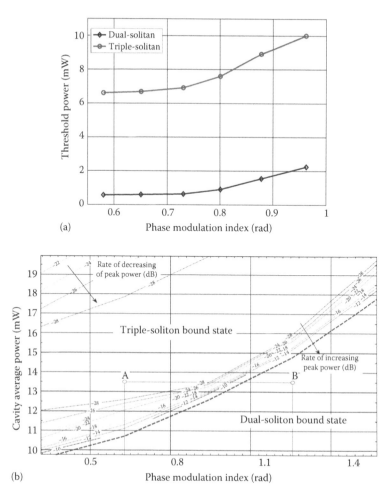

**FIGURE 4.16** (a) Experimentally measured variation of threshold splitting power with the phase modulation index of the bound soliton states and (b) simulated variation of the peak power of the triple-bound state with the phase modulation index.

power when the chirping is increased. An increase in the chirp rate requires a higher-threshold power to maintain a specific number of pulses in the bound state. When the phase modulation index increases, the expansion of the signal spectrum is enhanced. This is advantageous to not only shorten the pulse width but also to more tightly bind the pulses due to the increase in the number of phase-locked modes. However, the increase in the chirping rate also causes stronger energy transfer to the sidebands, which can degrade the soliton content of the pulse sequence. Although the up-chirped pulse is further compressed by anomalous dispersion of the fiber ring at a higher modulation index, larger frequency chirp at pulse edges also may require higher energy contained in the pulses. This tendency agrees with the theoretical analysis of soliton production from the chirped pulses in [22,23].

At a low phase modulation index, the chirping is almost unaffected by the threshold power due to weak binding of the bound solitons. Consequently, the waveforms of the bound states are noisier and more sensitive to phase fluctuations caused by the amplification stimulated emission (ASE) noise or the random polarization variation of the guided medium. A simulation result of the power limit of the cavity to maintain the bound state is shown in Figure 4.16b. The rates of increasing and decreasing of peak power of bound solitons are indicated showing two distinct regions of operations, the triple and the double bound separated by the dashed bold line. This limit line (dashed bold line) is obtained when a minimum power level circulating in the cavity is required to maintain the bound state that divides the graph into the triple and double bound regions. At a specific modulation index, when the average power of the cavity decreases, the variation of the peak power is stronger. When the power is lower than the limit line, the higher-order soliton bound state switches to a lower-order soliton bound state. This tendency also indicates that the increase in chirping diminishes the number of solitons in the bound state at a specific optical power level, as shown in Figure 4.17a and b. These figures show the evolution of the triple-soliton bound state in the cavity at the same average power level but different phase modulation indices that correspond to the points A and B, respectively, in Figure 4.16b.

Thus, the phase modulation profile significantly affects multi-bound soliton states. In other words, multi-bound states can be modified by varying the modulation curve of the phase modulator. Enhancement of higher-order harmonics in the modulation signal is the simplest way to modify the phase modulation profile. Similar to rational harmonic mode-locking, the higher-order harmonics of modulation frequency are strongly enhanced by the saturation of the RF power amplifier. When the levels of the second- and the third-harmonics are sufficiently high to distort the phase modulation profile, multiple linear chirps in each modulation cycle can be created. In particular, there are two linear chirps in each modulation cycle created by an increase in the input power of the RF amplifier higher than 2 dBm in our setup. In fact a submodulation cycle would appear aside the main cycle due to the enhancement of the second harmonics in the modulation signal. When the laser is locked in harmonic mode-locking scheme instead of rational harmonic mode-locking at the same optical power level of various multi-bound solitons, the multi-bound solitons with smaller number of solitons are formed at two linear chirp positions as shown in Figure 4.18. The distortion of phase modulation profile splits higher-order bound solitons into two groups of lower-order bound solitons within the same cycle. Moreover, the change in the higher-order harmonics level also changes the distortion, which varies the chirp rate between two groups of solitons. Figure 4.19 shows a variation of time separation between two solitons split from the dual-bound soliton state as a function of the RF input power. Thus, the time separation reduces according to the reduction of the higher harmonic level in the modulation signal. When the RF input power is decreased to 2 dBm, two groups of bound soliton emerge together to form higher-order multi-bound soliton with the number of solitons being equal to the total of number solitons in two groups. The emergence of two bound soliton groups is due the near disappearance of the submodulation cycle generated by the second harmonic RF signal. It is interesting

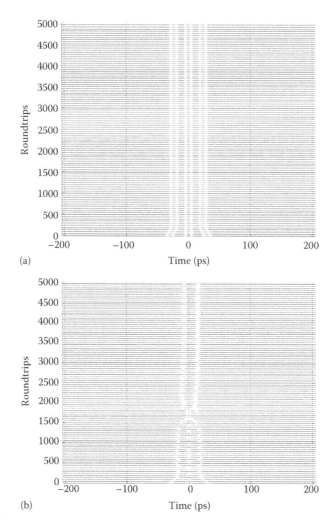

**FIGURE 4.17**    (a) and (b) Simulated evolution of the triple-soliton bound state at operation points A and B, respectively, in Figure 4.16b.

to see that the change in temporal separation is a complete linear function of the second harmonic RF power, as shown in Figure 4.20.

## 4.5  CONCLUDING REMARKS

In this chapter, we have demonstrated the influence of the EOPM on multi-bound solitons. The influence is reflected through two aspects of the phase modulator that consist of the inherent birefringence in the Ti:LiNbO$_3$ waveguide and the phase modulation profile or chirp rate caused by the modulation signal. Two typical phase modulators have been described and characterized to indicate a difference in structure between the lumped-type and the traveling-wave type that influence the operation

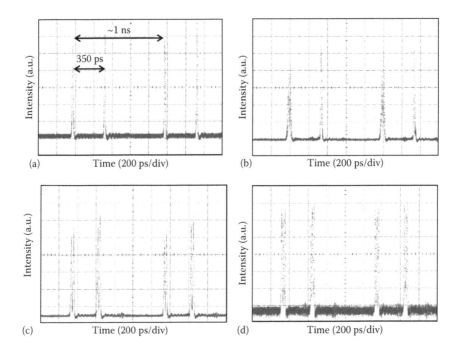

**FIGURE 4.18** Splitting of multi-bound solitons: (a) dual-bound soliton, (b) triple-bound soliton, (c) quadruple-bound soliton, and (d) sextuple-bound soliton into lower-order bound solitons.

modes of the mode-locked fiber laser. Because an artificial comb-like response can be formed in the ring cavity by the birefringence of the integrated phase modulators, a narrow FSR in the response for the phase modulator with long waveguide can limit pulse shortening and multi-bound soliton formation. Moreover, the ability of discrete wavelength tuning can be implemented in the fiber laser through simple tuning of the modulation frequency.

Mode-locking and multi-bound soliton operation in the actively frequency modulation-mode-locked fiber ring lasers is supported by linear chirping that can be approximated by a sinusoidal modulation. Besides the temporal separation of adjacent pulses, the optical power threshold for splitting the solitons and the number of solitons in the bound states can be influenced by the chirp rate induced by this phase modulation. When the modulation profile is modified by the enhancement of higher-order harmonic in the driving modulation signal, higher-order multi-bound solitons can be split into lower-order multi-bound solitons with controllable temporal separation between them. More importantly, the experimental results have demonstrated the controllability through manipulating and incorporating active devices such as the EOPM. This property can lead to several important practical applications.

Recent research and development of integrated Si photonics would offer greater opportunities in highly compact multi-bound soliton generators [24].

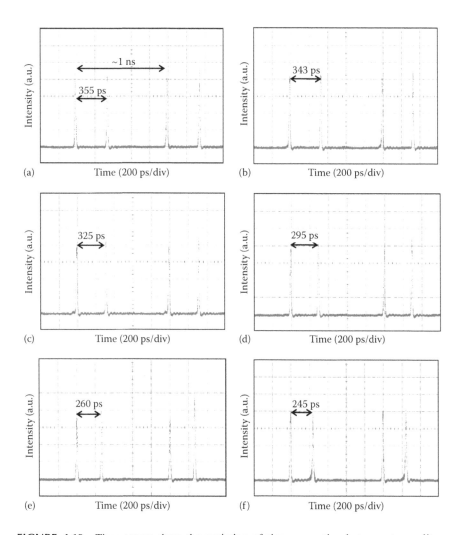

**FIGURE 4.19** Time traces show the variation of time separation between two solitons split from dual-bound soliton versus the change in RF input power: (a) 7 dBm, (b) 6 dBm, (c) 5 dBm, (d) 4 dBm, (e) 3 dBm, and (f) 2.5 dBm.

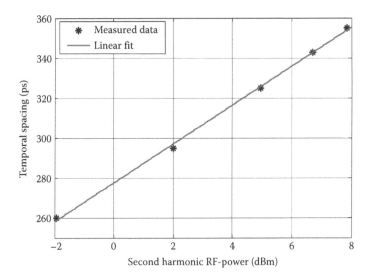

**FIGURE 4.20** Correlation between the temporal spacing between two groups of bound solitons and the second harmonic RF power.

## REFERENCES

1. L. N. Binh, *Photonic Signal Processing: Techniques and Applications*, CRC Press, Boca Raton, FL, 2008.
2. K. Kawano, T. Kitoh, O. Mitomi, T. Nozawa, and H. Jumonji, A wide-band and low-driving-power phase modulator employing a Ti:LiNbO$_3$ optical waveguide at 1.5 mm, *IEEE Photon. Technol. Lett., IEEE.*, 1, 33–34, 1989.
3. K. Kawano, T. Kitoh, H. Jumonji, T. Nozawa, and M. Yanagibashi, New travelling-wave electrode Mach-Zehnder optical modulator with 20 GHz bandwidth and 4.7 V driving voltage at 1.52 mm wavelength, *Electron. Lett.*, 25, 1382–1383, 1989.
4. K. Noguchi, O. Mitomi, H. Miyazawa, and S. Seki, A broadband Ti:LiNbO$_3$ optical modulator with a ridge structure, *J. Lightwave Technol.*, 13, 1164–1168, 1995.
5. R. Tench, Performance evaluation of waveguide phase modulators for coherent systems at 1.3 and 1.5181 mm, *J. Lightwave Technol.*, 5, 492–501, 1987.
6. Y. Shi, L. Yan, and A. E. Willner, High-speed electrooptic modulator characterization using optical spectrum analysis, *J. Lightwave Technol.*, 21, 2358, 2003.
7. R. Regener and W. Sohler, Loss in low-finesse Ti:LiNbO$_3$ optical waveguide resonators, *Appl. Phys. B: Lasers Opt.*, 36, 143–147, 1985.
8. E. Chan and R. A. Minasian, A new optical phase modulator dynamic response measurement technique, *J. Lightwave Technol.*, 26, 2882–2888, 2008.
9. W. Leeb, A. L. Scholtz, and E. Bonek, Measurement of velocity mismatch in traveling-wave electrooptic modulators, *IEEE J. Quant. Electron.*, 18, 14–16, 1982.
10. E. H. W. Chan and R. A. Minasian, Sagnac-loop-based equivalent negative tap photonic notch filter, *IEEE Photon. Technol. Lett.*, 17, 1740–1742, 2005.
11. R. Stubbe, G. Edwall, B. Sahlgren, L. Svahn, P. Granestrand, and L. Thylen, Polarization selective phase modulator in LiNbO$_3$, *IEEE Photon. Technol. Lett.*, 2, 187–190, 1990.
12. G. Shabtay, E. Eidinger, Z. Zalevsky, D. Mendlovic, and E. Marom, Tunable birefringent filters—Optimal iterative design, *Opt. Express*, 10, 1534–1541, 2002.

13. C. O'Riordan, M. J. Connelly, I. Evans, P. M. Anandarajah, R. Maher, and L. P. Barry, Actively mode-locked multiwavelength fibre ring laser incorporating a lyot filter, hybrid gain medium and birefringence compensated LiNbO$_3$ modulator, in *9th International Conference on Transparent Optical Networks, 2007. ICTON '07*, Rome, Italy, 2007, pp. 248–251.

14. G. J. Sellers and S. Sriram, Manufacturing of lithium niobate integrated optic devices, *Opt. News*, 14, 29–31, 1988.

15. S. P. Li and K. T. Chan, Electrical wavelength tunable and multiwavelength actively mode-locked fiber ring laser, *Appl. Phys. Lett.*, 72, 1954–1956, 1998.

16. Y. Zhao, C. Shu, J. H. Chen, and F. S. Choa, Wavelength tuning of 1/2-rational harmonically mode-locked pulses in a cavity-dispersive fiber laser, *Appl. Phys. Lett.*, 73, 3483–3485, 1998.

17. D. Lingze, M. Dagenais, and J. Goldhar, Smoothly wavelength-tunable picosecond pulse generation using a harmonically mode-locked fiber ring laser, *J. Lightwave Technol.*, 21, 930–937, 2003.

18. T. Aizawa, T. Sakai, A. Wada, and R. Yamauchi, Effect of spectral-hole burning on multi-channel EDFA gain profile, in *Optical Fiber Communication Conference, 1999, and the International Conference on Integrated Optics and Optical Fiber Communication. OFC/IOOC '99. Technical Digest*, San Jose, CA, vol. 2, 1999, pp. 102–104.

19. M. Bolshtyansky, Spectral hole burning in erbium-doped fiber amplifiers, *J. Lightwave Technol.*, 21, 1032, 2003.

20. D. Kovsh, S. Abbott, E. Golovchenko, and A. Pilipetskii, Gain reshaping caused by spectral hole burning in long EDFA-based transmission links, in *Optical Fiber Communication Conference, 2006 and the 2006 National Fiber Optic Engineers Conference. OFC 2006*, Anaheim, CA, 2006, p. 3.

21. L. Rapp and J. Ferreira, Dynamics of spectral hole burning in EDFas: Dependency on pump wavelength and pump power, *IEEE Photon. Technol. Lett.*, 22, 1256–1258, 2010.

22. M. Desaix, L. Helczynski, D. Anderson, and M. Lisak, Propagation properties of chirped soliton pulses in optical nonlinear Kerr media, *J. Phys. Rev. E*, 65, 2002.

23. J. E. Prilepsky, S. A. Derevyanko, and S. Turitsyn, Conversion of a chirped Gaussian pulse to a soliton or a bound multisoliton state in quasi-lossless and lossy optical fiber spans, *J. Opt. Soc. Am. B*, 24, 1254–1261, 2007.

24. L. Vivien and L. Pavesi (ed.), *Handbook of Silicon Photonics*, Series in Optics and Optoelectronics, CRC Press, Boca Raton, FL, 2013.

# 5 Bound-Soliton Bispectra and Nonlinear Photonic Signal Processing

In analog or digital signal processing techniques, power spectrum estimation is commonly used to obtain the power distribution of the signal in the frequency domain. It is a useful and popular tool to analyze or characterize a signal or processing system. However, under this processing mechanism the phase information of the spectral components is suppressed, especially under 1-D Fourier transform. Multi-bound solitons are generated via the phase control of the lightwaves transporting the soliton envelopes. It is very important to understand the phase evolution of the lightwave carrier transporting these bound solitons. The bispectrum technique is a very powerful transformation technique to obtain the amplitude and phases of the solitons distributed in two dimensions (2-D). This chapter is dedicated to introduce this bispectrum transform technique and its applications in the identification of the properties of multi-bound solitons.

## 5.1 BISPECTRUM OF MULTI-BOUND SOLITONS

### 5.1.1 BISPECTRUM

In signal processing, the power spectrum estimation showing the distribution of power in the frequency domain is a useful and popular tool to analyze or characterize a signal or process; however, the phase information between the frequency components is suppressed in the power spectrum. Therefore it is necessarily useful to exploit higher-order spectra known as multidimensional spectra instead of the power spectrum in some cases, especially in nonlinear processes or systems [1,2]. Different from the power spectrum, the Fourier transforms of the autocorrelation, multidimensional spectra are known as the Fourier transforms of higher-order correlation functions, hence they provide us not only the magnitude information but also the phase information.

In particular, the two-dimensional spectrum also called bispectrum is by definition the Fourier transform of the triple correlation or the third-order statistics [2]. For a signal $x(t)$ its triple correlation function $C_3$ is defined as

$$C_3(\tau_1, \tau_2) = \int x(t)x(t - \tau_1)x(t - \tau_2)\, dt \qquad (5.1)$$

where $\tau_1$, $\tau_2$ are the time-delay variables. Thus, the bispectrum can be estimated through the Fourier transform of $C_3$ as follows

$$B_i(f_1, f_2) \equiv F\{C_3\} = \iint C_3(\tau_1, \tau_2) \exp(-2\pi j(f_1\tau_1 + f_2\tau_2)) d\tau_1 d\tau_2 \qquad (5.2)$$

where

$F\{\}$ is the Fourier transform

$f_1, f_2$ are the frequency variables

From the definitions (Equations 5.1 and 5.2), both the triple correlation and the bispectrum are represented in a 3D graph with two variables of time and frequency, respectively. Figure 5.1 shows the regions of power spectrum and bispectrum, respectively, and their relationship. The cut-off frequencies are determined by the intersection between the noise and spectral lines of the signal. These frequencies also determine the distinct areas that are basically bounded by a hexagon in the bispectrum. The area inside the hexagon shows the relationship between frequency components of the signal only, otherwise the area outside shows the relationship between the signal components and noise. Due to a two-dimensional representation in the bispectrum, the variation of the signal and the interaction between signal components can be easily identified.

Because of the unique features of the bispectrum, it is really useful in characterizing the non-Gaussian or nonlinear processes and is applicable in various fields such as signal processing [1,2], biomedicine [3], and image reconstruction [4,5]. Extending the number of representation dimensions makes the bispectrum to more easily and significantly represent different types of signals and differentiate various processes, especially nonlinear processes such as doubling and chaos. Hence multidimensional spectra is proposed as a useful tool to analyze the behaviors of signals generated from these systems such as multi-bound solitons, especially transition states in the formation process of multi-bound solitons.

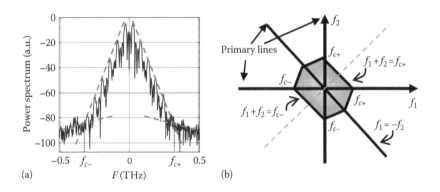

**FIGURE 5.1**  (a) Power spectrum regions, power versus frequency, and (b) corresponding bispectrum regions, 3D graph amplitude versus frequency components in plane distribution.

## 5.1.2 Various States of Bound Solitons

For multi-bound solitons, the optical power spectrum is modulated with the appearance of main lobes and sub-lobes due to the phase relationship between bound pulses as demonstrated in Chapter 3. Therefore, various multi-bound states generated from the FM mode-locked fiber laser can be distinguished by the optical power spectrum analysis. However, as a result the bispectrum of each multi-bound soliton state obviously exhibits a distinct structure for characterization. In this section, various multi-bound solitons in the steady state obtained from simulation are used to estimate their bispectra. Moreover, some interesting information of the MBS can be obtained from the bispectrum.

First, the triple correlations of various multi-bound solitons are estimated as shown in Figure 5.2. The triple correlation of a pulse can be represented by the elliptical contour lines corresponding to its magnitude, as shown in Figure 5.2a.

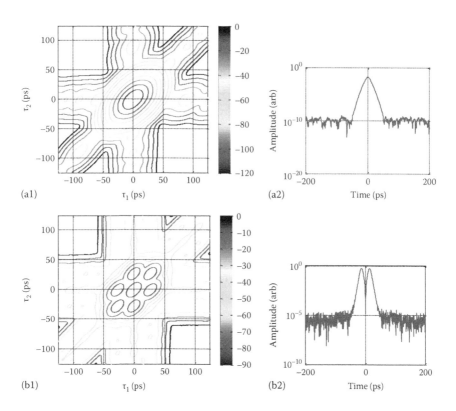

**FIGURE 5.2**  Triple correlations in contour plot view of various bound soliton states: (a1) single soliton, (b1) dual-bound solitons, (c1) triple-bound solitons, and (d1) quadruple-bound solitons. Insets (Right) (a2), (b2), (c2), and (d2): Corresponding temporal waveforms in amplitude logarithm scale showing the enhancement of the pedestal in higher-order multi-bound solitons.                                                                                    (*Continued*)

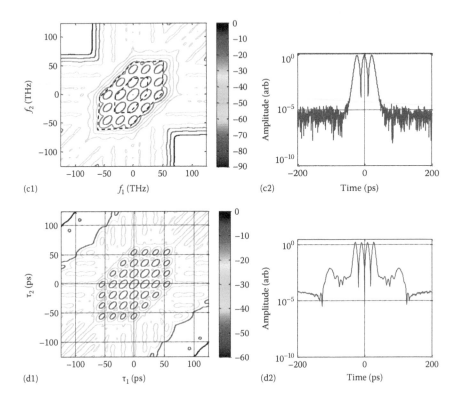

(c1)   $f_1$ (THz)                                    (c2)   Time (ps)

(d1)   $\tau_1$ (ps)                                    (d2)   Time (ps)

**FIGURE 5.2 (Continued)** Triple correlations in contour plot view of various bound soliton states: (a1) single soliton, (b1) dual-bound solitons, (c1) triple-bound solitons, and (d1) quadruple-bound solitons. Insets (Right) (a2), (b2), (c2), and (d2): Corresponding temporal waveforms in amplitude logarithm scale showing the enhancement of the pedestal in higher-order multi-bound solitons.

Depending on the number of solitons in the bound state, the triple correlation of the multi-bound soliton is represented by the layers of elliptical contour lines bounded in a hexagon, as shown in Figure 5.2b through d. The quality and symmetry of pulses can be reflected by the uniformity of the ellipsis in the same layer. In the triple correlation, the presence of pedestals, which is commonly characterized by a high dynamic range autocorrelation measurement [6], can be easily observed by the contour lines outside the hexagon. When the gain is enhanced in the cavity for the higher multi-bound state, the pulses are not only shortened but also possibly degraded by the increase in pedestal energy which is proportional to the level of the lines outside the hexagon area.

The Fourier transform of the triple correlations results in the bispectra of multi-bound solitons that consist of the magnitude and the phase representations. Figure 5.3 shows the magnitude in contour plot view of different multi-bound soliton states with $\pi$ phase difference that circulate stably in the FM mode-locked fiber laser cavity. In the single-soliton state, the magnitude bispectrum contains only closed contour lines inside the area bounded by the hexagon as shown in Figure 5.3a. While the bispectra

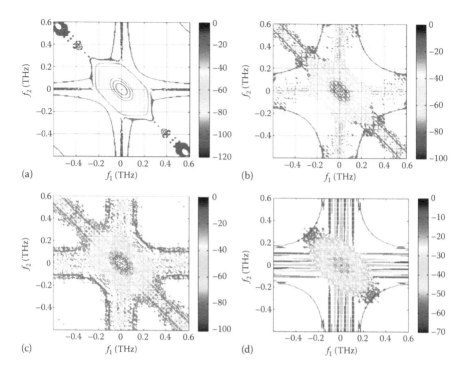

**FIGURE 5.3** The magnitude bispectra in the logarithm scale of different multi-bound soliton states: (a) single soliton, (b) dual-, (c) triple, and (d) quadruple-bound solitons.

of higher-order bound solitons from dual- to quadruple-bound states exhibit a periodic structure inside this area, as shown in Figure 5.3b through d. Depending on the number of pulses in the bound states, the periodic structures of both phase and magnitude spectra can vary correspondingly. When the number of pulses in the bound state increases, the periodic structure of its bispectra is much more complex because of the interactive relationship between sidelobes or the various frequency components. Moreover, the hexagonal area in the bispectrum that is broader at higher-order bound states reflects the bandwidth of the signal that is broadened by the pulse compression due to the enhancement of the gain or power in the fiber cavity. Similar to the observation in the triple correlation, the stability of the bound states is also clearly exhibited in the bispectrum. A lower stability of higher-order bound states is shown by the blurred contour lines inside the hexagon and the enhancement of the contour ridges outside the hexagon.

The phase bispectra provide additional information about status and quality of the bound state. Figure 5.4 shows the phase bispectra of various soliton states from the single pulse to quadruple-pulse. For the phase bispectrum, with periodic phase variation of a stable state in the frequency domain is easily exhibited in two dimensions. The sharply periodic structure of the phase bispectrum inside the hexagon with two distinct regions, which are complex conjugate of each other and separated by the line $f_1 = -f_2$ as shown in Figure 5.4a, indicates high stability of the single soliton state. On the other hand, the periodic structure of the stable multi-bound state is corrupted by the change in operating conditions of the multi-bound state such as gain and the noise level in the fiber cavity. At higher-order bound soliton states, the phase variation in the bispectrum is more blurred in the hexagon; however, two conjugate regions can still be clearly identified.

### 5.1.3 TRANSITIONS IN MULTI-BOUND SOLITON FORMATION

One of important advantages of the bispectrum representation is to differentiate the linear and nonlinear responses or processes; therefore it is especially useful in analyzing the transition processes from the noise to the steady state or from the unstable state to the stable one. In transition processes of multi-bound solitons, the signal experiences strong fluctuations in phase and magnitude like a chaotic state before

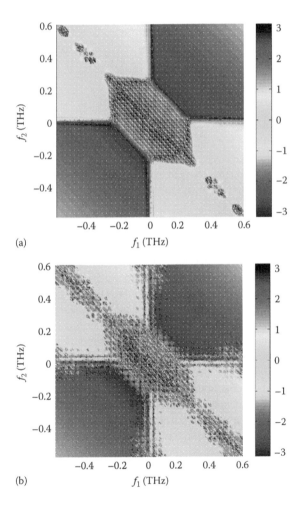

**FIGURE 5.4** The phase bispectra of different multi-bound soliton states: (a) single soliton, (b) dual-bound. *(Continued)*

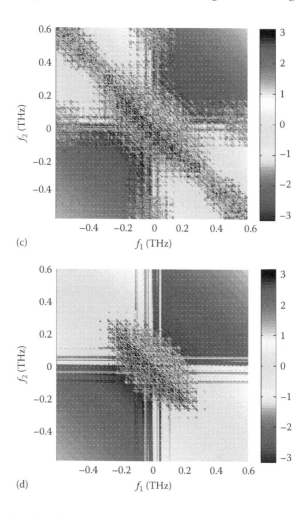

**FIGURE 5.4 (*Continued*)** The phase bispectra of different multi-bound soliton states: (c) triple- and (d) quadruple-bound solitons.

reaching a new steady state. Therefore, the average bispectrum of these processes also exhibits clearly the variation in terms of both phase and magnitude, which can be used to analyze multi-bound solitons at this stage. One of the most important transition processes is the formation of multi-bound solitons from the noise in the fiber cavity, which was numerically investigated in Chapter 3. The evolution of the signal from the noise into the multi-bound solitons can be split into three stages in which the first stage is the process that the noise builds up into the pulses, yet the amplitude and phase of the signal fluctuate strongly during circulating inside the cavity like a chaotic state. In order to analyze this stage, the bispectrum is averaged over the first 1000 round trips in the evolution of the multi-bound soliton formation, which was simulated in Chapter 3. Figure 5.5 shows the magnitude and phase bispectra of the dual-bound soliton and the triple-bound soliton evolutions, respectively. For the case of dual-bound soliton, the bispectra in Figure 5.5a and b show a nonperiodic structure in a hexagon with a small area that corresponds to the rapid variation of magnitude and phase of the signal from round trip to round trip. In particular, the phase structure shows a uniform distribution indicating the independence of the frequency components in the signal envelope because the signal behavior in this stage is similar to a chaotic state. While the magnitude bispectrum for the case of the triple-bound soliton shows the closed contours in a low-frequency region and that the phase bispectrum has a periodic structure due to the existence of two dominant short

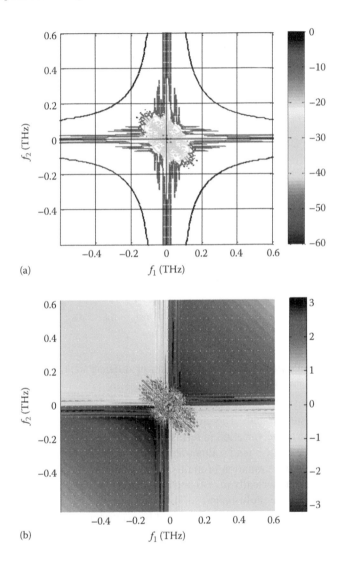

(a)

(b)

FIGURE 5.5 Magnitude and phase bispectra averaged over 1000 round trips of evolution of the signal at the first stage in the formation process from the noise of various multi-bound solitons: (a–b) dual-bound soliton. (*Continued*)

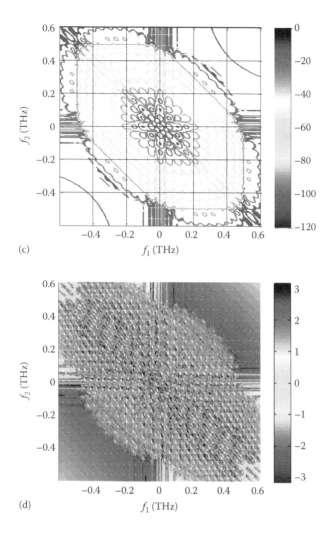

(c)

(d)

**FIGURE 5.5 (*Continued*)** Magnitude and phase bispectra averaged over 1000 round trips of evolution of the signal at the first stage in the formation process from the noise of various multi-bound solitons: (c–d) Triple-bound soliton.

pulses in this first stage, the state of these two pulses is unstable, which is shown by the ridges in the hexagon of the magnitude bispectrum and the nonclarity of the two distinct regions in the phase structure of the bispectrum.

The second stage in the next 1000 round trips is when the built-up pulses interact and bind together to form multi-bound solitons. For the evolution of dual-bound soliton, the elliptical contours in magnitude bispectrum and the periodic structure in phase bispectrum in this stage appear obviously because of the presence of periodic components as shown in Figure 5.6a and b. The structure of closed contours in the magnitude bispectrum becomes finer and extended due to the appearance of three solitons in the evolution of triple-bound soliton, as shown in Figure 5.6c. However, the bound state in this stage is unstable during circulating in the cavity because the operating condition, the gain factor, is not optimized. Hence, the contamination in phase bispectra still exists and the structure of ridges is dominant in the hexagon of the magnitude spectra. The transition process between different bound states due to the change in operating conditions behaves similar to the evolution in this stage, which is easily identified by the bispectral representation. When the fluctuations of multi-bound soliton occurs, the periodic lines of bispectrum at that process is commonly varied and smeared, before another periodic structure of the bispectrum is formed when a new stable state is established.

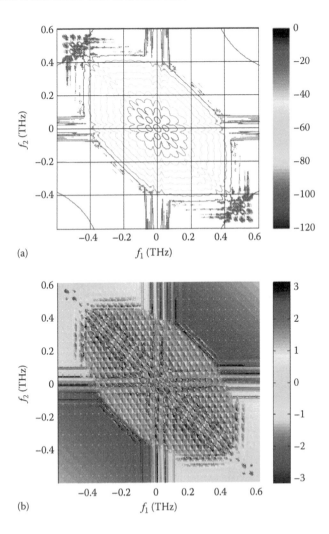

(a)

(b)

**FIGURE 5.6** Magnitude and phase bispectra averaged over 1000 round trips of evolution of the signal at the second stage in the formation process of various multi-bound solitons: (a–b) Dual-bound soliton. *(Continued)*

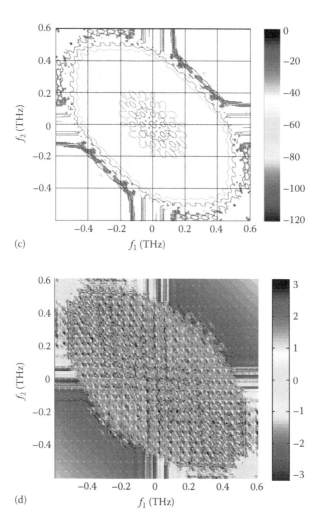

**FIGURE 5.6 (*Continued*)** Magnitude and phase bispectra averaged over 1000 round trips of evolution of the signal at the second stage in the formation process of various multi-bound solitons: (c–d) Triple-bound soliton.

The last stage in the evolution of multi-bound soliton is the process in which the pulses do self-adjustments until the multi-bound soliton reaches a steady state after the gain of the cavity is optimally adjusted. Although small fluctuations in the phase and amplitude still exist in this process, the pulses in the bound states are well defined in periodic evolution. Therefore, the magnitude bispectra in both the dual- and triple-bound solitons show the periodic structure with the closed elliptical contours dominant in the hexagon, as shown in Figure 5.7a and c. Figure 5.7b and d shows the phase bispectra with a periodic relationship between the frequency components of pulses and two clearly distinct conjugate regions separated by the diagonal $f_1 = -f_2$. On the other hand, the periodic structure in the bispectra is progressively widened during the transition from the unstable to the stable state.

Another state of multi-pulse in the fiber cavity can be obtained when the gain factor is not optimized for higher multi-bound soliton states. Therefore, the magnitude and the phase fluctuate strongly during circulation in the cavity. Nonuniformity and deviation of the phase difference between pulses from the value of π result in the collision of pulses in this state and the formation of new pulses. Figure 5.8 shows both the magnitude and the phase bispectra of this state averaged over 2000 round trips. The presence of both the closed contours and the ridges in a jumble indicate the presence of multipulsing operation and its dynamic state with a rapid variation in time due to the

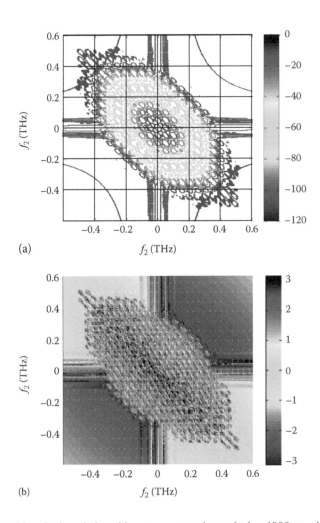

(a)

$f_2$ (THz)

(b)

$f_2$ (THz)

**FIGURE 5.7** Magnitude and phase bispectra averaged over the last 1000 round trips of evolution of the signal at the third stage in the formation process of various multi-bound solitons: (a–b) dual-bound soliton. *(Continued)*

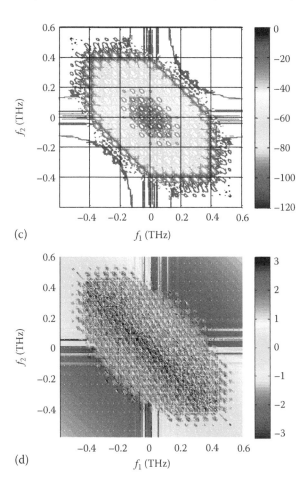

**FIGURE 5.7 (*Continued*)**   Magnitude and phase bispectra averaged over the last 1000 round trips of evolution of the signal at the third stage in the formation process of various multi-bound solitons: (c–d) Triple-bound soliton.

collision and the interaction of pulses in the cavity. The chaos in this state also exhibits through the uniform distribution of phase in the bispectrum, as observed in Figure 5.8b.

## 5.2  THIRD-ORDER NONLINEARITY FOUR-WAVE MIXING FOR PHOTONIC SIGNAL PROCESSING

When the processing speed is over that of the electronic limit or requires massive parallel and high-speed operations, the processing in the optical domain offers significant advantages. Thus all-optical signal processing is a promising technology for future optical communication networks. An advanced optical network requires a variety of signal processing functions including optical regeneration, wavelength conversion, optical switching, and signal monitoring. An attractive way to realize

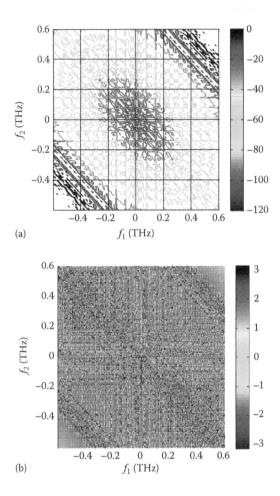

(a)

(b)

**FIGURE 5.8** (a) Magnitude spectrum and (b) phase bispectra averaged over 2000 round trips of the unstable multipulse state in the FM fiber ring cavity.

these processing functions in transparent and high-speed mode is to exploit the third-order nonlinearity in optical waveguides, particularly parametric processes.

### 5.2.1 FOUR-WAVE MIXING IN NONLINEAR WAVEGUIDES

Four-wave mixing (FWM) is a parametric process of the third-order susceptibility $\chi^{(3)}$. In the FWM process, the superposition and generation of the propagating of the waves with different amplitudes $A_k$, frequencies $\omega_k$, and wave numbers $k_k$ through the waveguide can be represented as

$$A = \sum_k A_k e^{\left[j(k_k z - \omega_k \tau)\right]} \quad \text{with } k = 1, \dots, 4 \tag{5.3}$$

The propagation and interaction between different waves in nonlinear waveguides can be described by the system of coupled differential equations, which is derived by the introduction of Equation 5.3 into the nonlinear Schrödinger equation (NLSE) ignoring some linear and scattering effects for simplification [7]. The coupled differential equations, each of which is responsible for one distinct wave in the optical waveguide are as follows:

$$\frac{\partial A_1}{\partial z} + \frac{\alpha}{2} A_1 = j\gamma A_1 \left[ |A_1|^2 + 2\sum_{k \neq 1} |A_k|^2 \right] + j\gamma 2 A_3 A_4 A_2^* \exp(-j\Delta k_1 z) \tag{5.4}$$

$$\frac{\partial A_2}{\partial z} + \frac{\alpha}{2} A_2 = j\gamma A_2 \left[ |A_2|^2 + 2\sum_{k \neq 2} |A_k|^2 \right] + j\gamma 2 A_3 A_4 A_1^* \exp(-j\Delta k_2 z) \tag{5.5}$$

$$\frac{\partial A_3}{\partial z} + \frac{\alpha}{2} A_3 = j\gamma A_3 \left[ |A_3|^2 + 2\sum_{k \neq 3} |A_k|^2 \right] + j\gamma 2 A_1 A_2 A_4^* \exp(-j\Delta k_3 z) \tag{5.6}$$

$$\frac{\partial A_4}{\partial z} + \frac{\alpha}{2} A_4 = j\gamma A_4 \left[ |A_4|^2 + 2\sum_{k \neq 4} |A_k|^2 \right] + j\gamma 2 A_1 A_2 A_3^* \exp(-j\Delta k_4 z) \tag{5.7}$$

where $\Delta k = k_1 + k_2 - k_3 - k_4$ is the wave vector mismatch. In Equations 5.4, the last term represents the interaction that can generate new waves. For three waves with different frequencies, a fourth wave can be generated at frequency $\omega_4 = \omega_1 + \omega_2 - \omega_3$. The waves at frequencies $\omega_1$ and $\omega_2$ are called pump waves, whereas the wave at frequency $\omega_3$ is the signal and the wave generated at $\omega_4$ is called idler wave, as shown in Figure 5.9a.

If all three waves have the same frequency $\omega_1 = \omega_2 = \omega_3$, the interaction is called a degenerate FWM with the new wave at the same frequency. If only two of the three waves are at the same frequency ($\omega_1 = \omega_2 \neq \omega_3$), the process is called partly degenerate FWM, which is important for some applications like wavelength converter and parametric amplifier.

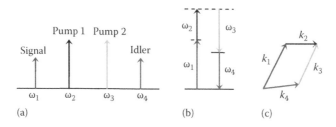

**FIGURE 5.9** (a) Position and notation of the distinct waves, (b) diagram of energy conservation, and (c) diagram of momentum conservation in FWM.

### 5.2.2 Phase Matching

In parametric nonlinear processes such as FWM, the conservation of energy and momentum must be satisfied to obtain a high efficiency of the energy transfer, as shown in Figure 5.9b and c. The phase matching condition for the new wave requires

$$\Delta k = k_1 + k_2 - k_3 - k_4 = \frac{1}{c}(n_1 \omega_1 + n_2 \omega_2 - n_3 \omega_3 - n_4 \omega_4) = 2\pi \left( \frac{n_1}{\lambda_1} + \frac{n_2}{\lambda_2} - \frac{n_3}{\lambda_3} - \frac{n_4}{\lambda_4} \right)$$

(5.8)

where $n_i$ is the refractive index of the wave at the wavelength $\lambda_i$. During propagation in optical waveguides, the relative phase difference $\theta(z)$ between the four waves is determined by [8,9]

$$\theta(z) = \Delta k z + \phi_1(z) + \phi_2(z) - \phi_3(z) - \phi_4(z) \tag{5.9}$$

where $\phi_k(z)$ relates to the initial phase and the nonlinear phase shift during propagation. An approximation of the phase matching condition can be given as follows [10]:

$$\frac{\partial \theta}{\partial z} \approx \Delta k + \gamma (P_1 + P_2 - P_3 - P_4) = \kappa \tag{5.10}$$

where
  $P_k$ is the power of the waves
  $\kappa$ is the phase mismatch parameter

Thus the FWM process has maximum efficiency for $\kappa = 0$. The mismatch comes from the frequency dependence of the refractive index and the dispersion of the guided medium, the optical waveguides. Depending on the dispersion profile of the nonlinear waveguides, it is very important to ensure that the phase mismatch parameter is minimized when selecting pump wavelengths.

### 5.2.3 Simulink® Model for FWM in Optical Waveguides

To model the parametric FWM process between multi-waves, basic propagation equations are used. There are two approaches to simulate the interaction between waves. The first approach is to use the coupled equations system (Equation 5.4) which is called the separating channels approach. In this approach, interactions between different waves are obviously modeled by certain coupling terms in each coupled equation. Thus each optical wave considered as one separated channel is represented by a vector. The coupled equations system will be solved for solutions of the FWM process. The outputs of the nonlinear waveguide are also represented by separated vectors; hence the desired signal can be extracted without using a filter, as described in Figure 5.10a.

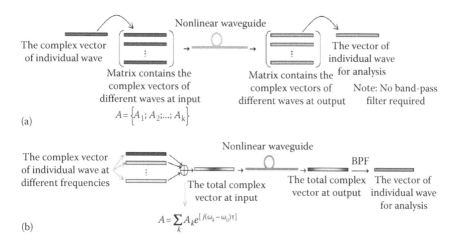

(a)

(b)

**FIGURE 5.10**  A description of two approaches in modeling the parametric FWM process in nonlinear waveguides: (a) The separating wave approach and (b) the total complex field approach.

The second or alternative approach is to use the NLSE, as described in Chapter 2, which can allow us to simulate all evolution effects of optical waves in the nonlinear waveguides. In this approach, a total field is used instead of separated waves. The total complex envelope $A$ is represented by only one vector, which is a summation of individual complex amplitudes of different waves:

$$A = \sum_{k=1}^{N_c} A_k e^{[j(\omega_k - \omega_0)\tau]}$$ (5.11)

where
  $\omega_0$ is the defined angular central frequency
  $A_k$, $\omega_k$ are the complex envelope and the carrier frequency of individual waves, respectively

Hence, various waves at different frequencies are combined into only a total signal vector that facilitates integration of the nonlinear waveguide model into the Simulink® platform. By this approach, the NLSE is numerically solved by the split-step method and then the block of nonlinear waveguide is developed by an embedded MATLAB® program. Because only complex envelopes of the optical waves are considered in the simulation, each optical wave is shifted by a frequency difference between the central frequency and the frequency of the wave to allocate the wave in the frequency band of the total field. Then the summation of individual waves, which is equivalent to the combination process at the optical coupler, is processed in the block placed in front of that representing nonlinear waveguide, as shown in Figure 5.10b. The output of nonlinear waveguide will be selected by an optical band-pass filter (BPF). In this way, the model of the nonlinear waveguide can easily connect to other Simulink blocks available in the platform for the simulation of optical fiber communication systems [11].

Based on the Simulink platform for optical fiber communication systems, a range of signal processing applications based on parametric processes will be demonstrated through simulations in which the model of nonlinear waveguide is used as a functional block. Details of the Simulink models are presented in Appendix C.

## 5.3 APPLICATIONS OF FWM IN PHOTONIC SIGNAL PROCESSING

### 5.3.1 SIGNAL PROCESSING BASED ON PARAMETRIC AMPLIFICATION

One of the important applications of the $\chi^{(3)}$ nonlinearity is parametric amplification. Optical parametric amplifiers (OPAs) offer a wide gain bandwidth, high differential gain, and optional wavelength conversion and operation at any wavelength [9,10,12,13]. OPA have these important features because their parametric gain process does not rely on energy transitions between energy states but they are based on highly efficient FWM in which two photons at one- or two-pump wavelengths will interact with a signal photon. A fourth photon, the idler, will be formed with a phase such that the phase difference between the pump waves and the signal and idler waves satisfies a phase matching condition. A typical scheme of the fiber-based parametric amplifier is shown in Figure 5.11a.

For a parametric amplifier using one pump source, from the coupled Equations 5.4 with $A_1 = A_2 = A_p$, $A_3 = A_s$ and $A_4 = A_i$, it is possible to derive three coupled equations for the complex field amplitude of the three waves $A_{p,s,i}$:

$$\frac{\partial A_p}{\partial z} = -\frac{\alpha}{2} A_p + i\gamma A_p \left[ \left|A_p\right|^2 + 2\left( \left|A_s\right|^2 + \left|A_i\right|^2 \right) \right] + i2\gamma A_s A_i A_p^* \exp(-i\Delta kz) \quad (5.12)$$

$$\frac{\partial A_s}{\partial z} = -\frac{\alpha}{2} A_s + i\gamma A_s \left[ \left|A_s\right|^2 + 2\left( \left|A_p\right|^2 + \left|A_i\right|^2 \right) \right] + i\gamma A_p^2 A_i^* \exp(-i\Delta kz) \quad (5.13)$$

$$\frac{\partial A_i}{\partial z} = -\frac{\alpha}{2} A_i + i\gamma A_i \left[ \left|A_i\right|^2 + 2\left( \left|A_s\right|^2 + \left|A_p\right|^2 \right) \right] + i\gamma A_p^2 A_s^* \exp(-i\Delta kz) \quad (5.14)$$

The analytical solution of these coupled equations determines the gain of the amplifier [8]:

$$G_s(L) = \frac{\left|A_s(L)\right|^2}{\left|A_s(0)\right|^2} = 1 + \left[ \frac{\gamma P_p}{g} \sinh(gL) \right]^2 \quad (5.15)$$

where
  $L$ is the length of the highly nonlinear fiber (HNLF)/waveguide
  $P_p$ is the pump power
  $g$ is the parametric gain coefficient

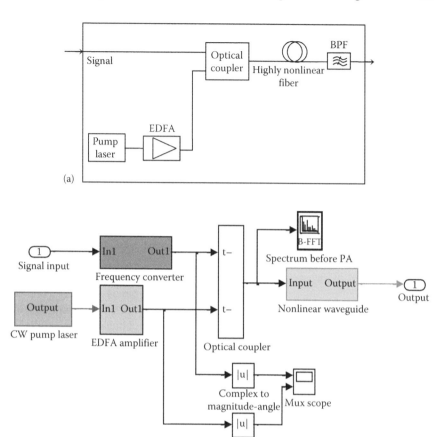

**FIGURE 5.11** (a) A typical setup of an optical parametric amplifier [9] and (b) Simulink®
model of optical parametric amplifier.

$$g^2 = -\Delta k \left( \frac{\Delta k}{4} + \gamma P_p \right) \tag{5.16}$$

where the phase mismatch $\Delta k$ can be approximated by extending the propagation
constant in a Taylor series around $\omega_0$:

$$\Delta k = -\frac{2\pi c}{\lambda_0^2} \frac{dD}{d\lambda} (\lambda_p - \lambda_0)(\lambda_p - \lambda_s)^2 \tag{5.17}$$

where $dD/d\lambda$ is the slope of the dispersion at the zero-dispersion wavelength, $\lambda_k = 2\pi c / \omega_k$ is the optical wavelength.

A setup of the 40 Gbit/s RZ transmission system using a parametric amplifier has
been implemented in the Simulink platform. The setup contains a 40 Gbit/s optical

**TABLE 5.1**

**Important Parameters of the Parametric Amplifier in a 40 Gbit/s System**

**RZ 40 Gbit/s Transmitter**

$\lambda_s = 1520$ nm – 1600 nm, $\lambda_0 = 1559$ nm

Modulation: RZ-OOK, $P_s = 0.01$ mW (peak), $B_r = 40$ Gbit/s

**Parametric Amplifier**

Pump source: $P_p = 1$ W (after EDFA), $\lambda_p = 1560.07$ nm

HNLF: $L_f = 500$ m, $D = 0.02$ ps/km/nm, $S = 0.09$ ps/nm²/km, $\alpha = 0.5$ dB/km, $A_{eff} = 12$ µm²,

    $\gamma = 13$ W⁻¹ km⁻¹.

BPF: $\Delta\lambda_{BPF} = 0.64$ nm

RZ transmitter, an optical receiver for monitoring, a parametric amplifier block and a band-pass filter that filters the desired signal from the total field output of the amplifier. Details of the parametric amplifier block can be seen in Figure 5.11b. The block setup of the parametric amplifier consists of a continuous wave (CW) pump laser source, an optical coupler to combine the signal and the pump, and a HNLF block that contains the embedded MATLAB model for nonlinear propagation. The important simulation parameters of the system are shown in Table 5.1.

Figure 5.12 shows the signals before and after the amplifier in the time domain. The time trace indicates the amplitude fluctuation of the amplified signal as a noisy source from the wave mixing process. Their corresponding spectra are shown in Figure 5.13. The noise floor of the output spectrum of the amplifier shows the

(a)

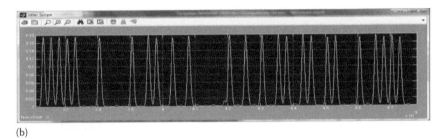

(b)

**FIGURE 5.12**   Time traces of the 40 Gbit/s signal at location: before (a) and after (b) the parametric amplifier.

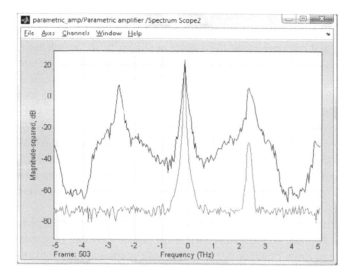

**FIGURE 5.13**    Optical spectra at the input (light gray) and the output (dark gray) of the OPA.

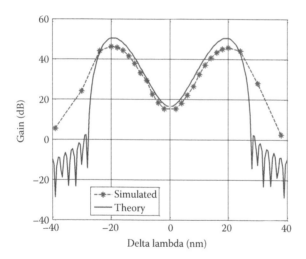

**FIGURE 5.14**    Calculated and simulated gain of the OPA at $P_p = 30$ dBm.

gain profile of OPA. Simulated dependence of OPA gain on the wavelength difference between the signal and the pump is shown in Figure 5.14, together with the theoretical gain obtained from Equation 5.10. The plot shows an agreement between theoretical and simulated results. The peak gain is achieved at the phase-matched condition where the linear phase mismatch is compensated by the nonlinear phase shift.

In fact, the application of parametric amplifiers is not popular in transmission systems because of some disadvantages [9]. However, the structure of a parametric amplifier is perceptively applied to signal processing functions such as wavelength conversion, which can be obtained from the idler, after the wave mixing process. Due to very fast response of the third-order nonlinearity, the wavelength conversion based on this effect is transparent to the modulation format and the bit rate of signals. For the design of a flat wideband converter, which is a key device in wavelength-division multiplexing (WDM) networks, a short length HNLF with a low dispersion slope is required. Selecting a suitable pump wavelength optimizes the wavelength converter to obtain a bandwidth of 200 nm [12]. Therefore, the wavelength conversion between bands such as C and L bands can be performed in WDM networks. Figure 5.15 shows an example of the wavelength conversion for four WDM channels at the C-band. The important parameters of the wavelength converter are shown in Table 5.2. The WDM signals are converted into L-band with a conversion efficiency of −12 dB.

Another important application with the same setup is nonlinear phase conjugation (NPC). A phase conjugated replica of the signal wave can be generated by the FWM process. From Equation 5.9, the idler wave is approximately given in the case of degenerate FWM for simplification: $E_i \sim A_p^2 A_s^* e^{-j\Delta kz}$ or $E_i \sim r A_s^* e^{[j(-kz-\omega\tau)]}$ with the signal wave $E_s \sim A_s e^{[j(kz-\omega\tau)]}$. Thus the idler field is a complex conjugate of the signal field. In appropriate conditions, optical distortions can be compensated by using NPC, and optical pulses propagating in the fiber link can be recovered. The basic principle of distortion compensation with NPC refers to spectral inversion. When an optical pulse propagates in an optical fiber, its shape will be spread in time and distorted by the group velocity dispersion. The phase-conjugated replica of the pulse is generated in the middle point of the transmission link by a nonlinear effect. On the other hand, the pulse is spectrally inverted, where spectral components in the lower frequency range are shifted to a higher frequency range and vice versa. If the pulse propagates in the second part of the link in the same manner as in the first part, it is inversely distorted again, which can cancel the distortion in the first part and recover the pulse shape at the end of the transmission link. By using NPC for distortion compensation, a 40%–50% increase in transmission distance compared to a conventional transmission link can be obtained [14–16]. However, NPC for distortion compensation has been mostly implemented by exploiting the second-order susceptibility of nonlinear crystals such as KTP and periodically poled LiNbO$_3$ (PPLN) [15]. Recently, planar nonlinear waveguides based on AsS$_3$ glass have emerged as promising devices for ultrahigh-speed photonic processing [17–19]. These nonlinear waveguides offer a lot of advantages such as no free-carrier absorption, stable at room temperature, no requirement of quasi-phase matching, and the possibility of dispersion engineering (Table 5.3).

Details of the Simulink setup of a long-haul 40 Gbit/s transmission system demonstrating the distortion compensation using NPC are shown in Appendix D. The fiber transmission link of the system is divided into two sections by an NPC based on a parametric amplifier. Each section consists of five spans with 100 km standard single mode fiber (SSMF) in each span. Figure 5.16a shows an eye diagram of the signal after propagating through the first fiber section. After the NPC at the mid-link, the idler signal, a phase conjugated replica of the original signal, is filtered for transmission in the next section. The signal in the second section suffers the same

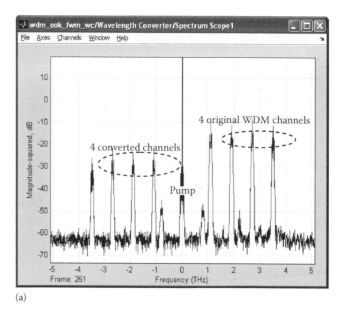

(a)

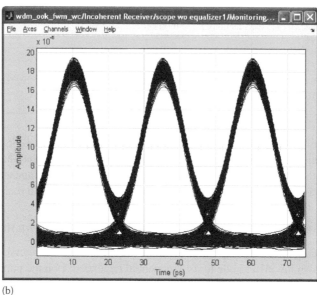

(b)

**FIGURE 5.15**   (a) The wavelength conversion of four WDM channels, (b) Eye diagram of the converted 40 Gbit/s signal after BPF.

**TABLE 5.2**

**Important Parameters of the Wavelength Converter**

**RZ 40 Gbit/s Signal**

$\lambda_0 = 1559$ nm, $\lambda_s = \{1531.12, 1537.4, 1543.73, 1550.12\}$ nm

$P_s = 1$ mW (peak), $B_r = 40$ Gbit/s

**Parametric Amplifier**

Pump source: $P_p = 100$ mW (after EDFA), $\lambda_p = 1560.07$ nm

HNLF: $L_f = 200$ m, $D = 0.02$ ps/km/nm, $S = 0.03$ ps/nm²/km, $\alpha = 0.5$ dB/km, $A_{eff} = 12$ μm², $\gamma = 13$ 1/W/km

BPF: $\Delta\lambda_{BPF} = 0.64$ nm, $\lambda_i = \{1587.91, 1581.21, 1574.58, 1567.98\}$ nm

---

**TABLE 5.3**

**Important Parameters of the Long-Haul Transmission System Using NPC for Distortion Compensation**

**RZ 40 Gbit/s Transmitter**

$\lambda_s = 1547$ nm, $\lambda_0 = 1559$ nm

Modulation: RZ-OOK, $P_s = 1$ mW (peak), $B_r = 40$ Gbit/s

**Fiber Transmission Link**

SMF: $L_{SMF} = 100$ km, $D_{SMF} = 17$ ps/nm/km, $\alpha = 0.2$ dB/km

EDFA: Gain = 20 dB, $NF = 5$ dB;

Number of spans: 10 (5 in each section), $L_{link} = 1000$ km

**NPC Based on OPA**

Pump source: $P_p = 500$ mW (after EDFA), $\lambda_p = 1560.07$ nm

NW: $L_w = 7$ cm, $D = 28$ ps/km/nm, $S = 0.03$ ps/nm²/km, $\alpha = 5$ dB/cm, $A_{eff} = 1$ μm², $\gamma = 10{,}000$ W⁻¹ km⁻¹

BPF: $\Delta\lambda_{BPF} = 0.64$ nm

---

dispersion as in the first section. At the end of the transmission link, the optical signal is totally regenerated as shown in Figure 5.16b. It is noted that due to change in wavelength of the signal after the NPC, a tunable dispersion compensator will be required to compensate the residual dispersion after transmission in real systems.

## 5.3.2 Ultrahigh-Speed OTDM Demultiplexing (Optical Switching)

When the pump is an intensity modulated signal instead of a CW signal, the gain of the OPA is also modulated due to its exponential dependence on the pump power in a phase-matched condition. The width of gain profile in the time domain is inversely proportional to the product of the gain slope ($S_p$) or the nonlinear coefficient and the length of the nonlinear waveguide ($L$) [8]. Therefore, an OPA with a high gain or a large $S_pL$ operates as an optical switch with an ultrahigh bandwidth, which is very important in some signal processing applications such as pulse compression or short pulse generation [20,21]. A Simulink setup for a 40 GHz short pulse generator is built with the configuration shown in Figure 5.17. In this setup, the input signal is a CW

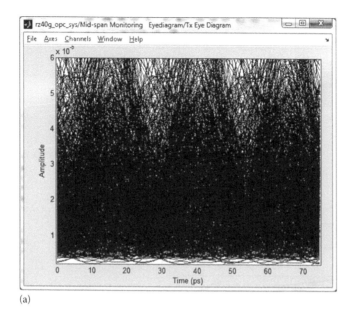

(a)

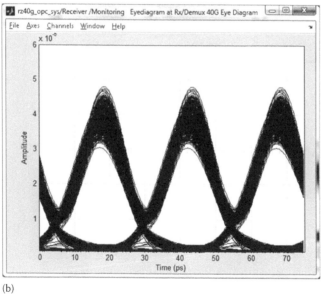

(b)

**FIGURE 5.16** Eye diagrams of the 40 Gbit/s signal at the end of (a) the first section and (b) the transmission link.

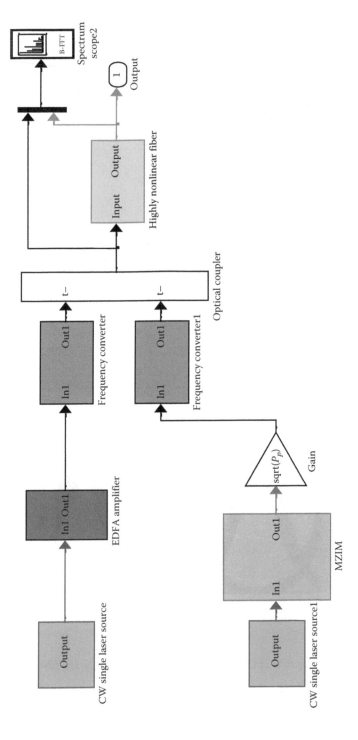

**FIGURE 5.17** Simulink setup of the 40 GHz short pulse generator to demonstrate ultrahigh-speed switching based on parametric amplification. The difference compared to previous applications is the use of an intensity-modulated source for pump.

source with low power and the pump is amplitude-modulated by a Mach–Zehnder intensity modulator which is driven by an RF sinusoidal wave at 40 GHz. The waveform of the modulated pump is shown in Figure 5.18a. The important parameters of the FWM-based short pulse generator are shown in Table 5.4. Figure 5.18b shows the generated short pulse sequence with a pulse width of 2.6 ps at the signal wavelength after the optical BPF.

Another important application of the optical switch based on the FWM process is the demultiplexer, a key component in ultrahigh-speed optical time division multiplexing (OTDM) systems. OTDM is a key technology for Tb/s Ethernet transmission, which can meet the increasing demand of traffic in future optical networks. A typical scheme of OTDM demultiplexer to generate short pulses for control, in which the pump is a mode-locked laser (MLL), is shown in Figure 5.19a. The working principle of the FWM-based demultiplexing is described as follows: The control pulses generated from a MLL at tributary rate are pumped and copropagated with the OTDM signal through a nonlinear medium such as HNLF or nonlinear waveguide. The mixing process between the control pulses and the OTDM signal during propagation through the nonlinear medium converts the desired tributary channel to a new idler wavelength. Then the demultiplexed signal at the idler wavelength is extracted by a band-pass filter before it reaches a receiver, as shown in Figure 5.19a.

Using HNLF is relatively popular in structures of OTDM demultiplexer [13,22]. However its stability, especially the walk-off problem, is still a serious obstacle. With the same operational principle, planar waveguide-based OTDM demultiplexers are very compact and suitable for photonic integrated solutions [19]. Figure 5.19b shows the Simulink setup of an FWM-based demultiplexer of the on-off keying (OOK) 40 Gbit/s signal from the 160 Gbit/s OTDM signal using a highly nonlinear waveguide instead of HNLF. Important parameters of the OTDM system used in the simulation are listed in Table 5.5. Figure 5.20a shows the spectrum at the output of the nonlinear waveguide. Then the demultiplexed signal is extracted by the bandpass filter as shown in Figure 5.20b. The developed model of OTDM demultiplexer can be applied not only to the conventional OOK format but also to advanced modulation formats such as DQPSK, which increases the data load of the OTDM system without increasing the bandwidth of the signal. By using available blocks developed for the DQPSK system [11], a Simulink model of DQPSK-OTDM system is also setup for demonstration. The bitrate of the OTDM system is doubled to 320 Gbit/s with the same pulse repetition rate. Figure 5.21 shows the simulated performance of the demultiplexer in both 160 Gbit/s OOK-OTDM and 320 Gbit/s DQPSK-OTDM systems. The bit error rate (BER) curve in case of the DQPSK-OTDM signal shows a low error floor, which may be a result of the influence of nonlinear effects on the phase-modulated signals in the waveguide.

### 5.3.3  FWM-Based Triple Correlation and Bispectrum

One of the promising possibilities exploiting the $\chi^{(3)}$ nonlinearity is implementation of triple correlation in the optical domain. To implement triple correlation in the optical domain, the product of three signals including different delayed versions of

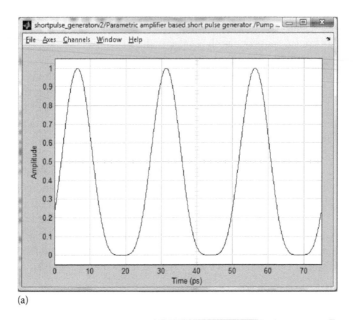

(a)

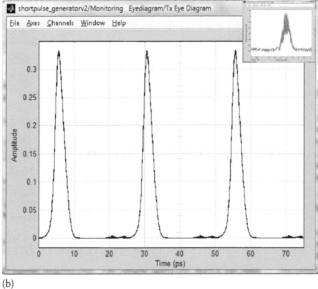

(b)

**FIGURE 5.18**  Time traces of (a) the sinusoidally amplitude-modulated pump and (b) the generated short pulse sequence (Inset: the corresponding pulse spectrum).

**TABLE 5.4**

**Parameters of a 40 GHz Short Pulse Generator**

**Short Pulse Generator**

Signal: $P_s = 0.7$ mW, $\lambda_y = 1535$ nm, $\lambda_0 = 1559$ nm

Pump source: $P_p = 1$ W (peak), $\lambda_p = 1560.07$ nm, $f_m = 40$ GHz

HNLF: $L_f = 500$ m, $D = 0.02$ ps/km/nm, $S = 0.03$ ps/nm2/km

$\alpha = 0.5$ dB/km, $A_{eff} = 12$ µm², $\gamma = 13$ W⁻¹ km⁻¹. BPF: $\Delta\lambda_{BPF} = 3.2$ nm

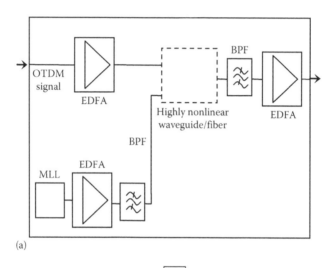

(a)

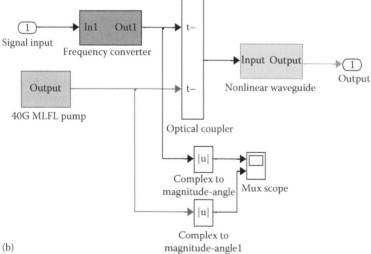

(b)

**FIGURE 5.19**    (a) A typical setup of the FWM-based OTDM demultiplexer and (b) Simulink model of the OTDM demultiplexer.

**TABLE 5.5**

**Important Parameters of the FWM-Based OTDM
Demultiplexer Using a Nonlinear Waveguide**

**OTDM Transmitter**

MLL: $P_0 = 1$ mW, $T_p = 2.5$ ps, $f_m = 40$ GHz

Modulation formats: OOK and DQPSK; OTDM multiplexer: $4 \times 40$ Gsymbols/s

**FWM-Based Demultiplexer**

Pumped control: $P_p = 500$ mW, $T_p = 2.5$ ps, $f_m = 40$ GHz, $\lambda_p = 1556.55$ nm

Input signal: $P_s = 10$ mW (after EDFA), $\lambda_s = 1548.51$ nm

Waveguide: $L_w = 7$ cm, $D_w = 28$ ps/km/nm, $S_w = 0.003$ ps/nm²/km

$\alpha = 0.5$ dB/cm, $\gamma = 10^4$ W⁻¹ km⁻¹

BPF: $\Delta\lambda_{BPF} = 0.64$ nm

original signal need to be generated and then detected by an optical photodiode to perform the integral operation. From the representation of nonlinear polarization vector (see Appendix A), this triple product can be generated by the $\chi^{(3)}$ nonlinearity. One conventional way to generate the triple correlation is based on third harmonic generation (THG), where the generated new wave containing the triple product is at a frequency three times the original carrier frequency. Thus, if the signal wavelength is in the 1550 nm band, the new wave need to be detected at around 517 nm. The triple optical autocorrelation based on single-stage THG has been demonstrated in direct optical pulse shape measurement [23,24]. However, in this method it is hard to obtain high efficiency in the wave mixing process due to the difficulty in the phase-matching between three signals. Moreover, the triple product wave is in 517 nm where wideband photodetectors are not available for high-speed communication applications. Therefore, a possible alternative to generate the triple product is based on other nonlinear interactions such as FWM. From Equation 5.4, the fourth wave is proportional to the product of the three waves $A_4 \sim A_1 A_2 A_3^* e^{-j\Delta kz}$. If $A_1$ and $A_2$ are the delayed versions of the signal $A_3$, the mixing of the three waves results in the fourth wave $A_4$, which is clearly the triple product of the three signals. As mentioned in Section 5.2.1, all three waves can have the same frequency; however, these waves should propagate into different directions to possibly distinguish the new generated wave in a diverse propagation direction, which requires a strict arrangement of the signals in the spatial domain for phase matching. An alternative way we propose is to convert the three signals into different frequencies ($\omega_1$, $\omega_2$, and $\omega_3$). Then the triple product wave can be extracted at the frequency $\omega_4 = \omega_1 + \omega_2 - \omega_3$, which is still in the 1550 nm band.

A Simulink model for the triple correlation estimation based on FWM in nonlinear waveguide is setup for investigation. The structural block consists of two variable delay lines to generate delayed versions of the original signal and frequency converters to convert the signal into three different waves before combining at the optical coupler to launch into the nonlinear waveguide, as shown in Figure 5.22a. Then the fourth wave signal generated by FWM is extracted by the band-pass filter. To verify

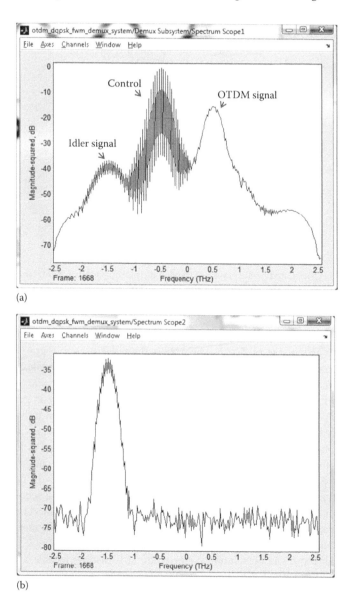

(a)

(b)

**FIGURE 5.20**    Spectra at the outputs of (a) nonlinear waveguide and (b) BPF.

the triple product based on FWM, another model, shown in Figure 5.22b, to estimate the triple product by a mathematical function block is also implemented for comparison. The integration of the generated triple product signal is then performed with a photodetector in the optical receiver to estimate the triple correlation of the signal. Repetitive signals with different patterns are generated for investigation. Important parameters of the setup are shown in Table 5.6. The triple correlation is represented by a 3D plot, an image with warm colors corresponding to the intensity of the triple

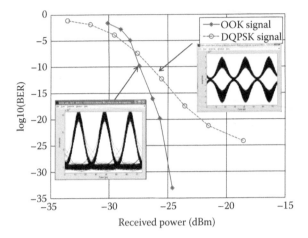

**FIGURE 5.21** Simulated performance of the demultiplexed signals for 160 Gbit/s OOK and 320 Gbit/s DQPSK-OTDM systems (Insets: eye diagrams at the receiver).

correlation. The $x$ and $y$ axes are the time-delay variables ($\tau_1$ and $\tau_2$) in terms of samples with a step-size of $T_m/32$, where $T_m$ is the pulse period. Figure 5.23 shows the normalized frequency domain of the triple correlation of the single-pulse signal (a) and (b) theoretical estimation and (c) FWM in nonlinear waveguide.

In the simple case, a single Gaussian pulse sequence at 5 GHz is generated for investigation. Figure 5.23 shows the triple correlations of the signal in both models, which are based on the FWM process and the theory for comparison with the same resolution. The intensity of the triple correlation based on the FWM is focused in a spot similar to that shown in the figure. By changing into another pattern, the triple correlations of the dual-pulse with equal and unequal amplitudes in both models are obtained as shown in Figure 5.24. Because of the presence of a second pulse in the pattern, a hexagonal layer of spots appears in the triple correlation based on the FWM process. When the amplitude of the second pulse is lower, the intensity of the spots outside is also reduced.

In general, the triple correlation patterns are still distinguishable as compared to the theory. However, FWM-based triple correlations are contaminated by noise that is mainly generated by other mixing processes and crosstalk from neighboring wavelengths similar to those in WDM systems. If the wavelength spacing between three waves is equal, the noise level from undesired mixing processes is enhanced, as demonstrated in Figure 5.25a. Furthermore, if the bandwidth of the signal is broadened, the crosstalk noise increases correspondingly as demonstrated in Figure 5.25b. In all triple correlations obtained by the FWM, the noise is deterministic, which is enhanced only at certain vertical lines in the image. We note that the noise can be reduced by the optical band-pass filter; however, the bandwidth of optical filter can distort the triple correlation. Depending on the bandwidth of the signal, the wavelength positions as well as the bandwidth of the filter need to be carefully chosen to obtain the best results.

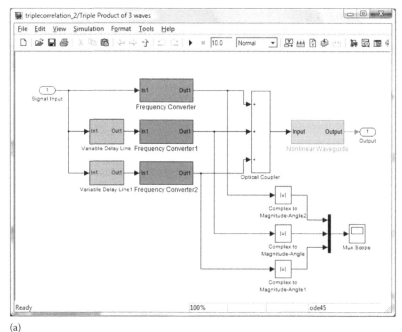

(a)

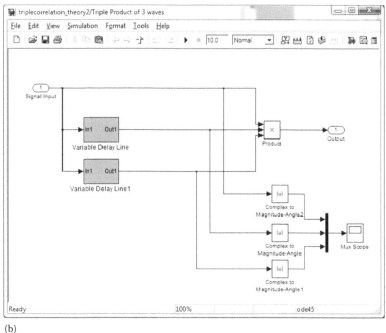

(b)

FIGURE 5.22 (a) Simulink setup of the FWM-based triple product generation and (b) Simulink setup of the theory-based triple product generation.

**TABLE 5.6**

**Parameters of the FWM-Based Triple Product Generator Using a Nonlinear Waveguide**

**Signal Generator**

Single-pulse: $P_0 = 200$ mW, $T_p = 5$ ps, $f_m = 5$ GHz

Dual-pulse: $P_p = 200$ mW, $T_p = 5$ ps, $f_m = 5$ GHz

**FWM-Based Triple Product Generator**

Original signal: $\lambda_{s1} = 1552.52$ nm

Delayed $\tau_1$ signal: $\lambda_{s2} = 1553.32$ nm, Delayed $\tau_2$ signal: $\lambda_{s3} = 1551.5$ nm

Waveguide: $L_w = 7$ cm, $D_w = 28$ ps/km/nm, $S_w = 0.003$ ps/nm²/km

$\alpha = 0.5$ dB/cm, $\gamma = 10^4$ W$^{-1}$ km$^{-1}$

BPF: $\Delta\lambda_{BPF} = 0.64-1$ nm

## 5.4 CONCLUDING REMARKS

Bispectrum estimation, a Fourier transform of triple correlation, has been proposed to characterize various states of multi-bound solitons. Owing to the distinct representation of the bispectrum, dynamic aspects of multi-bound solitons can be identified. Specially, the stability and the transition from a chaotic state to a steady state or vice versa were clearly distinguished by variation of the periodic structure in the magnitude bispectrum and the phase distribution in the phase bispectrum.

Triple correlation based on the FWM process has been examined by the Simulink model, which was developed for signal processing using nonlinear waveguides. The validity of the model has been demonstrated using important applications of the FWM process such as wavelength conversion, nonlinear phase conjugation, and ultrafast switching. Although it is noisy due to other mixing processes, the obtained

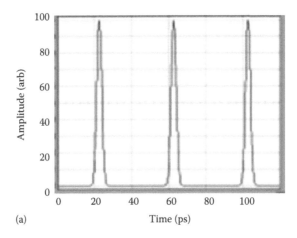

(a)

**FIGURE 5.23** Triple correlation of the single-pulse signal based on (a) single soliton sequence. *(Continued)*

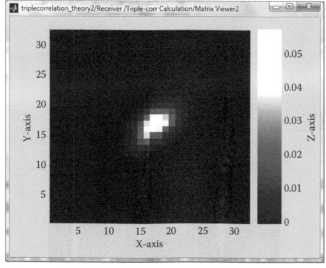

(b)

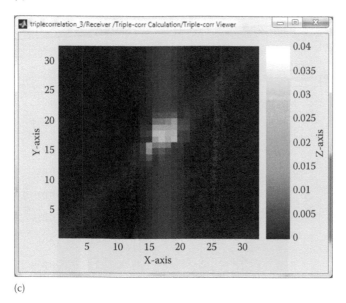

(c)

**FIGURE 5.23 (*Continued*)** Triple correlation of the single-pulse signal based on (b) theoretical estimation, and (c) FWM in nonlinear waveguide.

results showed that the triple correlation based on FWM is still distinguishable for signal processing. However, the complexity of properly selecting the parameters, such as in this technique, may prevent its use in practical applications, especially when real-time processing is required. This remains the principal issue for further research in the near future.

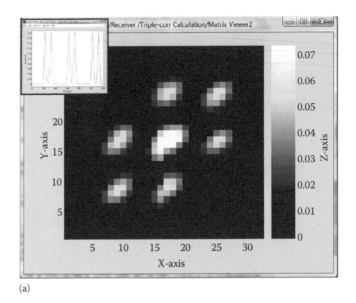

(a)

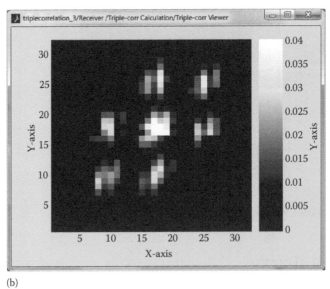

(b)

**FIGURE 5.24** Triple correlations based on (a) theoretical estimation and (b) FWM in the nonlinear waveguide of the dual-pulse signal with equal amplitudes.          (*Continued*)

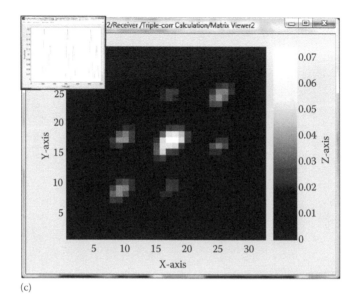

(c)

(d)

**FIGURE 5.24 (*Continued*)**   Triple correlations based on (c) theoretical estimation and (d) FWM in the nonlinear waveguide of the dual-pulse with unequal amplitudes (peak power of the second pulse $P_2 = 2/3P_p$). (Insets: the dual-pulse patterns.)

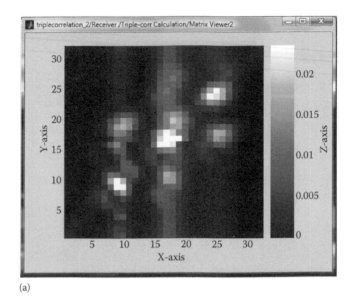

(a)

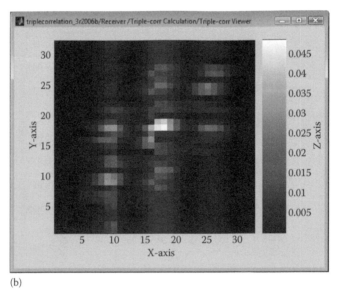

(b)

**FIGURE 5.25**   Triple correlations of the dual-pulse signal based on the FWM process in various conditions: (a) with double the bandwidth of the signal and (b) with equal wavelength spacing.

## REFERENCES

1. C. L. Nikias and M. R. Raghuveer, Bispectrum estimation: A digital signal processing framework, *Proc. IEEE*, 75, 869–891, 1987.
2. C. L. Nikias and J. M. Mendel, Signal processing with higher-order spectra, *IEEE Signal Proces. Mag.*, 10, 10–37, 1993.
3. J. M. Lipton, K. P. Dabke, J. F. Alison, H. Cheng, L. Yates, and T. I. Brown, Use of the bispectrum to analyse properties of the human electrocardiograph, *Aust. Phys. Eng. Sci. Med.*, 21, 1–10, 1998.
4. A. W. Lohmann, G. Weigelt, and B. Wirnitzer, Speckle masking in astronomy: Triple correlation theory and applications, *Appl. Opt.*, 22, 4028–4037, 1983.
5. M. J. Northcott, G. Ayers, and J. Dainty, Algorithms for image reconstruction from photon-limited data using the triple correlation, *J. Opt. Soc. Am. A*, 5, 986–992, 1988.
6. I. D. Jung, F. X. Kärtner, J. Henkmann, G. Zhang, and U. Keller, High-dynamic-range characterization of ultrashort pulses, *Appl. Phys. B: Lasers Opt.*, 65, 307–310, 1997.
7. G. P. Agrawal, *Nonlinear Fiber Optics*, 3rd edn., Academic Press, San Diego, CA, 2001.
8. M. E. Marhic, N. Kagi, T.-K. Chiang, and L. G. Kazovsky, Broadband fiber optical parametric amplifiers, *Opt. Lett.*, 21, 573–575, 1996.
9. J. Hansryd, P. A. Andrekson, M. Westlund, L. Jie, and P. Hedekvist, Fiber-based optical parametric amplifiers and their applications, *IEEE J. Sel. Top. Quant. Electron.*, 8, 506–520, 2002.
10. R. Stolen and J. Bjorkholm, Parametric amplification and frequency conversion in optical fibers, *IEEE J. Quant. Electron.*, 18, 1062–1072, 1982.
11. L. N. Binh, *Optical Fiber Communications Systems: Theory and Practice with MATLAB® and Simulink® Models*, CRC Press, Boca Raton, FL, 2010.
12. M.-C. Ho, K. Uesaka, M. Marhic, Y. Akasaka, and L. G. Kazovsky, 200-nm-bandwidth fiber optical amplifier combining parametric and Raman gain, *J. Lightwave Technol.*, 19, 977, 2001.
13. P. O. Hedekvist, M. Karlsson, and P. A. Andrekson, Fiber four-wave mixing demultiplexing with inherent parametric amplification, *J. Lightwave Technol.*, 15, 2051–2058, 1997.
14. R. M. Jopson, A. H. Gnauck, and R. M. Derosier, Compensation of fiber chromatic dispersion by spectral inversion, *Electron. Lett.*, 29, 576–578, 1993.
15. S. L. Jansen, D. vandenBorne, P. M. Krummrich, S. Spalter, G. D. Khoe, and H. deWaardt, Long-haul DWDM transmission systems employing optical phase conjugation, *IEEE J. Sel. Top. Quant. Elect.*, 12, 505–520, 2006.
16. C. Lorattanasane and K. Kikuchi, Design theory of long-distance optical transmission systems using midway optical phase conjugation, *J. Lightwave Technol.*, 15, 948–955, 1997.
17. M. D. Pelusi, V. G. Ta'eed, L. B. Fu, E. Magi, M. R. E. Lamont, S. Madden, D.-Y. Choi, D. A. P. Bulla, B. Luther-Davies, and B. J. Eggleton, Applications of highly-nonlinear chalcogenide glass devices tailored for high-speed all-optical signal processing, *IEEE J. Sel. Top. Quant. Electron.*, 14, 529–539, 2008.
18. M. D. Pelusi, F. Luan, S. Madden, D. Y. Choi, D. A. Bulla, B. Luther-Davies, and B. J. Eggleton, Wavelength conversion of high-speed phase and intensity modulated signals using a highly nonlinear chalcogenide glass chip, *IEEE Photon. Technol. Lett.*, 22, 3–5, 2010.
19. D. V. Trung, H. Hu, M. Galili, E. Palushani, J. Xu, L. K. Oxenlowe, S. J. Madden et al., Photonic chip based 1.28 Tbaud transmitter optimization and receiver OTDM demultiplexing, San Diego, CA, 2010, p. PDPC5.

20. T. Yamamoto and M. Nakazawa, Active optical pulse compression with a gain of 29.0 dB by using four-wave mixing in an optical fiber, *IEEE Photon. Technol. Lett.*, 9, 1595–1597, 1997.

21. J. Hansryd and P. A. Andrekson, Wavelength tunable 40 GHz pulse source based on fibre optical parametric amplifier, *Electron. Lett.*, 37, 584–585, 2001.

22. R. Slavik, L. K. Oxenlowe, M. Galili, H. C. H. Mulvad, Y. Park, J. Azana, and P. Jeppesen, Demultiplexing of 320-Gb/s OTDM data using ultrashort flat-top pulses, *IEEE Photon. Technol. Lett.*, 19, 1855–1857, 2007.

23. B. Wirnitzer, Measurement of ultrashort laser pulses, *Opt. Commun.*, 48, 225–228, 1983.

24. T.-M. Liu, Y.-C. Huang, G.-W. Chern, K.-H. Lin, C.-J. Lee, Y.-C. Hung, and C.-K. Sun, Triple-optical autocorrelation for direct optical pulse-shape measurement, *Appl. Phys. Lett.*, 81, 1402–1404, 2002.

# 6 Solitons and Multi-Bound Solitons in Passive Mode-Locked Fiber Lasers

## 6.1 INTRODUCTORY REMARKS

Since the first demonstration of the propagation of optical pulses of constant incident amplitude in a passive nonlinear optical cavity could exhibit multi-stability and chaos in response [1], extensive studies on the dynamics in both the passive and active nonlinear optical cavities have been carried out. It has been shown that period-doubling bifurcation and period-doubling route to chaos are generic features of light traversing a nonlinear cavity (see also Appendix A).

However, majority of the researches have been focused on the continuous-wave (CW) operation of the cavities [2]. As a matter of fact, apart from the CW operation, short pulse operation of the cavities is also available. Moreover, determined by the cavity parameters, various types of solitons can even be formed in a nonlinear cavity. Solitons as a special nonlinear wave that can propagate long distances in dispersive materials without distorting their shapes have been found in a wide range of physical systems in thermodynamics, plasma physics, condensed matter physics and optics. As solitons can be considered as a nonlinear wave packet intrinsically stable against perturbations, it would be of much interest to study the dynamical features of the cavities and their manifestations under the ultrashort pulse operation, in particular when the ultrashort pulse is itself an optical soliton.

Soliton operation has been generated by various ultrashort pulse mode-locked fiber lasers [3–10], as well as by mode-locked bulk solid-state lasers [11,12]. In the case of fiber lasers, besides the cavity-incorporated components necessary for mode-locking and for generating laser soliton output sequences, its cavity path is mainly made up of single-mode optical fibers (SMF). Due to the small core size of the single-mode fibers and the long propagating distance, strong nonlinear phase shift can be accumulated in ultra-short pulses circulating around the fiber laser cavity. Therefore, a mode-locked pulse could be easily shaped into an optical soliton in a fiber laser.

It has been shown that despite some actions of the other discrete cavity components, the average dynamics of the formed solitons in a fiber laser can be well described by the well-known nonlinear Schrödinger equation (NLSE; see also Appendix B). A soliton formed in a laser is inherently different from those generated in a single mode fiber (SMF) due to their influences by both the optical loss and gain. It has been shown both experimentally and numerically that under certain conditions, the generated solitons in a fiber laser could exhibit deterministic dynamics [13–22], such as the soliton period-doubling bifurcations, soliton intermittency, multi-bound solitons, and soliton quasi-periodicity. Moreover, the appearance of the soliton deterministic dynamics can be independent on the concrete laser cavity structure. This indicates a general feature of the system.

This chapter describes mainly the deterministic dynamics of solitons in fiber lasers under passive mode-locking (see Appendix A for some fundamental mathematic representations). Our discussions focus on fiber soliton lasers formed by passive mode-locking using the nonlinear polarization rotation (NLPR) technique. In fiber lasers the scalar solitons are normally formed. We also investigated the vector soliton and multi-bound solitonic dynamics in fiber lasers mode-locked by the semiconductor saturable absorber mirrors (SESAMs) and the dissipative solitons dynamics in fiber lasers with normal cavity dispersion. This chapter is organized as follows: Section 6.2 presents the general theoretical background for NLPR mode-locking and soliton generation in fiber lasers. Section 6.3 reviews the deterministic dynamics of various solitons in different fiber lasers, which is followed by the corresponding numerical simulations in Section 6.4. Section 6.5 describes the cavity-induced modulation instability effect and its relation to the soliton deterministic dynamics. We show that the various forms of the soliton deterministic dynamics observed could be related to the cavity-induced soliton modulation instability effect, namely, their physical origin could be traced back to the self-induced nonlinear resonant wave coupling in the laser cavity. Section 6.6 outlines some concluding remarks.

## 6.2 SOLITON GENERATION BY PASSIVELY MODE-LOCKED FIBER LASERS

Soliton formation in anomalous dispersion cavity fiber lasers is mainly due to the natural balance between the cavity dispersion and the fiber nonlinear optical Kerr effect. An optical pulse can be routinely generated in a fiber laser by various mode-locking techniques. In terms of passive mode-locking, these include the nonlinear optical loop mirror method [23], Figure of Eight cavity method [24], the NLPR technique [25], and the SESAM method [26]. Among these, the NLPR technique is the most widely used. The technique exploits the nonlinear birefringence of single-mode optical fibers for the generation of an artificial saturable absorber effect in the laser cavity. It is the artificial saturable absorber effect that initiates a self-started mode-locking in the laser. To understand the operating principle of the technique, it is important to note that generally the polarization state of light passing through a piece of birefringent fiber varies linearly with the fiber birefringence and length. However, when the light intensity is strong, the nonlinear optical Kerr effect of the fiber introduces an extra change to the light polarization. As the extra polarization change is proportional to the light intensity, if a polarizer is placed behind the fiber, then by appropriately selecting the orientation of the polarizer, a feature where light with higher intensity experiences larger transmission through the polarizer could be obtained. Such an effect is known as an artificial saturable absorber effect. Incorporating such an artificial saturable absorber effect in a fiber laser is equivalent to inserting a saturable absorber in the laser cavity. Under effect of a saturable absorber, the operation of a laser can automatically become mode locked. In the

NLPR mode-locked fiber laser, nonlinear phase shift is required to achieve mode-locking. If the nonlinear phase modulation of the pulse could balance the pulse width broadening caused by cavity dispersion, a soliton is then formed in the laser.

### 6.2.1 PULSE PROPAGATION IN SINGLE-MODE FIBERS

When an optical pulse propagates in a birefringent single-mode fiber, the fiber dispersion would broaden the pulse width. At the same time, the different polarized pulses would have different group velocities that in turn broaden the pulse width and/or even split the pulse. The nonlinear fiber Kerr effect not only generates self-phase modulation (SPM) on each polarization, but also creates associate effects such as cross-phase modulations (XPM) between the two polarized components. Coherent coupling between the two polarization components also leads to a degenerated four-wave mixing (FWM) term, whose strength is related to that of the linear fiber birefringence [27].

In a fiber laser system, one of the essential components is the active medium or optical amplification element, which is generally generated from energy pumping on rare earth ions doped on the fiber, for example, $Er^{3+}$-doped silica fiber for 1550 nm gain region. When an optical pulse propagates in the optical fiber amplifier, the effect of light amplification must be considered. A fiber amplifier is characterized by its small signal gain, gain spectral bandwidth, gain saturation, and its noise figure. Therefore, to precisely describe pulse evolution in a fiber laser, it is necessary to consider the cavity dispersion, fiber nonlinearity (SPM and XPM), fiber birefringence, FWM, gain and loss, gain dispersion effects, etc.

Optical pulse propagation in fiber segments in a passive fiber laser can be described by the extended Ginzburg–Landau equation (GLE) (see also Chapter 2) [28]:

$$\frac{\partial A_x}{\partial z} = i\beta A_x - \delta \frac{\partial A_x}{\partial t} - \frac{i}{2}\beta_2 \frac{\partial^2 A_x}{\partial t^2} + i\gamma \left( |A_x|^2 + \frac{2}{3}|A_y|^2 \right) A_x + \frac{i\gamma}{3} A_y^2 A_x^*$$

$$+ \frac{g}{2} A_x + \frac{g}{2\Omega_g^2} \frac{\partial^2 A_x}{\partial t^2}$$

$$\frac{\partial A_y}{\partial z} = -i\beta A_y + \delta \frac{\partial A_y}{\partial t} - \frac{i}{2}\beta_2 \frac{\partial^2 A_y}{\partial t^2} + i\gamma \left( |A_y|^2 + \frac{2}{3}|A_x|^2 \right) A_y + \frac{i\gamma}{3} A_x^2 A_y^*$$

$$+ \frac{g}{2} A_y + \frac{g}{2\Omega_g^2} \frac{\partial^2 A_y}{\partial t^2}$$

(6.1)

where
   $A_x$ and $A_y$ are the two normalized slowly varying pulse envelopes along the slow and fast axes
   $A_x^*$ and $A_y^*$ are their conjugates, respectively
   $2\beta = 2\pi(n_x - n_y)/\lambda = 2\pi/L_b$ is the wave number difference or the beat length
   $L_b$, $n_x - n_y$ is the refractive index difference of the x- and y-polarized direction indices
   $2\delta = 2\beta\lambda/2\pi c$ is the inverse group-velocity difference.

The equations are written in the coordinate system that moves with the average group velocity $\bar{v}_g^{-1} = \bar{\beta}_1 = \beta_{1x} + \beta_{1y}/2$. The GLE is nonintegrable but numerical simulations are available for studying the pulse dynamics in such fiber lasers.

The last two terms on the right-hand side (RHS) of Equation 6.1 correspond to the gain and the gain dispersion effects when an optical pulse propagates in the active fiber segments of a fiber laser. $g$ is the peak gain coefficient, $\Omega_g$ is the gain bandwidth. When the gain saturation results from light along both polarizations, the saturated gain coefficient is calculated by

$$g = g_0 \exp\left(-\frac{1}{E_s}\int\left(|A_x|^2 + |A_y|^2\right)\right) \tag{6.2}$$

where
   $g_0$ is the small signal gain
   $E_s$ is the saturation energy

The typical value of $E_s$ for erbium-doped fibers (EDF) is about 10 µJ. As the pulse energies are normally much smaller than the saturation energy $E_s$, the variation of the gain saturation is negligible over the duration of a single pulse. However, in the case of a pulsed fiber laser, the pulse circulates in the laser cavity, then the average power of the light beam would saturate the gain.

We can observe that Equation 6.1 is two pair of coupled equations representing the mechanism of polarized pulse sequences circulating around the ring resonators. The energies of the dual sequences are exchanged and the coupled strength depends on the resonant condition that can be satisfied by the phase of the carrier of either polarized lightwaves. Due to this competition of energy of the system, which is saturated by optical amplifiers, frequency doubling of the generated solitons can occur. Hence, this can be followed by a multiple-soliton sequence. The observations and descriptions of these mechanisms are given in the following sections.

## 6.2.2 CAVITY TRANSMISSION OF NLPR MODE-LOCKED FIBER LASERS

Apart from the pulse propagation in various fiber segments, cavity boundary condition is another intrinsic feature that needs to be considered for soliton generation in fiber lasers. A fiber laser mode-locked with the NLPR technique can always be equivalently simplified into three parts, as shown in Figure 6.1, where a polarization controller $P_1$ is at the beginning and an analyzer at the end of the fiber, which is a weakly birefringent fiber. The two principal polarization axes of the birefringent fiber are $x$ (horizontal) and $y$ (vertical) axis, and we consider that the birefringent axes of the different segments of fibers are the same. Let $\theta$ be the angle formed between the fast axis of $P_1$ and the $y$-axis of the polarized direction of the fiber. Similarly $\varphi$ is the angle between the transmission axis of the analyzer and the $y$-axis of the fiber polarization, as shown in Figure 6.1.

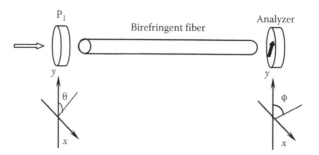

**FIGURE 6.1**  Equivalent physically simplified laser cavity for fiber lasers using the NLPR mode-locking.

In brief, the pulse evolution in the fiber laser starts with an arbitrary weak pulse and it circulates in the laser cavity until a steady pulse evolution state is established. Whenever the pulse encounters an individual intra-cavity component except the fiber segments, the Jones matrix of the component is multiplied to the optical field of the pulse.

Starting from a linearly polarized weak pulse $F_n$, two polarization components are obtained when $F_n$ travels through $P_1$:

$$\begin{cases} u = F_n \sin\theta \exp(i\Delta\Phi) \\ v = F_n \cos\theta \end{cases} \tag{6.3}$$

where $\Delta\Phi$ is the phase shift between the wave components in the two orthogonal birefringent axes $x$ and $y$.

Then the propagation of the two polarization components in the fiber segments is governed by the coupled GLEs (Equation 6.1) as given in Section 6.2.1. The light propagates along the laser cavity and finally projects on the transmission axis of the analyzer:

$$F_{n+1} = u' \sin\varphi + v' \cos\varphi \tag{6.4}$$

where $u'$ and $v'$ are the two orthogonal polarization components of light after propagation in all the fiber segments.

The transmittance/transmissivity $T$ of lightwaves passing through cavity can be expressed by [29]

$$T = \frac{I_{out}}{I_{in}} = \sin^2\theta\sin^2\phi + \cos^2\theta\cos^2\phi + \frac{1}{2}\sin(2\theta)\sin(2\phi)\cos(\Delta\Phi_F) \tag{6.5}$$

where $\Phi_F$ is the phase delay generated between the two light polarization components when they traverse the birefringent fiber. The phase delay in fact consists of two parts: a linear phase delay $\Delta\Phi_1$ prolonged because of the linear birefringence of the fiber, this part always exists, and a nonlinear part $\Delta\Phi_{nl}$ generated by the nonlinear effects of the fiber. If the light intensity is weak, the nonlinear part becomes zero. Hence, one can divide the light transmission into the linear transmission and the nonlinear transmission. The linear intensity transmission can be written as follows:

$$T_l = \sin^2\theta\sin^2\varphi + \cos^2\theta\cos^2\varphi + \frac{1}{2}\sin(2\theta)\sin(2\varphi)\cos(\Delta\Phi_l) \qquad (6.6)$$

Once the orientations of the polarizer and analyzer with respect to the fiber polarization principal axes are fixed, the linear transmission of the setup is a sinusoidal function of the linear phase delay between the two polarization components. Although the relative orientation between the polarizer and the analyzer could be arbitrarily selected, in order to possibly achieve a stable mode-locked operation, it is preferred to choose the relative orientation between the polarizer and the analyzer 90°, as under such a selection the linear transmission is possible to set to zero.

To illustrate the effect of the NLPR on intensity transmission, we assume that the nonlinear fiber birefringence only introduces a small extra phase delay $\Delta\Phi_{nl}$. Then the intensity transmission, or the transmittance or transmissivity, can be written as

$$T = T_l - \frac{1}{2}\sin(2\theta)\sin(2\varphi)\sin(\Delta\Phi_l)\Delta\Phi_{nl} \qquad (6.7)$$

The nonlinear phase delay introduces a transmission change relative to the value of the linear transmission $T_l$. The change is not only the nonlinear phase delay $\Delta\Phi_{nl}$, and therefore, light intensity dependent, but also the orientation of the polarizer, making it linear phase delay dependent as well. As a result, in order to generate an artificial saturable absorber effect through the nonlinear fiber birefringent effect, it is necessary to select an appropriate combination of all these parameters.

It is instructive to consider a special case to illustrate it. Suppose the polarizer and the P1 are set orthogonal. In this case the linear intensity transmission is

$$T_l = \sin^2 2\theta\frac{[1 - \cos(\Delta\Phi_l)]}{2} \qquad (6.8)$$

Therefore, the maximum linear intensity transmission is limited by $\sin^2 2\theta$. The intensity transmission under the existence of a nonlinear phase delay $\Delta\Phi_{nl}$ is

$$T = T_l + \frac{1}{2}\sin^2(2\theta)\sin(\Delta\Phi_l)\Delta\Phi_{nl} \qquad (6.9)$$

The orientation of the polarizer determines the projection of light on the two polarization axes of the fiber, therefore, determining whether a positive or a negative $\Delta\Phi_{nl}$ would be generated. Assuming that a negative $\Delta\Phi_{nl}$ is generated, then depending further on the linear phase delay, the NLPR could cause either an increase or decrease in the intensity transmission. The magnitude of $\Delta\Phi_{nl}$ is always proportional to the intensity of light. This means that with an increase in the light intensity, either an increase or a decrease in the transmission could be generated, which is purely determined by the linear phase delay selection.

Strictly speaking, the nonlinear phase delay between the two polarization components of an optical pulse traversing a piece of birefringent fiber can only be determined by numerically solving the coupled GLEs (Equation 6.1). However, in the case of CW light propagation, and ignoring the effect of energy exchange between the two polarization components, this can be calculated explicitly as follows [28]:

$$\Delta\Phi_{nl} = \frac{\gamma P_0 L}{3}\cos(2\theta) \tag{6.10}$$

where
$P_0$ is the power of the light
$L$ is the length of the fiber

Using the nonlinear phase delay, the intensity transmission can be written as [30]

$$T = T_l + \kappa|E_{in}|^2 \tag{6.11}$$

where $\kappa = -\dfrac{\gamma L}{12}\sin(4\theta)\sin(2\theta)\sin(\Delta\Phi_l)$. As far as $\kappa>0$, a saturable absorber effect can be obtained. Based on the formula, one could estimate the optimum selection for the $\theta$ value so that the strongest saturable absorber effect could be achieved [31]. In this case, it occurs at $\theta=27.7°$.

In practice, it is difficult to set the linear phase delay just within the range where the saturable absorption effect is achieved. Therefore, a PC is normally inserted in the setup to efficiently control the value of the linear phase delay $\Delta\Phi_l$. Mathematically, this is equivalent to adding a variable linear phase delay bias term in the intensity transmission formula. Hence, the intensity transmission can be further written as

$$T = \frac{I_{out}}{I_{in}} = \sin^2\theta\sin^2\varphi + \cos^2\theta\cos^2\varphi + \frac{1}{2}\sin(2\theta)\sin(2\varphi)\cos(\Delta\Phi_{PC} + \Delta\Phi_F) \tag{6.12}$$

where $\Phi_{PC}$ is the phase delay bias introduced by the PC and it is continuously tunable. With the insertion of a PC in the setup, it is always possible to achieve an artificial saturable absorber effect through changing the linear birefringence of the fiber.

In summary, the soliton generation in fiber lasers should be the stable pulse propagation that can both fulfill pulse propagation in fiber segments as described by Equation 6.1 and the cavity transmittance/transmissivity as indicated by Equation 6.12.

## 6.3 SOLITON DYNAMICS IN DUAL-POLARIZATION MODE-LOCKED FIBER LASERS

### 6.3.1 EXPERIMENTAL CONFIGURATION

The deterministic soliton dynamics have been experimentally studied in passively mode-locked fiber lasers under various cavity design and parameters. Figure 6.2 shows a typical fiber laser we used. The fiber laser has a ring cavity comprising a segment of the EDF used as the laser gain medium. The erbium fiber is pumped by a high-power fiber Raman laser source (BWC-FL-1480-1) of a wavelength of 1480 nm. The pump light is coupled into the cavity through a fiber wavelength division multiplexer (WDM). A fiber pigtailed isolator is inserted in the cavity to force the unidirectional operation of the cavity. We used the NLPR technique for the self-started mode-locking of the laser. To this end, two PC, one consisting of two quarter-wave plates and the other two quarter-wave plates and one half-wave plate, are used to adjust the polarization of the light. A cubic polarization beam splitter is used to output the laser pulses and set the polarization at the position of the cavity. The PCs

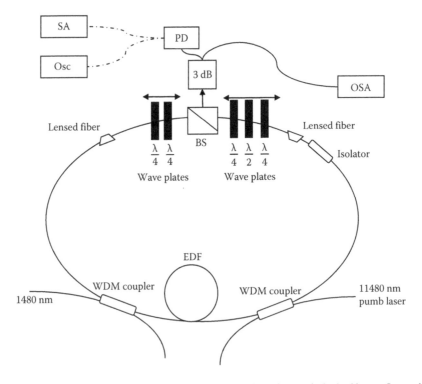

**FIGURE 6.2** A schematic of the experimental setup of passive mode-locked lasers. Legend: λ/4 = quarter-wave plate; λ/2 = half-wave plate; BS, beam splitter; WDM, wavelength-division multiplexer; EDF, erbium doped fiber; OSA, optical spectrum analyzer; PD, photodetector; Osci., Oscilloscope; SA, Spectrum Analyzer. RF, Radio Frequency. FC, fiber coupler. ppm, parts per million.---, dotted lines = electrical lines.

and the beam splitter are mounted on a 7 cm long fiber bench. The soliton output of the laser is monitored with an optical spectrum analyzer (OSA) (Model Ando AQ-6315B), a 26.5 GHz RF SA (Model Agilent E4407B ESA-E), and a 350 MHz oscilloscope (Model Agilent 54641A), together with a 5 GHz photodetector. A commercial optical autocorrelator (model autocorrelator pulsescope) is used to measure the soliton pulse width.

In the experiments, optical fibers are employed with various parameters, such as different fiber group velocity dispersion (GVD), different lengths and different doping concentrations of the gain fiber. The purpose of using fibers of different properties and lengths is to change the laser cavity parameters, so possibly different soliton dynamics in the cavity could be observed. Experimentally, we found that with the appropriate selection on the orientations of the waveplates, self-started mode-locking of the fiber lasers could always be achieved, and as far as the cavity nonlinearity is not too strong, a stable soliton operation of the fiber lasers can be observed. Such stable soliton operation of the fiber lasers has been extensively investigated and reported previously [29]. However, it has also been found that when the peak power of the formed solitons reaches a certain high level in a fiber laser, the deterministic dynamics of the optical solitons could also be observed.

### 6.3.2  SOLITON DETERMINISTIC DYNAMICS

Suggested by the average soliton theory of lasers [32], it is generally believed that the output of the fiber soliton lasers is a uniform soliton pulse train. However, Kim et al. found theoretically that depending on the strength of the fiber birefringence and the alignment of the polarizer with the fast- and slow-polarization axes of the fiber, the output pulse train of a fiber laser mode locked with the NLPR technique could exhibit periodic fluctuations in pulse intensity and polarization [33,34]. Nevertheless, the nonuniformity of the pulse train could be diminished by aligning the polarizer with either of the fast or slow axis of the fiber. The output property of a passively mode-locked fiber soliton ring laser was also experimentally investigated by the NLPR technique [35]. It was found that the soliton pulse nonuniformity is in fact an intrinsic feature of the laser, whose appearance is independent on the orientation of the polarizer in the cavity but closely related to the pump power. Based on numerical simulations it can be shown that depending on the linear cavity phase delay bias, the nonlinear polarization switching NPS effect could play an important role in the soliton dynamics of the lasers. When the linear cavity phase delay bias is set close to the NPS point and the pump power is strong, the soliton pulse peak intensity could be increased to high enough for the generated NLPR to cross over the NPS point, and consequently can drive the laser cavity from the positive feedback regime to the negative feedback regime. Eventually, the competition between the soliton pulses with the linear waves in the cavity such as the dispersive waves or CW laser emission then causes the amplitude of the soliton pulses to vary periodically. There exist two methods to suppress such periodical intensity fluctuations: the first is to reduce the pump power so that the peak intensity of the solitons is below the "polarization switch" threshold; the other is to increase the polarization switching power of a laser. However, the latter method needs to appropriately adjust the linear phase delay bias of the cavity.

Experimentally, lasers with the same cavity structure but different cavity lengths, different fiber birefringence, and fiber dispersion properties have been set up and investigated [13,15,17,19,20]. If the pump power is set beyond the mode-locking threshold in all lasers, and the linear cavity phase delay is appropriately selected, self-started mode-locking can always be obtained. However, depending on the laser parameter selection, deterministic dynamics of the solitons may not necessarily appear. In the following we summarize the observed typical soliton deterministic dynamics.

### 6.3.2.1  Period-Doubling Bifurcation and Chaos of Single-Pulse Solitons

Provided that the orientations of the PC are appropriately set, self-started soliton operation of the lasers is automatically obtained by simply increasing the pump power beyond the mode-locking threshold. Multiple solitons are initially obtained. However, if the pump power is decreased only that state with one soliton existing in the cavity can be achieved. Starting from a stable soliton operation state, experimentally it was noticed that turning the orientation of one of the quarter-wave plates to one direction, which theoretically corresponds to shifting the linear cavity phase delay bias away from the NPS point, the peak power of the soliton pulse formed in the cavity increases. Consequently, the strength of the nonlinear interaction of the soliton pulses with the cavity components such as the optical fiber and the gain medium increases. At a certain level of the nonlinear interaction, it was observed that the output soliton intensity pattern of the laser experiences period-doubling bifurcations and a period-doubling route to chaos. Figure 6.3 shows as example an experimentally observed period-doubling route to chaos. The results were obtained with

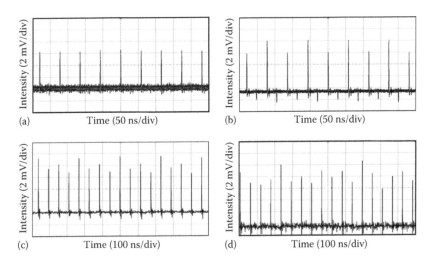

**FIGURE 6.3**  Period-doubling bifurcation to chaos of the soliton trains: (a) period-one state, (b) period-two state, (c) period-four state, (d) chaotic state. The pump intensity was increased from (a) to (d).

fixed linear cavity phase delay bias but increasing pump power. At a relatively weak pump power, a stable soliton pulse train with uniform pulse intensity was obtained. The pulses repeat themselves with the cavity fundamental repetition rate (Figure 6.3a). The laser output power has been experimentally measured when it is operating in such a state. With a pump power of about 26 mW an average output power of about 140 μW was obtained, which shows that the single soliton pulse energy is about 8.05 pJ. Carefully increasing the pump power further, the intensity of the soliton pulse is no longer uniform, but alternates between two values (Figure 6.3). Although the round trip time of the solitons circulating in the cavity is still the same, the pulse energy returns only every two round trips, forming a so-called period-doubled state as compared with that in Figure 6.3a. Increasing the pump power slightly further results in a period-quadrupled state (Figure 6.3c). Eventually the process ends in a chaotic soliton pulse energy variation state (Figure 6.3d).

With fixed pump power, all the states shown are stable. Provided there are no huge disturbances that can last for several hours. It was also confirmed experimentally by the combined use of an autocorrelator (PulseScope model, scanning range from 50 fs to 50 ps) and a high-speed oscilloscope (Agilent 86100A bandwidth 50 GHz) that there is only one soliton in the cavity. Limited by the resolution of our autocorrelator, the measured autocorrelation traces show no unusual features of the soliton pulses and thus give no evidence of the behavior of period-doubling bifurcations, which is similar to what was observed by Sucha et al. [36]. In all states, the average soliton duration measured was about 316 ± 10 fs. The experimental results demonstrate that contrary to the general understanding of mode-locked lasers, after one round trip the mode-locked pulse does not return to its original value, but returns only after every two or four cavity round trips in the stable periodic states. Depending on the strength of the nonlinear interaction between the pulse and the cavity components, the pulse could even never return to its original state in the chaotic state. To exclude any possibility of the artificial digital sampling effect of the oscilloscope, the pulse intensity alternation of the various periodic states by using a high-speed sampling oscilloscope (Agilent 86100A bandwidth 50 GHz) has been confirmed. Figure 6.4 shows the result corresponding to a period-two state. In obtaining this figure, the soliton

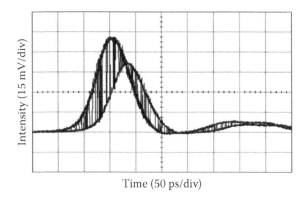

**FIGURE 6.4**   Captured oscilloscope traces of a period-two state of laser emission.

pulse is employed as the trigger pulse as well as the signal input, or self-triggering, fed into the oscilloscope ports. A high oscilloscope resolution of 50 ps/div is used. Due to the high scan speed of the oscilloscope, one can clearly see that the individual soliton pulse trace on the screen becomes broader. Therefore, no sampling antibiasing problem exists. Triggered by different pulses, the oscilloscope traces formed have two distinct peak intensities, indicating that the solitons in the laser output have indeed two different pulse energies.

The optical spectra of the solitons corresponding to the period-one and period-two states are shown in Figure 6.5a and b, respectively. While the spectral curve shown in Figure 6.5a is smooth, the spectral curve shown in Figure 6.5b exhibits clear modulations. The spectral curve shown in Figure 6.5a possesses typical features of the soliton spectra of the passively mode-locked fiber lasers, characterized by the existence of sidebands superposing on the soliton spectrum. As in a period-one state, solitons are identical in a soliton train, Figure 6.5a shows the optical spectrum of a captured sole soliton. In contrast, the optical spectrum shown in Figure 6.5b is that of the average spectra of two different solitons, each with different energy and frequency chirps. Based on Figure 6.5b, we can draw a conclusion that after a period-doubling bifurcation, the solitons possess different frequency properties as that of the solitons prior to the bifurcation.

The intensity-modulated soliton train has been monitored on a radio frequency (RF) SA. If period-doubling does occur, there should appear a new frequency component in the RF spectrum locating exactly at half of the fundamental cavity repetition rate position. Figure 6.6 shows the RF spectra of the laser output measured. As expected, after a period-doubling bifurcation, a new frequency component of about 8.7 MHz appears in the spectrum. The amplitude of the new frequency component is quite strong compared to the fundamental frequency component, which vividly suggests that the soliton peak intensity alternates between two values with a large difference. When period quadrupling occurs, in the RF spectrum we found that the amplitude of the new frequency component decreased, however, the frequency components corresponding to period quadrupling were too weak to be clearly distinguished from the background noise.

It should be noted that if the linear cavity phase delay bias is selected close to the NPS point of the cavity, although the single soliton operation of the laser can still be obtained. Because the peak intensity of the soliton pulses is limited by the NPS power, which is weak under the linear cavity phase delay bias selection, no matter how strong the pump power is, no period-doubling bifurcation could be observed. Thus, a threshold level exists there for the period-doubling bifurcation. Only when the linear cavity phase delay bias, which determines the stable soliton peak intensity, meets certain resonant conditions at which the stable soliton peak intensity exceeds a certain value, period-doubling bifurcation can be achieved. The experimental result further confirms that the appearance of period-doubling bifurcations and period-doubling route to chaos is soliton pulse intensity dependent and that it is a nonlinear dynamic feature of the laser. We can, in our experiment, confirm that for this effect to occur, the soliton pulse energy or peak power must be high. In this case the nonlinear interactions between light and the gain medium and light and the nonlinear laser cavity become strong. It is well-known that as a result of the

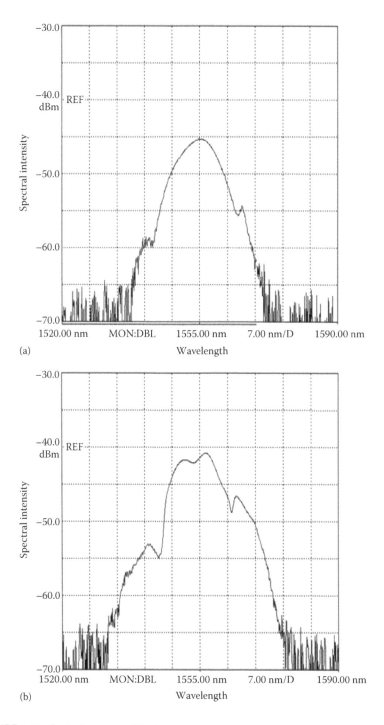

**FIGURE 6.5**   Optical spectra of the laser measured in (a) the period-one state and (b) the period-two state.

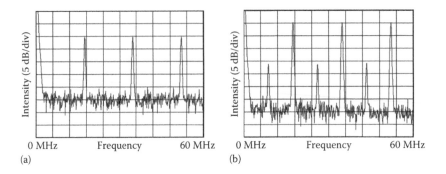

**FIGURE 6.6**   RF spectra corresponding to (a) the period-one and (b) the period-two state.

strong nonlinear interaction between light and the gain medium in the laser cavity, a laser operating in the CW or Q-switched mode can exhibit a period-doubling route to chaos. Our experimental observation demonstrates this phenomenon appearing in such a mode-locked soliton laser. Finally, we can point out that Côté et al. have reported period-doubling in a femtosecond Ti–sapphire laser by total mode-locking of the $TEM_{00}$ and $TEM_{01}$ modes in an effective confocal cavity [37]. They believe that the gain saturation is a likely mechanism to support the transverse mode-locking and the period-doubling. However, in our laser only one transverse mode of the SMF exists.

### 6.3.2.2   Period-Doubling and Quadrupling of Bound Solitons

The full route of soliton period-doubling bifurcation to chaos was only observed under appropriately selected laser cavity parameters. In most cases, only one or two soliton period-doubling bifurcations could be obtained. A common feature of all soliton fiber lasers is multiple soliton generation under strong pumping. And in the steady states all the solitons generated have identical properties, a result of the soliton energy quantization effect [38]. Interaction between the multiple solitons has been extensively investigated previously. It was found that various modes of the multiple soliton operation could be formed [39]. A special situation also experimentally observed is that under appropriate laser cavity conditions, the solitons in the cavity could automatically bind together and form states of bound solitons [40,41]. Depending on the strength of the soliton binding, a certain state of the bound solitons can even become the only stable state in a laser. In such a state the bound solitons as an entity exhibit features that are in close similarity to those of the single pulse solitons [40,42,43]. It was speculated that such a state of bound solitons could even be regarded as a new form of multipulse solitons in the lasers [44]. We found experimentally that period-doubling bifurcations could also occur on these bound solitons. Like the conventional single pulse solitons, the bound solitons as an entity can also exhibit complicated deterministic dynamics.

Figure 6.7a through c show, for example, an experimentally observed period-doubling bifurcation route of a state of bound solitons. In the current case, multiple solitons are tightly bound and "move" together at the cavity fundamental repetition frequency. When the peak intensity of the bound solitons is strong, the total intensity of the multi-bound solitons exhibits the period-doubling bifurcation. The response time of the photo-detector used in our experiment is about 140 ps, which can clearly resolve the pulse train but not the intensity profile of the bound solitons. Therefore, the measured oscilloscope traces shown in Figure 6.7 have no distinctions to those of the period-doubling bifurcations of the single-pulse solitons [13]. Figure 6.7d through f depict the simultaneously measured optical spectra corresponding to the Figure 6.7a through c, respectively. The features of the optical spectral modulations exist on the spectra, which show that there are more

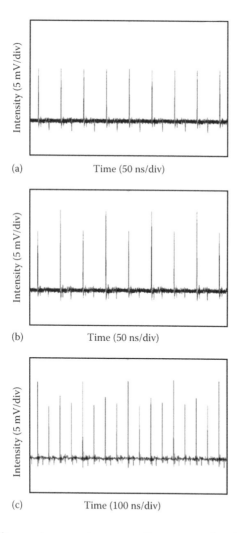

(a)     Time (50 ns/div)

(b)     Time (50 ns/div)

(c)     Time (100 ns/div)

**FIGURE 6.7** Oscilloscope traces and corresponding spectra of period-doubling bifurcations of a bound-soliton pulse train: (a and d) period-one state, (b and e) period-doubled state, (c and f) period-quadrupled state.                              (*Continued*)

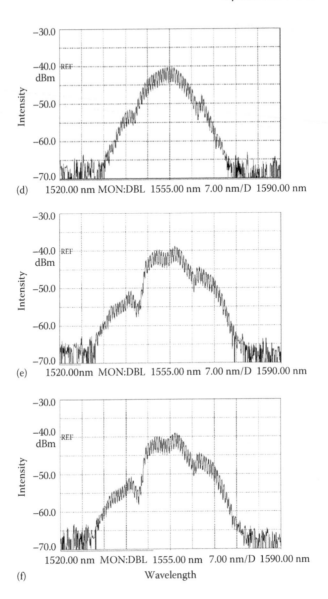

**FIGURE 6.7 (Continued)** Oscilloscope traces and corresponding spectra of period-doubling bifurcations of a bound-soliton pulse train: (a and d) period-one state, (b and e) period-doubled state, (c and f) period-quadrupled state.

than one soliton in the cavity and they are closely spaced. The oscillating peaks of the optical spectra have a peak separation of 0.70 nm. Assuming that the central wavelength of the soliton pulse is 1550 nm, based on the Fourier transformation, the solitons have a peak separation of about 11.5 ps in the time domain. The bound soliton nature of the state is also confirmed by the simultaneous autocorrelation trace measurement, as shown in Figure 6.8. Limited by our autocorrelator

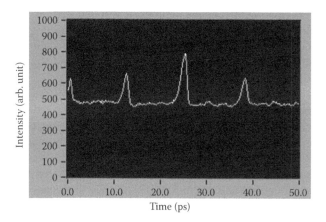

**FIGURE 6.8**    Measured autocorrelation trace of a bound-soliton.

resolution, the measured autocorrelation traces show no distinguishable features among different period-doubled states. The measured single soliton duration of the state is about $326 \pm 12$ fs, and the soliton separation between neighboring solitons is about 11.8 ps, which agrees well with the 0.70 nm period of optical spectral modulation within the experimental errors. The soliton separations under different period-doubled states are also the same. From the autocorrelation traces it is seen that there are at least three solitons with exactly the same soliton separations binding together in the bound-soliton state. Due to the limited scan range of our autocorrelator, we could not identify how many solitons are actually in the state. Nevertheless, it is confirmed that all the solitons in the state are equally spaced, and the bound soliton as an entity exhibit period-doubling bifurcations. We note that the state of bound solitons shown in Figure 6.7 is the only pulse pattern existing in the cavity, which is confirmed by the combined monitoring of the autocorrelation trace and a high-speed oscilloscope trace.

Figure 6.7a through c are obtained with continuously increased pump power while keeping all the other cavity parameters fixed. The chaotic state of the bound solitons could be obtained with further increasing pump power from the state shown in Figure 6.7c. However, the chaotic state was not stable. It quickly evolved into a chaotic state of the single-pulse soliton. The change of the laser emission from a chaotic state of bound solitons to that of single-pulse solitons was experimentally identified by the change of the optical spectra. It was observed that whenever the bound-soliton became chaotic, the modulations on the optical spectrum, which is a direct indication of close soliton separation in the time domain, disappeared, and subsequently the optical spectrum had exactly the same profile as that of the chaotic state of the single-pulse solitons. Considering that in a chaotic state the energy of solitons varies randomly, which may affect the binding energy between the solitons and destroy their binding, this result seems also plausible. The states of period-doubled and period-quadrupled bound solitons are stable. Nevertheless, compared with those states of the single pulse solitons [13], they are more sensitive to environmental perturbations. Analyzing the optical spectra of the period-doubled bound solitons, it is found that the overall spectral profile in each state is similar to those of the single-pulse soliton that underwent period-doubling, except that it is now modulated [13]. The feature of the optical spectra indicates that the individual solitons within a bound-soliton are still the same as the single-pulse soliton of the laser.

RF spectra of the bound-soliton pulse train also disclose different period-doubled states. The emergence of a new frequency component at the position of half of the fundamental cavity repetition frequency clearly shows that the repetition rate of the bound solitons is doubled. The amplitude of the new frequency component is nearly half of that of the fundamental frequency component, which vividly suggests that the total peak intensity of the bound solitons alternates between two values with large difference. Same as observed in Ref. [13], the new frequency component corresponding to the period quadrupling is not distinguishable from the background noise. Limited by the resolution of our measurement system, the detailed intensity variations of each of the solitons under period-doubling bifurcations could not be resolved. There are two possible ways for a two-pulse bound-soliton exhibiting a period-doubling intensity pattern. Either the two solitons simultaneously experience the period-doubling, or only one soliton experiences the period-doubling while the other one still remains stable. With more solitons binding together, the process could become more complicated.

Experimentally, the period-doubling bifurcations of bound solitons have been observed with different soliton separations. Bound solitons with different pulse separations can be easily distinguished by their optical spectra—as different spectral modulation periods correspond to different pulse separations in the time domain. This suggests that the appearance of the phenomenon should be a generic feature of the laser, which is independent on the concrete property of the optical pulses. Period-doubling route to chaos is a well-known nonlinear dynamic phenomenon widely investigated. Period-doubling route to chaos of the CW and the Q-switched lasers as a result of strong nonlinear interaction between the light field and the gain medium have already been reported [45,46]. Except that the laser modes are phase locked, physically the interaction between the light and the gain medium in a mode-locked laser is still the same. Therefore, it is not surprising that under the existence of strong mode-locked pulses, a period-doubling route to chaos on the pulse repetition rate could still be obtained. Nevertheless, it was a slightly unexpected that a bound-soliton can exhibit period-doubling bifurcations so intuitively that the dynamic bifurcation of the laser state could easily damage the binding between the solitons. In our experiment, a dispersion-managed laser cavity was used. The purpose of using a dispersion managed cavity is to possibly make the energy of the formed solitons strong, so in average, a strong nonlinear interaction between the pulses with the cavity components could be achieved.

### 6.3.2.3   Period-Doubling of Multiple Solitons

It is well-known that a passively mode-locked fiber laser can operate with multiple solitons in cavity, and depending on the soliton interaction, the multiple solitons can either form soliton bunch or randomly distribute with stable relative soliton separations [47]. With more than two solitons coexisting in the cavity, soliton interaction cannot be ignored. In Section 6.3.2.2, the period-doubling bifurcation of bound solitons is described. In that case two or more solitons coexist in the cavity. However, due to that the solitons are strongly coupled, the state of bound solitons actually behaves like a single pulse soliton. No difference in their period-doubling bifurcations was observed compared to those of the single-pulse soliton. Different from the case of the bound solitons, the solitons are now distributed far apart in the cavity with stable relative separations. We show that even with multiple solitons coexisting in a fiber cavity, with obvious gain competition between them, period-doubling bifurcation can still occur in the laser, and specifically each soliton in the cavity experiences period-doubling bifurcation. However, the intensity variation of the individual solitons is not necessarily synchronized.

Figure 6.9 shows, for example, a typical state of the multiple soliton operation observed in a fiber laser. The laser has a cavity round trip time of about 38 ns. It is to see that four solitons coexist in the cavity and locate far apart from each other. In the case of current soliton operation, all solitons have the same pulse height in the oscilloscope trace, indicating that after every cavity round trip the energy of each soliton returns to its previous value. Such a multiple soliton operation has also been reported by other authors, which is a typical case of the conventional soliton fiber laser operation [6]. Changing the orientation of one of the waveplates in the laser

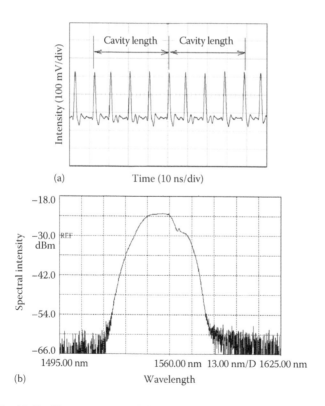

**FIGURE 6.9**    (a) Oscilloscope trace and (b) optical spectrum of a period-one state of the randomly distributed multiple solitons (four solitons in the cavity).

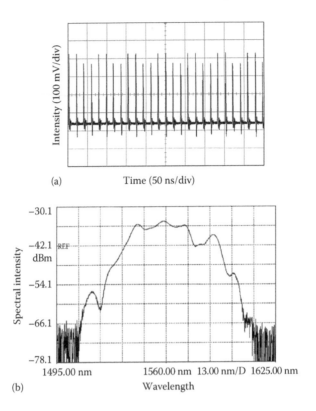

**FIGURE 6.10** (a) Oscilloscope trace and (b) optical spectrum of a period-doubled multiple-soliton state (two solitons in the cavity). Note the central peak indicating residual carrier and two side peaks for in-phase between adjacent pulses.

cavity, which corresponds to shifting the linear cavity phase delay bias away from the cavity polarization switching point, and therefore, increasing the formed soliton peak power [29], the period-doubling bifurcation occurs in the laser. Figure 6.10 shows a case of the period-doubled state of the laser with two solitons coexisting in the cavity. Comparing with the oscilloscope trace shown in Figure 6.9, the pulse height of each soliton in the oscilloscope returns to its value after every two cavity roundtrips. It is obvious that each individual soliton in the cavity experiences a period-doubling bifurcation, suggesting that each of them has the same bifurcation threshold and behaves identically in the cavity. For the sake of completeness, the corresponding optical spectrum of the solitons is shown in Figures 6.9b and 6.10b. As the solitons have large separations in the cavity, the spectrum of the multiple solitons is the same as that of the single soliton. Again, note that after the period-doubling bifurcation, extra spectral structures appear on the soliton spectrum, indicating the existence of dynamical sideband generation [18].

As there are only two solitons in the cavity, it is difficult to judge from the oscilloscope trace whether the intensity variations of the solitons are synchronized or not. To clarify it, we show in Figure 6.11a, another period-doubled state of the multiple

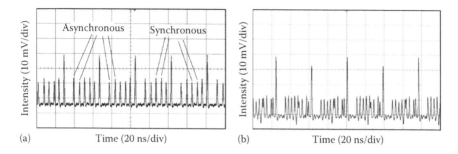

**FIGURE 6.11**   Oscilloscope traces of period-doubled multiple-soliton states: (a) low pulse amplitude and (b) higher amplitude pulses with nearly twice the intensity caused by partial superposition of two closely spaced solitons.

solitons in which six pulses are randomly scattered in the cavity. Again after the period-doubling bifurcation, the period of each pulse in the cavity becomes doubled, but it is now clear that the intensity variation of the pulses is not all synchronized. In the same cavity round trip, some pulses are in their high power state while others in their low power state. Note that one pulse in the oscilloscope trace has significantly larger pulse height than the others. It is actually due to that two solitons are too close in the cavity that our detector cannot resolve them. Therefore, in the oscilloscope trace, they appear as one pulse. It is important to note that due to the opposite intensity variation of the two solitons in the current state, their total pulse height in the oscilloscope exhibits no change. It is only an experimental artificial appearance. In the experiment, such state can also be obtained as shown in Figure 6.11b, where the strong pulse in the oscilloscope trace also exhibits period-doubling, indicating that the two solitons that form the pulse have synchronized intensity variation. Using the commercial autocorrelator, the soliton pulse width of the laser is measured, which is about 1.54 ps assuming a Gaussian pulse shape.

It was shown that even when multiple solitons are present in the cavity, the laser can still experience period-doubling bifurcation. In particular, each soliton in the cavity exhibits the same period-doubling bifurcation. However, the detailed intensity variation of the solitons could be unsynchronized, indicating that their period-doubling is actually not related.

### 6.3.2.4   Period-Doubling of Dispersion-Managed Solitons at Point near Cavity Zero-Dispersion

Although a dispersion-managed cavity consists of fibers with either positive or negative dispersion, and on average the pulse peak power is lower than that of the pulses formed in the equivalent uniform dispersion cavity, as far as the net cavity GVD is negative and large, conventional solitary waves (solitons with clear sidebands) could still be formed in the lasers. Furthermore, in the regime of near zero net cavity GVD, a different type of solitary waves known as the dispersion-managed solitons [48] could also be formed. Comparing with the conventional solitons formed in a laser, the optical spectra of the dispersion-managed solitons have a Gaussian profile without spectral sidebands. Dispersion-managed solitons in the fiber transmission lines have been extensively investigated [49–53]. It was shown that such a soliton could even exist in a system with positive near zero net GVD [49–52].

It is well-known that cavity dispersion is wavelength related, and the central wavelength of the mode-locked pulses shifts with the experimental conditions. It is difficult to accurately determine the net cavity dispersion under various laser mode-locking states. However, one can roughly estimate the dispersion of a laser cavity by simply accumulating the dispersion of each cavity component, and then fine tune it through cutting back the SMF length.

The period-doubling bifurcation of the dispersion-managed solitons has also been observed. After having obtained a uniform dispersion-managed soliton pulse train as shown in Figure 6.12a, we then carefully tune one of the waveplates to the direction that causes the pulse peak power to increase while keeping all other laser parameters unchanged. Physically, this action corresponds to shift the linear cavity phase delay bias to the direction that causes the peak power of the optical pulse to be clamped at a higher level by the cavity [14]. To a certain level of the pulse peak power, it was observed that the dispersion-managed solitons exhibited period-doubling bifurcation as shown in Figure 6.12b. Obviously, in a period-two state the pulse intensity returned to its original value after every two cavity round trips. Figure 6.13b shows the optical spectrum of the dispersion-managed solitons corresponding to Figure 6.12b. The fundamental repetition rate of the laser is 21.3 MHz. Limited by the resolution of our autocorrelator, the autocorrelation trace measured under a

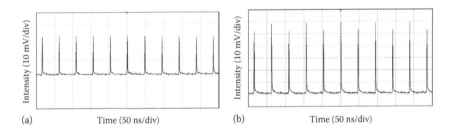

(a) Time (50 ns/div)      (b) Time (50 ns/div)

**FIGURE 6.12** Oscilloscope traces of dispersion-managed solitons: (a) period-one state and (b) period-doubled state.

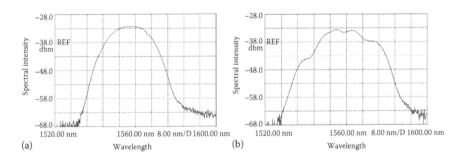

(a)     Wavelength      (b)     Wavelength

**FIGURE 6.13** Dispersion-managed solitons of the laser: (a) optical spectrum of the period-one state and (b) optical spectrum of the period-doubled state.

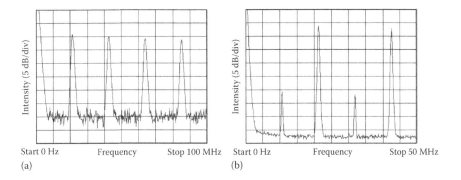

**FIGURE 6.14**    RF spectrum of the dispersion-managed solitons: (a) period-one state and (b) period-doubled state.

period-doubled state has no obvious difference to that of the uniform soliton pulse train. However, the optical spectrum of the state shows clear modulation as compared with that of the period-one state (Figure 6.13a). The measured optical spectrum is an average of the pulse spectra. As in the period-doubled state the laser emits alternately between two mode-locked pulse states, it is expected that the resultant spectrum is different from that of the period-one state. Monitored by a high-speed sampling oscilloscope, we confirmed experimentally that in the state shown there is only one soliton pulse propagating in the cavity.

Figure 6.14 further shows the RF spectra of the laser outputs. Obviously, in the period-doubled state, a new frequency component appears at the half of the cavity repetition rate, while there is no such frequency component in the case of period-one state.

### 6.3.2.5    Period-Doubling of Gain-Guided Solitons with Large Net Normal Dispersion

Although the dynamics of the lasers with large positive cavity dispersion is still determined by the extended GLE, the formed soliton pulses in these lasers have distinct features from those of the solitons formed in lasers with negative cavity dispersion. While the solitons formed in the latter are dominantly a result of the balanced interaction between the cavity negative dispersion and the fiber nonlinear Kerr effect, the solitons formed in lasers with positive cavity dispersion are due to the spectral filtering of the limited gain bandwidth and the cavity nonlinearity. To highlight their differences, solitons formed in fiber lasers with positive cavity dispersion were also called gain-guided solitons (GGSs) [54]. The GGSs belong to the family of dissipative solitons [55]. They are a localized, stable chirped nonlinear wave. It was found that the GGSs could also exhibit deterministic dynamics despite the fact that GGSs have large chirp and broad pulse width.

GGSs can be characterized by their steep spectral edges and pump-power-dependent spectral bandwidth. They could be easily obtained in a mode-locked fiber laser with large positive cavity dispersion. To observe period-doubling bifurcation in GGSs, one starts from a stable GGS operation state. If the saturable absorption strength of the cavity is gradually increased, this could be done through shifting the linear cavity phase delay bias away from the polarization switching point [29] of the laser, the peak power of the mode-locked pulses increase. It is to note that accompanying the increase of the saturable absorption strength, the pump power should also be carefully increased in order to maintain a stable GGS operation. To a certain point of the pulse peak power, it could be observed that some spectral spikes suddenly appear on the long wavelength side of the soliton spectrum as shown in Figure 6.15a. However, the

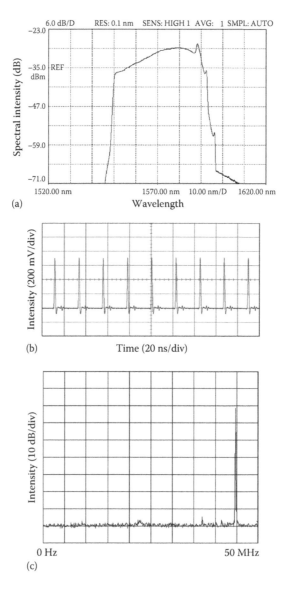

**FIGURE 6.15** Period-one/doubling of GGSs in the laser: (a)/(d) optical spectrum, (b)/(e) oscilloscope trace, and (c)/(f) RF spectrum. (*Continued*)

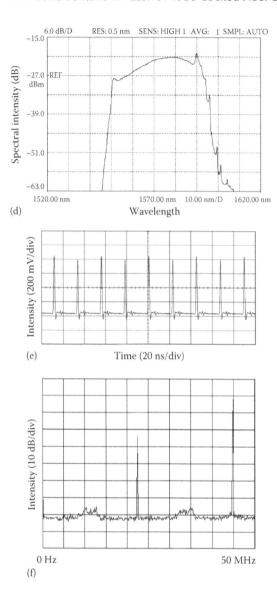

(d)

(e)

(f)

**FIGURE 6.15 (*Continued*)**    Period-one/doubling of GGSs in the laser: (a)/(d) optical spectrum, (b)/(e) oscilloscope trace, and (c)/(f) RF spectrum.

appearance of soliton spikes does not affect stable GGS operation. Figure 6.15b and c show the oscilloscope trace and the RF spectrum measured immediately after the appearance of the spectral spikes. Obviously, the laser still emitted uniform pulses. The laser cavity length was about 4.5 m, which matched the soliton repetition rate of 44.8 MHz shown in Figure 6.15c. According to the autocorrelation measurements, the GGSs of the laser have a pulse duration of about 3.18 ps if a Gaussian pulse profile is assumed. After the spectral spikes are obtained, if the pump power is further increased but with all other laser operation conditions fixed, a period-doubling bifurcation of the GGS is then observed, as shown in Figure 6.15d through f. Associated with the period-doubling bifurcation, more spectral spikes appear on the soliton spectrum (Figure 6.15d). The period-doubling of the soliton is clearly visible on the oscilloscope trace, where after every two cavity roundtrips the pulse energy returned, and on the RF spectrum, where a new spectral component appears at the position of half cavity fundamental repetition frequency. The above process is stable and repeatable in a laser. Nevertheless, in our experiments no further period-doubling bifurcation but a noise-like state was observed when the pump power was further increased.

### 6.3.2.6   Period-Doubling of Vector Solitons in a Fiber Laser

A characteristic of the NLPR mode-locked fiber lasers is that a polarizer is inserted in the laser cavity for the generation of an artificial saturable absorption effect in the cavity [56]. As the intracavity polarizer fixes light polarization at the cavity position, solitons formed in the lasers are considered as scalar solitons.

Due to technical limitation in fabricating perfectly circular cross section core fibers and/or random mechanical stresses, in practice a single-mode fiber always supports two polarization eigenmodes, or, in other words, it possesses weak birefringence. It has been shown that without a polarizer in cavity, solitons formed in a fiber laser could exhibit complicated polarization dynamics. Cundiff et al. have demonstrated the formation of vector solitons in a fiber laser mode locked with a SESAM [57]. A vector soliton differs from a scalar soliton by that it consists of two orthogonal polarization components, and the two polarization components are nonlinearly coupled. Depending on the features of their coupling, there are different types of vector solitons, the polarization-locked vector solitons (PLVSs), the group velocity-locked vector solitons (GVLVSs), and the polarization rotation-locked vector solitons (PRLVSs).

A vector soliton fiber laser can also exhibit deterministic dynamics. Using a cavity configuration similar to the one reported in Ref. [58], the period-doubling bifurcation in a vector soliton fiber laser has been first experimentally observed. Briefly, the vector soliton fiber laser has a ring cavity with a length of about 9.40 m, which consists of 2.63 m Er-doped fiber (StockerYale EDF-1480-T6) and all other fibers used are standard SMF. With the help of a 3-port polarization-independent circulator, a SESAM is introduced in the ring cavity of the laser. A 1480 nm Raman fiber laser with a maximum output of 220 mW is used to pump the laser. Backward pump scheme is adopted to avoid the CW overdriving of the SESAM by the residual pump strength. The laser outputs through a 10% fiber coupler. A fiber-based PC is inserted in the cavity to control the cavity birefringence. The SESAM used has a saturable absorption of 8%, and a recovery time of 2 ps.

As no explicit polarization discrimination components are used in the cavity, and all the fibers used have weak birefringence, vector solitons are easily obtained in the laser by simply increasing the pump power above the mode-locking threshold. Determined by the detailed laser operation conditions, various types of vector solitons, such as the PLVS, incoherently coupled vector soliton, and polarization rotating vector solitons are obtained in the fiber laser. In particular, it is found that the polarization rotation of the formed polarization rotating vector solitons could be locked to the laser cavity round trip times [58].

In the parameter regime of a PRLVS, a state of the period-doubling of the vector solitons, as shown in Figure 6.16a, is also observed. The vector soliton output of the fiber laser is directly monitored by a photodetector with 2 GHz bandwidth. Although the pulse intensity difference between two adjacent cavity round trips is weak, the intensity period-doubling of the soliton pulse train is evident. The RF spectrum of the laser emission is also measured, as shown in Figure 6.16b. A weak but clearly visible frequency component appears at the position of the half of the cavity fundamental repetition frequency. Measured after passing through an external polarizer, the polarization rotation state with polarization rotation locked to twice of the cavity round trip time is the same as shown in Figure 6.16. However, such a state is not a period-doubled one because the observed pulse intensity alternation displayed by the oscilloscope trace is due to the polarization rotation of the vector soliton. The period-doubling shown in Figure 6.16 is formed due to the intrinsic dynamic feature of the laser.

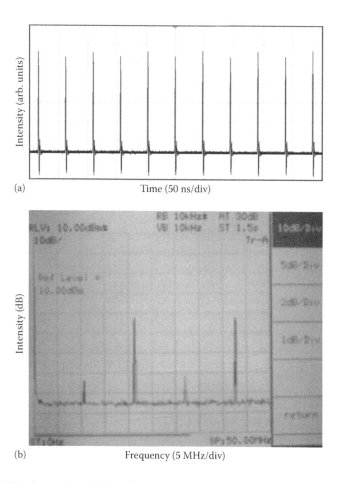

(a)                      Time (50 ns/div)

(b)                      Frequency (5 MHz/div)

**FIGURE 6.16**  Screenshot of (a) oscilloscope trace and (b) corresponding RF spectrum of direct intensity period-doubling.

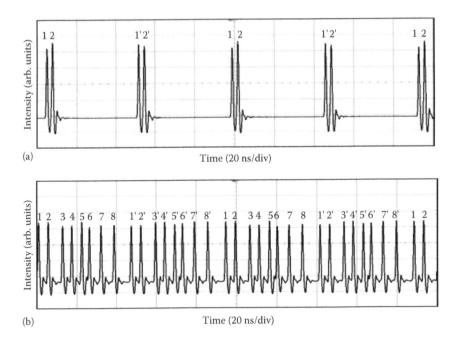

**FIGURE 6.17** Period-doubling of multiple vector solitons: (a) dual vector solitons and (b) octave vector solitons.

After the period-doubling state is achieved, keeping all other laser operation parameters fixed but increasing the pump power increases the number of vector solitons circulating in the cavity. The new vector solitons generated exhibit the same period-doubling feature. Figure 6.17a and b show example cases where two vector solitons and eight vector solitons coexist in the cavity, respectively. Period-doubling of the vector solitons can be clearly identified. Decreasing pump power reduces the number of vector solitons in the cavity. Varying pump power could also change the soliton pulse intensity within a small range. However, no period-one state could be obtained by simply decreasing the pump power. Period-one state could be obtained only if the cavity birefringence is changed. Similarly, after a period-one state is obtained, the period-doubling state could not be obtained by simply increasing the pump strength. Experimentally it is found that in a period-doubled state the polarization rotation of the vector solitons is locked to twice of the cavity round trip time.

## 6.4 SOLITON DETERMINISTIC DYNAMICS IN FIBER LASERS: SIMULATION

### 6.4.1 ROUND TRIP MODEL OF SOLITON FIBER LASERS

The soliton propagation in a fiber laser is characterized by the soliton circulation in the nonlinear ring cavity and the periodical interaction with the cavity components. Therefore, in order to simulate soliton dynamics, in particular to gain an insight into the deterministic dynamics of the various types of solitons in mode-locked fiber lasers, a so-called round-trip model is used. Concretely, we start a simulation with an arbitrarily small light pulse and let it circulate in the cavity. We follow the pulse circulation in the laser cavity. Whenever the pulse encounters a discrete cavity component, for example, the cavity output coupler, the action of the cavity component on the light pulse is considered by multiplying the transfer-matrix of the discrete cavity component to the pulse. The pulse propagation in the fiber segments of the cavity is described by Equation 6.1. After one round of circulation in the cavity, the result of the previous round of calculation is then used as an input of the next round of calculation, until a steady state is reached, which is denoted as a stable soliton state of the laser.

The round-trip model has several advantages. The calculation is made following the pulse propagation in the cavity. Therefore, the detailed pulse evolution within one cavity round trip can be studied. Within each step of calculation, the pulse variation is always small, even if the change of a pulse within one cavity round trip is big. As there is no limitation on the pulse change within one cavity roundtrip, dynamical processes of soliton evolution, such as the process of new soliton generation in the cavity, soliton interaction, and soliton collapse, etc., can be investigated. In addition, the effect of discrete cavity components on the soliton, the influence of the dispersive waves and the different order of the cavity components on the soliton property are automatically included in the calculation.

The soliton operation of a fiber laser is simulated using exactly the laser cavity parameters whenever they are known. Numerically we found that independent of the concrete laser cavity design, as far as the intensity of the formed soliton is strong, soliton quasi-periodicity and period-doubling bifurcations can be obtained. In some laser designs, soliton intermittency could also be numerically observed. Nevertheless, in order to obtain a full period-doubling route to chaos, cavity parameters such as fiber dispersion and fiber lengths must be appropriately selected.

### 6.4.2 Deterministic Dynamics of Solitons in Different Fiber Lasers

Based on the round trip model, the soliton operation of the passively mode-locked fiber ring lasers mode locked with the NLPR technique were numerically simulated. Properties of the soliton pulses formed in the lasers, influence of different laser cavity parameters on the soliton properties, as well as the intrinsic laser cavity effect on the soliton operation were numerically studied. In all of the numerical simulations if it is not explicitly pointed out, the simulation parameters used were as follows:

Fiber nonlinearity—$\gamma = 3$ $W^{-1}$ $m^{-1}$
Fiber dispersion: $-12.8$ $ps^2/km$ for the EDF $-23.0$ $ps^2/km$ for single-mode Fiber, $-2.6$ $ps^2/km$ for the dispersion-shifted fiber
Fiber beat length: $L_b = L/2$
The orientation of the intracavity polarizer to the fiber fast birefringent axis: $\Psi = 0.152\pi$
Gain saturation energy: $E_{sat} = 300$ pJ

Gain bandwidth: $\Omega_g = 16$ nm

Laser cavity length: $L = 6$ m

The orientations of the polarizer and the analyzer with the fast axis of the bire-fringent fiber: $\theta = \pi/8$ and $\varphi = \pi/2 + \pi/8$.

### 6.4.2.1  Period-Doubling Route to Chaos of Single Pulse Solitons

By appropriately choosing the linear cavity phase delay bias, which corresponds in the experiment to appropriately selecting the orientations of the PCs, soliton opera-tion can always be obtained in our simulations. With a fixed linear cavity phase delay bias but different values of gain, as long as the generated peak power of the soliton pulse is weaker than that of the polarization switching power of the cavity [29], stable uniform soliton pulse train can be obtained. Figure 6.18a shows the soliton profiles of the laser under different gain coefficients when the linear cavity phase delay bias is fixed at $\delta\phi = 1.2\pi$. Figure 6.18b is the corresponding soliton spectra. The exact soliton parameters, such as the pulse width and peak power, are determined by the laser parameter settings and the laser operation condition such as the gain value. Under a larger pump power, the solitons generated have higher peak power and nar-rower pulse width. With the current linear cavity phase delay bias selection, NPS threshold of the cavity is low. Therefore, the maximum peak power of the solitons reachable is clamped by the NPS effect. Once the soliton peak is clamped, increasing the gain will further increase the soliton peak power and a new soliton is generated, and consequently a multiple soliton operation state of the laser can be obtained. As a laser operating in the state has weak linear cavity loss, mode-locking can be easily achieved. Therefore, in practice a laser will always start the soliton operation from the state. However, our numerical simulations show that no soliton period-doubling bifurcation could occur at such a linear cavity phase delay bias as the soliton peak power is weak.

When the linear cavity phase delay bias is chosen as $\delta\phi = 1.6\pi$, which corresponds to raising the NPS threshold of the cavity higher so that the soliton pulse formed could have higher peak power, stable soliton operation can still be obtained. Now, as the soliton peak power can reach a very high value under strong pumping, it is observed that when the peak power of the soliton increases to a certain value, period-doubling bifurcations and period-doubling route to chaos of the solitons as observed in the experiments automatically appear. Figure 6.19 shows, for example, a numerically calculated period-doubling route to chaos of a laser. When the gain coefficient was set at $G = 800$, a stable and uniform high-intensity soliton train is obtained (Figure 6.19a). Increasing the value of G and keeping all the other param-eters fixed, the soliton repetition period in the cavity is then doubled at $G = 850$ (Figure 6.19b). At $G = 902$, it doubles again (Figure 6.19c), and further doubles at $G = 908$ (Figure 6.19d). Eventually the soliton repetition in the laser becomes cha-otic (Figure 6.19e). Figure 6.19f through j shows the corresponding optical spec-tra of the solitons in Figure 6.19a through e. Associated with the soliton intensity variation, the soliton spectrum also exhibits period-doubling changes. Figure 6.20 further shows the soliton spectral variation within one period of the period-four state. In this state, after every four cavity round trips the soliton intensity and pro-file return to the original value and shape. From Figure 6.20, we can see that the

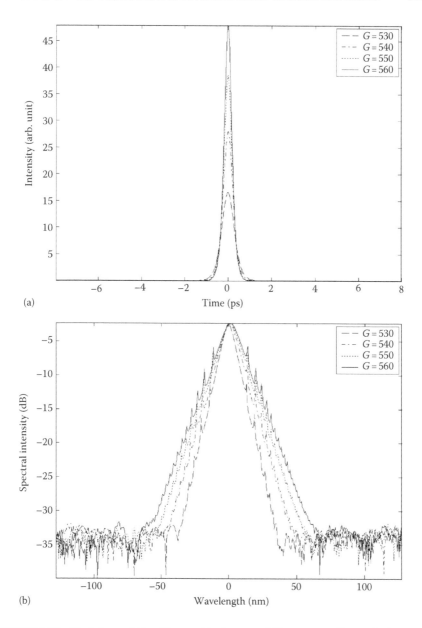

**FIGURE 6.18** Variations of (a) soliton profile shapes and (b) corresponding optical spectra as a function of gain factor under a linear cavity phase delay bias of $\delta\varphi = 1.2\pi$.

soliton spectrum in each round trip is different. Corresponding to the soliton of the highest peak power, the soliton spectrum (Figure 6.20a) also exhibits the strongest sidebands and spectral modulations, indicating the existence of strong nonlinear SPM on the pulse. The change from one soliton operation state to the other is abrupt. At the bifurcation point, when the gain coefficient is slightly increased, the soliton

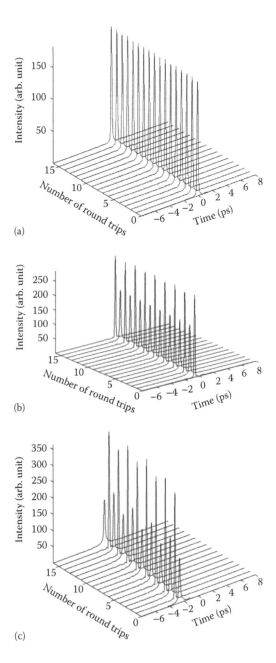

**FIGURE 6.19** Soliton pulse evolution and the corresponding optical spectra numerically calculated under different pump strengths. The linear cavity phase delay bias is set as $\delta\phi = 1.6\pi$: (a)/(f) period-one soliton state, $G=800$; (b)/(g) period-two soliton state, $G=850$; (c)/(h) period-four soliton state, $G=902$; (d)/(i) period-eight soliton state, $G=908$; and (e)/(j) chaotic soliton state, $G=915$.                                   *(Continued)*

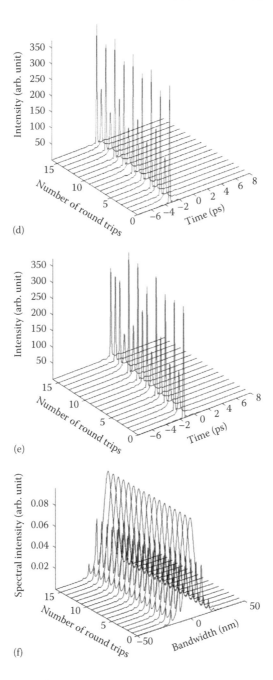

(d)

(e)

(f)

**FIGURE 6.19 (Continued)**   Soliton pulse evolution and the corresponding optical spectra numerically calculated under different pump strengths. The linear cavity phase delay bias is set as $\delta\phi=1.6\pi$: (a)/(f) period-one soliton state, $G=800$; (b)/(g) period-two soliton state, $G=850$; (c)/(h) period-four soliton state, $G=902$; (d)/(i) period-eight soliton state, $G=908$; and (e)/(j) chaotic soliton state, $G=915$.    (*Continued*)

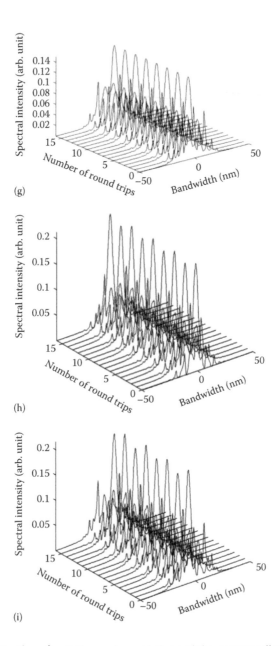

**FIGURE 6.19 (Continued)**   Soliton pulse evolution and the corresponding optical spectra numerically calculated under different pump strengths. The linear cavity phase delay bias is set as $\delta\phi = 1.6\pi$: (a)/(f) period-one soliton state, $G=800$; (b)/(g) period-two soliton state, $G=850$; (c)/(h) period-four soliton state, $G=902$; (d)/(i) period-eight soliton state, $G=908$; and (e)/(j) chaotic soliton state, $G=915$.                                    (Continued)

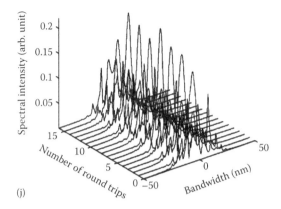

(j)

**FIGURE 6.19 (*Continued*)**   Soliton pulse evolution and the corresponding optical spectra numerically calculated under different pump strengths. The linear cavity phase delay bias is set as $\delta\phi = 1.6\pi$: (a)/(f) period-one soliton state, $G = 800$; (b)/(g) period-two soliton state, $G = 850$; (c)/(h) period-four soliton state, $G = 902$; (d)/(i) period-eight soliton state, $G = 908$; and (e)/(j) chaotic soliton state, $G = 915$.

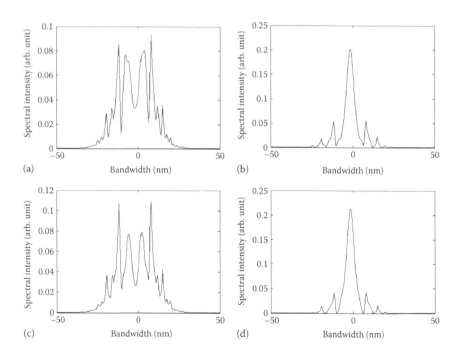

(a)

(b)

(c)

(d)

**FIGURE 6.20**   Soliton spectral variation within one period-four state (soliton evolves from (a) to (d)).

quickly jumps to another state with doubled periodicity, exhibiting the universal characteristic of the period-doubling bifurcation and route to chaos of the nonlinear dynamics systems.

The appearance of the period-doubling bifurcation is independent of the concrete laser cavity design. Under different cavity parameter settings, as far as the soliton peak power is unlimited by the NPS effect of the cavity, we could always obtain the phenomenon in our simulations. Even the period-doubling route to chaos of a period-three state, which in terms of the nonlinear dynamics theory is known as a periodic window within the chaotic regime, has also been numerically revealed. However, it is to point out that due to the coexistence of other effects in the laser, for example, the soliton peak nonuniformity [35] and soliton collapse [59], in order to obtain a full period-doubling route to chaos in a laser, the cavity parameters must be appropriately selected. In some of our numerical simulations frequently only certain period-doubling bifurcations, for example, the period-one to period-two bifurcation and then to chaos, or the period-one to chaotic state could be obtained. If the linear cavity phase delay bias is not set large enough, the NPS effect could also limit the peak power of the solitons. Consequently, only bifurcations to a certain periodic state, for example, the period-four state could be reached. Further increase of the gain would cause the generation of a new soliton rather than further bifurcation to chaos, which clearly shows the direct relation of the period-doubling bifurcation to the soliton peak power.

Previous studies on the synchronously pumped passive ring cavities have also revealed period-doubling cascade to chaos in the sequence of pulses emerging from the cavities [60,61]. It was shown that the bifurcations and route to chaos of the system was caused by repetitive interference between the input pulse and the pulse that has completed a round trip in the cavity. As the pulse traveling in the cavity suffers nonlinear phase shift, which itself is pulse intensity dependent, the transmission of the cavity is a nonlinear function of the pulse intensity. It is an intrinsic property of such a nonlinear cavity that under larger nonlinear phase shift of the pulse, its output exhibits the period-doubling route to chaos [2]. We note that a similar repetitive interference process exists in the fiber soliton lasers. Due to the birefringence of the laser cavity, the pulse propagation in the laser actually comprises two orthogonal polarization components. Although there are nonlinear couplings between the two orthogonally polarized pulses as can be seen from Equation 6.1, after one round trip they experience different linear and nonlinear phase shifts. The interference between them at the intracavity polarizer results in a state where the effective cavity transmission is a nonlinear function of the soliton intensity. Based on the studies on the synchronously pumped passive ring cavities, it is safe to assume that under strong soliton peak intensity, period-doubling bifurcation and route to chaos could also appear in the lasers. Indeed, our numerical simulation shows that the effect could only occur in the fiber lasers when the linear cavity phase delay bias is set away from the NPS point. In this case, the soliton peak power is unclamped by the cavity and can increase to a very high value with the increase of the pump strength. A high soliton peak power generates large nonlinear phase shift difference between the two polarization components, which causes an intrinsic instability of the system.

To obtain a full route to chaos, the cavity parameters such as the fiber dispersion and fiber lengths must be appropriately selected. In one of our numerical simulations even the period-doubling route to chaos of a period-three state, which in the nonlinear dynamics theory is known as a periodic window within the chaotic regime, has also been obtained as shown in Figure 6.21.

### 6.4.2.2   Period-Doubling Route to Chaos of Bound Solitons

Period-doubling bifurcations and route to chaos have also been numerically revealed for the bound solitons. Figure 6.22 shows for example one of such results obtained. In calculating the state we used the same laser parameters as those for obtaining the single-pulse soliton period-doubling route to chaos, only the pump strength and initial state are different. Figure 6.22a shows that two solitons coexist in the cavity and bind together. Note that due to the close separation between the solitons, their optical spectra have strong intensity modulations, but the modulation patterns do not change during each round trip, which indicates that the phase difference between the solitons is fixed as well. The binding nature of the solitons is represented by the fixed soliton separation and phase difference even under the existence of soliton interaction between them. Increasing pump strength from the state of Figure 6.22b, both solitons experience simultaneously period-doubling bifurcations and route to chaos. Associated with the soliton period-doubling the soliton separation between the bound solitons also changes slightly. However, after the period-doubling bifurcation, the soliton separation then remains constant again. This soliton separation change suggests that the dynamic bifurcation of the system could affect soliton interaction.

### 6.4.2.3   Period-Doubling of Multiple Solitons

As have been shown previously, multiple soliton formation in the fiber lasers is a result of the cavity pulse peak clamping effect [29]. Therefore, by appropriately setting the linear cavity phase delay bias and the pumping strength, randomly distributed multiple solitons can be easily obtained numerically. The maximum achievable soliton pulse peak power is also determined by the setting of the linear cavity phase delay bias. If the linear cavity phase delay bias is set too close to the cavity polarization switching point, only low peak power of the solitons could be obtained. In this case, increasing the pumping strength will only increase the number of the soliton in the cavity and no period-doubling bifurcation of the solitons is observed. Therefore, to obtain soliton period-doubling bifurcation, it is necessary to set the linear cavity phase delay bias away from the polarization switching point. Numerically we find that with current simulation parameters only when the linear cavity phase delay bias is larger than $1.4\pi$, the phenomenon of soliton period-doubling could be achieved. In the simulation, $\delta\varphi = 1.5\pi$ is set. Depending on the initial state and pump strength, either synchronous or asynchronous period-doubled evolutions are obtained. Figure 6.23a and b shows the typical numerical results, where two solitons have a separation of 64 ps in the cavity. Figure 6.23a shows that both solitons are period-doubled and have exactly the same intensity evolution with the cavity round trips, while Figure 6.23b shows a case where the solitons have unsynchronized pulse intensity evolution with the cavity round trips. Numerically, the states of period-doubling of multiple solitons are obtained and experimentally observed with more than two solitons.

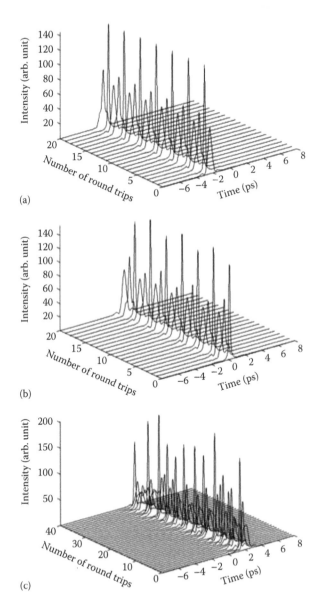

(a)

(b)

(c)

**FIGURE 6.21**  Soliton profiles and corresponding optical spectra numerically calculated: (a)/(d) State of period-three, $G=730$; (b)/(e) State of period-six, $G=735$; and (c)/(f) Chaotic state, $G=750$.                                                                                    (*Continued*)

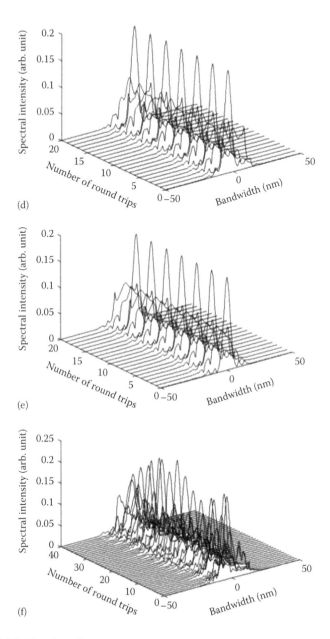

**FIGURE 6.21 (Continued)** Soliton profiles and corresponding optical spectra numerically calculated: (a)/(d) State of period-three, $G=730$; (b)/(e) State of period-six, $G=735$; and (c)/(f) Chaotic state, $G=750$.

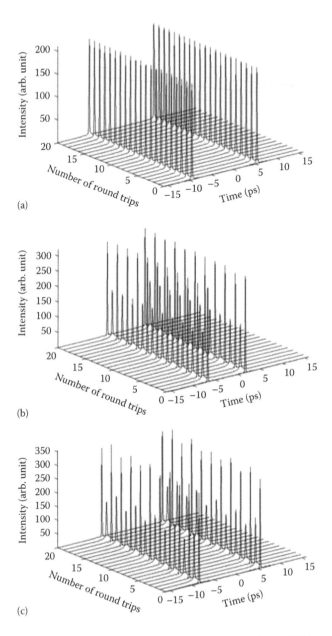

**FIGURE 6.22** Period-doubling route to chaos of the bound solitons: (a)/(e) state of stable bound solitons, $G = 1149$; (b)/(f) state of period-two of the bound solitons, $G = 1300$; (c)/(g) state of period-four of the bound solitons, $G = 1353$; (d)/(h) Chaotic state of the bound solitons, $G = 1358$.                                                                          (*Continued*)

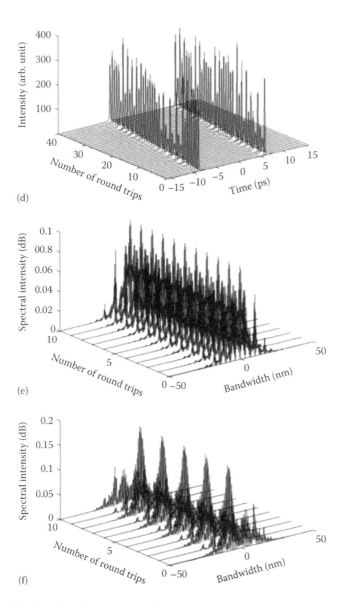

(d)

(e)

(f)

**FIGURE 6.22** (*Continued*)   Period-doubling route to chaos of the bound solitons: (a)/(e) state of stable bound solitons, $G = 1149$; (b)/(f) state of period-two of the bound solitons, $G = 1300$; (c)/(g) state of period-four of the bound solitons, $G = 1353$; (d)/(h) Chaotic state of the bound solitons, $G = 1358$.      (*Continued*)

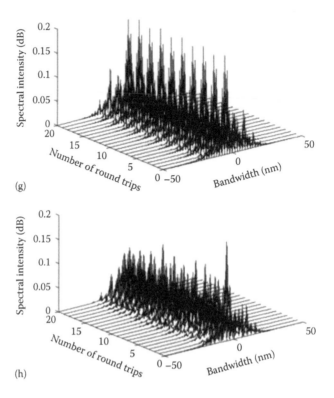

**FIGURE 6.22 (*Continued*)** Period-doubling route to chaos of the bound solitons: (a)/(e) state of stable bound solitons, $G = 1149$; (b)/(f) state of period-two of the bound solitons, $G = 1300$; (c)/(g) state of period-four of the bound solitons, $G = 1353$; (d)/(h) Chaotic state of the bound solitons, $G = 1358$.

### 6.4.2.4 Period-Doubling of Dispersion-Managed Solitons

The cavity of a mode-locked fiber laser is a nonlinear cavity. Therefore, it is expected that when the intensity of the pulse circulating in the fiber ring laser has strong peak power, the laser output could exhibit similar nonlinear dynamical behaviors. Furthermore, as the occurrence of the period-doubling is a property of the nonlinear laser cavity, no matter whether the pulse circulating in it is a conventional soliton or a dispersion-managed soliton, the same cavity dynamics should be observed.

Figure 6.24 shows an example of a numerically calculated dispersion-managed soliton for the positive near-zero cavity dispersion case. The optical spectrum of the dispersion-managed soliton shown in Figure 6.24a has a Gaussian-like spectral profile and no obvious sidebands, which agrees with the experimental observation. Figure 6.24b shows the corresponding temporal pulse profile. Again it is closer to the Gaussian shape than the sech²-form. Our numerical results clearly show that the dispersion-managed soliton could be formed in lasers with positive near-zero cavity dispersion just as experimentally observed.

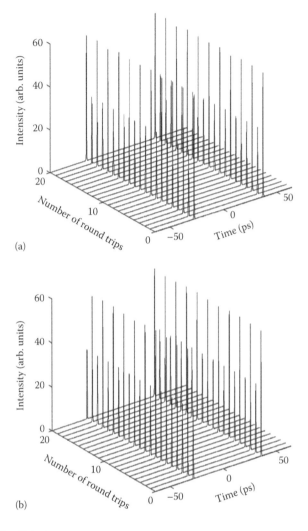

(a)

(b)

**FIGURE 6.23**    Calculated pulse intensity evolution with the cavity round trips of the period-doubled solitons: (a) synchronous evolution and (b) asynchronous evolution.

After a Gaussian-like stable uniform pulse train is obtained, with all the other simulation parameters fixed but the small signal gain coefficient increased, a period-doubled state as shown in Figure 6.24c is observed. This shows that, like the conventional solitons, the dispersion-managed solitons can experience period-doubling bifurcation as well (see also Appendix A). With our current laser cavity design, only a period-two state could be obtained. If the pump strength is further increased, then a new dispersion-managed soliton is generated instead of further period-doubling of the pulse [62]. Similar results were also obtained when the net cavity dispersion is set at negative near-zero or zero.

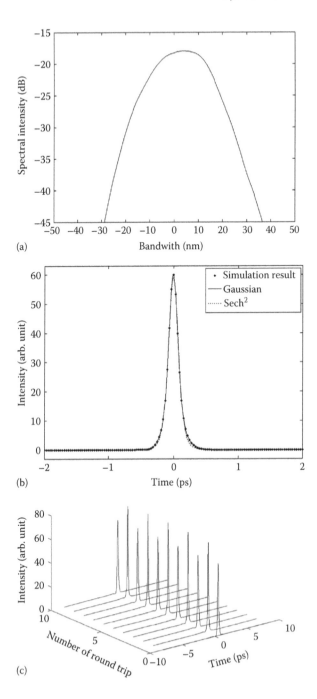

(a)

(b)

(c)

**FIGURE 6.24** Numerical simulations of the dispersion-managed soliton in the positive near zero net cavity GVD regime: (a) optical spectrum, (b) temporal profile of a dispersion-managed soliton—small signal gain coefficient $g_0 = 520$, and (c) a period-doubled state—small signal gain coefficient $g_0 = 550$.

### 6.4.2.5  Period-Doubling of Gain-Guided Solitons

The period-doubling bifurcation of the GGSs could also be numerically simulated. Figure 6.25 shows, for example, a numerically calculated period-doubling of a GGS with a pump strength of $G=4300$. Figure 6.25a shows the evolution of the calculated GGS with the cavity round trips. Period-doubling of the pulse is evidenced by that the pulse returns to its previous parameters at every two cavity round trips. Figure 6.25b shows the optical spectra of the soliton in two adjacent round trips and the averaged one. Clear differences between them are visible. However, no obvious spectral spikes were obtained. We believe the absence of the spectral spikes could be caused by the parabolic gain profile approximation used in our simulations. From the experimental results, the spikes only appeared on the edges of the spectrum, but where the parabolic gain profile artificially introduced large losses. Nevertheless, the numerical simulations have reasonably reproduced the essential features of the soliton period-doubling bifurcation, that

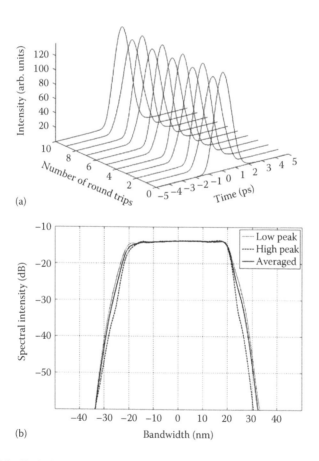

(a)

(b)

**FIGURE 6.25**  Period-doubling of the GGSs numerically calculated: (a) soliton evolution with the cavity roundtrips and (b) optical spectra of the soliton in two adjacent round trips and their average.

is, the simulated spectra have a flat and smooth top, and the spectral variations between the period-one and period-doubled states only occur on the edges of the spectrum, which are in agreement with the experimental observations and different from those of the soliton period-doubling observed in fiber lasers of negative cavity dispersion [13].

Figure 6.26 shows a comparison between the evolutions of the GGSs in cavity before and after the period-doubling bifurcation numerically calculated. Figure 6.26a and b shows the pulse evolution in the time domain along the cavity; Figure 6.26c and d shows the corresponding pulse width variation along the cavity; Figure 6.26e and f shows the evolution of the pulse spectrum corresponding to Figure 6.26a and b, respectively. We note that Figure 6.26 has shown pulse evolution in two cavity round trips in order to display the period-doubling effect. With all other parameters fixed, the period-one state could be obtained in a large pump strength range. In a period-one state, the chirped GGS is compressed in the SMF segments. The stronger the pump strength, the larger is the chirp accumulated in the EDF, and consequently narrower pulse and higher pulse peak power are obtained in the SMF. After this, the pulse peak power goes beyond a certain value and period-doubling of the pulse occurs. Numerically we found that in a period-doubled state, the GGS may be de-chirped to a transform-limited pulse in the SMF, and retain the shortest pulse width until entering the EDF, as shown in Figure 6.26d. However, a pedestal also associates with the compressed pulse, which could be understood as a result of the pulse breaking.

### 6.4.2.6  Period-Doubling of Vector Solitons

Due to the existence of a polarizer in the cavity, which fixes the polarization of light in the position of the cavity, solitons formed in fiber lasers mode locked with the NLPR technique are considered as scalar solitons. In fiber lasers, if there are no polarization-dependent components in the cavity, vector solitons could be formed. The formed vector solitons have two coupled orthogonal polarization components. Numerical simulations have shown that the vector solitons can also exhibit period-doubling bifurcations as observed in the experiments.

Similarly, in the experiments where a polarization insensitive SESAM was used instead of the NLPR mode-locking in order to obtain vector soliton operation of a fiber laser, the effect of a real saturable absorber is considered to numerically simulate the features of vector solitons formed in a fiber laser. Figure 6.27 shows the results of simulations obtained under different net cavity birefringence, the other laser parameters being the same. Numerically, it is found that when the cavity birefringence is selected as $L_b = L/0.4$, where $L$ is the cavity length, a period-one vector soliton state is obtained. When the cavity birefringence is selected as $L_b = L/0.09$, a period-doubled vector soliton is then obtained. Figure 6.27a and b show the calculated vector soliton evolution with cavity round trips. The soliton period-doubling of the state shown in Figure 6.27a in comparison with that shown in Figure 6.27b is evident. Figure 6.27c/e and d/f show further the evolution of the vector soliton components, of the period-doubled vector soliton and the period-one vector soliton, respectively. Each component of the period-doubled vector soliton also exhibits period-doubling. However, the pulse intensity variation between the

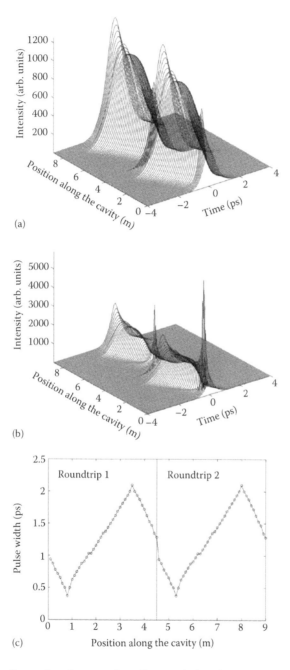

**FIGURE 6.26** Comparison between the soliton evolutions in cavity before and after the period-doubling bifurcation: (a)/(b) pulse evolution in time domain, (c)/(d) pulse width evolution, and (e)/(f) optical spectrum evolution in double cavity-length.     (*Continued*)

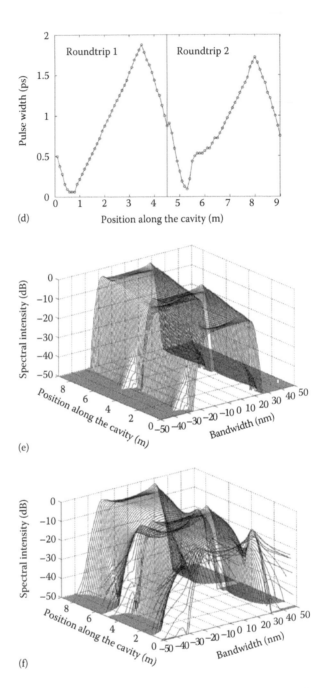

(d)

(e)

(f)

**FIGURE 6.26 (*Continued*)** Comparison between the soliton evolutions in cavity before and after the period-doubling bifurcation: (a)/(b) pulse evolution in time domain, (c)/(d) pulse width evolution, and (e)/(f) optical spectrum evolution in double cavity-length.

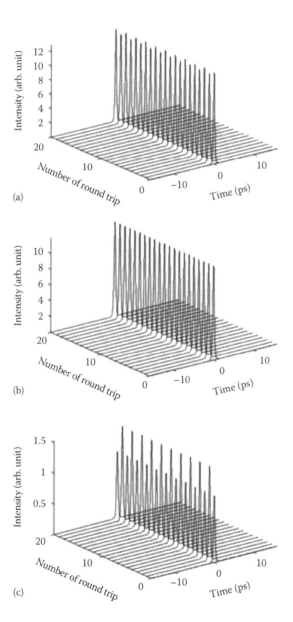

**FIGURE 6.27**  Numerically simulated results of period-doubling/period one of (a)/(b) the vector soliton, (c)/(d) the vector soliton along the horizontal birefringent axis, and (e)/(f) the vector soliton along the vertical birefringent axis with the cavity roundtrips.    (*Continued*)

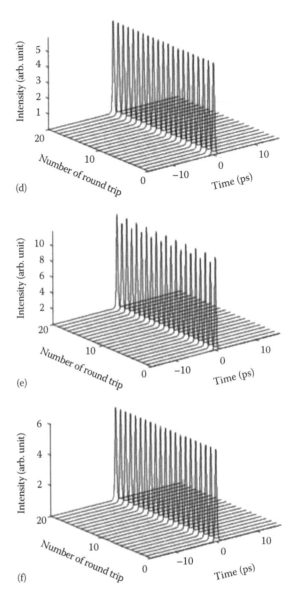

(d)

(e)

(f)

**FIGURE 6.27 (Continued)** Numerically simulated results of period-doubling/period one of (a)/(b) the vector soliton, (c)/(d) the vector soliton along the horizontal birefringent axis, and (e)/(f) the vector soliton along the vertical birefringent axis with the cavity roundtrips.

two orthogonal polarization components is antiphase. It is because of this antiphase pulse intensity evolution of the period-doubled vector soliton that only a weak period-doubling could be observed on the vector soliton intensity evolution, as shown in Figure 6.27a. The result agrees well with our experimental observations in Figure 6.16a.

## 6.5   CAVITY-INDUCED SOLITON MODULATION INSTABILITY EFFECT

Modulation instability (MI) is also a well-known phenomenon that destabilizes strong CW propagation in dispersive media (see also Appendix E). In optics, Tai et al. first reported the experimental observation of MI in light propagation in single mode fibers [63]. They found that MI is physically a special case of four-wave-mixing where the phase matching condition is self-generated by the nonlinear refractive index change and the anomalous dispersion of the fibers. Soliton propagation in single mode fibers is intrinsically stable against MI. However, it was theoretically shown that if dispersion of a fiber is periodically varied or the intensity of a solitary wave is periodically modulated, MI of solitons could still occur [62,63]. MI establishes resonant energy exchange between the soliton and dispersive waves, which results in a new type of spectral sideband generation on the soliton spectrum. Soliton propagation in a laser cavity experiences periodically fiber dispersion and/ or soliton energy variations. It is to expect that under certain conditions, such a soliton MI could appear. MI is an intrinsic feature of the solitons formed in a laser. The development of MI leads to strong resonant wave coupling between the soliton and dispersive waves, which further leads to the dynamic instability of the soliton. We believe that it is the dynamic instability of a soliton that ultimately limits the performance of a soliton laser. MI of a soliton in lasers is characterized by the formation of a new set of spectral sidebands on the soliton spectrum. Figure 6.28 shows a case of the soliton spectra numerically calculated for a period-doubling soliton state. In the soliton period-one state, the soliton spectrum has only the Kelly sidebands. The Kelly sidebands are generated by the constructive interference between the soliton and the dispersion waves in the cavity, whose positions are almost fixed

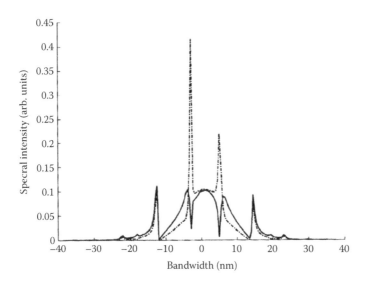

**FIGURE 6.28**   Soliton spectra of adjacent round trips in the period-doubling state.

for a fiber laser once its cavity parameters are fixed. We note that the formation of Kelly sidebands is a linear effect. However, after the soliton experienced a period-doubling bifurcation, it is observed that apart from the Kelly sidebands a new set of spectral sidebands also appears on the soliton spectrum. Positions of the new spectral sidebands vary with the laser cavity detuning. At certain cavity detuning settings, or when the new sidebands appear at certain positions on the soliton spectrum, the soliton period-doubling bifurcation occurs, while at other positions the soliton quasi-periodicity effect is observed. Carefully checking the soliton spectral evolution within one cavity round trip, it is further identified that in a soliton period-two state, while in one cavity round trip the new sideband is a spectral spike, in the subsequent cavity round trip, the new spectrum will become a spectral dip, and vice versa, which indicates that the new spectral sidebands are formed due to a parametric wave interaction process.

Figure 6.29 shows a comparison between the soliton spectra of the adjacent cavity round trips of a period-three state. Due to that the soliton period in the state is three cavity lengths the variation of the extra sidebands between two adjacent round trips is not exactly opposite, but $2\pi/3$ out of phase.

Based on the new spectral sideband formation and their relation to the observed soliton dynamics, we speculate that the appearance of the deterministic soliton dynamics is related to the cavity-induced soliton modulation instability. Independent of the cavity dispersion, soliton propagation in the cavity experiences a periodic intensity variation, which causes a phase matching of the soliton with certain particular frequencies of weak light or dispersive waves. Energy exchange between the soliton pulse and the phase-matched dispersive waves could be built up. When the phase difference between the soliton and the dispersive waves within one cavity

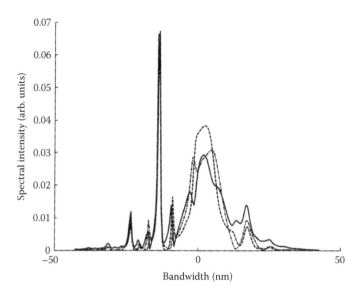

**FIGURE 6.29**   Soliton spectra of adjacent round trips in the period-tripling state.

round trip is $2\pi$, then a soliton period-two state is formed. As energy exchange between the soliton and weak light depends on their phase difference, within half of the period energy flows from soliton to the dispersive waves, while in the other half period it flows in reverse, resulting in a soliton period-two state. In the case that the phase difference between the soliton and dispersive waves is not a multiple of $2\pi$ but an arbitrary value, the quasi-periodicity state occurs. This could be why the quasi-periodic soliton operation state could be experimentally easily observed, while period-doubling and period-doubling route to chaos can only be observed under certain special laser conditions.

## 6.6 MULTISOLITON FORMATION AND SOLITON ENERGY QUANTIZATION IN PASSIVELY MODE-LOCKED FIBER LASERS

This section describes the numerical simulations of the generation of multiple soliton and soliton energy quantization in a soliton passively mode-locked fiber ring laser by using the NLPR technique. It is found by numerical method that the formation of multi-bound solitons in the laser is caused by a peak power limiting effect of the laser cavity. It is also the same effect that suppresses the soliton pulse collapse, an intrinsic feature of solitons propagating in the gain media, and makes the solitons stable in the laser. Furthermore, we show that the soliton energy quantization observed in the lasers is a natural consequence of the gain competition between multiple solitons. Enlightened by the numerical result, we speculate that the multisoliton formation and soliton energy quantization observed in other types of soliton fiber lasers could have similar mechanisms.

### 6.6.1 INTRODUCTORY REMARKS

Passively mode-locked fiber lasers as a simple and economic ultrashort pulse source have been extensively investigated over the past decade [7,10,39,64–69]. By implementing the soliton pulse shaping technique in the lasers it was demonstrated that optical pulses in the subpicosecond range could be routinely generated. Various passive mode-locking techniques, such as the nonlinear loop mirror method [3,4], the technique [5–7] and the semiconductor saturable absorber method [8,9], have been used to mode-lock the lasers. Independent of the concrete mode-locking techniques it was found that the soliton operation of all the lasers exhibited a common feature, namely, under strong pumping strength multiple soliton pulses are always generated in the laser cavity, and in the steady state all the solitons have exactly the same pulse properties: the same pulse energy and pulse width when they are far apart. The latter property of the solitons is also called the "soliton energy quantization effect" [38]. The multiple soliton generation and the soliton energy quantization effect limit the generation of optical pulses with larger pulse energy and narrower pulse width in the lasers. Therefore, to further improve the performance of the lasers it is essential to have a clear understanding of the physical mechanism responsible for these effects. It was conjectured that the soliton energy quantization could be an intrinsic property of the laser solitons, as solitons formed in a laser are intrinsically dissipative solitons, where the requirement of soliton internal energy balance ultimately determines the energy of a soliton [70].

However, this argument cannot explain the formation of multiple solitons in a laser cavity. In fact, multiple pulse generation has also been observed in other types of soliton lasers, for example, M. J. Lederer et al. reported the multipulse operation of a Ti–sapphire mode-locked fiber laser by an ion-implanted SESAM [71] and C. Spielmann et al. reported the breakup of single pulses into multiple pulses in a Kerr lens mode-locked Ti–sapphire laser [72]. Theoretically, Kärtner et al. have proposed a mechanism of pulse splitting for the multiple pulse generation in soliton lasers [73]. It was shown that when a pulse in a laser becomes so narrow that due to the effective gain bandwidth limit, the gain could no longer amplify the pulse but impose an extra loss on it, the pulse would split into two pulses with broader pulse width. Based on a similar mechanism and in the framework of a generalized complex GLE that explicitly takes into account the effect of a bandpass filter in the cavity, M.J. Lederer theoretically explained the multiple pulse operation of their laser [71]. However, we point out that this process of multipulse generation can be easily identified experimentally. In the case that no bandpass filter is in the cavity, a pulse splits into two pulses only when its pulse width has become so narrow that it is limited by the gain bandwidth. In the case of fiber lasers, no significant soliton pulse narrowing was ever observed before a new soliton pulse was generated, which obviously demonstrated that the multiple pulse generation in the soliton fiber lasers must have a different mechanism. G. P. Agrawal has also numerically shown multiple pulse formation when a pulse propagates in a strongly pumped gain medium [74]. Nevertheless, it can be shown that the multiple pulse formation has in fact the same mechanism as that described in [75]. Recently, P. Grelu et al. have numerically simulated the multiple pulse operation of the fiber soliton lasers [41,75]. By using a propagation model and also taking into account the laser cavity effect, they could quite well reproduce the multipulse states of the experimental observations. However, no analysis on the physical mechanism of the multi-pulse formation was given. In addition, in their simulations, the multi-bound soliton formation is only obtained for limited sets of parameters, which is not in agreement with the experimental observations. Recently, A. Komarov et al. have theoretically studied the multiple soliton operation and pump hysteresis of the soliton fiber lasers mode-locked by using the NLPR technique [76]. In their model, they have explicitly taken into account the nonlinear cavity effect so they can successfully explain the multisoliton formation and pump hysteresis based on the nonlinear cavity feedback. However, as they ignored the linear birefringence of the fiber and the associated linear cavity effects, their model could still not accurately describe the real laser systems, for example, in their model in order to obtain the multiple pulse operation, they have to add phenomenologically a frequency selective loss term. Physically, adding the term is like adding a bandpass filter in the laser cavity. In this section, we describe the numerical results of simulations on the soliton formation and soliton energy quantization in a fiber ring laser passively mode-locked by using the NLPR technique.

First we show that soliton formation is actually a natural consequence of a mode-locked pulse under strong pumping if a laser is operating in the anomalous total cavity dispersion regime. Especially we will show how the parameters of a laser soliton, such as the peak power and pulse width, vary with the laser operation conditions. Based on our numerical simulations, we further show that for the first time to our knowledge, the multiple soliton formation in the laser is caused by a peak power limiting effect of the laser cavity. It is also the effect of the cavity that suppresses the soliton collapse and makes the solitons stable in the laser even when the laser gain is very strong. Furthermore, we demonstrate numerically that the soliton energy quantization of the laser is a nature consequence of the gain competition between the solitons in the cavity.

### 6.6.2 EXPERIMENTAL OBSERVATIONS OF MULTI-BOUND SOLITONS

For the purpose of comparison and a better understanding of our numerical simulations, we present here again some of the typical experimental results on the multiple soliton operation and soliton energy quantization of the soliton fiber lasers. We note that although the results presented here were obtained from a particular soliton fiber ring laser as described in the following text, similar features were also observed in other lasers [39,65,66,69], which are in fact independent of the concrete laser systems. A schematic of the fiber soliton laser employed in the experiments is shown in Figure 6.30. It consists of (a) 1 m long dispersion-shifted fiber with a GVD of about −2 ps/nm km; (b) 4 m long EDF with a GVD of about −10 ps/nm km and 1 m long standard single-mode fiber (SSMF) with a GVD of about −18 ps/nm km; (c) two PCs, one consisting of two quarter-wave plates and the other one two quarter-wave plates and one half-wave plate, to control the polarization of the light in the cavity; (d) a polarization-dependent isolator was used to enforce the unidirectional operation of the laser and also determine the polarization of the light at the position; (e) a 10% output coupler was used to outlet the light. The soliton pulse width of the laser was measured with a commercial autocorrelator, and the average soliton output power

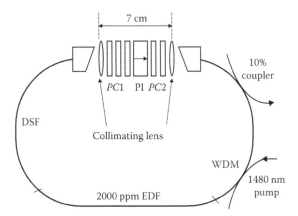

**FIGURE 6.30** Schematic of the soliton fiber laser. PI, polarization-dependent isolator. PC, polarization controller. DSF, dispersion shifted fiber. EDF, erbium-doped fiber. WDM, wavelength division multiplexer.

was measured with a power meter. The soliton pulse evolution inside the laser cavity is monitored with a high-speed detector and a sampling oscilloscope.

The soliton operation of the laser was extensively investigated previously [77–79] and various features such as pump power hysteresis, multiple soliton generation and various modes of multiple soliton operation, and bound states of solitons were observed. Worth mentioning here again is the pump hysteresis effect of the soliton operation. It was found experimentally that the laser always started mode-locking at a high pump power level, and immediately after the mode-locking, multiple solitons were formed in the cavity. After the soliton operation was obtained, the laser pump power could then be reduced to a very low level while the laser still maintained the soliton operation. This phenomenon of the laser soliton operation was known as the pump power hysteresis [80]. It was later found that the pump power hysteresis effect is related to the multiple soliton operation of the laser. Once multiple solitons are generated in the cavity, decreasing the pump power reduces the number of solitons. However, as long as one soliton remains in the cavity, the soliton operation state (and therefore, the mode-locking of the laser) is maintained. Not only the soliton operation of the laser exhibited pump power hysteresis, but also the generation and annihilation of each individual soliton in the laser exhibited pump power hysteresis [77]. Experimentally it was observed that if solitons existed in the cavity, then carefully increasing the pump power would generate new solitons in the cavity. As in this case the laser is already mode-locked, the generation of a new soliton only requires a small increase of the pump power.

An important characteristic of the multiple soliton operation of the laser is that as far as the solitons are far apart in the cavity, they all have exactly the same soliton parameters: the same pulse width, pulse energy, and peak power. To demonstrate the property, Figure 6.31 shows the oscilloscope trace of a typical experimentally measured multiple soliton operation state of the laser. The cavity round trip time of the laser is about 26 ns. There are six amplitude-equalized solitons coexisting in the cavity as observed on the oscilloscope trace. Although under the electronic detection

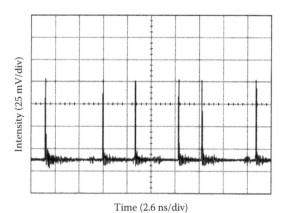

Time (2.6 ns/div)

**FIGURE 6.31** A typical "experimentally" observed oscilloscope trace of the multiple soliton operation of the laser.

system, the detailed pulse profile of the solitons cannot be resolved, nevertheless, the measured pulse height in the oscilloscope trace is directly proportional to the energy of each individual solitons. Based on the measured autocorrelation traces, and optical spectra, it was further identified that all the solitons indeed have the same parameters.

### 6.6.3  THEORETICAL MODELING

To find out the physical mechanism of the multiple soliton formation in the laser that has been confirmed and compared with those calculated from the conventional models of the fiber soliton lasers. Traditionally, the soliton operation of a laser was modeled by the GLE [81] or the Master equation [82], which also takes into account the gain, loss, and saturable absorber effects of a laser. However, a drawback of the model is that the laser cavity effect on the soliton is either ignored or not appropriately considered. Based on results of our experimental studies, we find that the cavity properties affect significantly the features of the solitons, as in the case of soliton lasers the solitons circulate inside a cavity. Therefore, we believe that in order to accurately model the soliton operation of a laser, the detailed cavity property must also be included in the model.

To this end, the conventional GLE model has been exploited by incorporating the cavity features. In previous papers, soliton sideband asymmetry [83] and subsideband generation has been simulated and experimentally observed [84]. Twin-pulse soliton [44] and soliton pulse train nonuniformity [85] have also been confirmed.

The basic idea of the model is fundamentally different to the conventional models. We do not make the assumption of small pulse variation approximation. Instead, we follow the dynamic circulation of the optical pulses in the laser cavity and take into considerations the influence and active action of all cavity components on the evolved pulses. Concretely, we describe the light propagation in the optical fibers by the NLSE, or the coupled NLSE (CNLSE) if the fiber is weakly birefringent. For the EDF, the gain effect is incorporated such as the light amplification and gain bandwidth limitation in the coupled Equation 6.1. Whenever the pulse encounters a discrete cavity component, for example, the output coupler, polarizer, we then account the effect of the cavity component by multiplying its transfer matrix to the light field. As the model itself is very complicated, we have to numerically solve it and find out the eigenstate of the laser under a certain operating condition. In our numerical simulations we always start the calculation with an arbitrary light field. After one round trip circulation in the cavity, we use the calculated result as the input of the next round of calculation until a steady state is obtained. The simulations always reach stable solutions, which correspond to a stable laser state under a certain operating condition.

We thus can present here the detailed procedure in simulating the soliton operation of the laser as follows. To describe light propagation in weakly birefringent fibers, we used the coupled complex NLSEs of the form (see also Equation 6.1 and Chapter 2):

$$\frac{\partial u}{\partial z} = i\beta u - \delta\frac{\partial u}{\partial t} - \frac{i\beta_2}{2}\frac{\partial^2 u}{\partial t^2} + i\gamma\left(|u|^2 + \frac{2}{3}|v|^2\right)u + \frac{i\gamma}{3}v^2 u* + \frac{g}{2}u + \frac{g}{2\Omega_g^2}\frac{\partial^2 u}{\partial t^2}$$

$$\frac{\partial v}{\partial z} = -i\beta v + \delta\frac{\partial v}{\partial t} - \frac{i\beta_2}{2}\frac{\partial^2 v}{\partial t^2} + i\gamma\left(|v|^2 + \frac{2}{3}|u|^2\right)v + \frac{i\gamma}{3}u^2 v* + \frac{g}{2}v + \frac{g}{2\Omega_g^2}\frac{\partial^2 v}{\partial t^2}$$

$$(6.13)$$

where

> $u$ and $v$ are the normalized envelopes of the optical pulse sequences projected on the two orthogonal polarization axes of the fiber
>
> $\beta = 2\pi\Delta n/\lambda$ is the wave-number difference between the two modes
>
> $2\delta = 2\beta\lambda/2\pi c$ is the inverse group velocity difference
>
> $\beta_2$ is the second-order dispersion factor
>
> $\beta_3$ is the third-order dispersion factor or dispersion slope as a function of wavelength (to be included if necessary)
>
> $\gamma$ represents the nonlinearity of the fiber
>
> $g$ is the saturable gain of the fiber
>
> $\Omega_g$ is the bandwidth of the laser gain.

For undoped fibers $g=0$ for the EDF, we further consider the gain saturation as

$$g = G\exp\left(-\frac{\int\left[|u|^2 + |v|^2\right]dt}{P_{sat}}\right)$$

$$(6.14)$$

where

> $G$ is the small signal gain coefficient
>
> $P_{sat}$ is the normalized saturation energy.

To possibly close to the experimental conditions of our laser, the following fiber parameters are used: $\gamma = 3\,W^{-1}km^{-1}$, $\beta_3 = 0.1\,ps^2/nm\,km$, $\Omega_g = 20\,nm$ gain saturation energy, $P_{sat} = 1000$, cavity length $L=6$ m, and the beat length of the fiber birefringence $L_b = L/4$. To simulate the cavity effect, we let the light circulate in the cavity. Starting from the intracavity polarizer, which has an orientation of $\theta = \pi/8$ to the fiber fast axis, the light then propagates on the various fibers, first through the 1 m dispersion-shifted fiber (DSF) whose GVD coefficient $\beta_2 = -2\,ps/nm\,km$, then the 4 m EDF whose GVD coefficient $\beta_2 = -10\,ps/nm\,km$, and finally the 1 m standard single mode fiber whose GVD coefficient $\beta_2 = -18\,ps/nm\,km$. Subsequently, the light passes through the waveplates, which cause a fixed polarization rotation of the light. Note that changing the relative orientations of the waveplates is physically equivalent to adding a variable linear cavity phase delay bias to the cavity. Certainly, the principal polarization axes of the waveplates are not aligned with those of the fibers, and in general the different fibers used in the laser cavity could also have different principal polarization axes. However, for the simplicity of numerical calculations, they have been treated with the same principal polarization axes. Furthermore, the effect caused by the principal polarization axis change has also been considered by assuming that the polarizer has virtually a different orientation to the fast axis of the fiber when it acts as an analyzer. In the simulations, the orientation angle of the analyzer to the fibers fast axis is set as $\phi = \pi/2 + \theta$. It is pointed out that in the real laser system, the analyzer has the same polarizer (PI). Therefore, the lightwaves passing through the analyzer remain the same after passing through the polarizer. Then the lightwaves can be used as the input for the next round calculation, and the iterative procedure is repeated until a steady state is achieved.

### 6.6.4 SIMULATION

The complex coupled NLSEs in Equation 6.1 can be numerically solved by using the split-step Fourier method (see Appendix E) [44]. We find that by appropriately setting the linear cavity phase delay bias of the cavity, so that an artificial saturable absorber effect is generated in the laser, self-started mode-locking can always be generated in our simulations through simply increasing the small signal gain coefficient, which corresponds to increasing the pump power in the experiments. Exactly like the experimental observations, multiple soliton pulses are formed in the simulation window immediately after the mode-locking. In the steady state and when the solitons are far apart, all the solitons obtained have exactly the same pulse parameters such as the peak power and pulse width. Figure 6.32 shows, for example, a numerically calculated multiple soliton operation of a laser. Like the experimental observations, the soliton operation of the laser and the generation and annihilation of each individual soliton in the cavity exhibit pump hysteresis. Decreasing numerically the pump power, the soliton number in the simulation window reduces one by one, while carefully increasing the pump strength, with at least one soliton already existing in the cavity, solitons can also be generated one by one as shown in Figure 6.33. All these numerically calculated results are in excellent agreement with the experimental observations.

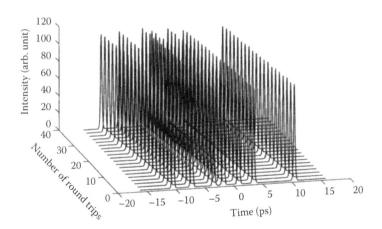

**FIGURE 6.32** Numerically simulated multiple soliton operation state of the laser. $\delta\Phi_l = 1.20\pi$, $G = 350$. Other parameters used are described in the text.

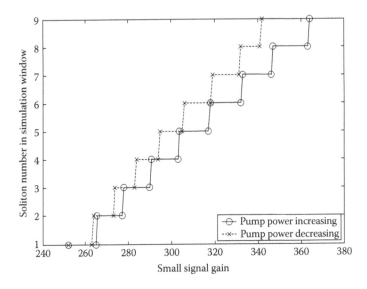

**FIGURE 6.33** Relationship between the soliton number in the simulation window and the pump strength. $\delta\Phi_l = 1.20\pi$.

In a practical laser due to the existence of laser output, fiber splices, etc., linear cavity losses are unavoidable. However, via numerical simulations we could artificially reduce the linear cavity losses and even make it to zero. We found numerically that the weaker the linear cavity loss, the smaller is the pump hysteresis of the soliton operation. With a very weak linear cavity loss we found numerically that a single soliton pulse could even be directly formed from a mode-locked pulse through increasing the pump strength. This numerical result clearly shows that the large pump hysteresis of the soliton operation of the laser is caused by the existence of large linear cavity loss of a practical laser. A large linear cavity loss makes the mode-locking threshold of a laser very high, which under the existence of cavity saturable absorber effect, causes an effective gain in the laser when mode-locking is very large. As will be shown succeeding text, when the peak power of a pulse is clamped, the large effective laser gain will result in the formation of multiple solitons immediately after the mode-locking of the laser. It has been mentioned in Section 6.1 the theoretical work of A. Komarov [79] on the multistability and hysteresis phenomena in passively mode-locked fiber lasers. In the framework of their model, they have explained these phenomena as caused by the competition between the positive nonlinear feedback and the negative phase modulation effect [79]. It is to note that in their model in order to obtain the multiple soliton operation, the cavity loss term caused by the frequency selective filter has to be added in, which from another aspect confirms our numerical result shown earlier. By making the linear cavity loss small, we can numerically investigate the process on how a soliton is formed in the laser cavity. Figure 6.34 shows the numerically simulated pulse shapes and corresponding spectra. In obtaining the result, the linear cavity phase delay bias is set to $\delta\Phi_l = 1.2\pi$. When $G$ is less than 251, there is no mode-locking. In the experiment, this corresponds to the case that the laser is operating

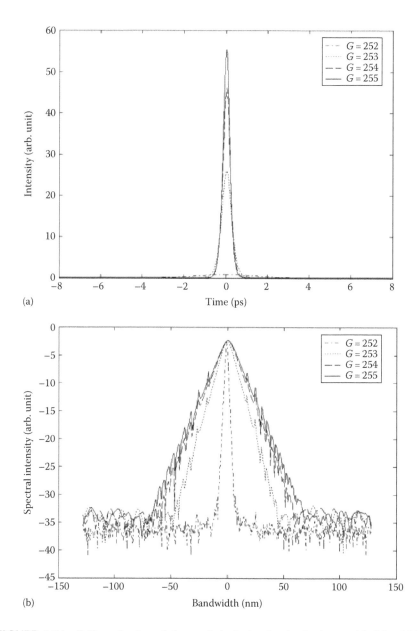

**FIGURE 6.34** Soliton shaping of the mode-locked pulse in the laser with $\delta\Phi_l = 1.20\pi$: (a) evolution of pulse profile with the pump strength and (b) evolution of the optical spectra with variation of the pump strength, hence the gain.

below the mode-locking threshold. When $G$ is equal to 252, a mode-locked pulse emerges in the cavity. The mode-locked pulse has weak pulse intensity and broad pulse width. Due to the action of the mode locker, which in the laser is the artificial saturable absorber, the mode-locked pulse circulates stably in the cavity, just like any mode-locked pulse in other lasers. Although such a mode-locked pulse has stable pulse profile during circulation in the cavity, we emphasize that it is not a soliton but a linear pulse. The linear nature of the pulse is also reflected in that its optical spectrum has no sidebands. When $G$ is further increased, the peak power of the pulse quickly increases. Associated with the pulse intensity increase, the nonlinear optical Kerr effect of the fiber also becomes strong and eventually starts to play a role. An effect of the pulse SPM is to generate a positive frequency chirp, which in the anomalous cavity dispersion regime counterbalances the negative frequency chirp caused by the cavity dispersion effect and compresses the pulse width. When the pulse peak power is strong enough, the nonlinear SPM effect alone can balance the pulse broadening caused by the cavity dispersion effect, even without the existence of the mode-locker, a pulse can propagate stably in the dispersive laser cavity. In this case, a mode-locked pulse then becomes a soliton. In the case of our simulation, this corresponds to the state of $G=253$. A soliton in the laser is also characterized by the appearance of the sidebands in the optical spectrum as shown in Figure 6.34b.

Once the laser gain is fixed, a soliton with fixed peak power and pulse width will be formed, which are independent of the initial conditions. The states shown in Figure 6.34 are stable and unique. This result confirms the auto-soliton property of the laser solitons [86]. However, if the pump power is continuously increased, solitons with even higher peak power and narrower pulse width will be generated. Associated with the soliton pulse width narrowing, the spectrum of the soliton broadens and, consequently, more sidebands become visible. However, the positions of the sidebands are almost fixed. The physical mechanism of sideband generation of laser solitons was extensively investigated previously and is well understood now [87]. It is widely believed that the sideband generation is a fundamental limitation to the soliton pulse narrowing in a laser [88]. However, our numerical simulations clearly show that the sideband generation is just an adaptive effect, whose existence does not limit the soliton pulse narrowing. As far as the pump power could balance the loss caused by the sidebands, soliton pulse width can still be narrowed. Based on our numerical simulation and if there is no other limitation as will be described later in this chapter, then the narrowest soliton pulse that can be formed in a laser should be ultimately only determined by the laser cavity dispersion property, including the net dispersion of all the cavity components and the dispersion of the gain medium. With already one soliton in the simulation window, we further increase the pump strength. Depending on the selection of the linear cavity phase delay bias, we found that the mechanism of the new soliton generation and the features of the multiple soliton operation in the laser are different. With the same laser parameters used earlier, we found that when the linear cavity phase delay bias is set small, say at about $\delta\Phi_l = 1.2\pi$, further increasing the pump power will initially increase the soliton pulse peak power and narrow its pulse width as expected. However, this will stop at a certain fixed value, and instead the background of the simulation window becomes unstable and weak background pulses become visible, as shown in Figure 6.35b. Increasing the pump

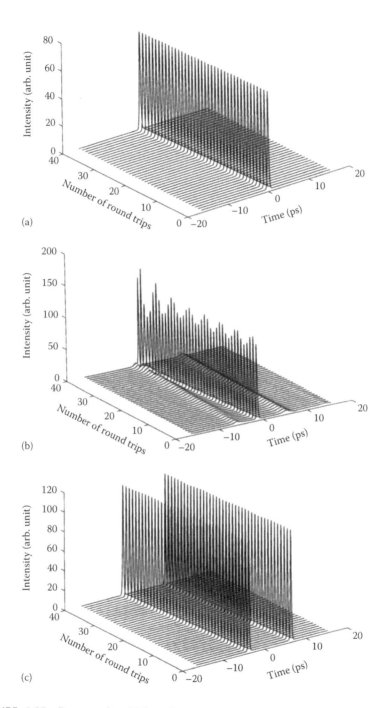

(a)

(b)

(c)

**FIGURE 6.35**  Process of multiple soliton generation in the laser under phase bias of $\delta\Phi_l = 1.20\pi$ with the variation of the gain factor (a) $G = 255$, (b) $G = 270$, (c) $G = 275$.

power slightly further quickly forms a new soliton in the cavity through the soliton shaping of one of the weak background pulses. As the weak background pulses are always initiated from the dispersive waves of the solitons, we have called the new soliton generation "soliton shaping of dispersive waves" [89]. In the steady state, both solitons have exactly the same pulse width and peak power as shown in Figure 6.35c.

When the pump power is further increased, new solitons are generated one by one in the simulation window in the same way and eventually a multiple-soliton state results after a number of round trip circulations as shown in Figure 6.32. This numerically simulated result is in agreement with experimental observations [80].

Due to multiple-soliton generation, the solitons formed in the laser cannot have large pulse energy. High peak power is achieved through increasing the pump power in order to enhance the gain factor. The larger the laser gain, the more solitons would be formed in the cavity. When the linear cavity phase delay bias is set at a very large value, say at about $\delta\Phi_l = 1.8\pi$, which is still in the positive cavity feedback range but close to the other end, no stable propagation of the solitons in cavity can be obtained. With the linear cavity phase delay bias selection, there is a big difference between the linear cavity loss and the nonlinear cavity loss. Therefore, if the gain of the laser is smaller than the dynamical loss that a soliton experiences, the soliton quickly dies out, as shown in Figure 6.36a.

While if the gain of the laser is even slightly larger than the dynamical loss that a soliton experiences, the soliton peak power will increase. Higher soliton peak power results in smaller dynamical loss and even larger effective gain, therefore, the soliton peak will continuously increase. Associated with the soliton peak increase, the soliton pulse width decreases, eventually the soliton breaks up into two solitons with weak peak power and broad pulse width as described by Kärtner et al. [41]. Once a soliton is broken into two solitons with weak peak power, the dynamical loss experienced by each of the solitons becomes very high.

Consequently, the gain of the laser cannot support them. The multiple solitons are then immediately destroyed as shown in Figure 6.36b. If very large gain is available in the laser, the multiple solitons may survive in the cavity temporally and each of them repeats the same process as shown in Figure 6.36b, and eventually a state as shown in Figure 6.36c is formed. Therefore, no stable soliton propagation is possible with an excessively large linear cavity phase delay setting in the laser. Even in the cases of stable multiple-soliton operation, depending on the selection of the linear cavity phase delay, the solitons obtained have different parameters. Figure 6.8 shows for comparison the multiple soliton operation obtained with the linear cavity phase delay bias set at $\delta\Phi_l = 1.55\pi$. It is to see that solitons with higher peak power and narrower pulse width can be formed with the linear cavity phase delay setting. Extensive numerical simulations have shown that the larger the linear cavity phase delay setting, the higher the soliton peak and the narrower the soliton pulse width achievable.

### 6.6.5 Multiple-Soliton Formation and Soliton Energy Quantization

Apparently, depending on the laser linear cavity phase delay bias setting, two different mechanisms exist for multiple soliton generation in the laser (Figure 6.37).

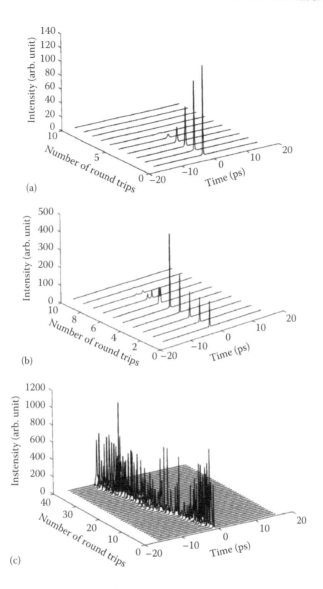

**FIGURE 6.36** Soliton evolutions calculated with $= 1.80\pi$. (a) $G=470$, (b) $G=478$, (c) $G=600$.

One is the soliton shaping of the unstable dispersive waves or the CW components, and the other is the well-known mechanism of pulse splitting. It is to see that in the laser the process of soliton splitting occurs only in the regime where the multiple solitons formed are practically unstable.

We have already reported previously the phenomenon of soliton genera-tion through unstable background in the lasers [89]. Here we further explain its physical origin. Our soliton fiber laser is mode locked using the NLPR technique.

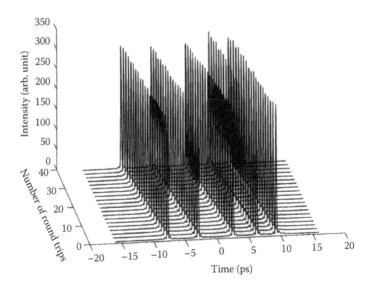

**FIGURE 6.37** Multiple soliton operation of the laser calculated with linear cavity phase delay bias set at $\delta\Phi_l = 1.55\pi$, $G = 465$.

The operating mechanism of the technique has already been analyzed by several authors [44,90,91]. It has been shown that through inserting a polarizer in the cavity and appropriately setting the linear cavity phase delay, the NLPR could generate an artificial saturable absorption effect in the laser. It is this effect that causes the self-started mode-locking of the laser. And after a soliton is formed in the laser cavity, it further stabilizes the soliton. Although previous studies have correctly identified the effects of NLPR and the saturable absorber in the laser, there is no further analysis on how and to what extent these effects affect the soliton parameters and soliton dynamics. Here we follow the description of Chen et al. [90] to understand it. The approach presented here involves first determining the linear and nonlinear cavity transmission of the laser and then, based on the results, finding out how they affect the solitons formed in the laser. Physically, the laser cavity shown in Figure 6.30 can be simplified to a setup as shown in Figure 6.38 for the purpose of determining its transmission property. Starting from the intracavity polarizer, which sets the initial polarization of light in relation to the birefringent axes of the fiber, the polarization of light after passing through the fiber is determined by both the linear and nonlinear birefringence of the fiber.

The lightwaves finally pass through the analyzer, which in the experimental system is the same intracavity polarizer. If we assume that the polarizer has an orientation of $\theta$ angle with respect to the fast axis of the fiber, then the analyzer has an angle of $\phi$, the phase delay between the two orthogonal polarization components caused by the linear fiber birefringence is $\Delta\Phi_l$, and that caused by the nonlinear birefringence is $\Delta\Phi_{nl}$, it can be shown that the transmission coefficient of the setup or the laser cavity is [90] as follows:

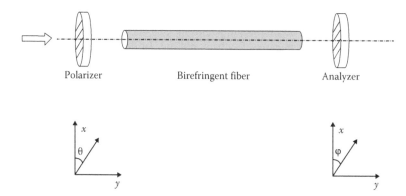

**FIGURE 6.38**  Equivalent setup of Figure 6.30 for determining the cavity transmission.

$$T = \sin^2\theta\sin^2\varphi + \cos^2\theta\cos^2\varphi + \frac{1}{2}\sin 2\theta\sin 2\varphi\cos(\Delta\Phi_l + \Delta\Phi_{nl}) \qquad (6.15)$$

Chen et al. [91] and Davey et al. [90] have already shown how to determine the orientations of the polarizer and the analyzer so that the cavity would generate an efficiently saturable absorption effect. In a previous paper [82] it was shown that the linear cavity transmission of the laser is a sinusoidal function of the linear cavity phase delay $\Delta\Phi_l$ with a period of $2\pi$. It is to point out that within one period of the linear cavity phase delay change, the laser cavity can provide positive (the saturable absorber type) cavity feedback only in half of the period, in the other half of the period it actually has negative feedback.

As shown in Equation 6.3, the actual cavity transmission for an optical pulse is also the nonlinear phase delay $\Delta\Phi_{nl}$ dependent. In the simulations, the orientation of the polarizer has an angle of $\theta = 0.125\pi$ to the fast axis of the fiber, so lightwaves propagation in the fiber will generate a negative nonlinear phase delay. The linear cavity beat length is ¼ of the cavity length, therefore, the maximum linear cavity transmission is at the positions of $(2n+1)\pi$ linear cavity phase delays, where $n = 0,1,2,\ldots$. Furthermore, when the linear cavity phase delay is biased within the range between the $(2n+1)\pi$ to $2(n+1)\pi$, the cavity will generate a positive feedback, as under the effect of the NLPR the actual cavity transmission increases. While if the linear cavity phase delay is located in the range from $2n\pi$ to $(2n+1)\pi$, the cavity will generate a negative feedback. The maximum linear cavity transmission point also marks the switching position of the two feedbacks. For the soliton operation, the laser is always initially biased in the positive cavity feedback regime. It is clearly seen that depending on the selection of the linear cavity phase delay and the strength of the nonlinear phase delay, it is possible for the cavity feedback to be dynamically switch from the positive feedback regime to the negative feedback regime. For the soliton operation of a laser, this cavity feedback switching consequently limits the peak of a soliton formed in the cavity. It is found that this soliton peak limits this effect resulting in multiple-soliton generation in both the soliton fiber laser and the soliton energy quantization. It is assumed that these remarkable phenomena are due to the peak power of a soliton being so strong that it switches the optical cavity from the positive to the negative feedback regime. In this case, although increasing the pump strength still causes the peak power of the soliton to increase, the higher the increase in the soliton peak power, the smaller the actual cavity transmission becomes. To a certain fixed level of the soliton peak power, which depends on the linear cavity phase delay, an enhancement of the soliton peak power would lead to less energy circulating in the actual cavity. Thus the soliton gains higher energy than in the linear cavity transmission level. At this point the soliton peak is clamped. Further increasing the laser gain would not amplify the soliton but enhance the background noise. If the background noise of a certain frequency fulfills the lasing condition, then it could also start to lase and form a CW component in the soliton spectrum. We note that the coexistence of solitons with CW is a generic effect of the soliton fiber lasers, and the phenomenon was reported by several authors [92,93]. Linear waves are intrinsically unstable in the cavity due to modulation instability. When they are strong enough, they become modulated. Furthermore, under the effect of saturable absorption, the strongest background pulse will be amplified and shaped into a multiple soliton. This was exactly what we have observed in the experiments on how a multiple soliton was generated. The dual solitons in the cavity share the same laser gain. As the cavity generates a positive feedback for the weak soliton and a negative one for the strong soliton, under the gain competition the two solitons have to adjust their strength so that the stronger one becomes weaker, and the weaker one becomes stronger, eventually they will stabilize at a state where both have exactly the same peak power. The soliton internal energy balance further determines their other parameters. Except that there are interactions between the solitons, they will always have identical parameters in the stable state. It turns out that the multiple soliton formation in the laser is in fact caused by the peak power clamping effect of the cavity. In addition, the soliton energy quantization observed is also a nature consequence of the gain competition between the solitons in the laser. Obviously, the maximum achievable soliton peak power in the laser is linear cavity phase delay dependent. When the linear cavity phase delay is set close to the cavity feedback switching point, solitons with relatively lower peak power could already dynamically switch the cavity feedback. Therefore, solitons obtained at this linear cavity phase delay setting have lower peak power and broader pulse width, as shown in Figure 6.35a.

While if the linear cavity phase delay is set far away from the switching point, soliton peak power is clamped at a higher value, then solitons with higher peak and narrower pulse width would then be obtained, as shown in Figure 6.37. In particular, if the linear cavity phase delay is set too close to the switching point, as the peak power of the pulse is clamped to too small a value, except the mode-locked pulses, then no soliton could be formed in the laser. While when the linear cavity phase delay is set too far away from the switching point as demonstrated numerically in Figure 6.36, before the soliton peak reaches the switching point, it has already become so high and so narrow that it splits, no stable soliton propagation could be obtained in the laser. Instead, only the state of so-called noise-like pulse emission will be observed [94].

Finally, we note that the multiple soliton operation and soliton energy quantization effect have also been observed in other passively mode-locked soliton fiber lasers, such as in the figure-of-eight lasers and the lasers passively mode-locked with semiconductor saturable absorbers [38,70]. Even in actively mode-locked fiber lasers [66] these phenomena have been observed. Despite the fact that those soliton lasers are not mode-locked with the NLPR technique, their detailed cavity transmission could not have the same feature as described by Equation 6.3. Enlightened by the result obtained in our laser, we conjugate that there must also be a certain pulse peak power limiting mechanism in the lasers, which causes their multiple pulse formation. Indeed, we found that for the figure-of-eight lasers, if the fiber birefringence of the nonlinear loop is further considered, it would also generate a similar pulse peak clamping effect in the laser. However, birefringence of fibers in the lasers is normally ignored. It was also reported that due to the two-photon absorption effect, the SESAM used for the passive mode-locking of fiber lasers has a pulse peak power limiting effect [95]. It is therefore, not surprising that soliton laser mode-locked with the material could also exhibit multiple solitons. For the actively mode-locked laser, in most cases the multiple soliton generation is due to the harmonic mode-locking. In this case, as too many solitons share the limited cavity gain, the energy of each pulse is weak. Therefore, even when the net cavity dispersion is negative, solitons can be normally difficult to form. Therefore, for an actively mode-locked fiber laser, if the cavity is not carefully designed with the right combination of the cavity birefringence and the modulator, which is a polarization-dependent device and a birefringence filter, then the unmatched combined system would further limit the peak power of the pulses formed in the lasers.

### 6.6.6 REMARKS

In this section, a numerical study on the mechanism of multi-bound soliton generation and soliton energy quantization in a passively mode-locked soliton fiber ring lasers using the NLPR technique is described. It has been identified that the multiple soliton generation in the laser is caused by the peak power clamping effect of the cavity. Depending on the linear cavity phase delay setting, the nonlinear phase delay generated by a soliton propagating in the fiber cavity could be so large that it switches the cavity feedback from the initially selected positive regime into a negative regime. As a result, due to the cavity feedback change, the maximum achievable soliton peak power is limited to a saturated level. In this case, increasing the laser pump power will not increase the peak power of the solitons, but generate a multiple soliton. Therefore, the multi-bound solitons can be formed in the laser. As the solitons share the same laser gain, gain competition between them combined with the cavity feedback feature further results, in that in the steady state they have exactly the same soliton parameters. The parameters of solitons formed in the laser are not fixed by the laser configuration but vary with the laser operating conditions, which are determined by the soliton internal energy balance between the shared laser gain and the dynamical losses of each soliton.

## 6.7 CONCLUDING REMARKS

In conclusion, in this chapter the deterministic dynamics of solitons formed in passively mode-locked fiber lasers has been investigated both experimentally and numerically. It has been shown experimentally that under strong soliton strength, deterministic soliton dynamics such as the soliton period-doubling bifurcations and soliton period-doubling route to chaos could appear. In particular, the deterministic soliton dynamics is independent of the number of solitons in the cavity and the types of solitons formed. We observed period-doubling route to chaos on the soliton repetition rate of both the single-pulse soliton and bound solitons, on the conventional solitons formed in anomalous dispersion cavity fiber lasers, on the dispersion-managed solitons generated in around net zero dispersion fiber lasers, and on the GGSs generated in normal dispersion fiber lasers. Period-doubling of vector solitons was also experimentally observed. Based on the coupled extended GLEs and the roundtrip model, the soliton operation of various fiber lasers has been numerically simulated. Successfully reproduced deterministic dynamics of solitons were experimentally observed. Moreover, based on results of both experimental studies and numerical simulations, we found that the appearance of these soliton dynamics could be related to the cavity-induced soliton modulation instability effect. We speculate that the cavity-induced modulation instability is also responsible for the deterministic dynamics of the nonlinear cavity under CW operation.

Multi-bound solitons have been proven, both by experiment and simulation, in passively mode-locked fiber ring lasers by polarization coupling and intercontrol of energy preserved at every polarization state of a linearly polarized fiber.

## REFERENCES

1. K. Ikeda, Multiple-value stationary state and its instability of the transmitted light by a ring cavity system, *Opt. Commun.*, 30, 257–261, 1979.
2. K. Ikeda, H. Daido, and O. Akimoto, Optical turbulence: Chaotic behavior of transmitted light from a ring cavity, *Phys. Rev. Lett.*, 45, 709–712, 1980.
3. D. U. Noske, N. Pandit, and J. R. Taylor, Subpicosecond soliton pulse formation from self-mode-locked erbium fiber laser using intensity dependent polarization rotation, *Electron. Lett.*, 28, 2185–2186, 1992.
4. V. J. Matsas, T. P. Newson, D. J. Richardson, and D. N. Payne, Self-starting passive mode-locked fiber ring soliton laser exploiting nonlinear polarization rotation, *Electron. Lett.* 28, 1391–1393, 1992.
5. I. N. Duling III, All-fiber ring soliton laser mode locked with a nonlinear mirror, *Opt. Lett.*, 16, 539–541, 1991.
6. D. J. Richardson, R. I. Laming, D. N. Payne, M. W. Phillips, and V. J. Matsas, 320 fs soliton generation with passively mode-locked erbium fiber laser, *Electron. Lett.*, 27, 730–732, 1991.
7. B. C. Collings, K. Bergman, S. T. Cundiff, S. Tsuda, J. N. Kutz, J. E. Cunningham, W. Y. Jan, M. Koch, and W. H. Knox, Short cavity erbium/ytterbium fiber lasers mode-locked with a saturable Bragg reflector, *IEEE J. Sel. Top. Quant. Electron.*, 3, 1065–1075, 1997.
8. S. Gray and A. B. Grudinin, Soliton fiber laser with a hybrid saturable absorber, *Opt. Lett.*, 21, 207–209, 1996.

9.  D. J. Jones, H. A. Haus, and E. P. Ippen, Subpicosecond solitons in an actively mode-locked fiber laser, *Opt. Lett.*, 21, 1818–1820, 1996.

10. R. P. Davey, N. Langford, and A. I. Ferguson, Interacting solitons in erbium fiber laser, *Electron. Lett.*, 27, 1257–1259, 1991.

11. Y. P. Tong, P. M. W. French, J. R. Taylor, J. O. Fujimoto, All-solid-state femtosecond source in the near infrared, *Opt. Commun.*, 136, 235–238, 1997.

12. A. M. Kowalevicz, Jr., A. Tucay Zare, F. Kartner, J. G. Fujimoto, S. Dewald, U. Morgner, V. Scheuer, and G. Angelow, Generation of 150-nJ pulses from a multiple-pass cavity Kerr-lens mode-locked Ti:Al$_2$O$_3$ oscillator, *Opt. Lett.*, 28, 1597–1599, 2003.

13. L. M. Zhao, D. Y. Tang, F. Lin, and B. Zhao, Observation of period-doubling bifurcations in a femtosecond fiber soliton laser with dispersion management cavity, *Opt. Express*, 12, 4573–4578, 2004.

14. D. Y. Tang, L. M. Zhao, and F. Lin, Numerical studies of routes to chaos in passively mode locked fiber soliton ring lasers with dispersion-managed cavity, *Europhys. Lett.*, 71(1), 56–62, 2005.

15. L. M. Zhao, D. Y. Tang, and B. Zhao, Period-doubling and quadrupling of bound solitons in a passively mode-locked fiber laser, *Opt. Commun.*, 252, 167–172, 2005.

16. L. M. Zhao, D. Y. Tang, and A. Q. Liu, Chaotic dynamics of a passively mode-locked soliton fiber ring laser, *Chaos*, 16, 013128, 2006.

17. L. M. Zhao, D. Y. Tang, T. H. Cheng, and C. Lu, Period-doubling of multiple solitons in a passively mode-locked fiber laser, *Opt. Commun.*, 273, 554–559, 2007.

18. D. Y. Tang, J. Wu, L. M. Zhao, and L. J. Qian, Dynamic sideband generation in soliton fiber lasers, *Opt. Commun.*, 275, 213–216, 2007.

19. L. M. Zhao, D. Y. Tang, T. H. Cheng, H. Y. Tam, C. Lu, and S. C. Wen, Period-doubling of dispersion-managed solitons in an Erbium-doped fiber laser at around zero dispersion, *Opt. Commun.*, 278, 428–433, 2007.

20. L. M. Zhao, D. Y. Tang, X. Wu, and H. Zhang, Period-doubling of gain-guided solitons in fiber lasers of large net normal dispersion, *Opt. Commun.*, 281, 3557–3560, 2008.

21. L. M. Zhao, D. Y. Tang, H. Zhang, X. Wu, C. Lu, and H. Y. Tam, Period-doubling of vector solitons in a ring fiber laser, *Opt. Commun.*, 281, 5614–5617, 2008.

22. D. Y. Tang, L. M. Zhao, X. Wu, and H. Zhang, Soliton modulation instability in fiber lasers, *Phys. Rev. A*, 80, 023806, 2009.

23. F. Ö. Ilday, F. W. Wise, and T. Sosnowski, High-energy femtosecond stretched-pulse fiber laser with a nonlinear optical loop mirror, *Opt. Lett.*, 27, 1531–1533, 2002.

24. M. J. Guy, D. U. Noske, and J. R. Taylor, Generation of femtosecond soliton pulses by passive mode locking of an ytterbium-erbium figure-of-eight fiber laser, *Opt. Lett.*, 18, 1447–1449, 1993.

25. D. Y. Tang, W. S. Man, H. Y. Tam, and M. S. Demokan, Modulational instability in a fiber siliton ring laser induced by periodic dispersion variation, *Phys. Rev. A* 61, 023804, 2000.

26. L. M. Zhao, D. Y. Tang, H. Zhang, X. Wu, and N. Xiang, Soliton trapping in fiber lasers, *Opt. Express*, 16, 9528–9533, 2008.

27. V. S. Letohkov, Generation of ultrashort light pulses in a laser with a nonlinear absorber, *Sov. Phys. JETP*, 28, 562–568, 1969.

28. G. P. Agrawal, *Nonlinear Fiber Optics*, 4th edn., Academic Press, Boston, MA, 2007.

29. D. Y. Tang, L. M. Zhao, B. Zhao, and A. Q. Liu, Mechanism of multisoliton formation and soliton energy quantization in passively mode-locked fiber lasers, *Phys. Rev. A*, 72, 043816, 2005.

30. M. Hofer, M. H. Ober, F. Haberl, and M. E. Fermann, Characterization of ultrashort pulse formation in passively mode-locked fiber lasers, *IEEE J. Quant. Electron.*, QE-28, 720–728, 1992.

31. R. P. Davey, N. Langford, and A. I. Ferguson, Role of polarization rotation in the mode-locking of an erbium fiber laser, *Electron. Lett.*, 29, 758–760, 1993.

32. S. M. J. Kelly, K. Smith, K. J. Blow, and N. J. Doran, Average soliton dynamics of a high-gain erbium fiber laser, *Opt. Lett.*, 16, 1337–1339, 1991.

33. A. D. Kim, J. N. Kutz, and D. J. Muraki, Pulse-train uniformity in optical fiber lasers passively mode-locked by nonlinear polarization rotation, *IEEE J. Quant. Electron.*, 36, 465–471, 2000.

34. K. M. Spaulding, D. H. Yong, A. D. Kim, and J. N. Kutz, Nonlinear dynamics of mode-locking optical fiber ring lasers, *J. Opt. Soc. Am. B* 19, 1045–054, 2002.

35. B. Zhao, D. Y. Tang, L. M. Zhao, and P. Shum, Pulse-train non-uniformity in a fiber soliton ring laser mode-locked by using the nonlinear polarization rotation technique, *Phys. Rev. A*, 69, 043808, 2004.

36. G. Sucha, S. R. Bolton, S. Weiss, and D. S. Chemla, Period doubling and quasi-periodicity in additive-pulse mode-locked lasers, *Opt. Lett.*, 20, 1794–1796, 1995.

37. D. Côté and H. M. van Driel, Period doubling of a femtosecond Ti:sapphire laser by total mode locking, *Opt. Lett.*, 23, 715–717, 1998.

38. A. B. Grudinin, D. J. Richardson, and D. N. Payne, Energy quantization in figure eight fiber laser, *Electron. Lett.*, 28, 67–68, 1992.

39. D. J. Richardson, R. I. Laming, D. N. Payne, V. J. Matsas, and M. W. Phillips, Pulse repetition rates in passive, femtosecond soliton fiber laser, *Electron. Lett.*, 27, 1451–1453, 1991.

40. D. Y. Tang, W. S. Man, H. Y. Tam, and P. D. Drummond, Observation of bound states of solitons in a passively mode-locked fiber laser, *Phys. Rev. A*, 64, 033814, 2001.

41. P. Grelu and J. M. Soto-Crespo, Multisoliton states and pulse fragmentation in a passively mode-locked fibre laser, *J. Opt. B Quant. Semiclass. Opt.*, 6, S271–S278, 2004.

42. B. Zhao, D. Y. Tang, P. Shum, Y. D. Gong, C. Lu, W. S. Man, and H. Y. Tam, Energy quantization of twin-pulse solitons in a passively mode-locked fiber ring laser, *Appl. Phys. B*, 77, 585–588, 2003.

43. B. Zhao, D. Y. Tang, P. Shum, W. S. Man, H. Y. Tam, Y. D. Gong, and C. Lu, Bound twin-pulse solitons in a fiber ring laser, *Phys. Rev. E*, 70, 067602, 2004.

44. D. Y. Tang, B. Zhao, D. Y. Shen, C. Lu, W. S. Man, and H. Y. Tam, Compound pulse solitons in a fiber ring laser, *Phys. Rev. A*, 68, 013816, 2003.

45. C. O. Weiss and H. King, Oscillation period doubling chaos in a laser, *Opt. Commun.*, 44, 59, 1982.

46. S. P. Ng, D. Y. Tang, L. J. Qin, X. L. Meng, and Z. J. Xiong, Polarization-resolved study of diode-pumped passively $Q$-switched Nd:GdVO$_4$ lasers, *Appl. Opt.*, 45(26), 6792–6797, 2006.

47. D. Y. Tang, B. Zhao, L. M. Zhao, and H. Y. Tam, Soliton interaction in a fiber ring laser, *Phys. Rev. E*, 72, 016616, 2005.

48. N. J. Smith, F. M. Knox, N. J. Doran, K. J. Blow, and I. Bennion, Enhanced power solitons in optical fibres with periodic dispersion management, *Electron. Lett.*, 32, 54–55, 1996.

49. J. H. B. Nijhof, N. J. Doran, W. Forysiak, and F. M. Knox, Stable soliton-like propagation in dispersion managed systems with net anomalous, zero and normal dispersion, *Electron. Lett.*, 33, 1726–1727, 1997.

50. J. N. Kutz and S. G. Evangelides, Jr., Dispersion-managed breathers with average normal dispersion, *Opt. Lett.*, 23, 685–687, 1998.

51. J. H. B. Nijhof, W. Forysiak, and N. J. Doran, Dispersion-managed solitons in the normal dispersion regime: A physical interpretation, *Opt. Lett.*, 23, 1674–1676, 1998.

52. Y. Chen and H. A. Haus, Dispersion-managed solitons in the net positive dispersion regime, *J. Opt. Soc. Am. B*, 16, 24–30, 1999.

53. V. S. Grigoryan, R.-M. Mu, G. M. Carter, and C. R. Menyuk, Experimental demonstration of long-distance dispersion-managed soliton propagation at zero average dispersion, *IEEE Photon. Technol. Lett.*, 12, 45–46, 2000.

54. L. M. Zhao, D. Y. Tang, T. H. Cheng, and C. Lu, Gain-guided solitons in dispersion-managed fiber lasers with large net cavity dispersion, *Opt. Lett.*, 31, 2957–2959, 2006.

55. N. Akhmediev and A. Ankiewicz (Eds.), *Dissipative Solitons*, Lecture Notes in Physics, 661 Springer, Heidelberg, Germany, 2005.

56. H. A. Haus, J. G. Fujimoto, and E. P. Ippen, Analytic theory of additive pulse and Kerr lens mode locking, *IEEE J. Quant. Electron.*, 28, 2086–2096, 1992.

57. S. T. Cundiff, B. C. Collings, and W. H. Knox, Polarization locking in an isotropic, mode-locked soliton Er/Yb fiber laser, *Opt. Express*, 1, 12–20, 1997.

58. L. M. Zhao, D. Y. Tang, H. Zhang, Wu, X., Polarization rotation locking of vector solitons in a fiber ring laser, *Opt. Express*, 16(14), 10053–10058, 2008.

59. S. K. Turitsyn, Theory of energy evolution in laser resonators with saturated gain and non-saturated loss, *Opt. Express*, 17(14), 11898–11904, 2009.

60. H. Nakatsuka, S. Asaka, H. Itoh, K. Ikeda, and M. Matsuoka, Observation of bifurcation to chaos in an all-optical bistable system, *Phys. Rev. Lett.*, 50, 109, 1983.

61. G. Steinmeyer, A. Buchholz, M. Hansel, M. Heuer, A. Schwache, and F. Mitschke, Dynamical pulse shaping in a nonlinear resonator, *Phys. Rev. A*, 52, 830, 1995.

62. F. Ilday, J. Buckley, and F. Wise, *Proceedings of the Nonlinear Guided Waves Conference*, Toronto, Ontario, Canada, 2004 (OSA, Washington, DC, 2004), Paper MD9.

63. K. Tai, A. Hasegawa, and A. Tomita, Observation of modulational instability in optical fibers, *Phys. Rev. Lett.*, 56, 135–138, 1986.

64. K. Smith, J. R. Armitage, R. Wyatt, and N. J. Doran, Erbium fiber soliton laser, *Electron. Lett.*, 26, 1149–1151, 1990.

65. M. Nakazawa, E. Yoshida, and Y. Kimura, Generation of 98 fs optical pulses directly from an erbium doped fiber ring laser at 1.57 um, *Electron. Lett.*, 29, 63–65, 1993.

66. V. J. Matsas, D. J. Richardson, T. P. Newson, and D. N. Payne, Characterization of a self-starting, passively mode-locked fiber ring laser that exploits nonlinear polarization evolution, *Opt. Lett.*, 18, 358–360, 1993.

67. K. Tamura, E. P. Ippen, H. A. Haus, and L. E. Nelson, 77-fs pulse generation from a stretched-pulse mode-locked all-fiber ring laser, *Opt. Lett.*, 18, 1080–1082, 1993.

68. M. Nakazawa, E. Yoshida, T. Sugawa, and Y. Kimura, Continuum suppressed, uniformly repetitive 136 fs pulse generation from an erbium doped fiber laser with nonlinear polarization rotation, *Electron. Lett.*, 29, 1327–1329, 1993.

69. L. A. Gomes, L. Orsila, T. Jouhti, and O. G. Okhotnikov, Picosecond SESAM-based ytterbium mode-locked fiber lasers, *IEEE J. Sel. Top. Quant. Electron.*, 10, 129–136, 2004.

70. N. N. Akhmediev, A. Ankiewicz, and J. M. Soto-Crespo, Multisoliton solutions of the complex Ginzburg-Landau equation, *Phys. Rev. Lett.*, 79, 4047–4051, 1997.

71. M. J. Lederer, B. Luther-Davies, H. H. Tan, C. Jagadish, N. N. Akhmediev, and J. M. Soto-Crespo, Multipulse operation of a Ti:sapphire laser mode locked by an ion-implanted semiconductor saturable-absorber mirror, *J. Opt. Soc. Am. B*, 16, 895–904, 1999.

72. C. Spielmann, P. F. Curley, T. Brabec, and F. Krausz, Ultrabroadband femtosecond lasers, *IEEE J. Quantum Electron.*, 30, 1100–1114, 1994.

73. F. X. Kärtner, J. Aus der Au, and U. Keller, Mode-locking with slow and fast saturable absorbers—What's the difference?, *IEEE J. Sel. Top. Quant. Electron.*, 4, 159–168, 1998.

74. G. P. Agrawal, Optical pulse propagation in doped fiber amplifiers, *Phys. Rev. A*, 44, 7493–7501, 1991.

75. P. Grelu, F. Belhache, F. Gutty, and J. M. Soto-Crespo, Relative phase locking of pulses in a passively mode-locked fiber laser, *J. Opt. Soc. Am. B*, 20, 863–870, 2003.

76. A. Komarov, H. Leblond, and F. Sanchez, Multistability and hysteresis phenomena in passively mode-locked fiber lasers, *Phy. Rev. A*, 71, 053809, 2005.

77. D. Y. Tang, W. S. Man, and H. Y. Tam, Stimulated soliton pulse formation and its mechanism in a passively mode-locked fiber soliton laser, *Opt. Commun.*, 165, 189–194, 1999.

78. B. Zhao, D. Y. Tang, P. Shum, Y. D. Gong, C. Lu, W. S. Man, and H. Y. Tam, Energy quantization of twin-pulse solitons in a passively mode-locked fiber ring laser, *Appl. Phys. B*, 77, 585–588, 2003.

79. D. Y. Tang, W. S. Man, H. Y. Tam, and P. Drummond, Observation of bound states of solitons in a passively mode-locked fiber soliton laser, *Phys. Rev. A*, 66, 033806, 2002.

80. M. Nakazawa, E. Yoshida, and Y. Kimura, Low threshold 290 fs erbium-doped fiber laser with a nonlinear amplifying loop mirror pumped by InGaAsP laser diodes, *Appl. Phys. Lett.*, 59, 2073–2075, 1991.

81. A. K. Komarov and K. P. Komarov, Multistability and hysteresis phenomena in passive mode-locked lasers, *Phys. Rev. E*, 62, R7607–R7610, 2000.

82. H. A. Haus, J. G. Fujimoto, and E. P. Ippen, Structures for additive pulse mode-locking, *J. Opt. Soc. B*, 8, 2068–2076, 1991.

83. W. S. Man, H. Y. Tam, M. S. Demonkan, P. K. A. Wai, and D. Y. Tang, Mechanism of intrinsic wavelength tuning and sideband asymmetry in a passively mode-locked soliton fiber ring laser, *J. Opt. Soc. B*, 17, 28–33, 2000.

84. D. Y. Tang, S. Fleming, W. S. Man, H. Y. Tam, and M. S. Demokan, Subsideband generation and modulational instability lasing in a fiber soliton laser, *J. Opt. Soc. Am. B*, 18, 1443–1450, 2001.

85. B. Zhao, D. Y. Tang, L. M. Zhao, P. Shum, and H. Y. Tam, Pulse train non-uniformity in a fiber soliton ring laser mode-locked by using the nonlinear polarization rotation technique, *Phys. Rev. A*, 69, 43808, 2004.

86. V. S. Grigoryan and T. S. Muradyan, Evolution of light pulses into autosolitons in nonlinear amplifying media, *J. Opt. Soc. Am. B*, 8, 1757–1765, 1991.

87. S. M. J. Kelly, Characteristic sideband instability of periodically amplified average soliton, *Electron. Lett.*, 28, 806–807, 1992.

88. M. L. Dennis and I. N. Duling III, Experimental study of sideband generation in femtosecond fiber lasers, *IEEE J. Quant. Electron.*, 30, 1469–1477, 1994.

89. W. S. Man, H. Y. Tam, M. S. Demonkan, and D. Y. Tang, Soliton shaping of dispersive waves in a passively node-locked fiber soliton ring laser, *Opt. Quant. Electron.*, 33, 1139–1147, 2001.

90. C. J. Chen, P. K. A. Wai, and C. R. Menyuk, Soliton fiber ring laser, *Opt. Lett.*, 17, 417–419, 1992.

91. R. P. Davey, N. Langford, and A. I. Ferguson, Role of polarization rotation in the modelocking of an Er fiber laser, *Electron. Lett.*, 29, 758–760, 1993.

92. S. Namiki, E. P. Ippen, H. A. Haus, and K. Tamura, Relaxation oscillation behavior in polarization additive pulse mode-locked fiber ring lasers, *Appl. Phys. Lett.*, 69, 3969–3971, 1996.

93. B. Zhao, D. Y. Tang, and H. Y. Tam, Experimental observation of FPU recurrence in a fiber ring laser, *Opt. Express*, 11, 3304–3309, 2003.

94. M. Horowitz, Y. Barad, and Y. Silberberg, Noise-like pulses with broadband spectrum generated from an erbium-doped fiber laser, *Opt. Lett.*, 22, 799–801, 1997.

95. E. R. Thoen, E. M. Koontz, M. Joschko, P. Langlois, T. R. Schibli, F. X. Kartner, E. P. Ippen, and L. A. Kolodziejski, Two-photon absorption in semiconductor saturable absorber mirrors, *Appl. Phys. Lett.*, 74, 3927–3929, 1999.

# 7 Multirate Multiplication Soliton Fiber Ring and Nonlinear Loop Lasers

This chapter presents the operational principles and implementation of mode-locked fiber lasers operating in nonlinear region via the optical saturated amplification or the photonic interactions of the pump sources and generated lightwaves. It is believed that multi-bound solitons can only be formed if at the initial state a single soliton is generated in a nonlinear resonator for the evolution to multi-bound soliton. The nonlinear fiber ring type is the typical resonance cavity in which the phase of the optical waves under the soliton envelope can be a basic structure for forming multi-bound solitonic waves.

Thus, this chapter describes simple harmonic and regenerative mode-locked types for 10 and 40 G-pulses/s employing the harmonic detuning technique for generating up to 200 G-pulses/s, and harmonic repetition rate multiplication using temporal diffraction. An ultrastable mode-locked laser operating at 10 GHz repetition rate has been designed, constructed, and tested. The laser generates optical pulse train of 4.5 ps pulse width when the modulator is biased at the phase quadrature quiescent region. Long-term stability of amplitude and phase noise indicates that the optical pulse source can produce an error-free pattern in a self-locking mode for more than 20 h, the most stable photonic fiber ring laser reported to date.

The repetition rate is demonstrated up to 200 G-pulses/s using harmonic detuning mechanism in a nonlinear fiber ring laser. In this system, the system operation under the rational harmonic mode-locking analyzed using the phase plane technique in control engineering is given. Furthermore, we examine the harmonic distortion contribution to this system performance. From the fundamental rate of 100 MHz pulse train, the multiplication factors of 660× and 1230× can be achieved at pulse rates of 66 and 123 GHz, respectively. The system behavior of group velocity dispersion (GVD) repetition rate multiplication is proven as one of the principal mechanisms. Stability and the transient response of the multiplied pulses are studied using the phase plane technique.

Furthermore, nonlinear fiber lasers can be used to generate bistable operations and generation of multi-bound solitons based on the nonlinearity in the ring cavity. Experimental and theoretical generation of multisoliton bound states in an active FM mode-locked fiber laser is described in Chapter 5, in which not only bound soliton pairs but also triple- and quadruple-soliton bound states can be generated.

In addition, nonlinear dynamics of nonlinear loop fibers are presented in Section 7.4 to describe the bistability, bifurcation, and chaos as analogies to electrical circuits described in Appendix A. Experimental demonstrations are also depicted to illustrate the nonlinear dynamics of such optical systems.

## 7.1  INTRODUCTION

Generation of ultrashort optical pulses with multiple gigabits repetition rate is critical for ultrahigh bit rate optical communications, particularly for the next generation of terabits/second optical fiber systems. As the demand for the bandwidth of the optical communication systems increases, the generation of short pulses with ultrahigh repetition rate becomes increasingly important in the coming decades.

The mode-locked fiber laser offers a potential source of such pulse trains. Although the generation of ultrashort pulses by mode-locking of a multimodal ring laser is well known, the applications of such short pulse trains in multi-gigabits/second optical communications challenge its designers on its stability and spectral properties. Recent reports on the generation of short pulse trains at repetition rates in the order of 40 Gbps, possibly higher in the near future [1], motivate us to design and experiment with these sources in order to evaluate whether they can be employed in practical optical communication systems.

Further, the interest of multiplexed transmission at 160 Gbps and higher in the foreseeable future requires us to experiment with optical pulse source having short pulse duration and high repetition rates. This report describes laboratory experiments of a mode-locked fiber ring laser (MLFRL), initially with a repetition rate of 10 GHz and preliminary results of higher multiple repetition rates up to 40 GHz. The mode-locked ring lasers reported hereunder adopt an active mode-locking scheme whereby partial optical power of the output optical waves is detected, filtered, and a clock signal is recovered at the desired repetition rate. It is then used as a radio frequency (RF) drive signal to the intensity modulator incorporated in the ring laser. A brief description of the principle of operation of the MLFRL is given in the next section followed by a description of the mode-locked laser experimental setup and characterization.

Active mode-locked fiber lasers remain as a potential candidate for the generation of such pulse trains. However, the pulse repetition rate is often limited by the bandwidth of the modulator used or the RF oscillator that generates the modulation signal. Hence, some techniques have been proposed to increase the repetition frequency of the generated pulse trains. Rational harmonic mode-locking is widely used to increase the system repetition frequency [1–3]. A 40 GHz repetition frequency has been obtained with fourth-order rational harmonic mode-locking using 10 GHz base band modulation frequency [2]. Wu and Dutta [3] reported 22nd-order rational harmonic detuning in the active mode-locked fiber laser, with 1 GHz base frequency, leading to 22 GHz pulse operation. This technique is simple and achieved by applying a slight deviated frequency from the multiple of fundamental cavity frequency. Nevertheless, it is well known that it suffers from inherent pulse amplitude instability as well as poor long-term stability. Therefore, pulse amplitude equalization techniques are often applied to achieve better system performance [3–5].

Other than this rational harmonic detuning, there are some other techniques that have been reported and used to achieve the same objective. The fractional temporal Talbot–based repetition-rate multiplication technique [6,7] uses the interference effect between the dispersed pulses to achieve the repetition rate multiplication.

The essential element of this technique is the dispersive medium, such as linearly chirped fiber grating (LCFG) [6,8] and dispersive fiber [9–11]. Intracavity optical filtering [12,13] uses modulators and a high finesse Fabry–Perot filter (FPF) within the laser cavity to achieve higher repetition rate by filtering out certain lasing modes in the mode-locked laser. Other techniques used in repetition rate multiplication include higher order FM mode-locking [14], optical time domain multiplexing [15], etc.

The stability of high repetition rate pulse train generated is one of the main concerns for practical multi-giga bits/second optical communication system. Qualitatively, a laser pulse source is considered stable if it is operating at a state where any perturbations or deviations from this operating point is not increased but suppressed. Conventionally the stability analyses of such laser systems are based on the linear behavior of the laser in which we can analyze the system behavior in both time and frequency domains. However, when the mode-locked fiber laser is operating under nonlinear regime, none of these standard approaches can be used, since direct solution of nonlinear different equation is generally impossible; hence, frequency domain transformation is not applicable. Some inherent nonlinearities in the fiber laser may affect its stability and performance, such as the saturation of the embedded gain medium, nonquadrature biasing of the modulator, nonlinearities in the fiber, etc.; hence, nonlinear stability approach should be used in any laser stability analysis.

In Section 7.2, we focus on the stability and transient analyses of the rational harmonic mode-locking in the fiber ring laser system using phase plane method, which is commonly used in the nonlinear control system. This technique has been previously used in Reference 11 to study the system performance of the fractional temporal Talbot–based repetition-rate multiplication systems. It has been shown that it is an attractive tool in system behavior analysis. However, it has not been used in the rational harmonic mode-locking fiber laser system. In Section 7.2, the rational harmonic detuning technique is briefly discussed.

Rational harmonic detuning [3,16,17] is achieved by applying a slight deviated frequency from the multiple of fundamental cavity frequency. A 40 GHz repetition frequency has been obtained by Wu and Dutta [3] using a 10 GHz base band modulation frequency with fourth-order rational harmonic mode-locking. This technique is simple in nature. However, this technique suffers from inherent pulse amplitude instability, which includes both amplitude noise and inequality in pulse amplitude; furthermore, it gives poor long-term stability. Hence, pulse amplitude equalization techniques are often applied to achieve better system performance [2,4,5]. Fractional temporal Talbot–based repetition-rate multiplication technique [4–8] uses the interference effect between the dispersed pulses to achieve the repetition rate multiplication. The essential element of this technique is the dispersive medium, such as LCFG [8,16] and single mode fiber [8,9]. This technique will be discussed further in Section 7.2. Intracavity optical filtering [13,14] uses modulators and a high finesse FPF the laser cavity to achieve higher repetition rate by filtering out certain lasing modes in the mode-locked laser. Other techniques used in repetition rate multiplication include higher order FM mode-locking [13], optical time domain multiplexing, etc.

Although Talbot–based repetition-rate multiplication systems are based on the linear behavior of the laser, there are still some inherent nonlinearities affecting its stability, such as the saturation of the embedded gain medium, nonquadrature biasing of the modulator, nonlinearities in the fiber, etc.; hence, nonlinear stability approach must be adopted. In Section 7.3, we focus on the stability and transient analyses of the group velocity dispersion (GVD)-multiplied pulse train using the phase plane analysis of nonlinear control analytical technique [2]. This section uses the phase plane analysis described in Chapter 2 to study the stability and transient performances of the GVD repetition-rate multiplication systems. In Section 7.2, the GVD repetition-rate multiplication technique is briefly given. Section 7.3 describes the experimental setup for the repetition rate multiplication. This section also investigates the dynamic behavior of the phase plane of GVD multiplication system, followed by some simulation results. Finally, some concluding remarks and possible future developments for this type of laser are given.

## 7.2 ACTIVE MODE-LOCKED FIBER RING LASER BY RATIONAL HARMONIC DETUNING

In this section, we investigate the system behavior of rational harmonic mode-locking in the fiber ring laser using the phase plane technique of the nonlinear control engineering. Furthermore, we examine the harmonic distortion contribution to this system performance. We also demonstrate 660x and 1230x repetition rate multiplications on 100 MHz pulse train generated from an active harmonically MLFRL, hence achieving 66 and 123 GHz pulse operations by using rational harmonic detuning, which is the highest rational harmonic order reported to date.

### 7.2.1 RATIONAL HARMONIC MODE-LOCKING

In an active harmonically mode-lock fiber ring laser, the repetition frequency of the generated pulses is determined by the modulation frequency of the modulator, $f_m = qf_c$, where $q$ is the $q$th harmonic of the fundamental cavity frequency, $f_c$, which is determined by the cavity length of the laser, $f_c = c/nL$, where $c$ is the speed of light, $n$ is the refractive index of the fiber, and $L$ is the cavity length. Typically, $f_c$ is in the range of several kHz or MHz. Hence, in order to generate GHz pulse train, mode-locking is normally performed by modulation in the states of $q \gg 1$ (that is, $q$ pulses circulating within the cavity), which is known as harmonic mode-locking. By applying a slight deviation or a fraction of the fundamental cavity frequency, $\Delta f = f_c/m$, where $m$ is the integer, the modulation frequency becomes

$$f_m = qf_c \pm \frac{f_c}{m} \tag{7.1}$$

This leads to $m$-times increase in the system repetition rate, $f_r = mf_m$, where $f_r$ is the repetition frequency of the system [2]. When the modulation frequency is detuned by a $m$ fraction, the contributions of the detuned neighboring modes are weakened, only every $m$th lasing mode oscillates in phase and the oscillation waveform maximums accumulate, hence achieving in $m$-times higher repetition frequency. However, the small but not negligible detuned neighboring modes affect the resultant pulse train, which leads to uneven pulse amplitude distribution and poor long-term stability. This is considered as harmonic distortion in our modeling, and it depends on the laser line width and amount of detuned, that is, a fraction $m$. The amount of the allowable detuneable range or rather the obtainable increase in the system repetition rate by this technique is very much limited by the amount of harmonic distortion. When the amount of frequency detuned is too small relative to the modulation frequency that is very high $m$, contributions of the neighboring lasing modes become prominent, thus reduce the repetition-rate multiplication capability significantly. In other words, no repetition frequency multiplication is achieved when the detuned frequency is unnoticeably small. Often in this case, it is considered as the system noise due to improper modulation frequency tuning. In addition, the pulse amplitude fluctuation is also determined by this harmonic distortion.

### 7.2.2 EXPERIMENT SETUP

In general, the experimental setup of the active harmonically MLFRL is similar to Figure 7.1. The principal element of the laser is an optical open loop with an optical gain medium, a Mach–Zehnder amplitude modulator (MZM), an optical amplifier to supply sufficient energy of photons, an optical polarization controller (PC), an optical band-pass filter (BPF), optical couplers, and other associated optics.

The gain medium used in our fiber laser system is an erbium-doped fiber amplifier (EDFA) with saturation power of 16 dBm. A polarization-independent optical isolator is used to ensure unidirectional lightwave propagation as well as to eliminate back reflections from the fiber splices and optical connectors. A free space filter with 3 dB bandwidth of 4 nm at 1555 nm is inserted into the cavity to select the operating wavelength of the generated signal and to reduce the noise in the system. In addition, it is responsible for the longitudinal modes selection in the mode-locking process. The birefringence of the fiber is compensated by a PC, which is also used for the polarization alignment of the linearly polarized lightwave before entering the planar structure modulator for better output efficiency. Pulse operation is achieved by introducing an asymmetric coplanar traveling wave 10 Gbps, lithium niobate, $Ti:LiNbO_3$

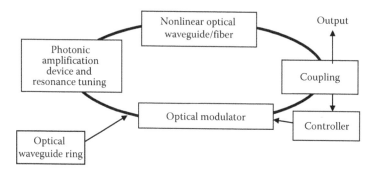

**FIGURE 7.1** Setup of a THz regenerative MLRL using nonlinear effects such as parametric or lumped sequence amplification and incorporating a feedback controller. The optical recirculating ring can be either fiber or planar type.

MZM into the cavity with half-wave voltage, $V_\pi$ of 5.8 V and insertion loss of $\leq$7 dB. The modulator is DC biased near the quadrature point and not more than the $V_\pi$ such that it operates around the linear region of its characteristic curve. The modulator is driven by a 100 MHz, 100 ps step recovery diode (SRD), which is in turn driven by an RF amplifier (RFA) a RF signal generator. The modulating signal generated by the SRD is a ~1% duty cycle Gaussian pulse train. The output coupling of the laser is optimized using a 10/90 coupler. Ninety percent of the optical field power is coupled back into the cavity ring loop, while the remaining portion is taken out as the output of the laser and analyzed.

### 7.2.3 PHASE PLANE ANALYSIS

Nonlinear system frequently has more than one equilibrium point. It can also oscillate at fixed amplitude and fixed period without external excitation. This oscillation is called limit cycle. However, limit cycles in nonlinear systems are different from linear oscillations. First, the amplitude of self-sustained excitation is independent of the initial condition, while the oscillation of a marginally stable linear system has its amplitude determined by the initial conditions. Second, marginally stable linear systems are very sensitive to changes, while limit cycles are not easily affected by parameter changes [31].

As illustrated in Figure 7.2, a phase plane analysis is a graphical method of representing second-order nonlinear systems, or the mutual dependence of the phases evolution of two variables of the system. The result is a family of system motion of trajectories on a two-dimensional plane, which allows us to visually observe the motion patterns of the system. Nonlinear systems can display more complicated patterns in the phase plane, such as multiple equilibrium points and limit cycles. In the phase plane, a limit cycle is defined as an isolated closed curve. The trajectory has to be both closed, indicating the periodic nature of the motion, and isolated, indicating the limiting nature of the cycle [31].

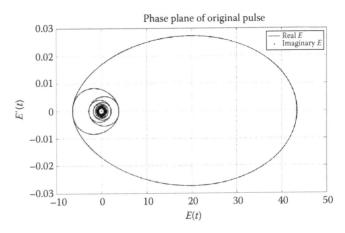

**FIGURE 7.2** Phase plane of a 10 GHz mode-locked pulse train (solid line—real part of the energy, dotted line—imaginary part of the energy, x-axis—$E(t)$, and y-axis—$E'(t)$).

The system modeling of the rational harmonic MLFRL system is implemented on the following assumptions: (1) detuned frequency is perfectly adjusted according to the fraction number required, (2) small harmonic distortion, (3) no fiber nonlinearity is included in the analysis, (4) no other noise sources are involved in the system, and (5) Gaussian lasing mode amplitude distribution analysis.

The phase plane of a perfect 10 GHz mode-locked pulse train without any frequency detune is shown in Figure 7.2 and the corresponding pulse train is shown in Figure 7.3a. The shape of the phase plane exposes the phase between the displacement and its derivative. From the phase plane obtained, one can easily observe that the origin is a stable node and the limit cycle around that vicinity is a stable limit cycle, hence leading to stable system trajectory. The 4× multiplication pulse trains (i.e., $m=4$) without and with 5% harmonic distortion are shown in Figure 7.3b and c. Their corresponding phase planes are shown in Figure 7.4a and b. For the case of 0% harmonic distortion, which is the ideal case, the generated pulse train is perfectly multiplied with equal amplitude and the phase plane has stable symmetry periodic trajectories around the origin too. However, for the practical case (i.e., with 5% harmonic distortion) it is obvious that the pulse amplitude is unevenly distributed, which can be easily verified with the experimental results obtained by Wu and Dutta [3]. Its corresponding phase plane shows more complex asymmetry system trajectories.

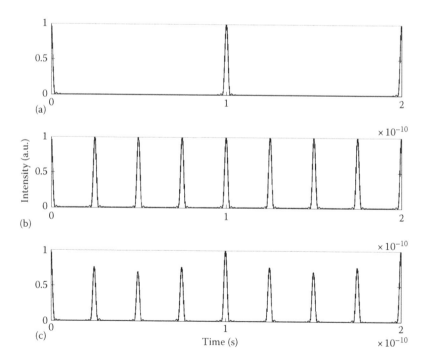

**FIGURE 7.3**  (a) Normalized pulse propagation of original pulse detuning fraction of 4, with (b) 0% and (c) 5% harmonic distortion noise.

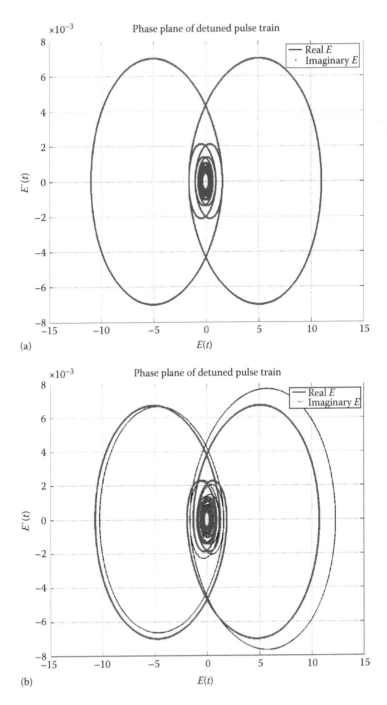

**FIGURE 7.4** Phase plane of detuned pulse train, $m=4$, with (a) 0% harmonic distortion and (b) 5% harmonic distortion (solid line—real part of the energy, dotted line—imaginary part of the energy, x-axes—$E(t)$, and y-axes—$E'(t)$).

One may naively think that the detuning fraction, $m$, could be increased to a very large number, so that a very small frequency, $\Delta f$, is deviated so as to obtain a very high repetition frequency. This is only true in the ideal world, if no harmonic distortion is present in the system. However, this is unreasonable for a practical mode-locked laser system.

We define the percentage fluctuation, %$F$ as follows:

$$\%F = \frac{E_{max} - E_{min}}{E_{max}} \times 100\% \qquad (7.2)$$

where $E_{max}$ and $E_{min}$ are the maximum and minimum peak amplitudes of the generated pulse train, respectively. For any practical mode-locked laser system, fluctuations above 50% should be considered as poor laser system design. Therefore, this is one of the limiting factors in a rational harmonic mode-locking fiber laser system. The relationships between the percentage fluctuation and harmonic distortion for three multipliers ($m = 2$, 4, and 8) are shown in Figure 7.5. Thus, the obtainable rational harmonic mode-locking is very much limited by the harmonic distortion of the system. For 100% fluctuation, it means no repetition rate multiplication, but with additional noise components; a typical pulse train and its corresponding phase plane are shown in Figures 7.6 (lower plot) and 7.7, respectively, with $m = 8$ and 20% harmonic distortion. The asymmetric trajectories of the phase graph explain the amplitude unevenness of the pulse train. Furthermore, it shows a more complex pulse formation system. Thus, it is clear that for any harmonic mode-locked laser system, the small side pulses generated are largely due to improper or not exact tuning of the modulation frequency of the system. An experiment result ($m = 8$) is depicted in Figure 7.6 for comparison.

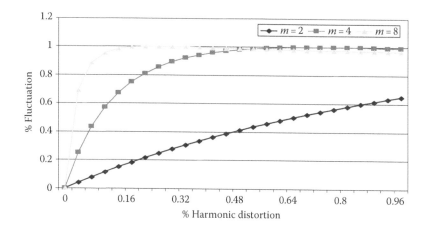

**FIGURE 7.5**   Relationship between the percentage amplitude fluctuation and the percentage harmonic distortion (diamond—$m = 2$, square—$m = 4$, and triangle—$m = 8$).

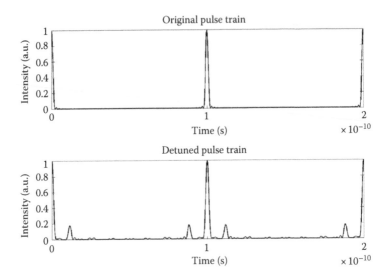

**FIGURE 7.6**  10 GHz pulse train (upper plot) and pulse train with $m=8$ and 20% harmonic distortion (lower plot).

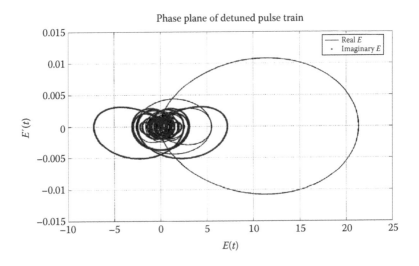

**FIGURE 7.7**  Phase plane of the pulse train with $m=8$ and 20% harmonic distortion.

### 7.2.4 DEMONSTRATION

By careful adjustment of the modulation frequency, polarization, gain level, and other parameters of the fiber ring laser, we are managed to obtain the 660th and 1230th order of rational harmonic detuning in the MLFRL with base frequency of 100 MHz, hence achieving 66 and 123 GHz pulse repetition frequency operation.

The autocorrelation traces and optical spectrums of the pulse operations are shown in Figure 7.8. With Gaussian pulse assumption, the obtained pulse widths of the operations are 2.5456 and 2.2853 ps, respectively. For the 100 MHz pulse operation, that is, without any frequency detune, the generated pulse width is about 91 ps. Thus, not only we achieved an increase in the pulse repetition frequency, but also a decrease in the generated pulse widths. This pulse-narrowing effect is partly due to the self-phase modulation (SPM) effect of the system as observed in the optical spectrums. Another reason for this pulse shortening is stated by Haus [18], where the pulse width is inversely proportional to the modulation frequency and given as

$$\tau^4 = \frac{2g}{M\omega_m^2\omega_g^2} \tag{7.3}$$

where
    $\tau$ is the pulse width of the mode-locked pulse
    $\omega_m$ is the modulation frequency
    $g$ is the gain coefficient
    $M$ is the modulation index
    $\omega_g$ is the gain bandwidth of the system

In addition, the duty cycle of our Gaussian modulation signal is ~1%, which is very much less than 50%, and this leads to a narrow pulse width too. Besides the uneven pulse amplitude distribution, high level of pedestal noise is also observed in the obtained results.

For the 66 GHz pulse operation, a 4 nm bandwidth filter is used in the setup, but it is removed for the 123 GHz operation. It is done so to allow more modes to be locked during the operation, thus, to achieve better pulse quality. In contrast, this increases the level of difficulty significantly in the system tuning and adjustment. As a result, the operation is very much determined by the gain bandwidth of the EDFA used in the laser setup [19–27].

The simulated phase planes for the given pulse operation are shown in Figure 7.9. They are simulated based on the 100 MHz base frequency, 10 round trips condition, and 0.001% of harmonic distortion contribution. There is no stable limit cycle in the phase graphs obtained; hence the system stability is hardly achievable, which is a known fact in the rational harmonic mode-locking [28–30]. Asymmetric system trajectories are observed in the phase planes of the pulse operations. This reflects the unevenness of the amplitude of the pulses generated. Furthermore, more complex pulse formation process is also revealed in the phase graphs obtained (Figure 7.9).

By a very small amount of frequency deviation, or improper modulation frequency tuning in the general context (see Figure 7.10), we can generate a pulse train with ~100 MHz modulation frequency with very short adjacent side pulses that is rather similar to Figure 7.6 (lower plot) despite the level of pedestal noise in the actual case. This is mainly because we do not consider other sources of noise in our modeling, except the harmonic distortion.

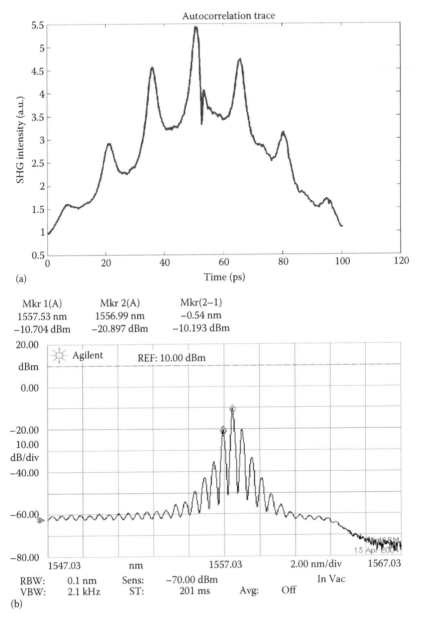

**FIGURE 7.8** Autocorrelation traces of (a) 66 GHz and (c) 123 GHz pulse operation and optical spectrums of (b) 66 GHz and (d) 123 GHz. *(Continued)*

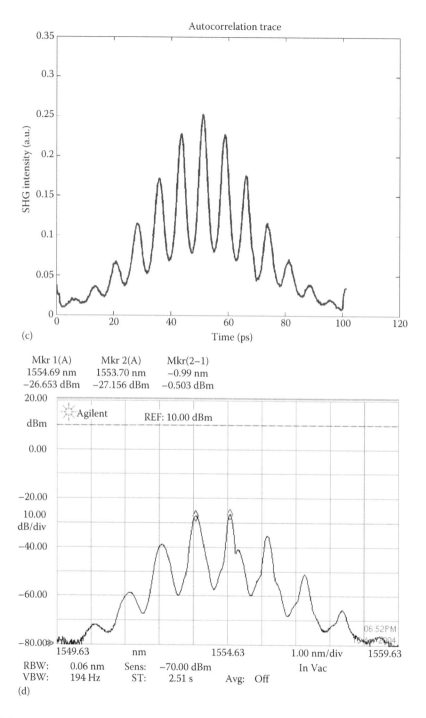

(c)

Mkr 1(A)        Mkr 2(A)        Mkr(2–1)
1554.69 nm      1553.70 nm      −0.99 nm
−26.653 dBm     −27.156 dBm     −0.503 dBm

(d)

**FIGURE 7.8 (Continued)**   Autocorrelation traces of (a) 66 GHz and (c) 123 GHz pulse operation and optical spectrums of (b) 66 GHz and (d) 123 GHz.

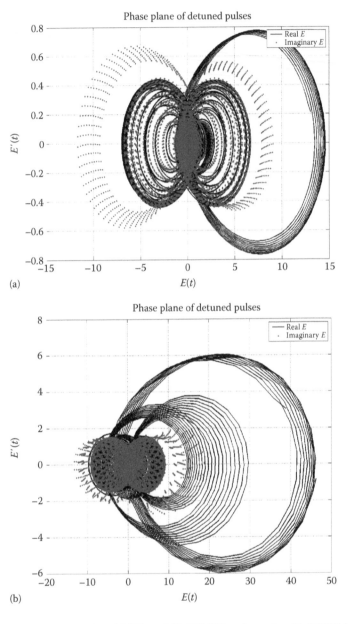

**FIGURE 7.9** Phase plane of (a) 66 GHz and (b) 123 GHz pulse train with 0.001% harmonic distortion noise.

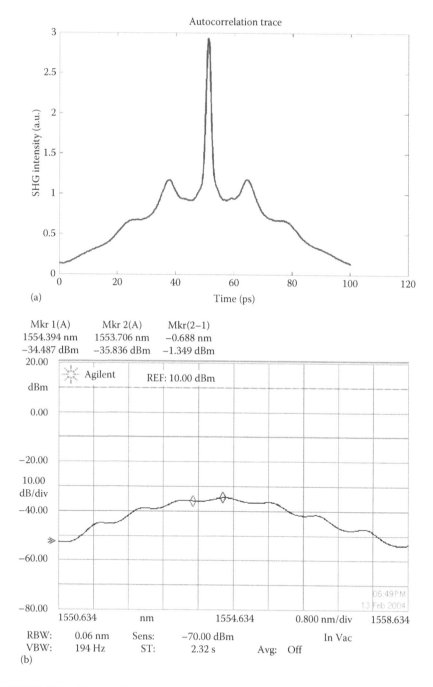

(a)

(b)

**FIGURE 7.10** Measured autocorrelation trace (a) and optical spectrum (b) of slight frequency detune in the mode-locked fiber ring laser.

### 7.2.5 Remarks

We have demonstrated 660th and 1230th order of rational harmonic mode-locking from a base modulation frequency of 100 MHz in the erbium-doped fiber ring laser (EDFRL), hence achieving 66 and 123 GHz pulse repetition frequency. To the best of our knowledge, this is the highest rational harmonic order obtained to date. Besides the repetition rate multiplication, we also obtain high pulse compression factor in the system, ~35× and 40×, relative to the nonmultiplied laser system.

In addition, we use phase plane analysis to study the laser system behavior. From the analysis model, the amplitude stability of the detuned pulse train can only be achieved either under negligible or no harmonic distortion condition, which is the ideal situation. The phase plane analysis also reveals the pulse forming complexity of the laser system.

## 7.3 REPETITION-RATE MULTIPLICATION RING LASER USING TEMPORAL DIFFRACTION EFFECTS

The pulse repetition rate of a mode-locked ring laser is usually limited by the bandwidth of the intracavity modulator. Hence, a number of techniques have to be used to increase the repetition frequency of the generated pulse train.

We investigate the stability and transient analyses of the GVD-multiplied pulse train using the phase plane analysis of nonlinear control analytical technique [2]. This is the first time, to the best of our knowledge that the phase plane analysis in the field of digital control can be used to analyze the stability and transient performances of the GVD repetition-rate multiplication systems.

The stability and the transient response of the multiplied pulses are studied using the phase plane technique of nonlinear control engineering. We also demonstrated 4× repetition rate multiplication on a 10 Gbps pulse train generated from the active harmonically MLFRL, hence achieving a 40 Gbps pulse train by using fiber GVD effect. It has been found that the stability of the GVD-multiplied pulse train, based on the phase plane analysis is hardly achievable even under the perfect multiplication conditions. Furthermore, an uneven pulse amplitude distribution is observed in the multiplied pulse train. In addition to that, the influences of the filter bandwidth in the laser cavity, nonlinear effect, and the noise performance are also studied in our analyses.

In Section 7.2, the GVD repetition-rate multiplication technique is briefly given. Section 7.3 describes the experimental setup for the repetition rate multiplication. Section 7.4 investigates the dynamic behavior of the phase plane of GVD multiplication system, followed by simulation and experimental results. Finally, some concluding remarks and possible future developments are given.

When a pulse train is transmitted through an optical fiber, the phase shift of $k$th individual lasing mode due to GVD is

$$\varphi_k = \frac{\pi \lambda^2 D z k^2 f_r^2}{c} \qquad (7.4)$$

where

λ is the center wavelength of the mode-locked pulses

$D$ is the fiber's GVD factor

$z$ is the fiber length

$f_r$ is the repetition frequency

$c$ is the speed of light in vacuum

This phase shift induces pulse broadening and distortion. At Talbot distance, $z_T = 2/\Delta\lambda f_r|D|$ [6], the initial pulse shape is restored, where $\Delta\lambda = f_r\lambda^2/c$ is the spacing between the Fourier-transformed spectra of the pulse trains. When the fiber length is equal to $z_T/2m$ (with $m = 2, 3, 4, \ldots$), every $m$th lasing modes oscillates in phase and the oscillation waveform maximums accumulate.

However, when the phases of other modes become mismatched, this weakens their contributions to pulse waveform formation. This leads to the generation of a pulse train with a multiplied repetition frequency with $m$-times. The pulse duration does not change that much even after the multiplication, because every $m$th lasing mode dominates in pulse waveform formation of $m$-times multiplied pulses. The pulse waveform therefore becomes identical to that generated from the mode-locked laser, with the same spectral property. Optical spectrum does not change after the multiplication process, because this technique utilizes only the change of phase relationship between lasing modes and does not use fiber's nonlinearity.

The effect of higher order dispersion might degrade the quality of the multiplied pulses, that is, pulse broadening, appearance of pulse wings, and pulse-to-pulse intensity fluctuation. In this case, any dispersive media to compensate the fiber's higher order dispersion would be required in order to complete the multiplication process. To achieve higher multiplications, the input pulses must have a broad spectrum and the fractional Talbot length must be very precise in order to receive high-quality pulses. If the average power of the pulse train induces the nonlinear suppression and experience anomalous dispersion along the fiber, solitonic dynamics would occur and prevent the linear Talbot effect from occurring.

The highest repetition rate obtainable is limited by the duration of the individual pulses, as pulses start to overlap when the pulse duration becomes comparable to the pulse train period, that is, $m_{max} = \Delta T/\Delta t$, where $\Delta T$ is the pulse train period and $\Delta t$ is the pulse duration.

The GVD repetition-rate multiplication system is used to achieve a 40 Gbps operation. The input to the GVD multiplier is a 10.217993 Gbps laser pulse source, generated from the active harmonically MLFRL, operating at 1550.2 nm wavelength.

The principal element of the active harmonically MLFRL is an optical closed loop with an optical gain medium, that is, the EDF under a 980 nm pump source, an optical 10 GHz amplitude modulator, optical BPF, optical fiber couplers, and other associated optics. The generic schematic construction of the active MLFRL is shown in Figures 7.1 and 7.11. In this case, the active MLFRL is based on a fiber ring cavity where the 25 m EDF with $Er^{3+}$ ion concentration of $4.7 \times 10^{24}$ ions/m³ is pumped by two diode lasers at 980 nm: SDLO-27-8000-300 and CosetK1116 with maximum forward pump power of 280 mW and backward pump power of 120 mW.

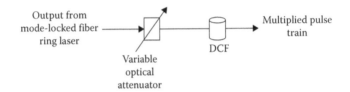

**FIGURE 7.11**    Experimental setup for GVD repetition-rate multiplication system.

The pump lights are coupled into the cavity by the 980/1550 nm WDM couplers with insertion loss for 980 and 1550 nm signals about 0.48 and 0.35 dB, respectively. A polarization-independent optical isolator ensures the unidirectional lasing. The birefringence of the fiber is compensated by a PC. A tunable Fabry–Perot (FP) filter with a 3 dB bandwidth of 1 nm and wavelength tuning range from 1530 to 1560 nm is inserted into the cavity to select the center wavelength of the generated signal as well as to reduce the noise in the system. In addition, it is used for the longitudinal modes selection in the mode-locking process. Pulse operation is achieved by introducing a JDS Uniphase 10 Gbps lithium niobate, Ti:LiNbO$_3$ MZM into the cavity with half-wave voltage, $V_\pi$ of 5.8 V. The modulator is DC biased near the quadrature point and not more than the $V_\pi$ such that it operates on the linear region of its characteristic curve and driven by the sinusoidal signal derived from an Anritsu 68347C synthesizer signal generator. The modulating depth should be less than unity to avoid signal distortion. The modulator has an insertion loss of ≤7 dB. The output coupling of the laser is optimized using a 10/90 coupler. Ninety percent of the optical field power is coupled back into the cavity ring loop, while the remaining portion is taken out as the output of the laser and is analyzed using a New Focus 1014B 40 GHz photodetector, Ando AQ6317B optical spectrum analyzer, Textronix CSA 8000 80E01 50 GHz communications signal analyzer (CSA), or Agilent E4407B RF spectrum analyzer.

One module of about 3.042 km of dispersion compensating fiber (DCF) with a dispersion value of −98 ps/nm/km was used in the experiment; the schematic of the experimental setup is shown in Figure 7.11. The variable optical attenuator used in the setup is to reduce the optical power of the pulse train generated by the MLFRL, hence to remove the nonlinear effect of the pulse. A DCF, that is, fiber of negative dispersion factor length for 4× multiplication factor on the ~10 GHz signal is required and estimated to be 3.048173 km. The output of the multiplier (i.e., at the end of DCF) is then observed using Textronix CSA 8000 80E01 50 GHz CSA.

### 7.3.1 Phase Plane Analysis

The system modeling for the GVD multiplier is done based on the following assumptions: (1) perfect output pulse from the MLFRL without any timing jitter, (2) the multiplication is achieved under ideal conditions (i.e., exact fiber length for a certain dispersion value), (3) no fiber nonlinearity is included in the analysis of the multiplied pulse, (4) no other noise sources are involved in the system, and (5) uniform or Gaussian lasing mode amplitude distribution.

## 7.3.2   Uniform Lasing Mode Amplitude Distribution

Uniform lasing mode amplitude distribution is assumed at the first instance, that is, ideal mode-locking condition. The simulation is done based on the 10 Gbps pulse train, centered at 1550 nm, with fiber dispersion value of –98 ps/km/nm, 1 nm flat-top BPF is used in the cavity of mode-locked fiber laser. The estimated Talbot distance is 25.484 km.

The original pulse (direct from the mode-locked laser) propagation behavior and its phase plane are shown in Figures 7.12a and 7.13a. From the phase plane obtained, one can observe that the origin is a stable node and the limit cycle around that vicinity is a stable limit cycle. This agrees very well with our first assumption: ideal pulse

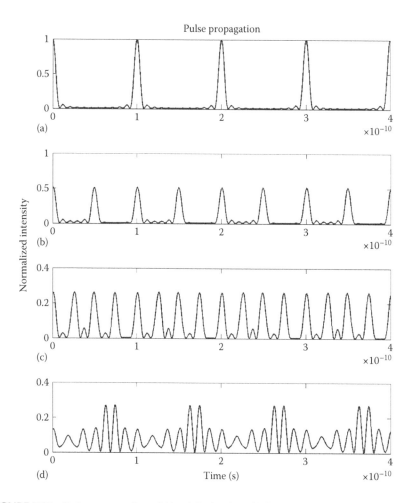

**FIGURE 7.12**   Pulse propagation of (a) original pulse, (b) 2× multiplication, (c) 4× multiplication, and (d) 8× multiplication with 1 nm filter bandwidth and equal lasing mode amplitude analysis.

Phase plane

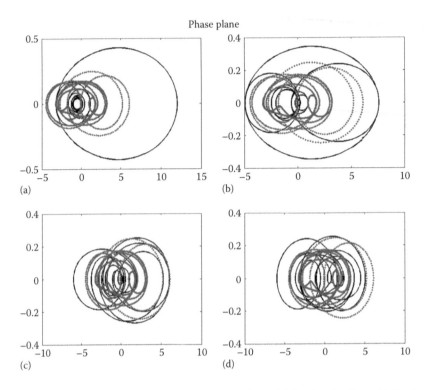

(a)

(b)

(c)

(d)

**FIGURE 7.13** Phase plane of (a) original pulse, (b) 2× multiplication, (c) 4× multiplication, and (d) 8× multiplication with 1 nm filter bandwidth and equal lasing mode amplitude analysis (solid line—real part of the energy, dotted line—imaginary part of the energy, $x$-axes—$E(t)$, and $y$-axes—$E'(t)$).

train at the input of the multiplier. Also, we present the pulse propagation behavior and phase plane for 2×, 4×, and 8× GVD multiplication system in Figures 7.12 and 7.13. The shape of the phase graph exposes the phase between the displacement and its derivative.

As the multiplication factor increases, the system trajectories are moving away from the origin. As for the 4× and 8× multiplications, there is neither stable limit cycle nor stable node on the phase planes even with the ideal multiplication parameters. Here we see the system trajectories spiral out to an outer radius and back to inner radius again. The change in the radius of the spiral is the transient response of the system. Hence, with the increase in multiplication factor, the system trajectories become more sophisticated. Although GVD repetition-rate multiplication system uses only the phase change effect in multiplication process, the inherent nonlinearities still indirectly affect its stability. Despite the reduction in the pulse amplitude, we observe uneven pulse amplitude distribution in the multiplied pulse train. The percentage of unevenness increases with the multiplication factor in the system.

Figure 7.14 displays (a) the pulse propagation of the original pulse, and then (b) 2× multiplication, (c) 4× multiplication, and (d) 8× multiplication sequence with 1 nm filter bandwidth. Figure 7.15 depicts the corresponding behavior in the phase plane of the time sequences of Figure 7.14 with (a) original pulse, (b) 2× multiplication, (c) 4× multiplication, and (d) 8× multiplication with 1 nm filter bandwidth

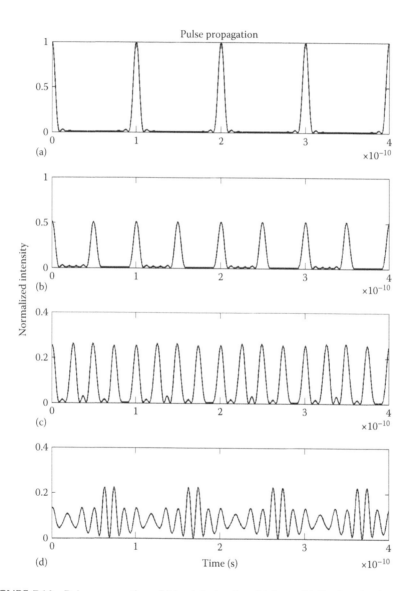

**FIGURE 7.14**   Pulse propagation of (a) original pulse, (b) 2× multiplication, (c) 4× multiplication, and (d) 8× multiplication with 1 nm filter bandwidth and Gaussian lasing mode amplitude analysis.

Phase plane

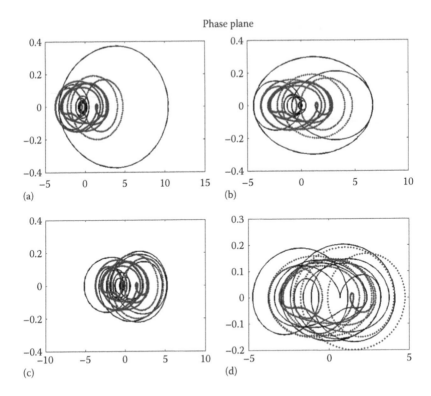

(a)

(b)

(c)

(d)

**FIGURE 7.15** Phase plane of (a) original pulse, (b) 2× multiplication, (c) 4× multiplication, and (d) 8× multiplication with 1 nm filter bandwidth and Gaussian lasing mode amplitude analysis (solid line—real part of the energy, dotted line—imaginary part of the energy, x-axes—$E(t)$, and y-axes—$E'(t)$).

and Gaussian lasing mode amplitude analysis shown in solid line (black). The real part of the energy is indicated in dotted line and the imaginary part of the energy in continuous gray lines.

Similarly, Figures 7.16 and 7.17 depict, respectively, the pulse propagation and their corresponding behavior in the phase plane of (a) original pulse, (b) 2× multiplication, (c) 4× multiplication, and (d) 8× multiplication with 3 nm filter bandwidth and Gaussian lasing mode amplitude analysis.

Figures 7.18 and 7.19 depict, respectively, the pulse propagation and phase plane dynamics of (a) original pulse, (b) 2× multiplication, (c) 4× multiplication, and (d) 8× multiplication with 3 nm filter bandwidth, Gaussian lasing mode amplitude analysis, and the input power of 1 W (30 dBm).

Figures 7.20 and 7.21 depict, respectively, the pulse propagation and corresponding dynamics in the phase plane of (a) the original pulse, (b) 2× multiplied pulse sequence, (c) 4× multiplication, and (d) 8× multiplications with 3 nm filter bandwidth, Gaussian lasing mode amplitude analysis, and 0 dB signal to noise ratio.

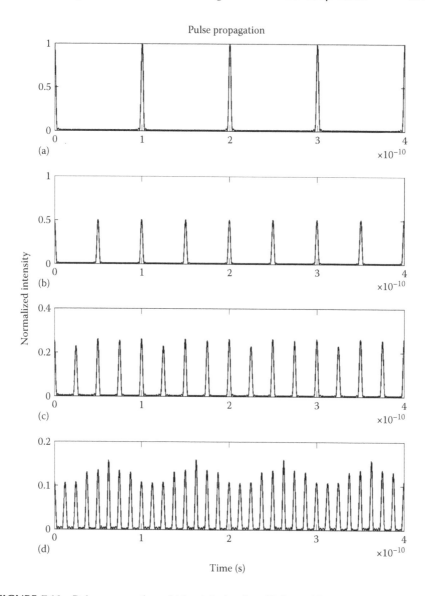

**FIGURE 7.16**   Pulse propagation of (a) original pulse, (b) 2× multiplication, (c) 4× multiplication, and (d) 8× multiplication with 3 nm filter bandwidth and Gaussian lasing mode amplitude analysis.

## 7.3.3   GAUSSIAN LASING MODE AMPLITUDE DISTRIBUTION

This set of the simulation models the practical filter used in the system. It gives us a better insight on the GVD repetition-rate multiplication system behavior. The parameters used in the simulation are exactly the same except the filter of the laser has been changed to 1 nm (125 GHz at 1550 nm) Gaussian-profile BPF. The spirals

Phase plane

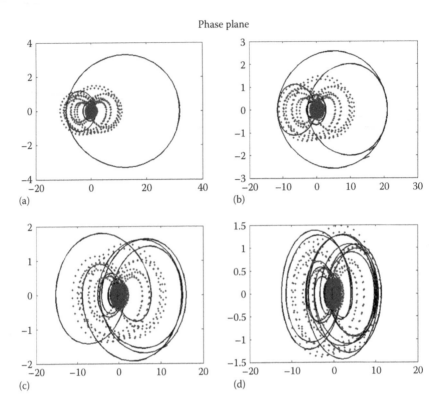

(a)

(b)

(c)

(d)

**FIGURE 7.17** Phase plane of (a) original pulse, (b) 2× multiplication, (c) 4× multiplication, and (d) 8× multiplication with 3 nm filter bandwidth and Gaussian lasing mode amplitude analysis (solid line—real part of the energy, dotted line—imaginary part of the energy, $x$-axes—$E(t)$, and $y$-axes—$E'(t)$).

of the system trajectories and uneven pulse amplitude distribution are more severe than those in the uniform lasing mode amplitude analysis.

### 7.3.4 EFFECTS OF FILTER BANDWIDTH

Filter bandwidth used in the MLFRL will affect the system stability of the GVD repetition-rate multiplication system as well. The analysis done above is based on 1 nm filter bandwidth. The number of modes locked in the laser system increases with the bandwidth of the filter used, which gives us a better quality of the mode-locked pulse train. The simulation results shown in the following are based on the Gaussian lasing mode amplitude distribution; 3 nm filter bandwidth used in the laser cavity and other parameters remain unchanged.

With wider filter bandwidth, the pulse width and the percentage pulse amplitude fluctuation decrease. This suggests a better stability condition. Instead of spiraling away from the origin, the system trajectories move inward to the stable node. However, this leads to a more complex pulse formation system.

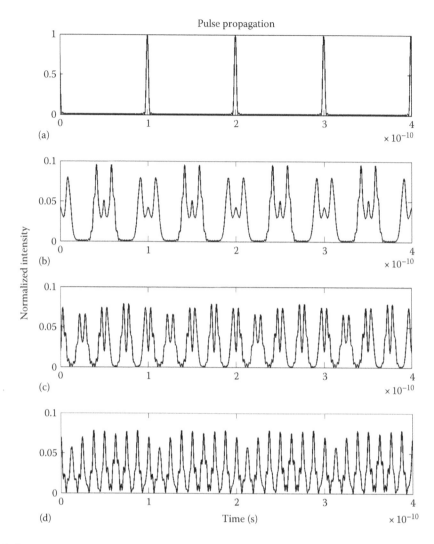

**FIGURE 7.18**    Pulse propagation of (a) original pulse, (b) 2× multiplication, (c) 4× multiplication, and (d) 8× multiplication with 3 nm filter bandwidth, Gaussian lasing mode amplitude analysis and input power = 1 W.

## 7.3.5  NONLINEAR EFFECTS

When the input power of the pulse train enters the nonlinear region, the GVD multiplier loses its multiplication capability as predicted. The additional nonlinear phase shift due to the high input power is added to the total pulse phase shift and destroys the phase change condition of the lasing modes required by the multiplication condition. Furthermore, this additional nonlinear phase shift also changes the pulse shape and the phase plane of the multiplied pulses.

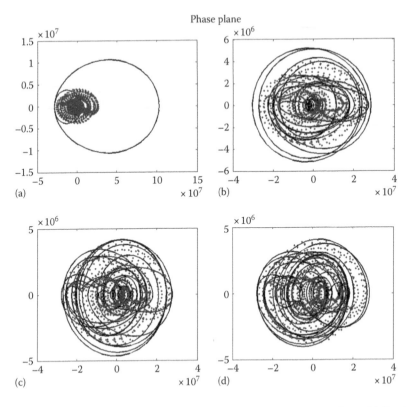

**FIGURE 7.19** Phase plane of (a) original pulse, (b) 2× multiplication, (c) 4× multiplication, and (d) 8× multiplication with 3 nm filter bandwidth, Gaussian lasing mode amplitude analysis and input power = 1 W (solid line—real part of the energy, dotted line—imaginary part of the energy, x-axes—$E(t)$, and y-axes—$E'(t)$).

### 7.3.6 NOISE EFFECTS

The above simulations are all based on the noiseless situation. However, in the practical optical communication systems, noises are always sources of nuisance that can cause system instability, therefore, it must be taken into the consideration for the system stability studies.

Since the optical intensity of the $m$-times multiplied pulse is $m$-times less than the original pulse, it is more vulnerable to noise. The signal is difficult to differentiate from the noise within the system if the power of multiplied pulse is too small. The phase plane of the multiplied pulse is distorted due to the presence of the noise, leading to poor stability of the soliton generator.

### 7.3.7 DEMONSTRATION

The obtained 10 GHz output pulse train from the MLFRL is shown in Figure 7.22. Its spectrum is shown in Figure 7.23. This output was then used as the input to the

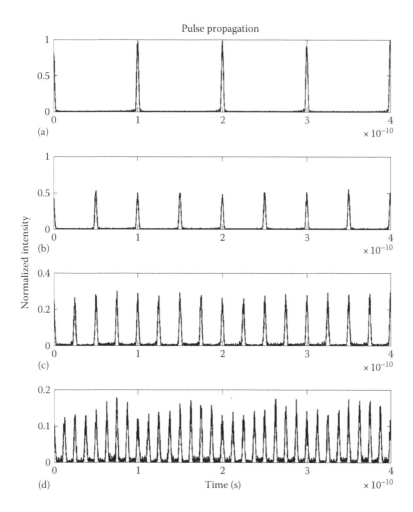

**FIGURE 7.20**  Pulse propagation of (a) original pulse, (b) 2× multiplication, (c) 4× multipli-cation, and (d) 8× multiplication with 3 nm filter bandwidth, Gaussian lasing mode amplitude analysis, and 0 dB signal-to-noise ratio.

DCF, which acts as the GVD multiplier in our experiment. The obtained 4× multi-plication by the GVD effect and its spectrum are shown in Figures 7.24 and 7.25.

The spectrums for both cases (original and multiplied pulse) are exactly the same since this repetition-rate multiplication technique utilizes only the change of phase relationship between lasing modes and does not use fiber's nonlinearity.

The multiplied pulse suffers an amplitude reduction in the output pulse train; however, the pulse characteristics should remain the same. The instability of the multiplied pulse train is mainly due to the slight deviation from the required DCF length (0.2% deviation). Another reason for the pulse instability, which derived from our analysis, is the divergence of the pulse energy variation in the vicinity around the origin as the multiplication factor gets higher. The pulse amplitude decreases

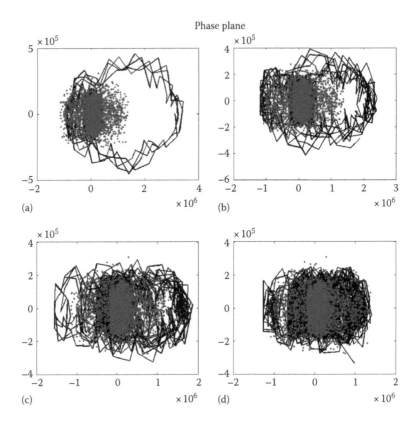

**FIGURE 7.21** Phase plane of (a) original pulse, (b) 2× multiplication, (c) 4× multiplication, and (d) 8× multiplication with 3 nm filter bandwidth, Gaussian lasing mode amplitude analysis, and 0 dB signal to noise ratio (solid line—real part of the energy, dotted line—imaginary part of the energy, $x$-axes—$E(t)$, and $y$-axes—$E'(t)$).

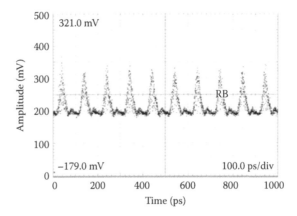

**FIGURE 7.22** 10 GHz pulse train from mode-locked fiber ring laser (horizontal scale:100 ps/div, vertical scale: amplitude 50 mV/div).

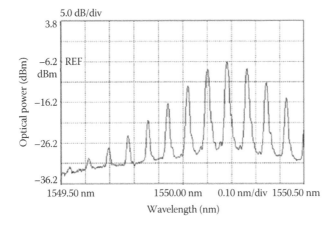

**FIGURE 7.23**    10 GHz pulse spectrum from mode-locked fiber ring laser.

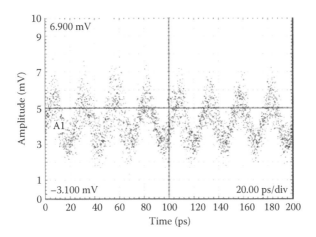

**FIGURE 7.24**    40 GHz multiplied pulse train (20 ps/div, 1 mV/div).

with the increase in multiplication factor; as the fact of energy conservation, when it reaches certain energy level, which is indistinguishable from the noise level in the system, the whole system will become unstable and noisy.

### 7.3.8   REMARKS

In this section, 4× repetition rate multiplication by using fiber GVD effect is demonstrated; hence, a 40 GHz pulse train can be obtained from a 10 GHz mode-locked fiber laser source. However, its stability is of great concern for the practical use in the optical communication systems. Although the GVD repetition-rate multiplication technique is linear in nature, the inherent nonlinear effects in such system may disturb the stability of the system. Hence any linear approach may not be suitable

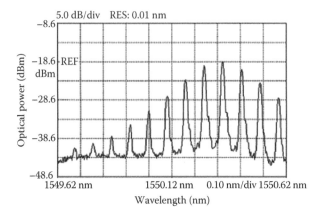

**FIGURE 7.25** 40 GHz pulse spectrum from GVD multiplier.

in deriving the system stability. Stability analysis for this multiplied pulse train has been studied by using the nonlinear control stability theory, which is for the first time, to the best of our knowledge, that phase plane analysis is being used to study the transient and stability performance of the GVD repetition rate multiplication system. Surprisingly, from the analysis model, the stability of the multiplied pulse train can hardly be achieved even under perfect multiplication conditions. Furthermore, we observed uneven pulse amplitude distribution in the GVD-multiplied pulse train, which is due to the energy variations between the pulses that cause some energy beating between them. Another possibility is the divergence of the pulse energy variation in the vicinity around the equilibrium point that leads to instability.

The pulse amplitude fluctuation increases with the multiplication factor. Also, with wider filter bandwidth used in the laser cavity, better stability condition can be achieved. The nonlinear phase shift and noises in the system challenge the system stability of the multiplied pulses. They not only change the pulse shape of the multiplied pulses, but also distort the phase plane of the system. Hence, the system stability is greatly affected by the SPM as well as the system noises.

This stability analysis model can further be extended to include some system nonlinearities, such as the gain saturation effect, non-quadrature biasing of the modulator, fiber nonlinearities, etc. The chaotic behavior of the system may also be studied by applying different initial phase and injected energy conditions to the model.

## 7.4   BISTABILITY, BIFURCATION, AND CHAOS IN NONLINEAR LOOP FIBER LASERS

All real physical systems are nonlinear in nature as briefly declared in Chapter 2. Apart from systems designed for linear signal processing, many systems have to be nonlinear by assumption, for instance flip-flops, modulators, demodulators, amplifiers, etc. In this section, we focus on the optical bistability, bifurcation, and chaos of a nonlinear system.

Bidirectional lightwaves propagation in an EDFRL and the behavior of optical bistability are studied. We exploit this commonly known undesirable bidirectional propagation of lightwaves for constructing fiber laser configurations based on nonlinear optical loop mirror (NOLM) and nonlinear amplifying loop mirror (NALM) structure. The lasers are operated in different regions. The switching capability of the laser based on its bistability characteristics is identified. Chaotic phenomena are also observed in these fiber lasers.

### 7.4.1 INTRODUCTION

The nonlinear phenomena of optical bistability have been studied in nonlinear resonators since 1976 by placing the nonlinear medium inside a cavity formed by using multiple mirrors [32]. As for the fiber-based devices, single mode fiber was used as the nonlinear medium inside a ring cavity in 1983 [33]. Since then, the study of nonlinear phenomena in fiber resonators has remained a topic of considerable interest.

Fiber ring laser is a rich and active research field in optical communications. Many fiber ring laser configurations have been proposed and constructed to achieve different objectives. It can be designed for CW or pulse operation, linear or nonlinear operation, fast or slow repetition rate, narrow or broad pulse width, etc., for various kinds of photonic applications.

The simplest fiber laser structure is an optical closed loop with a gain medium and some associated optical components such as optical couplers. The gain medium used can be any rare earth element–doped fiber amplifier, such as erbium and ytterbium, semiconductor optical amplifier, parametric amplifier, etc., as long as it provides the gain requirement for lasing. Without any mode-locking mechanism, the laser shall operate in the CW regime. By inserting an active mode-locker into the laser cavity, that is, either amplitude or phase modulator, the resulting output shall be an optical pulse train operating at the modulating frequency, when the phase conditions is matched, as discussed in Chapter 3. This often results in a high-speed optical pulse train, however, with broad pulse width. There is another kind of fiber laser, which uses the nonlinear effect in generating the optical pulses and is known as the passive mode-locking technique as described in Chapter 7. Saturable absorber, stretched pulse mode-locking [34], nonlinear polarization rotation [35], figure-eight fiber ring lasers, etc. can be grouped under this category. This type of laser generates shorter optical pulse sequences in exchange for high repetition frequency. Hence, this is the trade-off between the active and passive mode-locked systems.

Although there are different types of fiber laser, with different operating regimes, one common criterion is the unidirectionality, besides the gain and phase matching conditions. Unidirectional propagation has been proved to offer better lasing efficiency, less sensitive to back reflections, and good potential for single longitudinal mode operation [36], and can be achieved by incorporating an optical isolator within the laser cavity. Shi et al. [37] has demonstrated a unidirectional inverted S-type EDFRL without the use of optical isolator, but with optical couplers. In their laser system, the lightwaves are passed through in one direction and suppressed in the other using certain coupling ratios, hence, achieving unidirectional operation. However, only the power difference between the two CW lightwaves was studied and the works did not extend further into unconventional and nonlinear regions of operation.

In this chapter, we describe the bidirectional lightwaves propagation in an EDFRL and its behavior due to optical bistability, which has been well reported previously [38–40]. We exploit this in constructing a kind of fiber laser configuration based on NOLM and NALM structure. This laser configuration is similar to the one demonstrated by Shi et al. [37], but operating the laser in different regimes. We investigate the switching capability of the laser based on its bistability behavior. Furthermore, we focus on the nonlinear dynamics of the fiber laser and describe the optical bifurcation phenomenon.

### 7.4.2 Optical Bistability, Bifurcation, and Chaos

An optical bistable device is a device with two possible operation points. It will remain stable in any of the two optical states, one of high transmission and the other of low transmission, depending upon the intensity of the light passing through it. In this section, we will discuss the effect of optical nonlinearity together with the proper feedback, which can give rise to optical bistability and hysteresis. This is expected as these two effects are also observed in nonlinear electronic circuits with feedback, such as Schmitt trigger, as well as hybrid optical devices, such as an acousto-optic device with feedback [41].

A typical bistability curve is shown in Figure 7.26, with $Y_i$ and $Y_o$ as the input and output parameters, respectively. For an input value between $Y_1$ and $Y_2$, there are three possible output values. The middle segment, with negative slope, is known to be always unstable. Therefore, the output will eventually have two stable values. When two outputs are possible, one of the outputs that is eventually realized depends on the history of how the input is reached, and hence, the hysteresis phenomenon.

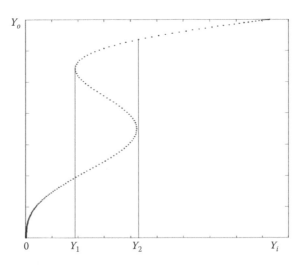

**FIGURE 7.26**  Bistability curve.

As the input value is increased from zero, the output will follow the lower branch of the curve until the input value reaches $Y_2$. Then it will jump up and follow the upper branch. However, if the input value is decreased from some points after the jump, the output will remain on the upper branch until the input value hits $Y_1$, and then the output will jump down and follow the lower branch. Hence, the bistability region observed is from $Y_1$ to $Y_2$.

In an optical ring cavity, the lightwaves can be split into two counterpropagating components that can mutually interact leading to optical gain competition. Hence, in a ring laser, the lightwave may propagate in one direction or another depending on the initial configuration and is thus a running wave in general. To eliminate this randomness, a device such as optical isolator can be inserted into the ring cavity to block the unwanted wave.

There are two different types of optical bistability, namely, absorptive bistability and dispersive bistability in a nonlinear ring cavity comprising a two-level gain medium as nonlinear medium. Absorptive bistability is the case when the incident optical frequency is close to or equal to the transition frequency of the atoms from one level to another. In other words, the system is in perfect resonance condition. In this case, the absorption coefficient becomes a nonlinear function of the incident frequency of the lightwaves.

On the other hand, if the frequencies are far apart, the gain medium behaves like a Kerr-type material and the system exhibits what is called dispersive bistability. A nonzero atomic detuning would introduce saturable dispersion in response to the medium. In this case, the material can be modeled by an effective nonlinear refractive index (the Kerr's effects), that is, its refractive index varies nonlinearly with respect to its intensity [41].

For optical bistability in a ring cavity, there is a possible instability due to the counterpropagating wave. Hence, it is interesting to know the lightwave behavior for both co- and counterpropagating components. The bidirectional operation of an optical bistable ring system has not yet been studied thoroughly. Although some studies have been done related to the bistability properties of the ring cavities, they focused on forcing the system to support only unidirectional operation [42]. The restriction to unidirectional propagation has been quite consistent with the experimental results; however, the exceptions were noted [43]. Unfortunately, no further investigation has been carried out since then.

It was discovered that the nonlinear response of a ring resonator could initiate a period-doubling (bifurcation) route to optical chaos [44]. In a dynamical system, a bifurcation is a period-doubling, -quadrupling, etc., that accompanies the onset of chaos. It represents a sudden appearance of a qualitatively different solution for a nonlinear system as some parameters are varied. A typical bifurcation map is shown in Figure 7.27 with $g$ and $y$ as the input and output system parameters, respectively. Period-doubling action starts at $g = {\sim}1.5$ and period-quadrupling at $g = {\sim}2$. Region beyond $g = {\sim}2.3$ is known as chaos.

Each of the local bifurcations may give rise to distinct route to chaos if the bifurcations appear repeatedly when changing the bifurcation parameter. These routes are important since it is difficult to conclude from experimental data alone whether irregular behavior is due to measurement noise or chaos. Recognition of

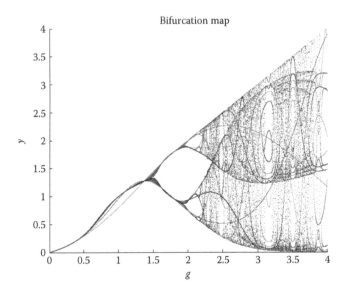

**FIGURE 7.27**  Typical bifurcation map.

one of the typical routes to chaos in experiments is a good indication that the dynamics may be chaotic [45].

1. Period-doubling route to chaos—when a cascade of successive period-doubling bifurcations occurs when changing the value of the bifurcation parameter, it is often the case that finally the system reached chaos.
2. Intermittency route to chaos—the route to chaos cased by saddle-node bifurcations. The common feature of which is a direct transition from regular motion to chaos.
3. Torus breakdown route to chaos—the quasiperiodic route to chaos results from a sequence of Hopf bifurcations.

It is well-known that self-pulsing often leads to optical chaos in the laser output, following a period-doubling or a quasiperiodic route. The basic idea is that the dynamics of the intracavity field is different from one round trip to another in a nonlinear fashion. The characteristics of a chaotic system [46] are as follows:

- Sensitive dependence on initial conditions—it gives rise to an apparent randomness in the output of the system and the long-term unpredictability of the state. Because the chaotic system is deterministic, the two trajectories that start from identical initial states will follow precisely the same paths through the state space.
- Randomness in the time domain—in contrast to the periodic waveforms, chaotic waveform is quite irregular and does not appear to repeat itself in any observation period of finite length.

- Broadband power spectrum—every periodic signal may be decomposed into Fourier series, a weighted sum of sinusoids at integers multiples of a fundamental frequency. Thus a periodic signal appears in the frequency domain as a set of spikes at the fundamental frequency and its harmonics. The chaotic signal is qualitatively different from the periodic signal. The aperiodic nature of its time-domain waveform is reflected in the broadband noise-like power spectrum. This broadband structure of the power spectrum persists even if the spectral resolution is increased to a higher frequency.

A typical example of the chaotic system is Lorenz attractor (see also Appendix A), or more commonly known as butterfly attractor, with the following system description and depicted in Figure 7.28, with $a = 10$, $b = 28$, and $c = 8/3$.

$$\frac{dx}{dt} = a(y - x) \tag{7.5}$$

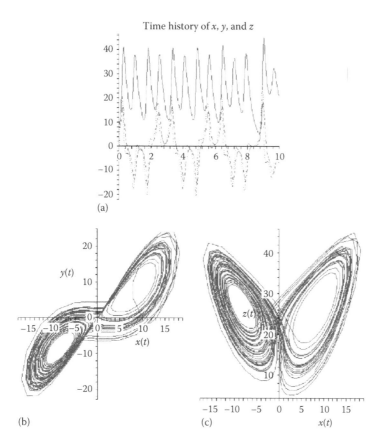

**FIGURE 7.28**   System response of Lorenz attractor (a) time traces of $x$—dash, $y$—dotted, and $z$—solid, (b) $y$–$x$ trajectory, and (c) $z$–$x$ trajectory.

$$\frac{dy}{dt} = x(b-z) - y \qquad (7.6)$$

$$\frac{dz}{dt} = xy - cz \qquad (7.7)$$

### 7.4.3 Nonlinear Optical Loop Mirror

The concept of NOLM was proposed by Doran and Wood [47]. It is basically a fiber-based Sagnac interferometer that uses the nonlinear phase shift of optical fiber for optical switching. This configuration is inherently stable since the two arms of the structure reside in the same fiber and same optical path lengths for the signals propagating in both arms, however, in opposite direction. There is no feedback mechanism in this structure since all lightwaves entering the input port exit from the loop after a single round trip. The NOLM in its simplest form contains a fiber coupler, with two of its output ports connected together, as shown in Figure 7.29, with $\kappa$ being the coupling ratio of the coupler, $E_1$, $E_2$, $E_3$, and $E_4$ are the fields at ports 1, 2, 3, and 4, respectively.

Under lossless condition, the input–output relationships of a coupler with an intensity coupling ratio $\kappa$ are

$$E_3 = \sqrt{\kappa}E_1 + j\sqrt{1-\kappa}E_2$$
$$E_4 = \sqrt{\kappa}E_2 + j\sqrt{1-\kappa}E_1 \qquad (7.8)$$

where $j^2 = -1$. If low intensity light is fed into port 1, that is, no nonlinear effect, the transmission, $T$ of the device is

$$T = \frac{P_{out}}{P_{in}} = 1 - 4\kappa(1-\kappa) \qquad (7.9)$$

If $\kappa = 0.5$, all light will be reflected back to the input, therefore, the name loop mirror; else, the counterpropagating lightwaves in the loop will have different intensities, and, thus, leads to different nonlinear phase shifts. This device can be designed to

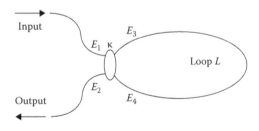

**FIGURE 7.29**   Nonlinear optical loop mirror.

transmit high power signal while reflecting it at low power levels, thus acting as an all-optical switch.

After $L$ propagation, the output signals $E_{3L}$ and $E_{4L}$ become as follows with non-linear phase shifts taken into account

$$E_{3L} = E_3 \exp\left( j \frac{2\pi n_2 L}{\lambda} |E_3|^2 \right)$$

$$E_{4L} = E_4 \exp\left( j \frac{2\pi n_2 L}{\lambda} |E_4|^2 \right)$$

(7.10)

where

$|E_3|^2$ and $|E_4|^2$ are the light intensities in the two arms
$n_2$ is the nonlinear coefficient
$L$ is the loop length
$\lambda$ is the operating wavelength of the signal
$E_1 = E_{01} \exp(j\omega_1 t)$ and $E_2 = E_{02} \exp(j\omega_2 t)$:

$$
\begin{aligned}
|E_{3L}|^2 &= |E_3|^2 \\
&= \kappa |E_1|^2 + (1-\kappa)|E_2|^2 + j\sqrt{\kappa(1-\kappa)}\left( E_1^* E_2 - E_1 E_2^* \right) \\
&= \kappa |E_{01}|^2 + (1-\kappa)|E_{02}|^2 - 2\sqrt{\kappa(1-\kappa)}E_{01}E_{02}\sin[(\omega_2 - \omega_1)t] \\
|E_{4L}|^2 &= |E_4|^2 \\
&= \kappa |E_2|^2 + (1-\kappa)|E_1|^2 + j\sqrt{\kappa(1-\kappa)}\left( E_2^* E_1 - E_2 E_1^* \right) \\
&= \kappa |E_{02}|^2 + (1-\kappa)|E_{01}|^2 + 2\sqrt{\kappa(1-\kappa)}E_{01}E_{02}\sin[(\omega_2 - \omega_1)t]
\end{aligned}
$$

(7.11)

The last term of Equation 7.11 represents the interference pattern between $E_1$ and $E_2$. Most literatures consider only simple case with single input at port 1, that is, $E_2 = 0$, and hence reduce the interference effect between the signals and the outputs at ports 1 and 2 are given as follows:

$$|E_{o1}|^2 = |E_{01}|^2 \left\{ 2\kappa(1-\kappa)\left[ 1 + \cos\left( (1-2\kappa)\frac{2\pi n_2 |E_{01}|^2 L}{\lambda} \right) \right] \right\}$$

$$|E_{o2}|^2 = |E_{01}|^2 \left\{ 1 - 2\kappa(1-\kappa)\left[ 1 + \cos\left( (1-2\kappa)\frac{2\pi n_2 |E_{01}|^2 L}{\lambda} \right) \right] \right\}$$

(7.12)

However, for more complete studies, we consider inputs at both ports; and the outputs can be expressed as

$$|E_{o1}|^2 = \kappa|E_{4L}|^2 + (1-\kappa)|E_{3L}|^2 + j\sqrt{\kappa(1-\kappa)}\left(E_{4L}^* E_{3L} - E_{4L}E_{3L}^*\right)$$

$$= [\kappa^2 + (1-\kappa)^2]|E_{02}|^2 + 2\kappa(1-\kappa)|E_{01}|^2 - 2(1-2\kappa)\sqrt{\kappa(1-\kappa)}$$

$$\times E_{01}E_{02}\sin[(\omega_2 - \omega_1)t] + j\sqrt{\kappa(1-\kappa)}\left(E_3 E_4^* \exp(j\Delta\theta) - E_3^* E_4 \exp(-j\Delta\theta)\right)$$

$$= |E_{01}|^2[2\kappa(1-\kappa)[1+\cos(\Delta\theta)]] + |E_{02}|^2[1-2\kappa(1-\kappa)[1+\cos(\Delta\theta)]]$$

$$- 2\sqrt{\kappa(1-\kappa)}E_{01}E_{02}[(1-2\kappa)\sin[(\omega_2 - \omega_1)t][1+\cos(\Delta\theta)]$$

$$+ \cos[(\omega_2 - \omega_1)t]\sin(\Delta\theta)] \tag{7.13}$$

$$|E_{o2}|^2 = \kappa|E_{3L}|^2 + (1-\kappa)|E_{4L}|^2 + j\sqrt{\kappa(1-\kappa)}\left(E_{3L}^* E_{4L} - E_{3L}E_{4L}^*\right)$$

$$= [\kappa^2 + (1-\kappa)^2]|E_{01}|^2 + 2\kappa(1-\kappa)|E_{02}|^2 + 2(1-2\kappa)\sqrt{\kappa(1-\kappa)}$$

$$\times E_{01}E_{02}\sin[(\omega_2 - \omega_1)t] + j\sqrt{\kappa(1-\kappa)}\left(E_4 E_3^* \exp(j\Delta\theta) - E_4^* E_3 \exp(-j\Delta\theta)\right)$$

$$= |E_{02}|^2[2\kappa(1-\kappa)[1+\cos(\Delta\theta)]] + |E_{01}|^2[1-2\kappa(1-\kappa)[1+\cos(\Delta\theta)]]$$

$$+ 2\sqrt{\kappa(1-\kappa)}E_{01}E_{02}[(1-2\kappa)\sin[(\omega_2 - \omega_1)t][1+\cos(\Delta\theta)]$$

$$+ \cos[(\omega_2 - \omega_1)t]\sin(\Delta\theta)] \tag{7.14}$$

where

$$\Delta\theta = \theta_3 - \theta_4 = 2\pi n_2 L\left\{(1-2\kappa)\left[\frac{|E_{02}|^2}{\lambda_2} - \frac{|E_{01}|^2}{\lambda_1}\right] - \frac{4\sqrt{\kappa(1-\kappa)}E_{01}E_{02}\sin[(\omega_2 - \omega_1)t]}{\sqrt{\lambda_1\lambda_2}}\right\}$$

$$\tag{7.15}$$

and $\lambda_1$ and $\lambda_2$ are the operating wavelengths of $E_1$ and $E_2$, respectively.

For $\lambda_1 = \lambda_2$ and $E_2 = pE_1$ where $p$ is a constant, the switching characteristics for (a) $\kappa = 0.45$ and (b) $\kappa = 0.2$ are shown in Figure 7.29. Switching occurs only for large energy difference between the two ports, that is, $p \gg 1$. Also, as can be seen from Figure 7.30, the switching behavior is better for coupling ratio close to 0.5.

### 7.4.4 NONLINEAR AMPLIFYING LOOP MIRROR

The structure of the NALM is somehow similar to the NOLM structure, and it is an improved exploitation of NOLM. For NALM configuration, a gain medium with gain coefficient, $G$ is added to increase the asymmetric nonlinearity within the loop [48]. The amplifier is placed at one end of the loop, closer to port 3 of the coupler, and is assumed short relative to the total loop length, as shown in

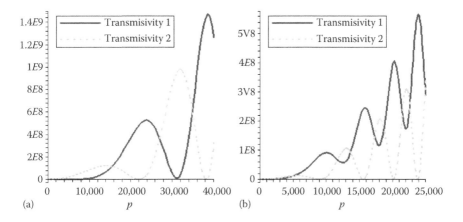

**FIGURE 7.30**    Switching characteristics for NOLM with (a) $\kappa = 0.45$ and (b) $\kappa = 0.2$.

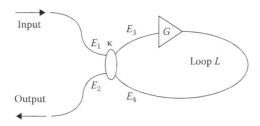

**FIGURE 7.31**    Nonlinear amplifying loop mirror.

Figure 7.31. One lightwave is amplified at the entrance to the loop, while the other experiences amplification just before exiting the loop. Since the intensities of the two lightwaves differ by a large amount throughout the loop, the differential phase shift can be quite large.

By following the analysis procedure stated in the previous section, we arrive at the input–output relationships as follows:

$$|E_{o1}|^2 = G \left\{ \begin{array}{l} \left[ |E_{01}|^2 [2\kappa(1-\kappa)[1+\cos(\Delta\theta)]] + |E_{02}|^2 \left[1-2\kappa(1-\kappa)[1+\cos(\Delta\theta)]\right] \right] \\ -2\sqrt{\kappa(1-\kappa)} E_{01} E_{02}[(1-2\kappa)\sin[(\omega_2-\omega_1)t][1+\cos(\Delta\theta)] \\ +\cos[(\omega_2-\omega_1)t]\sin(\Delta\theta)] \end{array} \right\} \quad (7.16)$$

$$|E_{o2}|^2 = G \left\{ \begin{array}{l} \left[ |E_{02}|^2 [2\kappa(1-\kappa)[1+\cos(\Delta\theta)]] + |E_{01}|^2 [1-2\kappa(1-\kappa)[1+\cos(\Delta\theta)]] \right] \\ +2\sqrt{\kappa(1-\kappa)} E_{01} E_{02}[(1-2\kappa)\sin[(\omega_2-\omega_1)t][1+\cos(\Delta\theta)] \\ +\cos[(\omega_2-\omega_1)t]\sin(\Delta\theta)] \end{array} \right\} \quad (7.17)$$

$$\Delta\theta = 2\pi n_2 L \left\{ \begin{array}{c} (1-\kappa-G\kappa)\left[\dfrac{|E_{02}|^2}{\lambda_2} - \dfrac{|E_{01}|^2}{\lambda_1}\right] \\ \\ -\dfrac{2(G+1)\sqrt{\kappa(1-\kappa)}E_{01}E_{02}\sin[(\omega_2-\omega_1)t]}{\sqrt{\lambda_1\lambda_2}} \end{array} \right\} \qquad (7.18)$$

### 7.4.5 NOLM–NALM Fiber Ring Lasers

The configuration of an NOLM–NALM fiber ring laser is shown in Figure 7.32. It is simply coupled loop mirrors, with NOLM (A–B–C–E–A) on one side and NALM (C–D–A–E–C) on the other side of the laser, with a common path in the middle section (A–E–C). One interesting feature about this configuration is the feedback mechanism of the fiber ring: one is acting as the feedback path to another, that is, NOLM is feedback part of the NALM's signal and vice versa.

As a matter of fact, the complex systems tend to encounter bifurcations, which when amplified, can lead to either order or chaos. The system can transit into chaos, through period-doubling, or order through a series of feedback loops. Hence, bifurcations can be considered as critical points in this system transitions.

In the formularization process, we ignore the nonlinear phase shift due to laser pulsation. The assumption is valid because of the saturation effect of gain medium as well as the energy stabilization provided by the filter.

Simulation of the laser behavior is conducted by combining the effects described in the previous section, and bifurcation maps are obtained based on CW operation as shown in Figure 7.8a and b, with $P_{o1}$ and $P_{o2}$ as the output powers at points $A$ and $C$, respectively, $\kappa_1 = 0.55$ and $\kappa_2 = 0.65$; $L_a$, $L_b$, and $L_c$ are 20, 100, and 20 m, respectively; nonlinear coefficient of $3.2 \times 10^{-20}$ m²/W; gain coefficient of 0.4 m⁻¹ at 1550 nm with

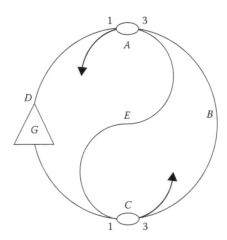

**FIGURE 7.32** Configuration of an NOLM–NALM fiber ring laser.

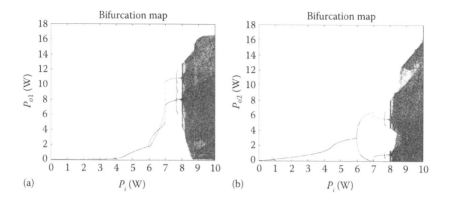

**FIGURE 7.33**  Bifurcation maps for (a) $P_{o1}$ and (b) $P_{o2}$ with $\kappa_1 = 0.55$ and $\kappa_2 = 0.65$ with only SPM consideration.

EDF length of 10 m and a saturation power of 25 dBm; and fiber effective area of 50 μm². $\kappa_1$ and $\kappa_2$ are the coupling ratios of couplers from port 1 to port 3 at points $A$ and $C$, respectively; $L_a$, $L_b$, and $L_c$ are the fiber lengths of left, mid, and right arm of the laser cavity, respectively. The maps are obtained with 200 iterations. Figure 7.33 shows the bifurcation maps with the contribution of SPM effect only, whereas, in Figure 7.34 both SPM and cross-phase modulation (XPM) effects are included.

In constructing the bifurcation maps of the system, we separate the lightwaves into clockwise and counterclockwise directions and similarly for the couplers involved in the system, that is, four couplers are used in simulating this bifurcation behavior. Initial conditions are set to be the pump power of the EDFA. We then propagate the lightwaves in both directions within the fiber ring with effects of various components taken into account, such as coupling ratios, gain, SPM, and XPM effects. The outputs of the system are then served as the input to the system for the next iteration. The process is repeated for 200×. Each iterated set of outputs is then combined together in constructing the bifurcation maps.

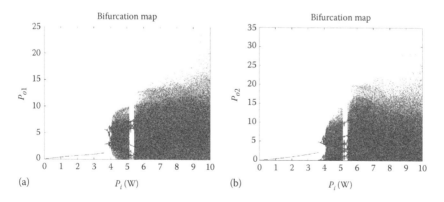

**FIGURE 7.34**  Bifurcation maps for (a) $P_{o1}$ and (b) $P_{o2}$ with $\kappa_1 = 0.55$ and $\kappa_2 = 0.65$ with both SPM and XPM considerations.

There are some similarities between Figures 7.33 and 7.34. Both of them indicate different operating regions at different power levels. With both nonlinear modulation effects (SPM and XPM) included (Figure 7.34), the transition from one operation state to another is faster, leading to earlier chaotic behavior with lower pump power because the strength of the XPM effect is twice that of the SPM effect. For simplicity, we discuss only the bifurcation behavior of Figure 7.33 in order to illustrate the transition behavior from one state to another.

From the bifurcation maps depicted in Figure 7.33, under this system setting, there are three operating regions within this NOLM–NALM fiber ring laser. The first is the linear operation region $(0 < P < 6\ \text{W})$, where there are single-value outputs and the output power increase with the input power. Period-doubling effect starts to appear when the input power reaches ~6 W. This is the second operating region $(6 \leq P < 8\ \text{W})$ of the laser, where double-periodic and quasiperiodic signals can be found. When the input power level reaches beyond 8 W, the laser enters into the chaotic state operation.

The Poincare map of the above system configuration with high pump power is shown in Figure 7.35. It shows the pattern of the attractor of the system when the laser is operating in the chaotic region. The powers required for the operations can be reduced by increasing the lengths of the fibers, $L_a$, $L_b$, and $L_c$.

By setting $\kappa_2 = 0$, the laser will behave like an NOLM. With an input pulse to port 1 of coupler A, we observe pulse compression at its port 2, as shown in Figures 7.36 and 7.37. Figure 7.37 presents the transmission capability of the setup at port 2 for various $\kappa_1$ values, when the input is injected to port 1 and $\kappa_2 = 0$. When $\kappa_1 = 0.5$, we observe no transmission at port 2 as the entire injected signal has been reflected back to port 1 where the mirror effect takes place. By changing the coupling ratio of $\kappa_2$,

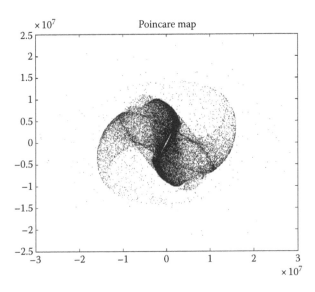

**FIGURE 7.35** Poincare map of the system under high pump power with $\kappa_1 = 0.55$ and $\kappa_2 = 0.65$.

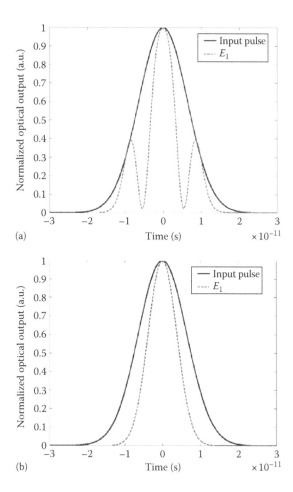

**FIGURE 7.36**  Input and output pulse at port 1 and port 2 comparison with $\kappa_2=0$ and (a) $\kappa_1=0.41$ and (b) $\kappa_1=0.49$ (solid line—input pulse and dotted line—output pulse at port 1).

we are able to change the zero transmission point away from $\kappa_1=0.50$. Figure 7.38 shows the transmissivities/transmittances of $P_{o1}$ and $P_{o2}$ for various sets of $\kappa_1$ and $\kappa_2$ and the complex switching dynamics of the laser for different sets of $\kappa_1-\kappa_2$, also with the pulse compression capability.

## 7.4.6  Experiment Setups and Analyses

### 7.4.6.1  Bidirectional Erbium-Doped Fiber Ring Laser

We start with a simple EDFRL; an optical closed loop with EDFA and some fiber couplers. It is used to study the bidirectional lightwave propagations behavior of the laser. The EDFA is made of a 20 m EDF and dual pumped by 980 and 1480 nm diode lasers with the saturation power of about 15 dBm. The slope efficiencies of the pump lasers are 0.45 mW/mA and 0.22 W/A, respectively. No isolator and filter are

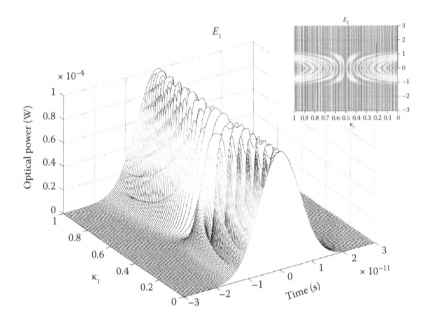

**FIGURE 7.37**   Output pulse behavior at various coupling coefficients $\kappa_1$ and when $\kappa_2=0$ for small input power (inset: top view of the pulse behavior).

used in the setup to eliminate the direction and spectra constraints. Two outputs are taken for examinations, that is, Output 1 in ccw direction and Output 2 in cw direction. All connections within the fiber ring are spliced together to reduce the possible reflections within the system. Output 1 (ccw) and Output 2 (cw) as shown in Figure 7.39 are taken for investigations.

The ASE spectrum of the laser covers the range from 1530 to 1570 nm. By increasing both the 980 and 1480 nm pumping currents to their maximum allowable values, we obtained bidirectional lasing, which is shown in Figure 7.40. The upper plot is the lightwave propagating in the cw direction while the other one in the ccw direction. To obtain bidirectional lasing, the losses and the gain must be balanced for the two lightwaves that propagate around the cavity in opposite directions [49]. The pump lasers used in the setup are not identical (in terms of power and pumping wavelength), and this gives rise to different lasing behaviors for both cw and ccw directions. Due to the laser diode controllers' limitation, the maximum pumping currents for both 980 and 1480 nm laser diodes are capped at 300 and 500 mA, respectively, which correspond to 135 and 110 mW in cw pump and ccw pump directions. This explains the domineering cw lasing as shown in Figure 7.39. One thing to note is that the lasings of the laser in both directions are not very stable due to the disturbance from the opposite propagating lightwave and the ASE noise contribution due to the absence of filter. Output 1 is mainly contributed by the back reflections from the fiber ends and connectors as well as some back-scattered noise. Since all the connections within the fiber ring are spliced together, the reflection due to the connections is the lowest. However, there are still some unavoidable reflections from the fiber ends, which contribute to the lightwaves in the opposite directions. Besides that, back-scattered noise also adds to the light-waves in the opposite directions.

We maintain a 980 nm pump current at certain value, and adjust the 1480 nm pump current upward and then downward to examine the bistability behavior of the laser. The bistable characteristics at a lasing wavelength of about 1562.2 nm for both cw and ccw directions are shown in Figures 7.41 and 7.42, with log and linear scales

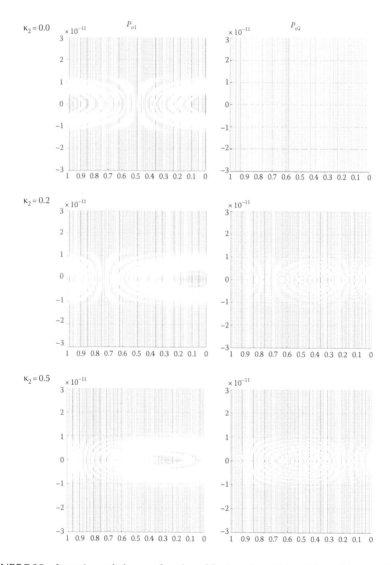

**FIGURE 7.38** Intensity variation as a function of $P_{o1}$ ($y$-axis—left hand side, arbitrary unit) and $P_{o2}$ ($y$-axis—right side figure, arbitrary unit) for different $\kappa_2$ values (indicated) as a function of $\kappa_1$ ($x$-axis—normalized).    (*Continued*)

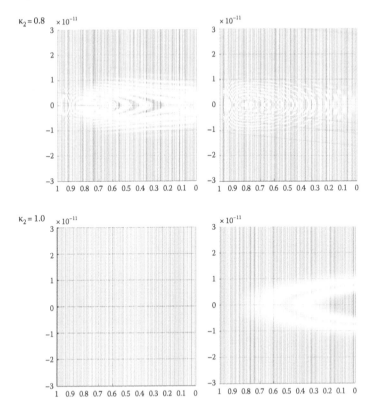

**FIGURE 7.38 (*Continued*)**   Intensity variation as a function of $P_{o1}$ (y-axis—left hand side, arbitrary unit) and $P_{o2}$ (y-axis—right side figure, arbitrary unit) for different $\kappa_2$ values (indicated) as a function of $\kappa_1$ (x-axis—normalized).

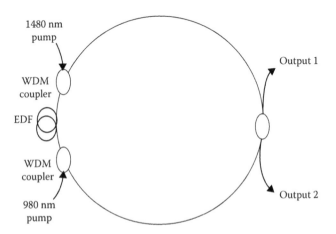

**FIGURE 7.39**   Bidirectional erbium-doped fiber ring laser.

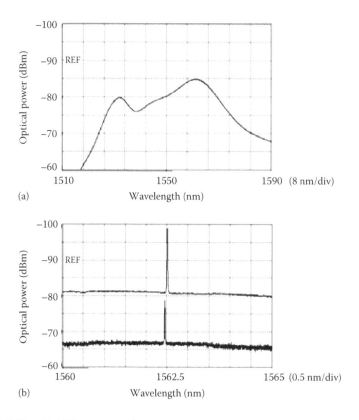

(a)

(b)

**FIGURE 7.40**    (a) ASE spectrum of the laser and (b) lasing characteristics in both directions (upper trace: Output 2; lower trace: Output 1).

for the vertical axes of Figures 7.41a and 7.42a and 7.41b and 7.42b, respectively. We obtain about 15 dBm difference between the two propagating lightwaves. The bistable region is obtained at a ~30 mA of pump current of a 1480 nm source and at a fixed value 100 mA for a 980 nm pump laser. No lasing is observed in the ccw lightwave propagation. This bistable region can be further enhanced by increasing the 980 nm pump power to a higher level. By maintaining the 980 nm pump current at ~175 mA, a bistable region can be observed as wide as at ~70 mA of the 1480 nm pump current. When the 980 nm pump current is maintained at a higher level (>175 mA), the lasing of Output 2 remains, even when the 1480 nm is switched off, as shown in Figure 7.42. This bistable behavior is mainly due to the saturable absorption of the EDF section.

### 7.4.6.2  NOLM–NALM Fiber Ring Laser

The experimental setup of an NOLM–NALM fiber ring laser is shown in Figure 7.43. It is simply a combination of NOLM on one side and NALM on the other side of the laser, with a common path in the middle section. The principal element of the laser is an optical close loop with an optical gain medium, two variable ratio couplers (VRCs),

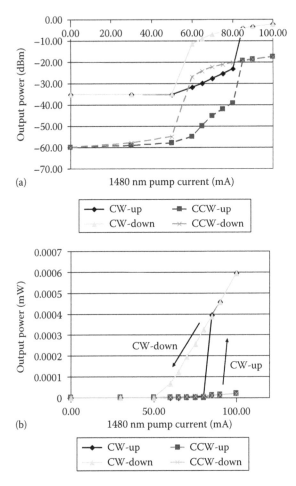

**FIGURE 7.41** Hysteresis loops of EDFRL at 980 nm pump laser at a driving current of 100 mA (a) log scale and (b) linear scale.

an optical BPF, optical couplers, and other associated optics. The gain medium used in our fiber laser system is the amplifier used in the preceding experiment. The two VRCs, with coupling ratios ranging from 20% to 80% and insertion loss of about 0.2 dB are added into the cavity at positions shown in Figure 7.43 to adjust the coupling power within the laser. They are interconnected in such a way that the output of one VRC is the input of the other. A tunable BPF with 3 dB bandwidth of 2 nm at 1560 nm is inserted into the cavity to select the operating wavelength of the generated signal and to reduce the noise in the system. It is noted that the lightwaves are traveling simultaneously in both directions, as there is no isolator used in the laser. Output 1 and Output 2 as shown in the figure are the outputs of the laser.

One interesting phenomenon observed before the BPF is inserted into the cavity is the wavelength tunability. The lasing wavelength is tunable from 1530 to 1560 nm

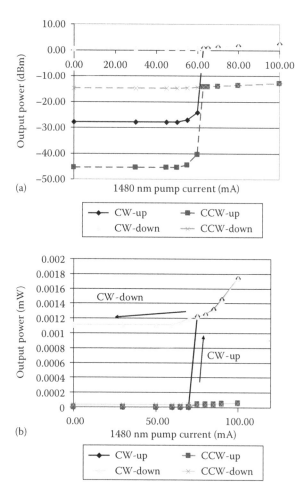

**FIGURE 7.42**  Hysteresis loops obtained from the EDFRL for 980 nm pump current of 200 mA (a) log scale and (b) linear scale.

(almost the entire EDFA C-band), by changing the coupling ratios of the VRCs. We believe that this wavelength tunability is due to the change in the traveling light-waves' intensities, which contributes to the nonlinear refractive index change, and in turn modifies the dispersion relations of the system, and, hence, the lasing wave-length. Therefore, the VRCs within the cavity not only determine the directionality of the lightwave propagation, but also the lasing wavelength.

For a conventional EDFRL, bistability is not observable when the pump current is far above the threshold value, where saturation starts to take place. However, a small hysteresis loop has been observed in our laser setup even with high pump current, that is, near saturation region, when changing the coupling ratio of one VRC while that of the other one remains unchanged, as shown in Figure 7.44.

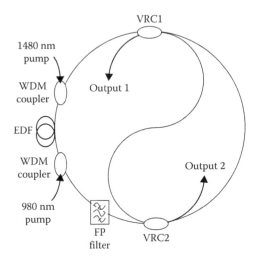

**FIGURE 7.43** Experimental setup of an NOLM–NALM fiber ring laser.

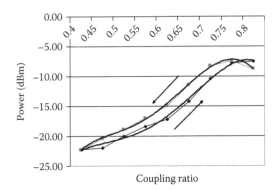

**FIGURE 7.44** Hysteresis loop observed; laser output power versus the coupling ratio of one VRC, while the other coupling remains unchanged operating under high pumping current.

Changing the coupling ratio of the VRC means directly altering the total power within the cavity, and hence modifying its gain and absorption behavior. As a result, a small hysteresis loop is observable even with a constant high pump power, which forms an additional member of the bistable state family. The power distribution of Output 1 and Output 2 of the NOLM–NALM fiber laser obtained experimentally is depicted in Figure 7.45. We are able to obtain the switching between the outputs by tuning the coupling ratios of VRC1 and VRC2. The simulation results for transmittance of various coupling ratios under linear operation are shown in Figure 7.46, since the available pump power of our experiment setup is insufficient to create high power within the cavity. Both the experimental and numerical results

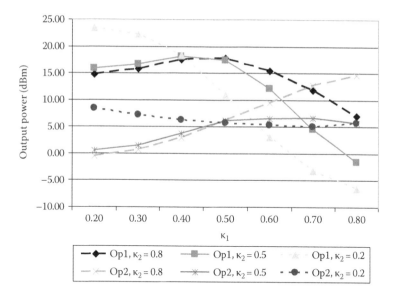

**FIGURE 7.45** Laser output power as function of coupling coefficient and experimental results for Output 1 (Op1) and Output 2 (Op2) with coupling ratios ($\kappa_1$—coupling ratio of VRC1, $\kappa_2$—coupling ratio of VRC2) as parameters.

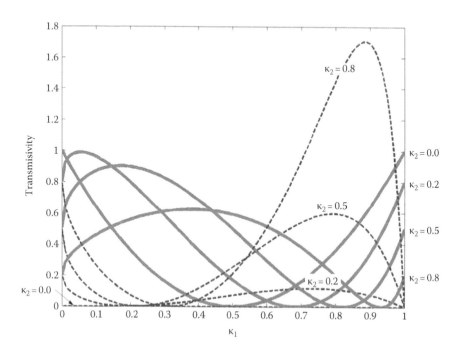

**FIGURE 7.46** Simulated transmittances of Output 1 (solid line) and Output 2 (dotted line) for various coupling ratios of VRC1 and VRC2.

have come to some agreements, but not all, since the model developed is simple and does not consider the polarization, the dispersion characteristics, etc., of the propagating lightwaves.

### 7.4.6.3 Amplitude-Modulated NOLM–NALM Fiber Ring Laser

The schematic of amplitude (AM)-modulated NOLM–NALM fiber ring laser is shown in Figure 7.47. A few new photonic components are added into the laser cavity. An asymmetric coplanar traveling wave 10 Gbps, Ti:LiNbO$_3$ MZM is used in the inner loop of the cavity with half-wave voltage, $V_\pi$ of 5.3 V. The modulator is DC biased near the quadrature point and not driven higher than $V_\pi$ such that it operates on the linear region of its characteristic curve to ensure minimum chirp imposing on the modulated lightwaves. It is driven by the sinusoidal signal derived from an Anritsu 68347C synthesizer signal generator. The modulator has an insertion loss of ≤7 dB. Two PCs are placed prior to the inputs of the modulator in both directions to ensure proper polarization alignment into the modulator. A wider bandwidth (i.e., 5 nm tunable FP filter) is used in this case to allow more longitudinal modes within the laser for possible mode-locking process.

With the insertion of AM modulator into the laser cavity, we are able to obtain the pulse operation from the laser by means of the active harmonic mode-locking technique. Both propagation lightwaves are observed. By proper adjustment of the modulation frequency, the PCs, and the VRCs, the unidirectional pulse operation at modulating frequency is obtained. However, it is highly sensitive to the environmental change. The direction of the lightwave propagation of the laser can be controlled by the VRCs. The unidirectional pulse train propagation obtained experimentally and numerically is shown in Figure 7.48. However, with a slight deviation to the system parameters, either in modulation

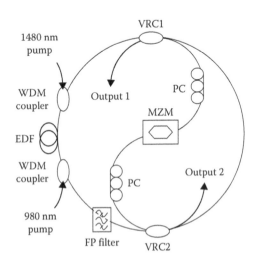

**FIGURE 7.47** Experimental setup for AM NOLM–NALM fiber laser (VRC—variable ratio coupler, PC—polarization controller, and MZM—Mach–Zehnder modulator).

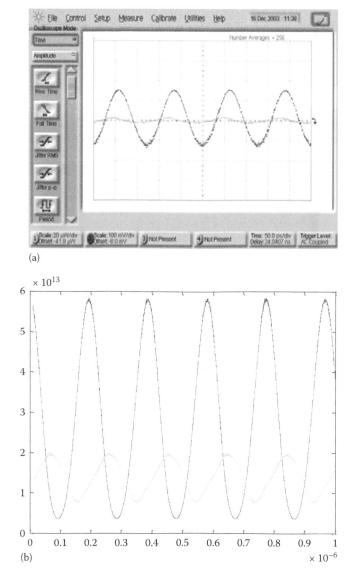

**FIGURE 7.48** Unidirectional pulse operation in AM NOLM–NALM fiber laser: (a) experimental results and (b) simulation results.

frequency detuning or polarization, period-doubling and quasiperiodic operations in the laser is recorded as shown in Figure 7.49. We believe that the effect is due the interference between the bidirectional propagation lightwaves, which have suffered nonlinear phase shifts in each direction, since we are operating in the saturation region of the EDFA. Another factor for this formation is that the lightwave in one direction is the feedback signal for another one. Furthermore, the intensity modulation of the optical modulator is not identical for co- and counter-interactions between the lightwaves and the traveling microwaves on the surface of the optical waveguide. This would contribute to the mismatch of the locking condition of the laser.

## 7.5 CONCLUDING REMARKS

We have successfully demonstrated in Sections 7.2 and 7.3 the mode-locked lasers operating under the open loop condition and with O/E RF feedback providing regenerative mode-locking. The O/E feedback can certainly provide a self-locking

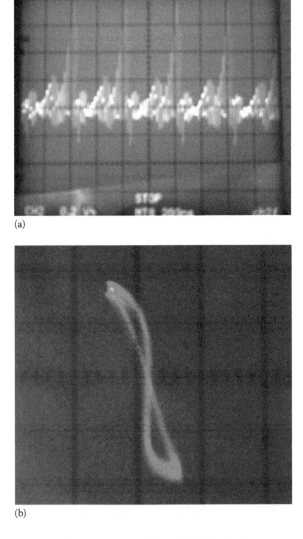

(a)

(b)

**FIGURE 7.49**   Quasiperiodic operation in AM NOLM–NALM fiber laser: (a) photograph of the oscilloscope trace and (b) XY plot of Outputs 1 and 2.                    (*Continued*)

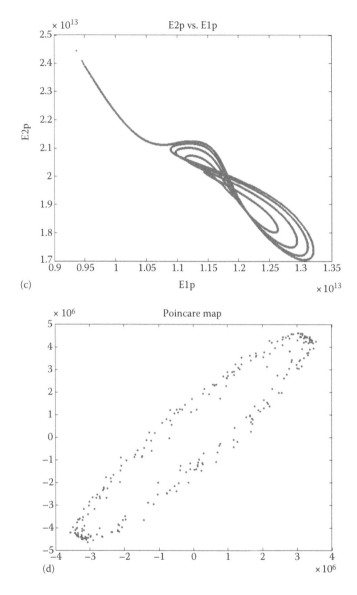

(c)

(d)

**FIGURE 7.49 (*Continued*)** Quasiperiodic operation in AM NOLM–NALM fiber laser: (c) simulated XY plot, and (d) simulated Poincare map.

mechanism under the condition that the polarization characteristics of the ring laser are manageable. The regenerative MLRL can self-lock even under the DC drifting effect of the modulator bias voltage (over 20 h).* The generated pulse

---

\* Typically the DC bias voltage of a LiNbO$_3$ intensity modulator is drifted by 1.5 V after 15 h of continuous operation.

trains of 4.5 ps duration can be, with minimum difficulty, compressed further to less than 3 ps for 160 Gbps optical communication systems.

We have also represented the 660th and 1230th order of rational harmonic mode-locking from a base modulation frequency of 100 MHz in the optically amplified fiber ring laser, hence achieving 66 and 123 GHz pulse repetition frequency. Besides the repetition rate multiplication, we also obtain high pulse compression factor in the system, ~35× and 40× relative to the nonmultiplied laser system. In addition, we use phase plane analysis to study the laser system behavior. From the analysis model, the amplitude stability of the detuned pulse train can only be achieved under negligible or no harmonic distortion condition, which is the ideal situation. The phase plane analysis also reveals the pulse forming complexity of the laser system.

Furthermore, $N$-times repetition rate multiplication by using fiber GVD effect, a linear diffraction mechanism, is achieved; hence, a 40 GHz pulse train is obtained from the 10 GHz mode-locked fiber laser source.

Currently the design and demonstration of multiwavelength mode-locked lasers to generate ultrashort and ultrahigh repetition-rate pulse sequences are under consideration by employing multispectral filter demultiplexers and multiplexers incorporated within the fiber ring. Furthermore, the locking in the THz region will also be reported. The principal challenge is the conversion from the THz photonic to electronic domain for stabilization. This can be implemented in either electronic or photonic sampling.

In Section 7.4, the important nonlinear dynamics, the bidirectional optical bistability, in a dual-pumped EDFRL without using isolator has been studied in both simulation and experiments. A ~70 mA 1480 nm pump current bistable region has also been obtained. With this bidirectional and bistable characteristics of the NOLM–NALM fiber lasers, we experimentally demonstrated and numerically simulated the switching and bifurcation behaviors. From the simulated bifurcations maps, three basic operation regions can be identified, namely, unidirectional, period-doubling, and chaos operations. The VRCs used in the setup not only control the lightwave directionality, but also its lasing wavelength. Unidirectional lightwave propagations, without isolator, were achieved in both CW and pulse operations by tuning the coupling ratios of the VRCs within the laser system. Bifurcation was also obtained from the AM NOLM–NALM fiber laser. However, chaotic operation was not observed experimentally due to the hardware limitation of our system, which required higher gain coefficient and input power as predicted in our simulation.

The configuration is somehow similar to optical flip-flop concept, which can be useful in optical communication systems. The optical flip-flop concept has been used in optical packet switches and optical buffers [50–52]. NOLM–NALM fiber lasers possess several interesting optical behaviors, which may be deduced for photonic signal processing, such as photonic flip-flops, optical buffer loop, optical pulse sampling devices, and secured optical communications.

## REFERENCES

1. K. Kuroda and H. Takakura, Mode-locked ring laser with output pulse width of 0.4 ps, *IEEE Trans. Inst. Meas.*, 48, 1018–1022, 1999.

2. G. Zhu, H. Chen, and N. Dutta, Time domain analysis of a rational harmonic mode-locked ring fibre laser, *J. Appl. Phys.*, 90, 2143–2147, 2001.

3. C. Wu and N. K. Dutta, High repetition rate optical pulse generation using a rational harmonic mode-locked fibre laser, *IEEE J. Quant. Electron.*, 36, 145–150, 2000.

4. K. K. Gupta, N. Onodera, and M. Hyodo, Technique to generate equal amplitude, higher-order optical pulses in rational harmonically mode locked fibre ring lasers, *Electron. Lett.*, 37, 948–950, 2001.

5. Y. Shiquan, L. Zhaohui, Z. Chunliu, D. Xiaoyi, Y. Shuzhong, K. Guiyun, and Z. Qida, Pulse-amplitude equalization in a rational harmonic mode-locked fibre ring laser by using modulator as both mode locker and equalizer, *IEEE Photon. Technol. Lett.*, 15, 389–391, 2003.

6. J. Azana and M. A. Muriel, Technique for multiplying the repetition rates of periodic trains of pulses by means of a temporal self-imaging effect in chirped fibre gratings, *Opt. Lett.*, 24, 1672–1674, 1999.

7. S. Atkins and B. Fischer, All optical pulse rate multiplication using fractional Talbot effect and field-to-intensity conversion with cross gain modulation, *IEEE Photon. Technol. Lett.*, 15, 132–134, 2003.

8. J. Azana and M. A. Muriel, Temporal self-imaging effects: Theory and application for multiplying pulse repetition rates, *IEEE J. Quant. Electron.*, 7, 728–744, 2001.

9. D. A. Chestnut, C. J. S. de Matos, and J. R. Taylor, 4x repetition rate multiplication and Raman compression of pulses in the same optical fibre, *Opt. Lett.*, 27, 1262–1264, 2002.

10. S. Arahira, S. Kutsuzawa, Y. Matsui, D. Kunimatsu, and Y. Ogawa, Repetition frequency multiplication of mode-locked using fibre dispersion, *J. Lightwave Technol.*, 16, 405–410, 1998.

11. W. J. Lai, P. Shum, and L. N. Binh, Stability and transient analyses of temporal Talbot-effect-based repetition-rate multiplication mode-locked laser systems, *IEEE Photon. Technol. Lett.*, 16, 437–439, 2004.

12. K. K. Gupta, N. Onodera, K. S. Abedin, and M. Hyodo, Pulse repetition frequency multiplication via intracavity optical filtering in AM mode-locked fibre ring lasers, *IEEE Photon. Technol. Lett.*, 14(3), 284–286, 2002.

13. K. S. Abedin, N. Onodera, and M. Hyodo, Repetition-rate multiplication in actively mode-locked fibre lasers by higher-order FM mode-locking using a high Finesse Fabry Perot filter, *Appl. Phys. Lett.*, 73, 1311–1313, 1998.

14. K. S. Abedin, N. Onodera, and M. Hyodo, Higher order FM mode-locking for pulse-repetition-rate enhancement in actively mode-locked lasers: Theory and experiment, *IEEE J. Quant. Electron.*, 35, 875–890, 1999.

15. W. Daoping, Z. Yucheng, L. Tangjun, and J. Shuisheng, 20 Gb/s optical time division multiplexing signal generation by fibre coupler loop-connecting configuration, in paper presented at *Fourth Optoelctronics and Communications Conference*, Beijing, China, 1999.

16. D. L. A. Seixasn and M. C. R. Carvalho, 50 GHz fibre ring laser using rational harmonic mode-locking, in paper presented at *Microwave and Optoelectronics Conference*, 2001.

17. R. Y. Kim, Fibre lasers and their applications, in paper presented at *Laser and Electro-Optics, CLEO/Pacific Rim'95*, 1995.

18. H. Zmuda, R. A. Soref, P. Payson, S. Johns, and E. N. Toughlian, Photonic beamformer for phased array antennas using a fibre grating prism, *IEEE Photon. Technol. Lett.*, 9, 241–243, 1997.

19. G. A. Ball, W. W. Morey, and W. H. Glenn, Standing-wave monomode erbium fibre laser, *IEEE Photon. Technol. Lett.*, 3, 613–615, 1991.

20. D. Wei et al., Multi-wavelength erbium-doped fibre ring laser with overlap-written fibre Bragg gratings, *Opt. Lett.*, 25, 1150–1152, 2000.

21. S. K. Kim, M. J. Chu, and J. H. Lee, Wideband multi-wavelength erbium-doped fibre ring laser with frequency shifted feedback, *Opt. Commun.*, 190, 291–302, 2001.

22. Z. Li, L. Caiyun, and G. Yizhi, A polarization controlled multi-wavelength Er-doped fibre laser, in presented at *APCC/OECC99*, 1999.

23. R. M. Sova, C. S. Kim, and J. U. Kang, Tunable dual-wavelength all-PM fibre ring laser, *IEEE Photon. Technol. Lett.*, 14, 287–289, 2002.

24. I. D. Miller et al., A $Nd^{3+}$-doped CW fibre laser using all-fibre reflectors, *Appl. Opt.*, 26, 2197–2201, 1987.

25. X. Fang and R. O. Claus, Polarization-independent all-fibre wavelength-division multiplexer based on a Sagnac interferometer, *Opt. Lett.*, 20, 2146–2148, 1995.

26. X. Fang et al., A compound high-order polarization-independent birefringence filter, *IEEE Photon. Technol. Lett.*, 19, 458–460, 1997.

27. X. P. Dong et al., Multi-wavelength erbium-doped fibre laser based on a high-birefringence fibre loop, *Electron. Lett.*, 36, 1609–1610, 2000.

28. D. Jones, H. Haus, and E. Ippen, Subpicosecond solitons in an actively mode locked fibre laser, *Opt. Lett.*, 21(22), 1818–1820, 1996.

29. X. Zhang, M. Karlson, and P. Andrekson, Design guideline for actively mode locked fibre ring lasers, *IEEE Photon. Technol. Lett.*, 51, 1103–1105, 1998.

30. A. E. Siegman, *Laser*, University Press, Mill Valley, CA, 1986.

31. J. J. E. Slotine and W. Li, *Applied Nonlinear Control*, Prentice-Hall, Englewood Cliffs, NJ, 1991.

32. H. M. Gibbs, S. L. McCall, and T. N. C. Venkatesan, differential gain and bistability using a sodium-filled Fabry-Perot interferometer, *Phys. Rev. Lett.*, 36(19), 1135–1138, 1976.

33. H. Nakatsuka, S. Asaka, H. Itoh, K. Ikeda, and M. Matsuoka, Observation of bifurcation to chaos in an all-optical bistable system, *Phys. Rev. Lett.*, 50(2), 109–112, 1983.

34. K. Tamura, E. P. Ippen, H. A. Haus, and L. E. Nelson, 77-fs pulse generation from a stretched-pulse mode locked all fiber ring laser, *Opt. Lett.*, 18(13), 1080–1082, 1993.

35. G. Yandong, S. Ping, and T. Dingyuan, 298 fs passively mode locked ring fiber soliton laser, *Microwave Opt. Technol. Lett.*, 32(5), 320–333, 2002.

36. A. E. Siegman, *Lasers*, University Science Books, Mill Valley, CA, 1986.

37. Y. Shi, M. Sejka, and O. Poulsen, A unidirectional $Er^{3+}$-doped fiber ring laser without isolator, *IEEE Photon. Technol. Lett.*, 7(3), 290–292, 1995.

38. J. M. Oh and D. Lee, Strong optical bistability in a simple L-Band tunable erbium-doped fiber ring laser, *J. Quant. Electron.*, 40(4), 374–377, 2004.

39. Q. Mao and J. W. Y. Lit, L-band fiber laser with wide tuning range bases on dual-wavelength optical bistability in linear overlapping grating cavities, *IEEE J. Quant. Electron.*, 39(10), 1252–1259, 2003.

40. L. Luo, T. J. Tee, and P. L. Chu, Bistability of erbium doped fiber laser, *Opt. Commun.*, 146, 151–157, 1998.

41. P. P. Banerjee, *Nonlinear Optics—Theory, Numerical Modeling, and Applications*, Marcel Dekker Inc., New York, 2004.

42. P. Meystre, On the use of the mean-field theory in optical bistability, *Opt. Commun.*, 26(2), 277–280, 1978.

43. L. A. Orozco, H. J. Kimble, A. T. Rosenberger, L. A. Lugiato, M. L. Asquini, M. Brambilla, and L. M. Narducci, Single-mode instability in optical bistability, *Phys. Rev. A*, 39(3), 1235–1252, 1989.

44. K. Ikeda, Multiple-valued stationary state and its instability of the transmitted light by a ring cavity, *Opt. Commun.*, 30(2), 257–261, 1979.

45. M. J. Ogorzalek, *Chaos and Complexity in Nonlinear Electronic Circuits*, World Scientific Series on Nonlinear Science, Series A No. 22, World Scientific Publishing Co. Pte. Ltd., Singapore, 1997.

46. M. P. Kennedy, Three steps to chaos—Part I: Evolution, *IEEE Trans. Circ. Syst.—I: Fundam. Theory Appl.*, 40(10), 640–656, 1993.

47. N. J. Doran and D. Wood, Nonlinear-optical loop mirror, *Opt. Lett.*, 13(1), 56–58, 1988.

48. M. E. Fermann, F. Haberl, M. Hofer, and H. Hochreiter, Nonlinear amplifying loop mirror, *Opt. Lett.*, 15(13), 752–754, 1990; M. E. Fermann, A. Galvanauskas, and G. Sucha, *Ultrafast Lasers—Technology and Applications*, Marcel Dekker Inc., New York, 2003.

49. M. Mohebi, J. G. Mejia, and N. Jamasbi, Bidirectional action of a titanium-sapphire ring laser with mode-locking by a Kerr lens, *J. Opt. Technol.*, 69(5), 312–316, 2002.

50. X. Zhang, M. Karlsson, and P. A. Andrekson, Design guidelines of actively mode-locked fiber ring lasers, *IEEE Photon. Technol. Lett.*, 10(8), 1103–1105, 1998.

51. R. Langenhorst, M. Eiselt, W. Pieper, G. Großkopf, R. Ludwig, L. Küller, E. Dietrich, and H. C. Weber, Fiber loop optical buffer, *J. Lightwave Technol.*, 14(3), 324–335, 1996.

52. A. Liu, C. Wu, Y. Gong, and P. Shum, Dual-Loop Optical Buffer (DLOB) based on a $3 \times 3$ collinear fiber coupler, *IEEE Photon. Technol. Lett.*, 16(9), 2129–2131, 2004.

# 8 Optical Multisoliton Transmission

Optical solitons have drawn considerable attention because of its fundamental nature and potential applications in optical fiber communications systems operating at ultrahigh bit rate and over extremely long distance. They are analyzed and discussed in this chapter.

The fundamental and higher-order solitons are simulated by using the numerical approach using beam propagation method (BPM) and the analytical approach using inverse scattering method (ISM) to solve the basic nonlinear propagation equation, also known as the nonlinear Schrodinger equation (NLSE). Solitons up to fifth order are observed, and the *soliton breakdown* effects are discussed. The interaction of solitons of two and three fundamental soliton pulses is simulated using the numerical approach, and its behavior is being observed. A number of methods are suggested to control the soliton interaction.

Furthermore, the optical generators and detectors from mode-locked fiber lasers (MLFL) are described; both periodic pulse sequence and multi-bound solitons are demonstrated. Experimental demonstration of generation of dual, triple, and quadruple solitons is presented. These experimental multi-bound solitons are transmitted through single mode optical fibers to prove that they have potential practical applications.

## 8.1 INTRODUCTION

A fascinating manifestation of the fiber nonlinearity occurs in the anomalous dispersion regime where the fiber can support optical solitons through interplay between the linear dispersive and nonlinear phase distortion effects. The term *soliton* refers to special kinds of waves that can propagate undistorted over long distances and remain unaffected after collision with each other. Solitons have been studied extensively in several fields of physics. In the context of optical fibers, solitons are not only of fundamental interest but also has the potential application in the field of optical fiber communications. The word soliton was coined [1] in 1965 to describe the particle-like properties of pulse envelopes in dispersive nonlinear media.

The basic concepts behind fiber solitons and its basic propagation equation, known as the NLSE, are introduced in Sections 8.3 and 8.4. Numerical approach using BPM (see also Appendix E) to solve the NLSE is discussed in detail in Section 8.3 and confirmed by experimental demonstration in Section 8.8.1.4. Analytical approach using ISM to solve the Schrodinger equation is discussed in detail in Section 8.4. The results obtained through these two approaches are compared and verified.

The results of the numerical approach are discussed in Section 8.5. In this section, the behavior of fundamental and higher-order solitons are being observed. In Section 8.6, solitons interactions are simulated and discussed. Techniques for controlling the solitons interaction effect are then suggested.

Section 8.7 revisits the ISM for simulation of optical solitons, in particular the interaction between sequential pulses for optical communications systems. Steps for interaction and reconstruction of solitons are described and proven to be the most accurate techniques for design and control of optical solitons pulses.

All results of our synthesis for optical solitons propagation and detection techniques are described for both bright and dark solitons that are summarized and suggestions are made. Section 8.8 presents the experimental demonstration of generation of dual, triple, and quadruple bound solitons in a MLFL and their dynamics when circulating around the fiber rings after several thousand times.

Section 8.8 then revisits the experimental generation of dual, triple, and quadruple solitons and then the experimental transmission of multi-bound solitons through single mode optical fibers is described to prove that they have potential practical applications.

## 8.2  FUNDAMENTALS OF NONLINEAR PROPAGATION THEORY

Similar to all electromagnetic phenomena, the propagation of optical fields either in linear or nonlinear regimes in fibers is governed by Maxwell's equations. From Maxwell's equations, one can obtain [2]

$$\nabla \times \nabla \times \mathbf{E} = -\frac{1}{c^2}\frac{\partial^2 \mathbf{E}}{\partial t^2} - \mu_0 \frac{\partial^2 \mathbf{P}}{\partial t^2} \tag{8.1}$$

where
  $\mathbf{E}$ is the electric field
  $\mathbf{P}$ is the induced polarization

The nonlinear effects in the optical fibers involve the use of short pulses with widths ranging from 10 ns to 10 fs. When such optical pulses propagate inside the fiber, both dispersive and nonlinear effects influence their shape and spectrum. Thus, Equation 8.1 can be further developed [2] to cater for these effects

$$\nabla^2 \mathbf{E} - \frac{1}{c^2}\frac{\partial^2 \mathbf{E}}{\partial t^2} = -\mu_0 \frac{\partial^2 \mathbf{P}_L}{\partial t^2} - \mu_0 \frac{\partial^2 \mathbf{P}_{NL}}{\partial t^2} \tag{8.2}$$

where $\mathbf{P}_L$ and $\mathbf{P}_{NL}$ are the linear and nonlinear part of the induced polarization, respectively. Using the method of separation of variables to solve Equation 8.2 and taking perturbation assumptions into consideration, we can obtain

$$\frac{\partial A}{\partial z} + \beta_1 \frac{\partial A}{\partial t} + \frac{j}{2}\beta_2 \frac{\partial^2 A}{\partial t^2} + \frac{\alpha}{2}A = j\gamma|A|^2 A \tag{8.3}$$

where the envelope $A$ is a slowly varying function of $z$. $\beta_1$ and $\beta_2$ are the first- and second-order derivatives of $\beta$ with respect to optical frequency $\omega$ as defined in Chapter 2. $\gamma$ is the fiber nonlinear coefficient, and $\alpha$ is the intrinsic fiber attenuation expressed in dB/km. Taking higher-order dispersion effect, stimulated inelastic scattering effect, and Raman gain effect into consideration together with perturbation approach leads to three additional terms in Equation 8.3. The generalized propagation equation takes the form

$$\frac{\partial A}{\partial z} + \frac{\alpha}{2} A + \frac{j}{2}\beta_2 \frac{\partial^2 A}{\partial T^2} - \frac{1}{6}\beta_3 \frac{\partial^3 A}{\partial T^3} = j\gamma \left\{ |A|^2 A + \frac{2j}{\omega_0} \frac{\partial}{\partial T}(|A|^2 A) - T_R A \frac{\partial |A|^2}{\partial T} \right\} \qquad (8.4)$$

where
$\omega_0$ is the carrier frequency
$T_R$ is related to the slope of the Raman scattering gain

Notice that $\beta_1$ has disappeared because of the transformation $T$, in which the observation frame is positioned on the soliton itself, hence the time frame is given by

$$T = t - \beta_1 z \qquad (8.5)$$

Before beginning to simulate solitons evolution using either numerical approach or analytical approach, a few important terms of solitons systems need to be clarified. It is useful to write Equation 8.4 in a normalized form by introducing the following:

$$\tau = \frac{t - \beta_1 z}{T_O} \qquad (8.6)$$

$$\xi = \frac{z}{L_D} \qquad (8.7)$$

$$U = \frac{A}{\sqrt{P_O}} \qquad (8.8)$$

where
$T_O$ is the soliton pulse width
$P_O$ is the soliton peak power

The dispersion length, $L_D$, is defined as

$$L_D = \frac{T_O^2}{|\beta_2|} \qquad (8.9)$$

where $\beta_2$ is obtained from the total dispersion factor of the optical fiber (see Chapter 2), which is given by

$$\beta_2 = -\left(\frac{D_T\lambda^2}{2\pi c}\right) \tag{8.10}$$

The order of the soliton, $N$, is defined as

$$N^2 = \frac{\gamma P_o T_o^2}{|\beta_2|} \tag{8.11}$$

The fundamental soliton corresponds to $N=1$. Thus, to maintain $N=1$, the soliton pulse needs to have its peak power so that the square root of Equation 8.11 will be an integer value. The fundamental soliton has a secant shape and its initial amplitude is given by

$$U(0,\tau) = \mathrm{sech}(\tau) \tag{8.12}$$

This pulse shape is to be launched into the fiber. Its shape remains unchanged during the propagation. However, for $N>1$, such input shape is recovered at each *soliton period* defined as follows:

$$z_O = \frac{\pi}{2}L_D = \frac{\pi}{2}\frac{T_o^2}{|\beta_2|} \tag{8.13}$$

The soliton period $z_O$ and the soliton order $N$ play an important role in the theory of optical solitons.

## 8.3 NUMERICAL APPROACH

### 8.3.1 BEAM PROPAGATION METHOD

The propagation Equation 8.4 is a nonlinear partial differential equation that does not generally lend itself to analytic solutions except for some specific cases in which the ISM (see Section 8.6) can be employed. A numerical approach is therefore, often necessary for an understanding of the nonlinear effects in optical fibers. Split-step Fourier model (SSFM) described in Appendix E and Chapter 2 is used to solve the pulse-propagation in the nonlinear guided wave medium. Thus, the evolution of solitons can be observed at any time span along the propagation distance. The BPM using split-step Method has been presented in detail in Appendix E. The SSFM is extensively implemented in this chapter for soliton transmission as demonstrated in Section 8.8.1.4 (see also Appendix E or SSFM).

## 8.3.2  ANALYTICAL APPROACH: ISM

The ISM is similar in spirit to the BPM that is commonly used to solve linear partial differential equations. The incident field at $z=0$ is used to obtain the initial scattering data whose evolution along $z$ is easily obtained by solving the linear scattering problem. The propagated field is reconstructed from the evolved scattering data. The details of the ISM are available in many texts [1,2], we describe only briefly how this method is used to solve NLSE given in Equation 8.4.

Using the normalized parameters in Equations 8.1 through 8.13, the parameter $N$ can be eliminated from Equation 8.4 by defining

$$u = NU = \sqrt{\left(\frac{\gamma T_o^2}{|\beta_2|}\right)} A \tag{8.14}$$

and the NLSE becomes

$$j\frac{\partial u}{\partial \xi} + \frac{1}{2}\frac{\partial^2 u}{\partial \tau^2} + |u|^2 u = 0 \tag{8.15}$$

In the ISM, the scattering problem associated with Equation 8.15 is

$$\frac{\partial v_1}{\partial \tau} + j\zeta v_1 = u v_2 \tag{8.16}$$

$$\frac{\partial v_2}{\partial \tau} - j\zeta v_2 = -u^* v_1 \tag{8.17}$$

where
   $v_1$ and $v_2$ are the amplitudes of the waves scattering in a potential $u(\xi, \tau)$
   $\zeta$ is the eigenvalue

The soliton order is characterized by the number $N$ of poles or eigenvalues $\zeta_j$ ($j=1-N$). The general solution is [3]

$$u(\xi, \tau) = -2 \sum_{j=1}^{N} \lambda_j^* \psi_{2j}^* \tag{8.18}$$

where

$$\lambda_j = \sqrt{c_j} \exp(j\zeta_j \tau + j\zeta_j^2 \xi) \tag{8.19}$$

$\psi_{2j}$ is obtained by solving the linear set of following equations:

$$\psi_{1j} + \sum_{k=1}^{N} \frac{\lambda_j \lambda_k^*}{\zeta_j - \zeta_k^*} \psi_{2k}^* = 0 \qquad (8.20)$$

$$\psi_{2j}^* - \sum_{k=1}^{N} \frac{\lambda_k \lambda_j^*}{\zeta_j^* - \zeta_k} \psi_{1k} = \lambda_j^* \qquad (8.21)$$

The eigenvalues $\zeta_j$ is given by

$$\zeta_j = i\eta_j \qquad (8.22)$$

Then Equation 8.19 becomes

$$\lambda_j = \sqrt{c_j} \exp\left(-\eta_j \tau - j\eta_j^2 \xi\right) \qquad (8.23)$$

Assuming the soliton is symmetrical about $\tau = 0$, the residues are related to the eigenvalues by the relation

$$c_j = \frac{\prod_{k=1}^{N} (\eta_j + \eta_k)}{\prod_{k\neq j}^{N} |(\eta_j - \eta_k)|} \qquad (8.24)$$

The fundamental soliton corresponds to the case of a single eigenvalue $\eta_1$ which has a value of 0.5 for $N=1$. In general, for $N>1$, the eigenvalues are given by

$$\eta_i = \frac{2i-1}{2} \qquad (8.25)$$

where $i = 1, 2, 3, \ldots, N$.

### 8.3.2.1   Soliton $N=1$ by Inverse Scattering

Fundamental soliton has only one eigenvalue which is shown in Equation 8.21. Thus, we obtain $\eta_1 = 0.5$. The residue that is related to the eigenvalue can be found by using Equation 8.20. Then, the eigenvalue and its residue are substituted in Equation 8.19 in order to obtain the eigenfunction $\lambda$. The complex eigenvalue can be obtained by using Equation 8.22. Eigenpotential $\psi_{21}$ shown in Equations 8.20 and 8.21 need to be solved simultaneously, and the process of solving this problem is described in the following. From Equations 8.20 and 8.21 we get the following

$$\psi_{11} + \frac{\lambda_1 \lambda_1^*}{\zeta_1 - \zeta_1^*} \psi_{21}^* = 0 \qquad (8.26)$$

$$\psi_{21}^* = \lambda_1^* + \frac{\lambda_1 \lambda_1^*}{\zeta_1^* - \zeta_1} \psi_{11} \qquad (8.27)$$

Substituting Equation 8.22 into 8.23, we obtain the following equation.

$$\psi_{21}^* = \frac{\zeta_1 - \zeta_1^*}{\lambda_1 \lambda_1^*} \left( \frac{\lambda_1 \lambda_1^{*2} \left( \zeta_1^* - \zeta_1 \right)}{\left( \zeta_1^* - \zeta_1 \right)\left( \zeta_1 - \zeta_1^* \right) + \left( \lambda_1 \lambda_1^* \right)^2} \right) \tag{8.28}$$

Thus, by substituting the eigenpotential and eigenfunction in Equation 8.18, we can obtained the fundamental soliton function $U(\xi, \tau)$.

### 8.3.2.2 Soliton $N = 2$ by Inverse Scattering

All the steps of solving the NLSE are the same as listed in Section 8.4.1 except that the eigenvalues, eigenfunction, and the eigenpotential have to be redefined. Solving the eigenvalues and the corresponding eigenfunction is straight forward. However, solving the eigenpotentials needs more effort. Again, from Equations 8.18 and 8.19, we obtain the following expressions:

$$\psi_{11} + \left( \frac{\lambda_1 \lambda_1^*}{\zeta_1 - \zeta_1^*} \psi_{21}^* + \frac{\lambda_1 \lambda_2^*}{\zeta_1 - \zeta_2^*} \psi_{22}^* \right) = 0 \tag{8.29}$$

$$\psi_{21}^* - \left( \frac{\lambda_1 \lambda_1^*}{\zeta_1^* - \zeta_1} \psi_{11} + \frac{\lambda_1 \lambda_2^*}{\zeta_1^* - \zeta_2} \psi_{12} \right) = \lambda_1^* \tag{8.30}$$

$$\psi_{12} + \left( \frac{\lambda_2 \lambda_1^*}{\zeta_2 - \zeta_1^*} \psi_{21}^* + \frac{\lambda_2 \lambda_2^*}{\zeta_2 - \zeta_2^*} \psi_{22}^* \right) = 0 \tag{8.31}$$

$$\psi_{22}^* - \left( \frac{\lambda_2 \lambda_1^*}{\zeta_2^* - \zeta_1} \psi_{11} + \frac{\lambda_2 \lambda_2^*}{\zeta_2^* - \zeta_2} \psi_{12} \right) = \lambda_2^* \tag{8.32}$$

There are four unknowns and four equations, thus it is possible to solve them simultaneously. We can represent Equations 8.29 through 8.32 in a matrix form as follows:

$$\begin{pmatrix} \psi_{11} \\ \psi_{12} \\ \psi_{21}^* \\ \psi_{22}^* \end{pmatrix} = \begin{pmatrix} 0 & 0 & \dfrac{\lambda_1 \lambda_1^*}{\zeta_1 - \zeta_1^*} \psi_{21}^* & \dfrac{\lambda_1 \lambda_2^*}{\zeta_1 - \zeta_2^*} \psi_{22}^* \\ 0 & 0 & \dfrac{\lambda_2 \lambda_1^*}{\zeta_2 - \zeta_1^*} \psi_{21}^* & \dfrac{\lambda_2 \lambda_2^*}{\zeta_2 - \zeta_2^*} \psi_{22}^* \\ -\dfrac{\lambda_1 \lambda_1^*}{\zeta_1^* - \zeta_1} \psi_{11} & -\dfrac{\lambda_1 \lambda_2^*}{\zeta_1^* - \zeta_2} \psi_{12} & 0 & 0 \\ -\dfrac{\lambda_2 \lambda_1^*}{\zeta_2^* - \zeta_1} \psi_{11} & -\dfrac{\lambda_2 \lambda_2^*}{\zeta_2^* - \zeta_2} \psi_{12} & 0 & 0 \end{pmatrix} \cdot \begin{pmatrix} \psi_{11} \\ \psi_{12} \\ \psi_{21}^* \\ \psi_{22}^* \end{pmatrix} + \begin{pmatrix} 0 \\ 0 \\ \lambda_1^* \\ \lambda_2^* \end{pmatrix} \tag{8.33}$$

Soliton $N = 1$: Numerical approach using beam propagation method

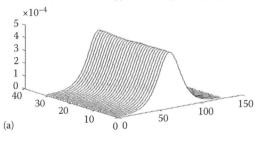

(a)

Soliton $N = 1$: Analytical approach using beam propagation method

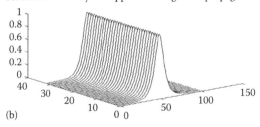

(b)

**FIGURE 8.1**    Comparison between (a) numerical and (b) analytic approach for soliton $N = 1$.

Soliton $N = 2$: Numerical approach using beam propagation method

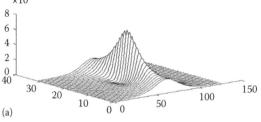

(a)

Soliton $N = 2$: Analytic approach using inverse scattering

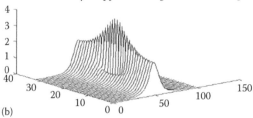

(b)

**FIGURE 8.2**    Comparison between (a) numerical and (b) analytic approach for soliton $N = 2$.

Thus, by substituting the eigenpotential and eigenfunction in Equation 8.22, we can obtained the second order of soliton function $U(\xi, \tau)$.

The results of both beam propagation (numerical approach) and ISM (analytical approach) are shown in Figures 8.1 and 8.2 for fundamental soliton and second-order soliton, respectively. From the results, we notice that the soliton pulse is recovered at one soliton period. The different pulse widths observed are due to different setting of the initial soliton pulse width. The results of the numerical approach agree with the analytical results, and hence, we conclude that our numerical algorithm in synthesizing solitons evolution has been a success.

## 8.4   FUNDAMENTAL AND HIGHER-ORDER SOLITONS

### 8.4.1   Soliton Evolution for $N = 1, 2, 3, 4$, and 5

The evolutionary process of optical soliton is shown in Figures 8.1 through 8.3 for the fundamental, second, third, fourth and fifth orders of soliton, respectively. Periodic evolution of soliton is very much depending on the soliton period $z_O$ and the soliton order $N$. As we can observe in those previous figures, the soliton pulse begins to evolve and its original shape will be recovered at every soliton period.

For the fundamental soliton, the pulse shape remains the same throughout the propagation distance. However, for soliton $N > 1$, the soliton pulse will be evolved into various shapes periodically depending on the order of the soliton. Eventually, these distorted pulses will come back to the shape of the fundamental soliton at every soliton period. As for soliton $N = 2$, the pulse begins to contract and the narrowing effect starts to take place and causes the peak to increase gradually. In the midpoint of the soliton period, the soliton peak has increased by three times (Figure 8.2—numerical results) and there are two minor components on each side of the main component of the soliton pulse.

For soliton $N = 3$, as the pulse propagates along the fiber, it first contracts to a fraction of its initial width, splits into two components, and then merges again to recover its original shape at the end of the soliton period. As for the soliton $N = 4$, the fundamental pulse shape splits into two components, and then further splitting into three components. After reaching the midpoint of the soliton period, the three-component pulse begins to merge into two components and finally recovers to its original shape at the end of the soliton period.

Each plot is taken at certain distance of propagation. Starting at 0 km, we have the fundamental soliton shape. Then it splits into two components at 400 km, further splitting into three components occurs at 600 km and when it reaches the midpoint of the soliton period which is at 1210 km, it splits into four components. After the midpoint, the four components begin to merge into three, and into two, and then into the original shape. Having analyzing the first five order of the soliton, we notice that the maximum number of component split ($N_C$) at the midpoint is given by

$$N_C = N - 1 \qquad\qquad (8.34)$$

Soliton order $N = 3$ and $T_o = 18$ ps over $Z_o$

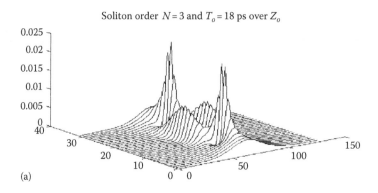

(a)

Soliton order $N = 4$ and $T_o = 18$ ps over $Z_o$

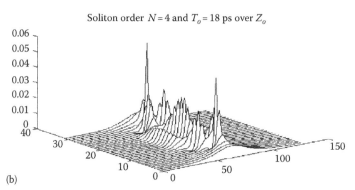

(b)

Soliton order $N = 5$ and $T_o = 30$ ps over $Z_o$

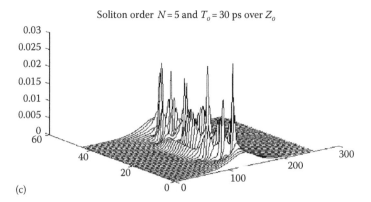

(c)

**FIGURE 8.3**    Soliton evolution for (a) $N=3$, (b) $N=4$, and (c) $N=5$.

where $N$ is the order of the soliton. For example, they are five-split components at the midpoint of the soliton period for a soliton of order $N=6$.

### 8.4.2   SOLITON BREAKDOWN

Another interesting observation of the soliton propagation is the soliton break-down effect. Referring back to Figure 8.4 for soliton $N=5$, we realized that the soliton pulse width of 30 ps is used, which is larger than the pulse width of the lower-order solitons (only 18 ps). In Figure 8.5, we purposely assign a narrow pulse width of 9 ps (instead of 18 ps) to the fourth-order soliton that the soliton break-down may be seen.

This proves that we need a substantial amount of soliton energy to support higher-order soliton in order for it to recover back to its own shape. Since the pulse width is related to the total power contained in the soliton, we would need to increase the pulse width for higher-order transmission. Otherwise, the soliton breakdown will occur where the soliton pulse will not recover to its original shape. At certain point in the propagation distance, the pulse will split and may not merge back again. All these effects can be seen in Figure 8.5. Hence, there is a trade-off between the order of soliton and the bit rate. For high bit rate transmission, we need a narrow soliton pulse width. However, this would limit the order of soliton that we can use.

## 8.5   INTERACTION OF FUNDAMENTAL SOLITONS

### 8.5.1   DUAL SOLITONS INTERACTION WITH DIFFERENT PULSE SEPARATION

It is necessary to determine how close two solitons can come closer to each other without interacting. The nonlinearity that bounds a single soliton introduces an inter-action force between the neighboring solitons. The amplitude of the soliton pair at the fiber input can be written in the following normalized form:

$$u(0, \tau) = \text{sech}(\tau - q_O) + r\,\text{sech}\{r(\tau + q_O)\}\exp(j\theta) \qquad (8.35)$$

where
$q_O$ is the initial pulse separation
$r$ is the relative amplitude

Thus, soliton interaction can be studied by solving NLSE using the numerical approach and the BPM discussed in Section 8.3.1. Figure 8.6 displays the evo-lution pattern showing periodic collapse of a soliton pair under various pulse separations.

The periodic collapse of neighboring solitons is undesirable from the system standpoint. One way to avoid the interaction problem is to increase the separation such that the collapse distance, $Z_p$, is much larger than the transmission distance $L_T$. From the results in Figure 8.6, we can measure the collapse distance $Z_p$ at each different pulse separation $q_O$ and, thus, draw a rough estimation by interpolating

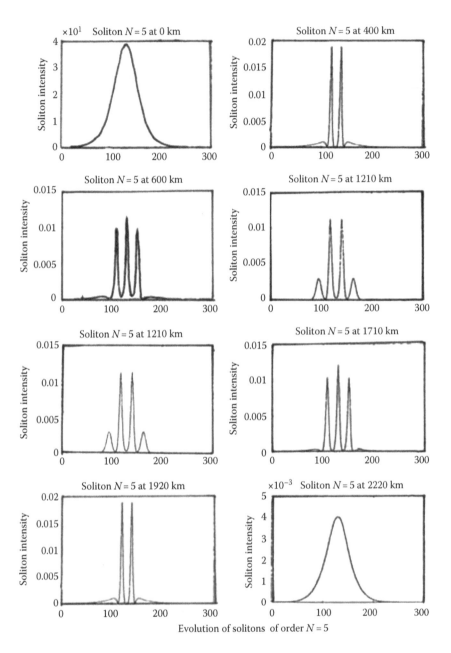

**FIGURE 8.4** Evolution of solution of fifth order with respect to different propagation distances as noted in each individual graph.

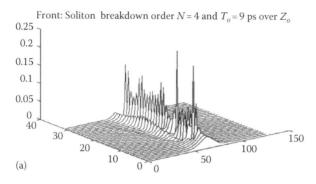

(a)

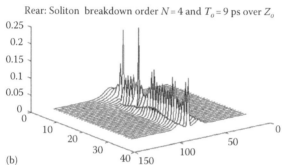

(b)

**FIGURE 8.5** Soliton breakdown phenomena: (a) front region and (b) rear region of the pulse sequence.

these data. All the data are tabulated in Table 8.1, and the interpolated curve is shown in Figure 8.7.

Data measured and collected from Figure 8.6. It is noted that $Z_p$ is defined as the distance where the first collapse of solitons happens. Thus, the curve shown in Figure 8.7 is useful as a guide line for selection of optimum pulse separation given a certain transmission distance. The pulse separation has to be minimized to achieve high bit rate transmission.

### 8.5.2 DUAL SOLITONS INTERACTION WITH DIFFERENT RELATIVE AMPLITUDE

Shown in Figure 8.8 is the two solitons interaction with different relative amplitudes. Referring to Equation 8.35, we can set different relative amplitude, via the parameter $r$, to the two pulses and the effect of setting asymmetrical pair of soliton can be examined in the following.

From Figure 8.8, we observe that setting a slight difference between the amplitudes of the soliton pulses does solve the soliton collapse problem effectively. Another observation we obtain is that as the relative amplitude increases, the periodical attraction between the two pulses happens more frequently. For example, in case of $r = 1.1$, the two pulses come closer to each other in three incidents, compared

Solitons interaction: $q = 1.5$, $r = 1$ and phase $= 0°$ over 1600 km

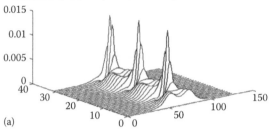

(a)

Solitons interaction: $q = 2.5$, $r = 1$ and phase $= 0°$ over 4700 km

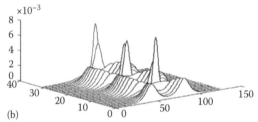

(b)

Solitons interaction: $q = 3.5$, $r = 1$ and phase $= 0°$ over 16,000 km

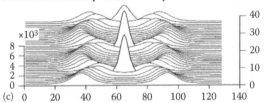

(c)

Solitons interaction: $q = 4.5$, $r = 1$ and phase $= 0°$ over 3400 km

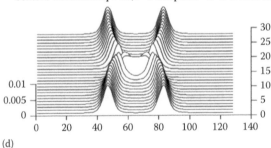

(d)

**FIGURE 8.6**  Dual solitons interaction at different pulse temporal separation and no phase difference (see notes accompanied each figure). (a) $q = 1.5$, $r = 1$, no phase difference, and propagation distance $L = 1600$ km. (b) $q = 2.5$, $r = 1$, no phase difference, and propagation distance $L = 4700$ km. (c) $q = 3.5$, $r = 1$, no phase difference, and propagation distance $L = 16,000$ km. (d) $q = 4.5$, $r = 1$, no phase difference, and propagation distance $L = 3400$ km.    (*Continued*)

Solitons interaction: $q = 5.5, r = 1$ no phase $= 0°$ over 10,000 km

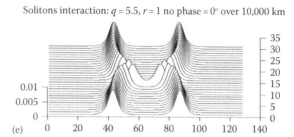

Solitons interaction: $q = 6.5, r = 1$ and phase $= 0°$ over 21,000 km

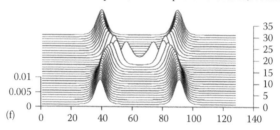

**FIGURE 8.6 (Continued)** Dual solitons interaction at different pulse temporal separation and no phase difference (see notes accompanied each figure). (e) $q=5.5$, $r=1$, no phase difference, and propagation distance $L=10,000$ km. (f) $q=6.5$, $r=1$, no phase difference, and propagation distance $L=21,000$ km.

---

**TABLE 8.1**

**Variation of Temporal Soliton Separation with Respect to Propagation Distance in km**

| $q_0$ | 1.5 | 2.5 | 3.5 | 4.5 | 5.5 | 6.5 |
|---|---|---|---|---|---|---|
| $Z_p$ (km) | 310 | 910 | 1,100 | 1,700 | 4,500 | 13,000 |

---

to four incidents for a given transmission distance in the case of $r=1.3$. Thus, by setting slightly different relative amplitude, we are able to overcome the problem of soliton pair collapse.

### 8.5.3 DUAL SOLITONS INTERACTION UNDER DIFFERENT RELATIVE PHASE

As we can observe from the results obtained in Figure 8.9, the strong attractive force turns into a repulsive force for $\theta \neq 0$ that the soliton pair is separated apart and may not merge back. Notice that relative phase of 180° could maintain the oscillation between the soliton pair without merging. The 90° relative phase soliton interaction is in fact the mirror image of the one with 270°. Apart from those relative phases mentioned earlier, they contribute to soliton pair repulsion. These characteristics are tabulated in Table 8.2.

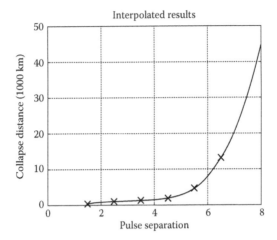

**FIGURE 8.7**  Interpolated data from Table 8.1.

In conclusion, by adjusting the pulse separation, relative amplitude, and relative phase of the soliton pair, we could increase the bit rate or the transmission capacity. The collapsing problems of soliton pair can be overcome effectively by varying those three parameters so that stable soliton transmission system can be achieved.

### 8.5.4  TRIPLE-SOLITON INTERACTION UNDER DIFFERENT RELATIVE PHASES

Similar to Equation 8.35, the amplitude of three solitons at the fiber input using can be rewritten in the following normalized form:

$$u(0, \tau) = \mathrm{sech}(\tau) + r\,\mathrm{sech}\{r(\tau - q_o) + r(\tau + q_o)\}\exp(j\theta) \tag{8.36}$$

As for the case of three solitons interaction, the observed characteristics are pretty much the same as those have been discussed for the soliton pair in terms of the relative phase. Again we define $r$ as the relative amplitude between the two solitons. In Figure 8.10, we notice that only at the relative phase of 180° between the center and the two side solitons, the three solitons would not merge. The oscillation of the three solitons would create severe problems at the receiving/detection subsystems. Thus, we must stabilize the oscillations by introducing a relative amplitude of $r = 1.5$ for the first and the third pulse described in Figure 8.11.

### 8.5.5  TRIPLE SOLITONS INTERACTION WITH DIFFERENT RELATIVE PHASE AND $r = 1.5$

From Figures 8.11 and 8.12, we can see that for a relative phase less than 180°, the side solitons are repelling away from the center soliton. The repulsive force reaches its maximum at 90° phase difference. Notice that the repulsive forces exerted in the case of 45° and 135° are the same. When the relative phase is adjusted to 180°,

Solitons interaction: $q = 2.5$, $r = 1.1$ and relative phase = 0° over 4700 km

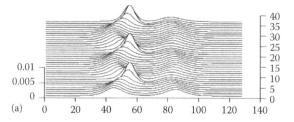

(a)

Solitons interaction: $q = 2.5$, $r = 1.2$ and relative phase = 0° over 4700 km

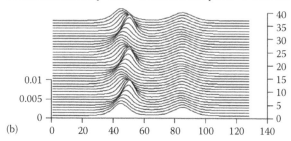

(b)

Solitons interaction: $q = 2.5$, $r = 1.3$ and relative phase = 0° over 4700 km

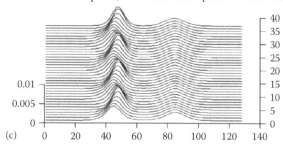

(c)

Solitons interaction: $q = 2.5$, $r = 1.5$ and relative phase = 0° over 4700 km

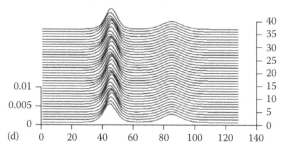

(d)

**FIGURE 8.8** Two solitons interaction of no phase difference at $L=4700$ km with different relative amplitude (ratio $r$): (a) $q=2.5$, $r=1.1$, no phase difference, $L=4700$ km. (b) $q=2.5$, $r=1.2$, no phase difference, $L=4700$ km. (c) $q=2.5$, $r=1.3$, no phase difference, $L=4700$ km. (d) $q=2.5$, $r=1.5$, no phase difference, $L=4700$ km.

Solitons interaction: $q = 3.5$, $r = 1$ and phase = $0°$ over 16,000 km

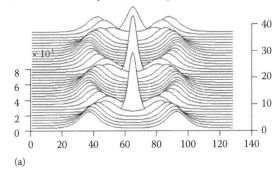

(a)

Solitons interaction: $q = 3.5$, $r = 1$ and phase = $45°$ over 16,000 km

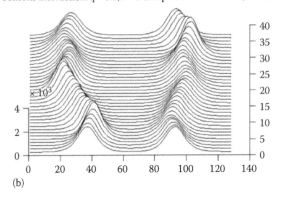

(b)

Solitons interaction: $q = 3.5$, $r = 1$ and phase = $90°$ over 16,000 km

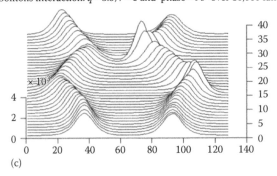

(c)

**FIGURE 8.9** Two solitons interaction under different relative phase of the carrier under the soliton envelope: (a) phase diff=0, (b) phase diff=$\pi/4$ (c) phase diff=$\pi/2$.        (*Continued*)

Solitons interaction: $q = 3.5$, $r = 1$ and phase = 135° over 16,000 km

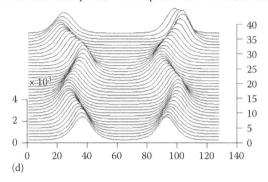

(d)

Solitons interaction: $q = 3.5$, $r = 1$ and phase = 180° over 16,000 km

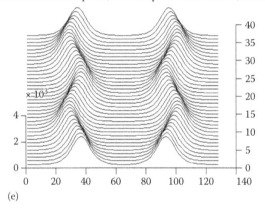

(e)

Solitons interaction: $q = 3.5$, $r = 1$ and phase = 270° over 16,000 km

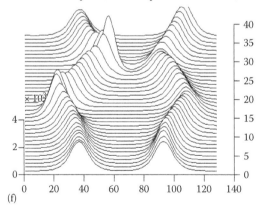

(f)

**FIGURE 8.9 (*Continued*)** Two solitons interaction under different relative phase of the carrier under the soliton envelope: (d) phase diff = $3\pi/4$, (e) $\pi$, and (f) $3\pi/2$.

## TABLE 8.2
## Characteristic of Soliton Pair due to the Relative Phase Angle

| Relative Phase | Observation and the Effect |
|---|---|
| 0° | Soliton pair merging periodically |
| 180° | Soliton pair oscillating periodically without merging |
| Others | Soliton pair in repulsion |

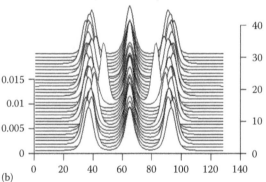

**FIGURE 8.10** Three solitons interaction at different relative phase between the adjacent solitons and the center soliton: (a) phase diff=0 rad, (b) phase diff=$\pi$/4.          *(Continued)*

Solitons interaction: $q = 7$, $r = 1$ and phase = 90° over 14,000 km

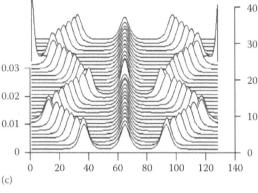

(c)

Solitons interaction: $q = 7$, $r = 1$ and phase = 180° over 14,000 km

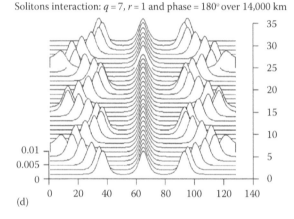

(d)

**FIGURE 8.10 (*Continued*)** Three solitons interaction at different relative phase between the adjacent solitons and the center soliton: (c) phase diff = $\pi/2$, and (d) phase diff = $\pi$.

there is neither attractive nor repulsive force at all. This is the same for the case when the relative phase = 0.

When the relative phase is greater than 180°, various interesting soliton dynamics can be observed. In Figure 8.12, we see that the two side solitons begin to merge together with the center one at a relative phase of 200°. Indeed, these solitons merge at the propagation distance of approximately 14,500 km (Figure 8.12). When the relative phase = 225°, the three solitons will be scattered apart at propagation distance of about 10,000 km (Figure 8.12). Higher relative phase would still merge the solitons, and there is no incidence of any repulsive separation.

## 8.6   SOLITON PULSE TRANSMISSION SYSTEMS AND ISM

This section deals with the design of a 4-bit soliton pulse train communication system. The dynamics of the optical solitons can be studied via the ISM. Of

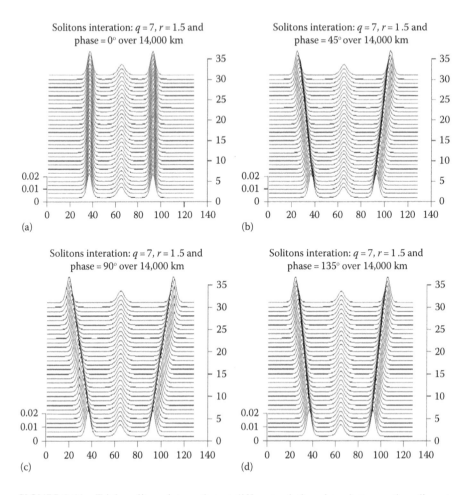

**FIGURE 8.11**   Triple solitons interaction at different relative phase between the adjacent and center solitons with different amplitude ratio of $r=1.5$: (a) phase diff$=0$ rad, (b) phase diff$=\pi/4$, (c) phase diff$=\pi/2$, and (d) phase diff$=3\pi/4$.

particular interest is the interaction between the neighboring solitons and its effect on the pulse separation. The design of an optimal 4-bit soliton pulse train can then be verified by simulation based on MATLAB®. Obtained results indicate that the system design is functioning well. However, the question as to whether the highest bit-rate is achieved or not remains unanswered. Furthermore, a re-examination of the phenomenon observed by the simulated results obtained in this section on the soliton breakdown due to "insufficient energy content" can be been proven to be false. The breakdown occurs due to improper adjustment of sampling parameters to accommodate higher-frequency components. A counter example is shown whereby the breakdown of a fourth-order soliton did not occur even with narrow pulse widths.

Solitons interaction: $q = 7$, $r = 1.5$ and phase = 200° over 14,000 km

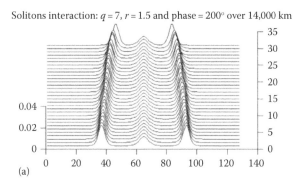

(a)

Solitons interaction: $q = 7$, $r = 1.5$ and phase = 250° over 14,000 km

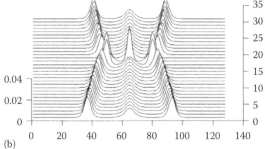

(b)

Solitons interaction: $q = 7$, $r = 1.5$ and phase = 315° over 14,000 km

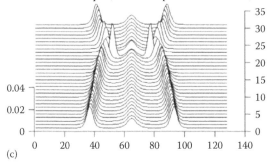

(c)

**FIGURE 8.12** Triple solitons interaction at different relative phase with relative amplitude difference of $r = 1.5$ observed at $L = 14,000$ km: (a) 200° of arc, (b) 250°, and (c) 315°.

We recall here again the wave equation governing optical soliton propagation (Equation 8.3). If this equation can be normalized to a simpler form and the third-order term neglected assuming that the operating wavelength is substantially far away from the zero dispersion wavelength of the optical fiber, then

$$j\frac{\partial U}{\partial \xi} - \text{sgn}(\beta_2)\frac{1}{2}\frac{\partial^2 U}{\partial \tau^2} + N^2 |U|^2 U = 0 \tag{8.37}$$

where $N$ denotes the order of the soliton solution. This also indicates how many solitons can be present in the solution.

### 8.6.1 ISM REVISITED

The ISM was developed to deal with many nonlinear wave equations encountered in the field of physics. One such equation studied was the NLSE (this equation differs slightly from Equation 8.37 by the roles of $t$ and $x$):

$$iq_t + q_{xx} + \chi |q|^2 q = 0 \tag{8.38}$$

A particular solution of this equation is the soliton wave given as follows:

$$u(x,t) = \sqrt{2\chi\eta}\, \frac{\exp[-4j(\xi^2 - \eta^2)t - 2j\xi x + j\varphi]}{\cosh[2\eta(x - x_0) + 8\eta\xi t]} \tag{8.39}$$

where
  the constant $\eta$ characterizes the amplitude as well as the width of the soliton
  $\xi$ is the velocity of the soliton (with respect to the group velocity)
  $x_0$ is the initial position of the soliton center
  $\varphi$ is its initial phase

For a more general solution, we can turn to the ISM.

Given the initial boundary constraints, the aim of the ISM is to reconstruct the potential $u$ in the NLSE from the asymptotic data that results from scattering processes. An outline of the method can be given in the following.

The route as presented shows that the ISM is broken up into three subproblems. This seems long and convoluted since from Figure 8.15, there can be a more direct approach. However, one must note that the three steps involved in the inverse method deals with linear processes only and this is the advantage of the ISM.

#### 8.6.1.1 Step 1: Direct Scattering

The scattering data $S_\pm(0)$ corresponding to the initial function $F(x)$ is first obtained by solving the NLSE for the Jost solutions, which are solutions with the asymptotic form $e^{-jkx}$ and $e^{+jkx}$ as $x \to \pm \infty$.

#### 8.6.1.2 Step 2: Evolution of the Scattering Data

Once the scattering data is obtained for $t=0$, the evolution of the scattering data is determined uniquely for arbitrary time $t$ by

$$C_1(\hat{L}^A)u_t + C_2(\hat{L}^A)u_x = 0, \quad x \in \Re, t \ge 0 \tag{8.40}$$

where
  $C_1, C_2$ are real analytic functions
  $L$ is basically a reformulation of the NLSE as a linear operator.

### 8.6.1.3 Step 3: Inverse Spectral Transform

The potential $u(x, t)$ is then reconstructed from the scattering data $S_{\pm}(t)$, usually by solving some system of linear integrals equations. Returning to the NLSE, Equation 8.38, we can associate with this equation the following Lax pair:

$$\hat{L} = i \begin{bmatrix} 1+p & 0 \\ 0 & 1-p \end{bmatrix} \frac{\partial}{\partial x} + \begin{bmatrix} 0 & u^* \\ u & 0 \end{bmatrix}, \quad \chi = \frac{2}{1-p^2}$$

$$\hat{A} = -p \begin{bmatrix} 1 & 0 \\ 0 & 1 \end{bmatrix} \frac{\partial^2}{\partial x^2} + i \begin{bmatrix} \dfrac{|u|^2}{1+p} & ju_x^* \\[2mm] -ju_x & \dfrac{-|u|^2}{1-p} \end{bmatrix} \tag{8.41}$$

and the $L$ operator evolves in time according to the commutation relation

$$\frac{\partial \hat{L}}{\partial t} = [\hat{A}, \hat{L}] \tag{8.42}$$

Here $L$ and $A$ are linear differential operators containing the sought function $u(x, t)$ in the form of a coefficient. By appropriate transformations of the eigenfunctions and rescaling of the eigenvalues, the pair could be transformed to resemble the form (see [4]):

$$\hat{L} = j \begin{bmatrix} \dfrac{\partial}{\partial x} & -U(x,t) \\[2mm] R(x,t) & -\dfrac{\partial}{\partial x} \end{bmatrix}$$

$$\hat{A} = \begin{bmatrix} A(x,t,k) & B(x,t,k) \\ C(x,t,k) & -A(x,t,k) \end{bmatrix} + f(k)\mathbf{I} \tag{8.43}$$

The key observation of Ablowitz, Kaup, Newell, and Segur is that a set of nonlinear equations can be obtained from the transformed Lax pair as

$$A_x - UC + RB = 0$$

$$U_t - B_x - 2AU + 2jkB = 0 \tag{8.44}$$

$$R_t - C_x + 2AR + 2jkC = 0$$

This set of equations, in turn, can generate several classes of nonlinear wave equations. A technique for obtaining some of the associated equations is to substitute the polynomials in $k$ for $A$, $B$, $C$ and solve recursively. This technique reproduces besides the NLSE, several other important equations. For example, one such equation is the $K\,dV$ equation.

$$U_t + 6UU_x + U_{xxx} = 0 \tag{8.45}$$

and this equation was generated with the polynomials $A$, $B$, $C$ having the following form:

$$A = -4ik^3 + 2jU - U_x,$$

$$B = 4Uk^2 + 2jU_xk - 2U^2 - U_{xx}, \tag{8.46}$$

$$C = -4k^2 + 2U, \quad R = -1$$

## 8.6.2 ISM Solutions for Solitons

Following the scheme as presented in Figure 8.13, we proceed with Step 1, the direct scattering problem.

### 8.6.2.1 Step 1: Direct Scattering Problem

Let us return to the $L$ operator associated with the NLSE. We consider the system of equations.

$$\hat{L}\psi = \lambda\psi, \quad \psi = \left\{ \begin{matrix} \psi_1 \\ \psi_2 \end{matrix} \right\} \tag{8.47}$$

Making the following change of variables:

$$\psi_1 = \sqrt{1-p}\,\exp\left(-j\frac{\lambda x}{1-p^2}\right)v_2, \quad \psi_2 = \sqrt{1+p}\,\exp\left(-j\frac{\lambda x}{1-p^2}\right)v_1 \tag{8.48}$$

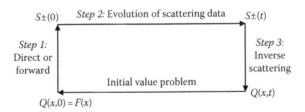

**FIGURE 8.13**   Steps of the ISM.

This equation can be rewritten in the following form:

$$\frac{\partial v_1}{\partial t} + j\zeta v_1 = qv_2, \quad \frac{\partial v_2}{\partial t} - j\zeta v_2 = -q^*v_1,$$

$$q = \frac{ju}{\sqrt{1-p^2}}, \quad \zeta = \frac{\lambda p}{1-p^2} \qquad (8.49)$$

We can define for real $\zeta = \xi$, the Jost functions $\varphi$ and $\psi$, as solutions of, with asymptotic values

$$\varphi \rightarrow \begin{Bmatrix} 1 \\ 0 \end{Bmatrix} \exp(-j\xi x), \quad x \rightarrow -\infty$$

$$\qquad (8.50)$$

$$\psi \rightarrow \begin{Bmatrix} 0 \\ 1 \end{Bmatrix} \exp(j\xi x), \quad x \rightarrow +\infty$$

The solution $\psi$ and its adjoint

$$\bar{\psi} = \begin{Bmatrix} \psi_2^* \\ -\psi_1^* \end{Bmatrix} \qquad (8.51)$$

form a complete system of solutions, and therefore,

$$\varphi = a(\xi)\bar{\psi} + b(\xi)\psi \qquad (8.52)$$

Note that if $\psi$ is a solution of the system at $\zeta = \xi + i\eta$, then its adjoint satisfies the system at $\zeta^* = \zeta - j\eta$. The points of the upper half-plane, Im $\zeta > 0$, $\zeta = \zeta_j$, $j = 1, \ldots, N$, at which $a(\zeta) = 0$, correspond to the eigenvalues of the problem. Here

$$\varphi(x, \zeta_j) = c_j \psi(x, \zeta_j), \quad j = 1, \ldots, N \qquad (8.53)$$

It can be shown that if $q$ is real, then the zeroes of $a(\zeta)$ lie on the imaginary axis.

### 8.6.2.2 Step 2: Evolution of the Scattering Data

Revisiting the commutation relation in Equation 8.49, we find that the eigenfunctions $\psi$ obey the following equation:

$$i\frac{\partial \psi}{\partial t} = \hat{A}\psi \qquad (8.54)$$

From this equation, the time dependencies of the coefficients $a(\xi, t)$, $b(\xi, t)$, $c_j(t)$ can be found as follows:

$$a(\xi, t) = a(\xi), \quad b(\xi, t) = b(\xi, 0)\exp(4i\xi^2 t), \quad c_j(t) = c_j(0)\exp(4i\zeta^2 t) \qquad (8.55)$$

### 8.6.2.3 Step 3: Inverse Scattering Problem

The task is now to reconstruct $u(x, t)$ from the scattering data $a(\xi)$, $b(\xi,0)$ for $-\infty < \xi < +\infty$ and $c_j(0)$, which we have denoted as $b(\zeta_j,0)$. The eigenfunctions are related to the residues at the poles $\zeta_j$ and the coefficients $a(\xi)$, $b(\xi)$ via the following system of equations. See [2] for complete derivation.

$$\phi_1 - c(x,\xi)\frac{1+J}{2}\phi_2^* = -c(x,\xi)\sum_{k=1}^{N}\frac{\exp\left(-i\zeta_k^* x\right)}{\xi - \zeta_k^*}c_k^*\psi_2^*(x,\zeta_k),$$

$$c^*(x,\xi)\frac{1-J}{2}\phi_1 + \phi_2^* = c^*(x,\xi) + c^*(x,\xi)\sum_{k=1}^{N}\frac{\exp(i\zeta_k x)}{\xi - \zeta_k}c_k\psi_1(x,\zeta_k),$$

$$\psi_1(x,\zeta_j)\exp(-i\zeta_j x) + \sum_{k=1}^{N}\frac{\exp\left(-i\zeta_k^* x\right)}{\zeta_j - \zeta_k^*}c_k^*\psi_2^*(x,\zeta_k) = \frac{1}{2}J\phi_2^*,$$

$$-\sum_{k=1}^{N}\frac{\exp(i\zeta_k x)}{\zeta_j^* - \zeta_k}c_k\psi_1(x,\zeta_k) + \psi_2^*(x,\zeta_j)\exp\left(i\zeta_j^* x\right) = 1 + \frac{1}{2}J\phi_1$$

(8.56)

where

$$(J\phi)(\xi) = \frac{1}{\pi i}\int_{-\infty}^{+\infty}\frac{\phi(\xi')}{\xi'-\xi}d\xi'$$

$$\phi(\xi) = \frac{b(\xi)}{a(\xi)}\exp(i\xi x)\psi(x,\xi)$$

The next step is to find an expression for $q$, because $u$ is related to $q$ by a simple constant as given in Equation 8.49. This can be done by comparing the asymptotic forms of the earlier system of equations.

$$\left\{\begin{array}{c}\psi_2(x,\zeta)\\-\psi_1(x,\zeta)\end{array}\right\}\exp(-i\zeta x) = \left\{\begin{array}{c}1\\0\end{array}\right\} + \frac{1}{\zeta}\left[\sum_{k=1}^{N}c_k^*\exp\left(-i\zeta_k^* x\right)\psi^*(x,\zeta_k) + \frac{1}{2\pi i}\int\phi^*(\xi)d\xi\right]$$

$$+O\left(\frac{1}{\zeta^2}\right), \quad \zeta \to \infty$$

(8.57)

and the differential equation in Equation 8.49 results in the following:

$$\psi(x,\zeta)\exp(-i\zeta x) = \left\{\begin{array}{c}0\\1\end{array}\right\} + \frac{1}{2i\zeta}\left\{\begin{array}{c}q(x)\\\int_x^\infty|q(s)|^2\,ds\end{array}\right\} + O\left(\frac{1}{\zeta^2}\right)$$

(8.58)

Thus, we get the following:

$$q(x) = -2i \sum c_k^* \exp\left(-i\zeta_k^* x\right) \psi_2^*(x, \zeta_k) - \frac{1}{\pi} \int \phi_2^*(\xi) d\xi,$$

$$\int_x^\infty |q(s)|^2 ds = -2i \sum c_k \exp(i\zeta_k x) \psi_1(x, \zeta_k) + \frac{1}{\pi} \int \phi_1(\xi) d\xi$$

(8.59)

Once $a$ and $b$ are obtained for an initial value, $u(x, t=0)$, then we can obtain $u(x, t)$ at an arbitrary instant through the Equations 8.57 through 8.59. We can now attempt to find an explicit formula for general solution of the NLSE.

### 8.6.3  N-SOLITONS SOLUTION (EXPLICIT FORMULA)

We consider the inverse scattering problem with the assumption $b(\xi, t) = 0$. Then $\varphi(\xi) \equiv 0$ and the system of Equation 8.58 reduces to

$$\psi_{1j} + \sum_{k=1}^{N} \frac{\lambda_j \lambda_k^*}{\zeta_j - \zeta_k^*} \psi_{2k}^* = 0,$$

$$-\sum_{k=1}^{N} \frac{\lambda_k \lambda_j^*}{\zeta_j^* - \zeta_k} \psi_{1k} + \psi_{2j}^* = \lambda_j^*,$$

(8.60)

$$\psi_j = \begin{Bmatrix} \psi_{1j} \\ \psi_{2j} \end{Bmatrix} = \sqrt{c_j} \psi(x, \zeta_j), \quad \lambda_j = \sqrt{c_j} \exp(i\zeta_j x)$$

Subsequently, the potential can also be simplified to

$$q(x) = -2i \sum_{k=1}^{N} \lambda_k^* \psi_{2k}^*, \quad \int_x^\infty |q(s)|^2 ds = -2i \sum_{k=1}^{N} \lambda_k \psi_{1k}$$

(8.61)

To see how a soliton can arise from this system of equations, we examine the case $N = 1$ and $a(\zeta)$ has only one zero in the upper-half plane. The following system of equations can be obtained.

$$\psi_1 + \frac{|\lambda|^2}{2i\eta} \psi_2^* = 0, \quad \frac{|\lambda|^2}{2i\eta} \psi_1 + \psi_2^* = \lambda^*$$

(8.62)

This system describes a soliton with amplitude $\eta$ and its initial center and phase given by

$$x_0 = \frac{1}{2\eta} \ln \frac{|\lambda(0)|^2}{2\eta}, \quad \varphi = -2 \arg \lambda(0)$$

(8.63)

Returning to the $N$-soliton case, in order to obtain the explicit solution for $q$ and ultimately $u$, we need to solve the system of Equations 8.62 for $\psi_j$. If we represent the aforementioned equations in matrix form, then the potential $u(x, t)$ can be shown [2] to be

$$\left|u(x,t)\right|^2 = \sqrt{2\chi}\,\frac{d^2}{dx^2}\ln\det\|\mathbf{A}\| = \sqrt{2\chi}\,\frac{d^2}{dx^2}\ln\det\left\|\mathbf{BB}^* + 1\right\|,$$

$$\tag{8.64}$$

$$\mathbf{B}_{jk} = \frac{\sqrt{c_j c_k^*}}{\zeta_j - \zeta_k^*}\exp\left[i\left(\zeta_j - \zeta_k^*\right)x\right]$$

The aforementioned result is a formula for reconstructing the potential $u(x, t)$. At the present moment, it does not describe a solution for any real physical problem, because we have not specified the initial values or boundary conditions. Therefore, let us examine the example from [5] where the initial potential has the form of a soliton, $A\,\text{sech}\,x$.

Example: $U(x,t) = A\,\text{sech}(x)$

Let us turn our attention to the NLSE. Note that in this example, the equation is slightly different from Equation 8.45, given as follows:

$$ju_t = \frac{1}{2}u_{xx} + \left|u\right|^2 u \tag{8.65}$$

We now consider the associated eigenvalue equation

$$jv_x + Uv = \zeta\sigma_3 v,$$

$$v = \begin{pmatrix} v_1 \\ v_2 \end{pmatrix}, \quad \sigma_3 = \begin{pmatrix} 1 & 0 \\ 0 & -1 \end{pmatrix}, \quad U = \begin{pmatrix} 0 & u \\ u^* & 0 \end{pmatrix} \tag{8.66}$$

Note that if $u$ is a solution of Equation 8.67, then the eigenvalue $\zeta$ is independent of time and $v$ evolves in time according to

$$iv_t = Av,$$

$$A = \begin{pmatrix} 1 & 0 \\ 0 & 1 \end{pmatrix}\left(\frac{1}{2}\frac{\rho^2 - 1}{\rho^2 + 1}\frac{\partial^2}{\partial x^2} - i\zeta\frac{\partial}{\partial x} + C\right) + \frac{1}{\rho^2 + 1}\begin{pmatrix} \rho^2\left|u\right|^2 & iu_x \\ -j\rho^2 u_x^* & -\left|u\right|^2 \end{pmatrix} \tag{8.67}$$

where $C$ is a constant independent of $x$. Proceeding with Step 1, we eliminate $v_2$ in Equation 8.67, and using the initial potential $u(x, t) = A\,\text{sech}\,x$, we have

$$s(1-s)\frac{d^2}{ds^2}v_1 + \left(\frac{1}{2}-s\right)\frac{d}{ds}v_1 + \left[A^2 + \frac{\zeta^2 + j\zeta(1-2s)}{4s(1-s)}\right]v_1 = 0, \quad s = \frac{1-\tanh x}{2} \tag{8.68}$$

Further transformation of the dependent variable $v_1$ into $s^\alpha(1-s)^\beta\omega_1$ reduces Equation 8.68 to the hypergeometric function forms:

$$v_{11}(s) = s^{j\zeta/2}(1-s)^{-j\zeta/2} F\left(-A, A, j\zeta + \frac{1}{2}; s\right),$$

$$v_{12}(s) = s^{(1/2)-j\zeta/2}(1-s)^{-j\zeta/2} F\left(\frac{1}{2} - j\zeta + A, \frac{1}{2} - j\zeta - A, \frac{3}{2} - j\zeta; s\right),$$

(8.69)

The solutions for $v$ can be obtained by replacing $\zeta$ with $-\zeta$ in the earlier equation,

$$v_{21}(s) = s^{-j\zeta/2}(1-s)^{j\zeta/2} F\left(-A, A, -j\zeta + \frac{1}{2}; s\right),$$

$$v_{12}(s) = s^{(1/2)+j\zeta/2}(1-s)^{j\zeta/2} F\left(\frac{1}{2} + j\zeta + A, \frac{1}{2} + j\zeta - A, \frac{3}{2} + j\zeta; s\right),$$

(8.70)

From the asymptotic requirements of Equation 8.50, we can find expressions for the coefficients as

$$\psi = \begin{pmatrix} \dfrac{A}{\xi + j/2} v_{12} \\ v_{21} \end{pmatrix}, \quad \overline{\psi} = \begin{pmatrix} v_{21}^* \\ -\dfrac{A}{\xi - j/2} v_{12}^* \end{pmatrix},$$

$$a(\xi) = \frac{[\Gamma(-j\xi + (1/2))]^2}{\Gamma(-j\xi + A + (1/2))\Gamma(-j\xi - A + (1/2))},$$

(8.71)

$$b(\xi) = \frac{j[\Gamma(j\xi + (1/2))]^2}{\Gamma(A)\Gamma(1-A)} = j\frac{\sin(\pi A)}{\cosh(\pi\xi)}.$$

The eigenvalues $\zeta_r$ are obtained from the zeroes of $a(\zeta)$.

$$\zeta_r = j\left(A - r + \frac{1}{2}\right)$$

(8.72)

$r$ must be positive integers satisfying $A - r + \frac{1}{2} > 0$. Step 2 gives

$$\lambda_k = \sqrt{\frac{b(\zeta_k)}{\partial a(\zeta_k)/\partial t}} \exp\left[j\zeta_k x - j\zeta_k^2 t\right], \quad c(x, \xi) = \frac{b(\xi)}{a(\xi)} \exp[2j\xi x - 2j\xi^2 t]$$

(8.73)

Note the slight difference in the aforementioned expressions when compared with Equation 8.24. Step 3 is an application of Equations 8.58 through 8.63.

### 8.6.4 SPECIAL CASE $A = N$

When $A = N$, a positive integer then by Equation 8.73, the coefficients $a(\zeta)$ reduces to

$$a(\zeta) = \prod_{r=1}^{N} \frac{(\zeta - \zeta_r)}{\zeta - \zeta_r^*} \qquad (8.74)$$

while $b(\xi) = 0$. At the eigenvalues,

$$b(\zeta_k) = j(-1)^{k-1} \qquad (8.75)$$

The coefficients $c(x, \xi)$ are defined by Equation 8.73 and subsequently, $\phi_1 = \phi_2^* = 0$. Substituting the results in Equation 8.58, we can solve the equations for $\psi_{k1}$ and $\psi_{k2}^*$. Explicit solutions for the cases $N = 1$ and $N = 2$ are given in the following:

$$u(x,t) = \exp\left(\frac{-jt}{2}\right)\operatorname{sech} x, \quad N = 1$$

$$\qquad (8.76)$$

$$u(x,t) = 4\exp\left(\frac{-jt}{2}\right)\frac{\cosh 3x + 3\exp(-4jt)\cosh x}{\cosh 4x + 4\cosh 2x + 3\cos 4t}, \quad N = 2$$

Of much interest to us is the long-term dynamic behavior of the soliton, and this will be explored in the next section.

### 8.6.5 N-SOLITON SOLUTIONS (ASYMPTOTIC FORM AS $\tau \to \pm\infty$)

The long-term behavior of the $N$-soliton solution is studied here for the special case where there are no two solitons with the same velocity, that is, no $\xi_j$ are the same. It is found that the $N$-soliton solution breaks up into diverging solitons as $t \to \pm\infty$. To verify this for the case $t \to +\infty$, let us arrange the $\xi_j$ in decreasing order, $\xi_1 > \xi_2 > \cdots > \xi_N$. From (8.57) we have the following:

$$\lambda_j(x,t) = \lambda_j(0)\exp\left[-\eta_j(x + 4\xi_j t) + j\left(\xi_j x + 2\left(\xi_j^2 - \eta_j^2\right)t\right)\right],$$

$$\qquad (8.77)$$

$$|\lambda_j(x,t)| = |\lambda_j(0)|\exp(-\eta_j y_j), \quad y_j = x + 4\xi_j t$$

Let us now consider the reference frame where $y_m = $ constant as $t \to +\infty$, that is, we are interested in a moving reference frame of one of the solitons. Then,

$$y_j \to 0, \quad |\lambda_j| \to 0 \quad \text{for } j < m,$$

$$\qquad (8.78)$$

$$y_j \to \infty, \quad |\lambda_j| \to \infty \quad \text{for } j > m,$$

It follows from the system of Equations 8.62 that $\psi_{1j}$, $\psi_{2j} \to 0$ when $j < m$, that is, the effect of these solitons is made negligible. The number of equations is therefore, reduced to $2(N - m - 1)$ given as follows:

$$\psi_{1m} + \frac{|\lambda_m|}{2i\eta_m} \psi_{2m}^* = -\lambda_m \sum_{k=m+1}^{N} \frac{1}{\zeta - \zeta_k^*} \phi_{2k}^*,$$

$$\frac{|\lambda_m|^2}{2i\eta_m} \psi_{1m} + \psi_{2m}^* = \lambda_m^* + \lambda_m^* \sum_{k=m+1}^{N} \frac{1}{\zeta^* - \zeta_k} \phi_{1k},$$

$$\sum_{k=m+1}^{N} \frac{1}{\zeta - \zeta_k^*} \phi_{2k}^* = -\frac{\lambda_m^*}{\zeta_j - \zeta_m^*} \psi_{2m}^*,$$

$$\sum_{k=m+1}^{N} \frac{1}{\zeta^* - \zeta_k} \phi_{1k} = -1 - \frac{\lambda_m}{\zeta^* - \zeta_m} \psi_{1m}$$

(8.79)

Solving first for $\phi_{1k}$ and $\phi_{2k}^*$, we obtain the following:

$$\phi_{1k} = a_k + \frac{2i\eta_m}{a_m} \frac{a_k}{\zeta_k - \zeta_m} \psi_{1m} \lambda_m,$$

$$\phi_{2k}^* = -\frac{2j\eta_m}{a_m^*} \frac{a_k^*}{\zeta_k^* - \zeta_m^*} \psi_{2m}^* \lambda_m^*,$$

(8.80)

$$a_k = \frac{\displaystyle\prod_{p=m+1}^{N} \zeta_k - \zeta_p^*}{\displaystyle\prod_{m < p \neq k}^{N} \zeta_k - \zeta_p}, \quad a_m = 2j\eta_m \prod_{p=m+1}^{N} \frac{\zeta_m - \zeta_p^*}{\zeta_m - \zeta_p}$$

After some manipulation (see [4] for details) we get the following:

$$\psi_{1m} + \frac{|\lambda_m^+|^2}{2i\eta_m} \psi_{2m}^* = 0,$$

(8.81)

$$\frac{|\lambda_m^+|^2}{2i\eta_m} \psi_{1m} + \psi_{2m}^* = \left(\lambda_m^+\right)^*, \quad \lambda_m^+ = \lambda_m \prod_{p=m+1}^{N} \frac{\zeta_m - \zeta_p}{\zeta_m - \zeta_p^*}$$

The soliton system represented by this Equation 8.81, can be considered much similar to that expressed in Equation 8.64 with a displaced position of the center $x_0^+$ and phase $\varphi^+$,

$$x_{0m}^+ - x_{0m} = \frac{1}{\eta_m} \sum_{p=m+1}^{N} \ln \left| \frac{\zeta_m - \zeta_p}{\zeta_m - \zeta_p^*} \right| < 0,$$

(8.82)

$$\varphi_m^+ - \varphi_m = -2 \sum_{p=m+1}^{N} \arg \left( \frac{\zeta_m - \zeta_p}{\zeta_m - \zeta_p^*} \right)$$

The calculations are similar for the case $t \to -\infty$. In the reference frames $y = x + \xi t$, where $\xi$ does not coincide with any of the $\xi_m$ as $t \to \pm\infty$, the reduced system tends to zero, hence proving the asymptotic breakdown of the $N$-soliton solution into solitons. When $t$ changes from $-\infty$ to $+\infty$, the corresponding changes in the soliton centers and phases are as follows:

$$\Delta x_{0m} = x_{0m}^+ - x_{0m}^- = \frac{1}{\eta_m}\left(\sum_{k=m+1}^{N} \ln\left|\frac{\zeta_m - \zeta_k}{\zeta_m - \zeta_k^*}\right| - \sum_{k=1}^{m-1} \ln\left|\frac{\zeta_m - \zeta_k}{\zeta_m - \zeta_k^*}\right|\right),$$

(8.83)

$$\Delta\varphi_m = \varphi_m^+ - \varphi_m^- = 2\sum_{k=1}^{m-1} \arg\frac{\zeta_m - \zeta_k}{\zeta_m - \zeta_k^*} - 2\sum_{k=m+1}^{N} \arg\frac{\zeta_m - \zeta_k}{\zeta_m - \zeta_k^*}$$

Formula can be interpreted by assuming that the solitons interact with each other pairwise. In each interaction, the faster soliton moves forward by the first product in formula and the slower one shifts backward by the second term in formula. The total soliton shift is equal to the algebraic sum of its shifts during the paired collisions or interactions. The potential thus takes the form

$$u(x,t) = \sqrt{2\chi}\sum_{j=1}^{N} \eta_j \operatorname{sech}[2\eta_j(x - x_{0j}) + 8\eta_j\xi_j t]\exp\left[-4j\left(\xi_j^2 - \eta_j^2\right)t - 2j\xi_j x + j\varphi_j\right]$$

(8.84)

which is the sum of all the individual modal solitons. Note that this conclusion was arrived at employing the assumption that $b(\xi)$ vanishes. This may not always be true, but Satsuma and Yajima [5] showed using the methods of complex analysis that the long-term solution still breaks down into individual solitons.

### 8.6.6  BOUND STATES AND MULTIPLE EIGENVALUES

From Equation 8.84, one can see that the rate of separation of a pair of solitons is proportional to the difference between the values of the parameters $\zeta$. At equal values of $\zeta$, the solitons do not separate but form a bound state. Consider the bound state of $N$ solitons with an initial potential $N \operatorname{sech}(x)$ as discussed earlier and suppose that their velocities $\zeta_j$ are all zero. Then the state of the soliton varies with time according to $\exp(-4j\eta_j^2 t)$ from the application of Equation 8.84. Thus, the state is a periodic variation with frequency $4\zeta_j^2$. For a bound state of $N$ solitons, there are $N$ frequencies, but we find that the behavior of the solitons is characterized by all the possible frequency differences or beat frequencies. For $N=2$, the bound state is characterized by the beat frequency $\omega = 4(\eta_1^2 - \eta_2^2)$. For $N=3$ or higher, the lowest beat frequency will give the main periodic variation. Referring to Figure 8.14, we can observe the effect of the periodic variation. The solitons appear to diverge and then coalesce together. This process repeats itself with a frequency determined by the lowest beat frequency.

Solitons interaction: $q = 7$, $r = 1.5$ and phase = 200° over 14,000 km

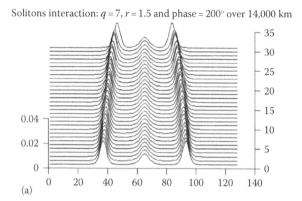

(a)

Solitons interaction: $q = 7$, $r = 1.5$ and phase = 200° over 16,000 km

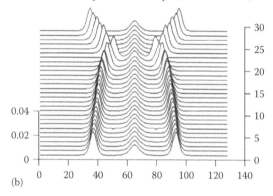

(b)

**FIGURE 8.14** Triple solitons interaction at (a) 14,000 km and (b) 16,000 km for 200° phase difference.

To study the effect of multiple eigenvalues, which is particularly important if we intend to study trains of equal sized solitons for communication purposes. Naturally, we would like to use the Equation 8.82 or 8.84, which have been developed earlier. Unfortunately, these formulas fail for solitons with equal amplitudes. In Reference 2, the authors have worked out an approximation for the limiting case when one eigenvalue approaches another. It was also found that the distance between the two solitons with equal amplitudes increases with time as $\ln(4\eta^2 t)$. Another recourse would be to turn to perturbation methods as developed by Lisak, Anderson, Karpman, and Solov'ev [8–10].

## 8.7 INTERACTION BETWEEN TWO SOLITONS IN AN OPTICAL FIBER

We now return to the NLSE describing the nonlinear propagation of light waves in optical fibers. The motivation for this study is the eventual design of a soliton communication system. One needs to know how two solitons affect one another.

The aim is to find suitable parameters for the soliton separation, widths, phases, or speeds so that the solitons do not diverge too much or coalesce together. Also, the shape and size of the solitons should change negligibly; otherwise, they may not be recognized by the soliton detector.

Desem and Chu [6] gave an exact derivation of a two-soliton solution with arbitrary initial phase and separation as follows:

$$q(x,\tau) = \frac{|\alpha_1|\cosh(a_1 + i\theta_1)e^{j\phi_2} + |\alpha_2|\cosh(a_2 + i\theta_2)e^{j\phi_1}}{\alpha_3 \cosh a_1 \cosh a_2 - \alpha_4[\cosh(a_1 + a_2) - \cos(\phi_2 - \phi_1)]},$$

$$\phi_{1,2} = \left[\frac{\left(\eta_{1,2}^2 - \xi_{1,2}^2\right)x}{2} - \tau\xi_{1,2}\right] + (\phi_0)_{1,2}, \quad a_{1,2} = \eta_{1,2}(\tau + x\xi_{1,2}) + (a_0)_{1,2},$$

$$|\alpha_{1,2}|e^{i\theta_{1,2}} = \pm\left\{\left[\frac{1}{\eta_{1,2}} - \frac{2\eta}{\Delta\xi^2 + \eta^2}\right] \pm j\frac{2\Delta\xi}{\Delta\xi^2 + \eta^2}\right\}, \quad \alpha_3 = \frac{1}{\eta_1\eta_2}, \quad \alpha_4 = \frac{2}{\eta^2 + \Delta\xi^2}$$

$$\zeta_{1,2} = \frac{\xi_{1,2} + j\eta_{1,2}}{2}, \quad \Delta\xi = \xi_2 - \xi_1, \quad \eta = \eta_1 + \eta_2 \tag{8.85}$$

The authors were interested in the initial condition of the form

$$q(0,\tau) = \text{sech}[\tau - \tau_0] + e^{j\theta}A\,\text{sech}[A(\tau + \tau_0)] \tag{8.86}$$

### 8.7.1 SOLITON PAIR WITH INITIAL IDENTICAL PHASES

Launching the solitons with equal phases, that is, $\theta = 0$ and hence, $\zeta_1 = \zeta_2 = 0$ will result in a bound system. If $\theta \neq 0$, then $\zeta_1 \neq \zeta_2$ and from the previous section on the asymptotic behavior of a general soliton solution, the solitons separate eventually. The resulting equation upon substituting $\theta = 0$ is simplified to

$$q(\tau,x) = Q\left\{\eta_1\,\text{sech}\,\eta_1(\tau + \gamma_0)e^{j\eta_1^2 x/2} + \eta_2\,\text{sech}\,\eta_2(\tau - \gamma_0)e^{j\eta_2^2 x/2}\right\},$$

$$Q = \frac{\eta_2^2 - \eta_1^2}{\eta_1^2 + \eta_2^2 - 2\eta_1\eta_2[\tanh a_1 \tanh a_2 - \text{sech}\,a_1\,\text{sech}\,a_2 \cos\psi]}, \tag{8.87}$$

$$a_{1,2} = \eta_{1,2}(\tau \pm \gamma_0), \quad \psi = \frac{\left(\eta_2^2 - \eta_1^2\right)x}{2}$$

The two-soliton solution (Equation 8.87) describes the interaction of two solitons with unequal amplitudes ($\eta_1 \neq \eta_2$) bound together by the nonlinear interaction. Here, owing to the slight variation in the coefficients of the NLSE in the optical fiber as compared to the one studied in the inverse scattering techniques, the beat frequencies are now

$$\omega = \left| \frac{\eta_2^2 - \eta_1^2}{2} \right| \tag{8.88}$$

and the two soliton pulses undergo a mutual interaction, which is periodic with $2\pi/\omega$.

### 8.7.2 SOLITON PAIR WITH INITIAL EQUAL AMPLITUDES

Here, the solitons are launched with equal amplitudes. The eigenvalues are

$$\eta_{1,2} \approx 1 + \frac{2\tau_0}{\sinh 2\tau_0} \pm \operatorname{sech}(\tau_0) \tag{8.89}$$

obtained by equating the first few terms of the Taylor series of Equation 8.85 and applying the first conserved quantity.

$$\int_{-\infty}^{\infty} |q|^2 d\tau = 2(\eta_1 + \eta_2) \tag{8.90}$$

Referring to Figure 8.15, the period of oscillation in terms of the initial separation $\tau_0$ is

$$\frac{\pi \sinh 2\tau_0 \cosh \tau_0}{2\tau_0 + \sinh 2\tau_0} \tag{8.91}$$

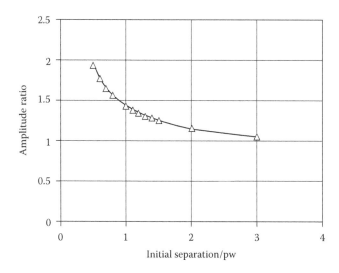

**FIGURE 8.15** Amplitude ratio plotted versus initial pulse separation relative to the pulse width (pw).

As the initial separation $\tau_0$ increases, the distance of coalescence increases exponentially. Therefore, to minimize interaction, the solitons should be widely separated.

### 8.7.3 Soliton Pair with Initial Unequal Amplitudes

The eigenvalues for the case of unequal amplitudes are

$$\eta_{1,2} \approx \frac{A+1}{2} + \frac{2\tau_0\sqrt{A}}{\sinh 2\tau_0\sqrt{A}} + (\sqrt{A}-1)\operatorname{sech} A\tau \pm \left(\frac{A-1}{2} + \operatorname{sech}(A_0)\right) \qquad (8.92)$$

and the separation parameter is now

$$\gamma_0 \approx \tau_0 - \left(1 - \frac{2\operatorname{sech} A\tau_0}{A}\right)\left(\frac{1+A}{2A}\right)\ln\left(\frac{1+A}{A-1}\right) \qquad (8.93)$$

One can see from Equation 8.93 that as the initial separation $\tau_0$ is increased, the eigenvalues $\eta_1,\eta_2$ approach $A$ and 1, respectively. This implies that the oscillation period will no longer be governed by the separation but by the amplitude ratio $A$. Another key observation: as $|\gamma_0|$ increases by increasing $A$, the value of the coefficient of $\cosh(\psi)$ in Equation 8.91 is decreased. This means that as the amplitude ratio $A$ is increased, the interaction is minimized. It can be seen that the interaction is still periodic, but at no stage do they coalesce to form one pulse. An approximate expression for the minimum separation, $\tau_{min}$, in units of the pulse width of the unity-amplitude solution is given by the following equation:

$$\tau_{min} = \tau_0 - \frac{A+1}{4A[1+4A\operatorname{sech} A\tau_0]}\ln\left[1+4\left(\frac{A+1}{A-1}\right)^2\exp\left(\frac{-4A\tau_0}{A+1}\right)\right] \qquad (8.94)$$

This finding suggests that by employing two solitons with unequal amplitudes, we would achieve a much more stable wave formation for long distances as well as a higher bit rate, since the pulses can be launched closer together. We can admit a higher bit rate, because there is a minimum separation for the unequal amplitudes case, whereas for the case of two solitons with equal amplitudes and zero phase difference, the soliton pulses do coalesce together, resulting in $\tau_{min}=0$.

A possibly troubling factor in employing unequal amplitudes is the fact that the amplitudes employed are not integers but can have fractions. Satsuma and Yajima [5] studied the behavior of nonintegral amplitudes and concluded that if $A=N+\alpha$, $|\alpha|<1/2$, then for large times.

$$\frac{\|u\|_\infty}{\|u\|_0} = 1 - \frac{\alpha^2}{(N+\alpha)^2} \qquad (8.95)$$

Numerical simulations show that the solitons oscillate around the limiting value and slowly settle toward the steady or limiting value.

## 8.7.4 Design Strategy

Using the findings reported by Desem and Chu [6], we consider two designs in which equiphase solitons correspond to the transmission of 4 digital "1" signals as this is the worst-case scenario that gives the most interaction. As mentioned earlier, a pair of nonequiphase solitons would result in an unbound system and the solitons will diverge. This is detrimental to the carrier system as the bit rate should be kept roughly constant. Also, the nonequiamplitude scheme is implemented as suggested by Desem and Chu [6,7], because it promises very little interaction among the soliton pulses.

Working in units of pulse width ($\pi\omega=2$ for a unity-amplitude soliton), let us impose a design constraint that the minimum separation $2\tau_{min}$ should not be less than $0.7 \cdot 2\tau_0$ ($2\tau_0$ is the initial separation). The reason is that we do not want the pulses to be too close together. Using Equation 8.94, we obtain for various values of $\tau_0$, the corresponding $A$ amplitude ratios and is plotted in Figure 8.15.

Another constraint that needs to be imposed is that amplitude ratio should not be too large; otherwise, the smaller pulse may be mistaken by the detector for a "0" pulse as the detector may have a certain threshold. Let us assume that the threshold is $0.7 \cdot A_{peak}^2$ where $A_{peak}$ is the largest pulse amplitude (it is assumed that the detector detects the energy or intensity of the wave). This means the largest amplitude ratio possible is 1.2 corresponding to $\tau_0 = 1.75$ pw for the first design and $A = 1.1$, $\tau_0 = 2.5$ pw for the second design. So, the first design seems to be superior to the second one.

Next, one has to check for the variation in the amplitude of the soliton pulses. The constraint is that the amplitude of any soliton pulse should not be less than ½ the amplitude of another pulse. One way is applying perturbation theory [9,10]. The other method is to substitute the values of $A$ and $\tau_0$ into Equations 10.94 and 10.95. Then, to find the variation of the soliton peaks, we let $\tau = -\gamma_0$ and $\tau = \gamma_0$ in Equation 8.89 to locate the peaks approximately. To estimate the extremum of the variation of the amplitude, we let $\cos\psi = \pm 1$. Note that this scheme gives a rough estimate only and no definite conclusion can be reached from this result. A MATLAB program for the investigation of the 4-bit soliton pulse train is developed. It includes an automatic checker that alerts the user if (1) the pulses can no longer be distinguished from one another that is, if we expect a signal of 4 "1"s, the pulses should be sufficiently far apart so that 4 "1"s can be distinguished. This happens when one pulse merges with another. (2) Pulse separation does not match the specification: 0.7/bit rate < pulse separation < 1.3/bit rate (the solitons drifts too much: drift > 0.35/bit rate; and the eye window is less than half the maximum pulse amplitude).

Finally, we have yet to ascertain the maximum range of the 4-bit soliton pulse train. From study on two-solitons interaction, the range can be extended indefinitely for nonequiamplitude soliton pulses. However, in the transmission scheme that I designed, there are scenarios where there may be interaction between equiamplitude solitons.

So, the maximum range is determined by the interaction between the equiamplitude solitons. The solitons coalesce at intervals defined by the soliton period. The formula for the soliton period is given by Desem and Chu [6]:

$$x_0 = \frac{\pi \sinh(2\tau_0)\cosh(\tau_0)}{2\tau_0 + \sinh(2\tau_0)} \qquad (8.96)$$

Substituting the aforementioned value of $\tau_0 = 3.5\pi\omega = 7$, we find that $x_0 = 1722.57$. Assuming that the coalescing process behaves like a cosine function, and if we demand that the separation should not be less than the 0.7 times the initial separation, then we have the maximum range is equal to $\cos^{-1}(0.7)/(\pi/2) * 1722.57{\sim}872$. This is of course in normalized units. In real units and using the normalization constants as used in the MATLAB program, we obtain a value of $872 * 157 = 58{,}404$ km. It turns out that for longer transmission distances, the second design is better as there is less chance of equiamplitude soliton interaction. But of course, the bit-rate allowed is lower.

In summary, the design process follows the steps. (1) Specify the minimum separation and amplitude threshold for a "1." (2) Use initial relative phases of zero between neighboring solitons. This is to prevent solitons spreading apart. Use the equations given by Desem and Chu [6] to obtain a suitable pulse separation to and amplitude ratio $A$. (3) Check for variation of soliton amplitudes. Determine maximum range of transmission. Simulate with numerical method to confirm.

## 8.8 GENERATION AND TRANSMISSION OF MULTI-BOUND SOLITONS: EXPERIMENTS

### 8.8.1 GENERATION OF BOUND-SOLITONS USING MODE-LOCKED FIBER RING RESONATORS

#### 8.8.1.1 Introductory Remarks

Bound solitons generated in actively mode-locked lasers enable new forms of pulse pairs and multiple pairs or groups of solitons as described in the earlier sections, in optical transmission or optical logic devices. In this section, we present the generation of stable bound states of multiple solitons in an active MLFL using continuous phase modulation for wideband phase matching. Not only that dual-soliton bound states but also the triple- and quadruple-soliton pulses can be established. Simulation of the generated solitons can be demonstrated in association with experimental demonstration. We can also prove by simulation that experimental relative phase difference and chirping caused by phase modulation of an optical modulator in the fiber loop significantly influences the interaction between the solitons and hence their stability as they circulate in the anomalous path-averaged dispersion fiber loop.

MLFLs are considered as important laser source for generating ultrashort soliton pulses. Recently, soliton fiber lasers have attracted significant research interests with experimental demonstration of bound states of solitons as predicted in some theoretical works [12–14]. These bound-soliton states have been observed mostly in passive MLFLs [15–18]. There are, however, few reports on bound solitons in active MLFLs. The active mode-locking offers significant advantage in the control of the repetition rate that would be critical for optical transmission systems. Observation of bound soliton pairs was first reported in a hybrid frequency modulation (FM)

MLFLs [19], in which a regime of bound-soliton pair harmonic mode-locking at 10 GHz could be generated. There are, however, no report on multiple-bound soliton states. Depending on the strength of soliton interaction, the bound solitons can be classified into two categories: loosely bound solitons and tightly bound solitons that are determined by the relative phase difference between adjacent solitons. The phase difference may take the value of $\pi$ or $\pi/2$ or any value depending on the fiber laser structures and mode-locking conditions.

In this section, we initially demonstrate the generations of bound states of multiple solitons using an active MLFL in which continuous phase modulation or FM mechanism is enforced. By tuning the parameters for the phase matching of the lightwaves circulating in the fiber loop, not only could we observe the dual-soliton bound state but also the triple- and quadruple-soliton bound states. Relative phase difference and chirping caused by phase modulation of the optical modulator, a $LiNbO_3$ type, in the fiber loop significantly influences the interaction between the solitons and hence their stability as they circulate in the anomalous path-averaged dispersion fiber loop.

### 8.8.1.2 Formation of Bound States in an FM Mode-Locked Fiber Laser

Although the formation of the stable bound soliton states, which are determined by the Kerr effect and anomalous averaged dispersion regime, has been discussed in some configurations of passive MLFLs [16–18], it can be quite distinct in our active MLFLs with the contribution of phase modulation of the $LiNbO_3$ modulator in stabilization of bound states. The formation of bound soliton states in an FM mode-locked laser can be experienced through two stages, which include a process of pulse splitting and stabilization of multisoliton bound states in the presence of a phase modulator in the cavity of mode-locked laser.

The first stage is splitting of a single pulse into multipulses, which occur when the power in the fiber loop increases above a certain mode-locking threshold [19,20]. At higher power, higher-order solitons can be excited and in addition, the accumulated nonlinear phase shift in the loop is so high that a single pulse breaks up into many pulses [21]. The number of split pulses depends on the optical power in the loop, so there is a specific range of power for each splitting level. The fluctuation of pulses may occur at region of power where there is a transition from the lower splitting level to the higher. Moreover, the chirping caused by phase modulator in the loop also makes the process of pulse conversion from a chirped single pulse into multipulses taking place more easily [22,23].

After splitting into multipulses, the multipulse bound states are stabilized subsequently through the balance of the repulsive and attractive forces between neighboring pulses while circulating in a fiber loop of anomalous-averaged dispersion. The repulsive force comes from direct soliton interaction depending on the relative phase difference between neighboring pulses [24] and the effectively attractive force comes from the variation of group velocity of soliton pulse caused by the frequency chirping. Thus, in an anomalous average dispersion regime, the locked pulses should be located symmetrically around the extreme of positive phase modulation half-cycle; in other words, the bound soliton pulses acquire an up-chirping when passing the phase modulator. In a specific MLFL setup, beside the optical power level and dispersion of the fiber cavity, the modulator-induced chirp or the phase modulation index determines not only the pulse width but also the time separation of bound-soliton pulses at which the interactive effects cancel each other.

The presence of a phase modulator in the cavity to balance the effective interactions among bound-soliton pulses is similar to the use of this device in a long-haul soliton transmission system to reduce the timing jitter; for this reason, the simple perturbation theory can be applied to understand the role of phase modulation on the mechanism of bound solitons formation. A multisoliton bound state can be described as following:

$$u_{bs} = \sum_{i=1}^{N} u_i(z,t) \tag{8.97}$$

and

$$u_i = A_i \operatorname{sech}\left\{ A_i \left[ \frac{t - T_i}{T_0} \right] \right\} \exp(j\theta_i - j\omega_i t) \tag{8.98}$$

where
  $N$ is number of solitons in the bound state
  $T_0$ is pulse width of soliton
  $A_i$, $T_i$, $\theta_i$, $\omega_i$ represent the amplitude, position, phase, and frequency of soliton, respectively.

In the simplest case of multisoliton bound state, $N$ is equal 2 or we consider the dual-soliton bound state with the identical amplitude of pulse and the phase difference of $\pi$ value ($\Delta\theta = \theta_{i+1} - \theta_i = \pi$), the ordinary differential equations for the frequency difference and the pulse separation can be derived by using the perturbation method.

$$\frac{d\omega}{dz} = -\frac{4\beta_2}{T_0^3} \exp\left[ -\frac{\Delta T}{T_0} \right] - 2\alpha_m \Delta T \tag{8.99}$$

$$\frac{d\Delta T}{dz} = \beta_2 \omega \tag{8.100}$$

where
  $\beta_2$ is the averaged GVD of the fiber loop
  $\Delta T$ is pulse separation between two adjacent solitons ($T_{i+1} - T_i = \Delta T$)
  $\alpha_m = m\omega_m^2/(2L_{cav})$
  $L_{cav}$ is the total length of the loop
  $m$ is the phase modulation index.

Equation 8.100 shows the evolution of frequency difference and position of bound solitons in the fiber loop in which the first term on the right-hand side represents the accumulated frequency difference of two adjacent pulses during a round trip of the fiber loop and the second one represents the relative frequency difference of these pulses when passing through the phase modulator. At steady state, the pulse separation is constant and the induced frequency differences cancel each other. On the other hand, if by setting Equation 8.99 to zero, we have

$$-\frac{4\beta_2}{T_0^3}\exp\left[-\frac{\Delta T}{T_0}\right]-2\alpha_m\Delta T = 0 \qquad (8.101)$$

or

$$\Delta T \exp\left[\frac{\Delta T}{T_0}\right] = -\frac{4\beta_2}{T_0^3}\frac{L_{cav}}{m\omega_m^2} \qquad (8.102)$$

We can see the effect of phase modulation on the pulse separation through the Equation 8.102, in addition $\beta_2$ and $\alpha_m$ must have opposite signs: in an anomalous dispersion fiber loop with negative value of $\beta_2$, the pulses should be up-chirped. With a specific setup of FM fiber laser, when the magnitude of chirping increases, the bound pulse separation decreases subsequently. The pulse width also reduces according to the increase in the phase modulation index and modulation frequency, so that the ratio $\Delta T/T_0$ can change not much. Thus, the binding of solitons in the FM MLFL is assisted by the phase modulator. Bound solitons in the loop experience periodically the frequency shift and hence their velocity in response to changes in their temporal positions by the interactive forces in equilibrium state.

### 8.8.1.3 Experimental Setup and Results

Figure 8.16 shows the experimental setup of the FM MLFL. Two erbium-doped fiber amplifiers (EDFAs) pumped at 980 nm are used in the fiber loop to control the optical power in the loop for mode-locking. Both are operating in saturated mode. A phase modulator driven in the region of 1 GHz modulation frequency assumes the role as a mode locker and controls the states of locking in the fiber ring. At input of the phase modulator, a polarization controller (PC) consisting of two quarter-wave plates and one half-wave plate is used to control the polarization of light, which relates to the nonlinear polarization evolution and influences multipulse operation in the formation of bound soliton states. A 50 m long Corning standard SMF-28 fiber is inserted after the phase modulator to ensure that the average dispersion in the loop is anomalous. The fundamental frequency of the fiber loop is 1.7827 MHz that is equivalent to the 114 m total loop length. The outputs of the mode locked laser from the 90:10 coupler are monitored by an optical spectrum analyzer (HP 70952B) and an oscilloscope (Agilent DCA-J 86100C) of an optical bandwidth of 65 GHz.

Under the normal conditions, the single pulse mode-locking operation is performed at the average optical power of 5 dBm and modulation frequency of 998.315 MHz (the harmonic mode-locking at the 560th order) as shown in Figure 8.17. The narrow pulses of 8–14 ps width depending on the radio frequency (RF)

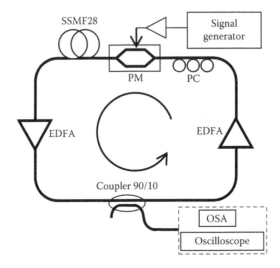

**FIGURE 8.16** Experimental setup of the FM mode-locked fiber laser. PM, phase modulator; PC, polarization controller; OSA, optical spectrum analyzer. Standard SMF-28 = standard single mode fiber—Corning type.

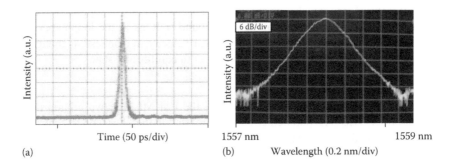

**FIGURE 8.17** (a) Oscilloscope trace of generated single soliton and (b) optical spectrum of a single soliton.

driving power of the phase modulator can be observed with the oscilloscope. The measured pulse spectrum has spectral shape of a soliton rather than a Gaussian pulse. By adjusting the polarization states of the PC wave plates at higher optical power, the dual-bound solitons or bound soliton pairs can be observed at an average optical power circulating inside the fiber loop of about 10 dBm. Figure 8.18a shows the typical time-domain waveform and Figure 8.18b the corresponding spectrum of the dual-bound soliton state. The estimated full width at half maximum (FWHM) pulse width is about 9.5 ps and the temporal separation between two bound pulses is 24.5 ps, which are correlated exactly to the distance between two spectral main lobes of 0.32 nm of the observed spectrum. When the average power inside the loop is increased to about 11.3 dBm and a slight adjustment of the polarization controller

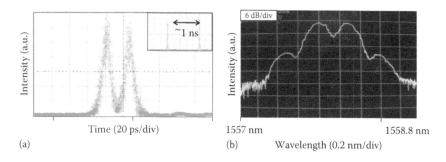

**FIGURE 8.18** (a) Oscilloscope trace of periodic bound soliton pairs in time-domain and (b) optical spectrum of soliton pair bound state.

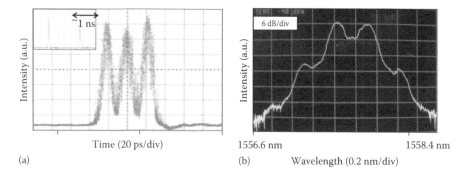

**FIGURE 8.19** (a) Oscilloscope trace of periodic groups of triple bound solitons in time-domain and (b) optical spectrum of triple-soliton bound state.

is performed, the triple-bound soliton state occurs as shown in Figure 8.19; the FWHM pulse width and the temporal separation of two close pulses are slightly less than 9.5 and 22.5 ps, respectively. Insets in Figures 8.18a and 8.19a show that the periodic sequence of bound solitons at the repetition rate are equal to the modulation frequency. This feature is quite different from that in a passive MLFL in which the positions of bound solitons are not stable and the direct soliton interaction causes a random movement and phase shift of bound soliton pairs [16]. On the other hand, it is advantageous to perform a stable periodic bound soliton sequence in an FM MLFL.

The symmetrical shapes of optical modulated spectrum in Figures 8.18b and 8.19b indicate clearly that the relative phase difference between two adjacent bound solitons is of $\pi$ value. At the center of spectrum, there is a dip due to the suppression of the carrier with $\pi$ phase difference in case of dual-soliton bound state. While there is a small hump in case of triple-soliton bound state because three soliton pulses are bound together with the first and the last pulses in phase and out of phase with the middle pulse. The shape of spectrum will change, which can be symmetrical or asymmetrical when this phase relationship varies. We believe that the bound state with the relative phase relationship of $\pi$ between solitons is the most stable in our experimental setup and this observation also agrees with the theoretical prediction

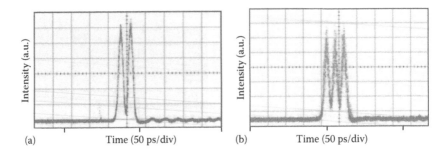

**FIGURE 8.20** Oscilloscope traces of the dual-soliton bound state (a) and the triple-soliton bound state (b) after propagating through 1 km SMF.

of stability of bound soliton pairs relating to photon-number fluctuation in different regimes of phase difference.

Similar to a single soliton, the phase coherence of bound solitons is still maintained as a unit when propagating through a dispersive medium. Figure 8.20 shows the dual-soliton bound state and triple-soliton state waveforms on the oscilloscope after they propagate through 1 km SSMF. There is also no change in their observed spectral shapes in both cases.

To prove that the multisoliton bound operation can be formed in an FM mode-locked fiber by operating at a critical optical power level and phase modulator-induced chirp in a specific fiber loop of anomalous average dispersion, we increased the average optical power in the loop to a maximum of 12.6 dBm and decrease the RF driving power of 15 dBm applied to the optical modulator. It is really amazing when the quadruple-soliton state is generated as observed in Figure 8.21. The bound state occurring at lower phase modulation index is due to maintaining a small enough frequency shifting in a wider temporal duration of bound solitons to hold the balance between interactions of the group of four solitons in the fiber loop. However, the optical power can still not be high enough to stabilize the bound state as well as

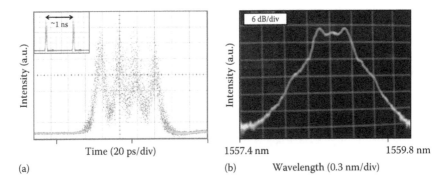

**FIGURE 8.21** (a) Oscilloscope trace and (b) optical spectrum of quadruple-soliton bound state.

to overcome the noise superimposed by the EDFAs, the optical amplifiers, so that the time-domain waveform of quadruple-soliton bound state is much noisier and its spectrum only can be seen clearly two main lobes being inversely proportional to the temporal separation of pulses which is about 20.5 ps in our experimental setup. The results show the pulse separation reduces when the number of pulses in the bound states is larger.

Thus through the experimental results, both the phase modulation index and the cavity's optical power influence the existence of the multisoliton bound states in the FM MLFL. The optical power determines the number of initially split pulses and maintains the pulse shape in the loop. Observations from the experiment show that there is a threshold of optical power for each bound level. At threshold value, the bound solitons show strong fluctuations in amplitude and oscillations in position considered as a transition state between different bound levels and the collisions of adjacent pulses even occur as shown in Figure 8.22. The phase difference of adjacent pulses can also change in these unstable states: the neighboring pulses are not out of phase anymore but in phase as seen through the measured spectrum in Figure 8.23. Although the decrease in phase modulation index is required to maintain the stability at higher bound soliton level, it increases the pulse width and reduces the peak power. So the waveform seen is noisier and more sensitive to ambient environment and its spectrum is not strongly modulated as shown in Figure 8.24.

### 8.8.1.4    Simulation of Dynamics of Bound States in an FM Mode-Locked Fiber Laser

*8.8.1.4.1    Numerical Model of an FM Mode-Locked Fiber Laser*

To understand the dynamics of FM bound soliton fiber laser, simulation technique is used to see the dynamic behavior of the formation and interactions of solitons in the FM mode-locked fiber loop. This is slightly different as compared with the analytical ISM described in Section 8.6. A recirculation model of the fiber loop is used to simulate propagation of the bound solitons in the fiber cavity. The simple model consists of basic components of an FM MLFL as shown in Figure 8.25; in other words, the cavity of FM MLFL is modeled as a sequence of different elements. The optical filter has a Gaussian transfer function with 2.4 nm bandwidth. The transfer function

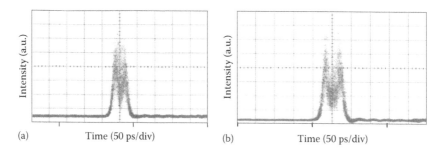

**FIGURE 8.22**  Oscilloscope traces of (a) the dual-soliton bound state at optical power threshold of 8 dBm and (b) the triple-soliton bound state at 10.7 dBm.

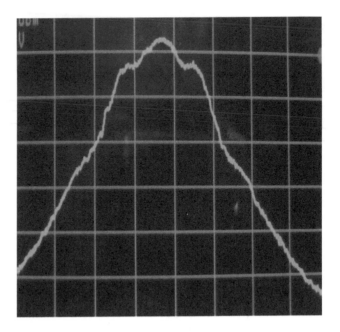

**FIGURE 8.23**  Spectrum of the bound soliton state with in-phase pulses. Center wavelength located at 1558 nm.

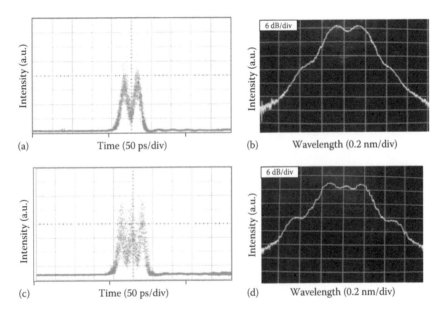

**FIGURE 8.24**  Oscilloscope traces of (a) the dual-soliton bound state and (c) the triple-soliton bound state and (b and d) their spectra, respectively, at low RF driving power of 11 dBm.

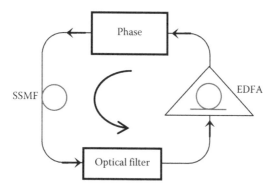

**FIGURE 8.25** A circulating model of MLFL for simulating the FM mode-locked fiber ring laser, SSMF: standard single mode fiber, EDFA: erbium-doped fiber amplifier.

of phase modulator is $u_{out} = u_{in}\exp[jm\cos(\omega_m t)]$, $m$ is the phase modulation index and $\omega_m = 2\pi f_m$ is angular modulation frequency, assumed to be a harmonic of the fundamental frequency of the fiber loop. The pulse propagation through the optical fibers is governed by the NLSE [24] (reiterated from Equation 8.15).

$$\frac{\partial u}{\partial z} + j\frac{\beta_2}{2}\frac{\partial^2 u}{\partial T^2} - \frac{\beta_3}{6}\frac{\partial^3 u}{\partial T^3} + \frac{\alpha}{2}u = j\gamma|u|^2 u \qquad (8.103)$$

where
  $u$ is the complex envelop of optical pulse sequence
  $\beta_2$ and $\beta_3$ account for the second- and third-order fiber dispersions
  $\alpha$ and $\gamma$ are the loss and nonlinear parameters of the fiber, respectively.

The amplification of signal including the saturation of the EDFA can be represented as [24]

$$u_{out} = \sqrt{G}u_{in} \qquad (8.104)$$

with

$$G = G_0 \exp\left(-\frac{G-1}{G}\frac{P_{out}}{P_{sat}}\right) \qquad (8.105)$$

where
  $G$ is the optical amplification factor
  $G_0$ is unsaturated amplifier gain
  $P_{out}$ and $P_{sat}$ are output power and saturation power, respectively

The difference of dispersion between the SSMF and the Er-doped fiber in the cavity arranges a certain dispersion map and the fiber loop gets a positive net dispersion or an anomalous average dispersion, which is important in forming the "soliton-like"

**TABLE 8.3**

**Parameter Values Used in the Simulations**

| | | |
|---|---|---|
| $\beta_2^{SMF} = -21$ ps²/km | $\beta_2 = 6.43$ ps²/km | $\Delta\lambda_{filter} = 2.4$ nm |
| $\gamma^{SMF} = 0.0019$ W⁻¹m⁻¹ | $\gamma^{ErF} = 0.003$ W⁻¹m⁻¹ | $f_m \approx 1$ GHz |
| $\alpha^{SMF} = 0.2$ dB/km | $P_{sat} = 7$–13 dBm | $m = 0.1\pi$–$1\pi$ |
| $L_{cav} = 115$ m | $NF = 6$ dB | $\lambda = 1558$ nm |

SSMF, Standard single mode fiber; ErF, Erbium doped fiber; NF, noise figure of the EDFA.

pulses in an FM fiber laser. Basic parameter values used in our simulations are listed in Table 8.3.

### 8.8.1.4.2  *Simulation of the Formation Process of the Bound Soliton States*

First, we simulate the formation process of bound states in the FM MLFL whose parameters are those employed in the experiments described earlier. The lengths of the Er-doped fiber and SSMF are chosen to get the cavity's average dispersion $\bar{\beta}_2 = -10.7$ ps²/km. The Schrodinger equation is applied with the initial conditions which are the amplitude of random optical noises of the amplification device as the initial amplitudes of the circulating optical waves circulating around the ring. This propagation is represented by this Schrodinger propagation equation. The optical power is built up with the phase matching conditions and hence the formation of the solitons.

Figure 8.26 shows a simulated dual-soliton bound state building up from initial Gaussian-distributed noise as an input seed over the first 2000 round-trips with $P_{sat}$ value of 10 dBm and $G_0$ of 16 dB and $m$ of 0.6$\pi$. The built-up pulse experiences transitions with large fluctuations of intensity, position, and pulse width during the first 1000 roundtrips before reaching to the stable bound steady state.

Figure 8.27 shows the time-domain waveform and spectrum of the output signal at the 2000th round trip. The bound states with higher number of pulses can be formed at higher gain of the cavity; hence, when the gain $G_0$ is increased to 18 dB, which is enhancing the average optical power in the loop, the triple-soliton bound steady state is formed from the noise seed via simulation as shown in Figure 8.28. In the case of higher optical power, the fluctuation of signal at initial transitions is stronger and it needs more round trips to reach to a more stable three pulses bound state. The waveform and spectrum of the output signal from the FM MLFL at the 2000th round trip are shown in Figure 8.29a and b, respectively.

Although the amplitude of pulses is not equal, which indicates the bound state can require a larger number of round trips before the effects in the loop balance, the phase difference of pulses accumulated during circulating in the fiber loop is approximately $\pi$ value that is indicated by strongly modulated spectra. In particular from the simulation result, the phase difference between adjacent pulses is 0.98$\pi$ in case of the dual-pulse bound state and 0.89$\pi$ in case of the triple-pulse bound state. These simulation results agree with the experimental results (shown in

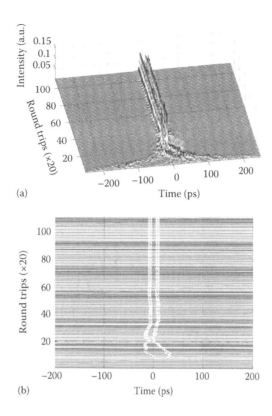

(a)

(b)

**FIGURE 8.26** (a) Numerically simulated dual-soliton bound state formation from noise and (b) the evolution of the formation process in contour plot view.

Figures 8.18b and 8.19b) discussed earlier to confirm the existence of multisoliton bound states in an FM MLFL

### 8.8.1.4.3 Simulation of the Evolution of the Bound Soliton States in an FM Fiber Loop

Stability of bound states in the FM MLFL strongly depends on the parameters of the fiber loop, which also determine the formation of these states. Beside the phase modulation and group velocity dispersion (GVD) as mentioned earlier (see also Appendix E), the cavity's optical power also influences the existence of the bound states. Using the same previous model, the effects of active phase modulation and optical power can be simulated to see the dynamics of bound solitons. Instead of the noise seed, the multisoliton waveforms following Equations 8.97 and 8.98 are used as the initial seed for simplification of our simulation processes. Initial bound solitons are assumed to be identical with the phase difference between adjacent pulses of $\pi$ value.

First, effect of the phase modulation index on the stability of the bound states is simulated through a typical example of the evolution of dual-soliton bound state over 2000 roundtrips in the loop at different phase modulation indexes with the same

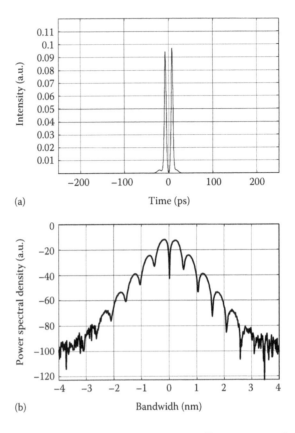

(a)

(b)

**FIGURE 8.27**  (a) The waveform and (b) the corresponding spectrum of simulated dual-soliton bound state at the 2000th roundtrip.

saturation optical power of 9 dBm as shown in Figure 8.30. The simulation results also indicate that the pulse separation decreases corresponding to the increase in the modulation index. In the first roundtrips, there is a periodic oscillation of bound solitons that is considered as a transition of solitons to adjust their own parameters to match the parameters of the cavity before reaching a finally stable state. Simulations in other multisoliton bound states also manifest this similar tendency. The periodic phase modulation in the fiber loop is not only to balance the interactive forces between solitons but also to maintain the phase difference of $\pi$ between them. At too small modulation index, the phase difference changes or reduces slightly after many round trips or in other words, the phase coherence is looser this leads to the amplitude oscillation due to the alternatively periodic exchange of energy between solitons as shown in Figure 8.31. The higher the number of solitons in bound state is, the more sensitive it is to the change in phase modulation index. The modulation index determines time separation between bound pulses, yet at too high modulation index, it is more difficult to balance the effectively attractive forces between solitons especially when the number of solitons in the bound state is larger. The increase in

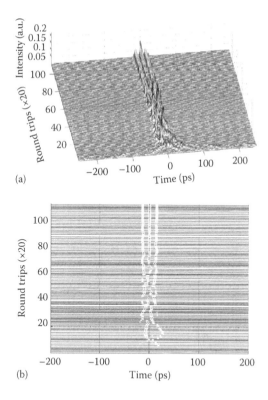

(a)

(b)

**FIGURE 8.28** (a) Numerically simulated triple-soliton bound state formation from noise in 2D view and (b) evolution of the formation process in contour plot view.

chirping leads to faster variation in group velocity of pulses when passing the phase modulator, which can create the periodic oscillation of pulse position in the time domain evolution.

Figure 8.32 shows the evolution of the triple-soliton bound state at the m value of $1\pi$ and the quadruple-soliton bound state at the $m$ value of $0.7\pi$. In case of quadruple-soliton bound state, solitons oscillate strongly and tend to collide together.

Another factor is the optical power of the fiber loop, it plays an important role not only in determination of multipulse bound states as in simulation of previous section, but also in stabilization of the bound states circulating in the loop. As mentioned in Section 8.8.1.3, each bound state has a specific range of operational optical power. In our simulations, dual-, triple-, and quadruple-soliton bound states are in stable evolution in the loop when the saturated power $P_{sat}$ is about 9, 11, and 12 dBm, respectively. When the optical power of the loop is not within these ranges, the bound states become unstable and they are more sensitive to the change of phase modulation index. At a power lower than the threshold, the bound states are out of bound and switched to lower level of bound state. In contrast, at high power level region, the generated pulses are broken into random pulse trains or decay into radiation.

Figure 8.33 shows unstable evolution of the triple-soliton bound states under nonoptimized operating conditions of the loop as an example. Different operating

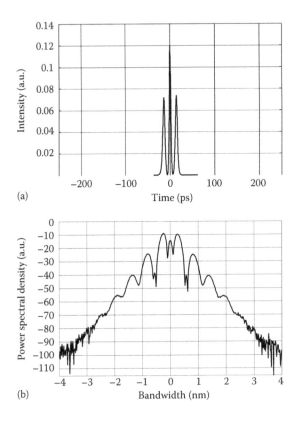

(a)

(b)

**FIGURE 8.29** (a) Waveform and (b) corresponding spectrum of simulated triple-soliton bound state at the end of 2000th roundtrip.

conditions of phase difference between bound solitons can also be simulated; the states of non-$\pi$ phase difference often behave unstably and can easily be destroyed as displayed in Figure 8.34. The simulation results agree well with the experimental observations as discussed earlier.

## 8.8.2   ACTIVE HARMONIC MODE-LOCKED FIBER LASER FOR SOLITON GENERATION

### 8.8.2.1   Experiment Setup

Figure 8.35 shows the schematic diagram of the active MLFL (MLFL). The fiber ring configuration incorporates an isolator to ensure unidirectional lasing. The gain media is obtained by using a 16 m erbium-doped fiber operating under bidirectional pump condition. The pump sources are two 980 nm laser diodes coupled to the ring through two 980/1550 WDM couplers. A FOL0906PRO-R17-980 laser diode from FITEL is used for the forward pump source and a COMSET PM09GL 980 nm laser diode for the backward pump source. A 18 m long dispersion shifted fiber (DSF) is used for shortening the locked pulse width. A 2 nm bandpass multilayer thin-film optical filter is used to select the operation wavelength of the laser and moderately

accommodate the bandwidth of the output pulses. A PC is also employed to maximize the coupling of lightwaves from fiber to the diffused waveguides of the Mach–
Zehnder intensity modulator (MZIM). Output pulse trains are extracted via a 90/10
coupler. Mode-locking is obtained by inserting into the ring a 20 GHz 3 dB bandwidth MZIM that periodically modulates the loss of the lightwaves traveling around
the ring. The modulator is biased at the quadrature point with a voltage of −1.5 V
and then driven by superimposing a sinusoidal signal derived from Anritsu 68347C
signal generator.

The output signal is then monitored using an Agilent 86100B wideband oscilloscope with optical input sampling module, an Ando 6317B optical spectrum analyzer with a resolution of 0.01 nm, and an Agilent E4407B electronic RF spectrum
analyzer. Figure 8.36 shows the picture of the whole system setup for experiment.

## 8.8.2.2 Tunable Wavelength Harmonic Mode-Locked Pulses

The total optical cavity length is 55.4 m, which corresponds to a fundamental frequency $f_R$ of 3.673 MHz (the measurement of $f_R$ is presented in Section 8.8.2.3).
When the modulation frequency is set at 2.695038 GHz or equivalently equal to
733 times of $f_R$, mode-locking laser pulse is obtained at the output as shown in

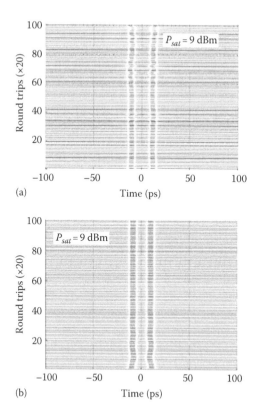

**FIGURE 8.30** Simulated evolution of dual-soliton bound state over 2000 roundtrips in the
fiber loop at different phase modulation indexes: (a) $m=0.1\pi$, (b) $m=0.4\pi$.    (*Continued*)

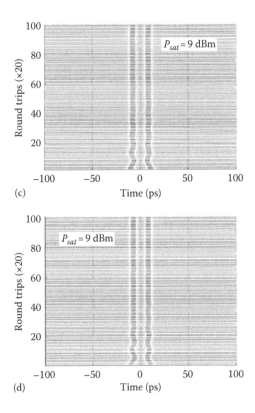

**FIGURE 8.30 (*Continued*)** Simulated evolution of dual-soliton bound state over 2000 roundtrips in the fiber loop at different phase modulation indexes: (c) $m = 0.8\pi$, and (d) $m = 1\pi$.

Figure 8.37a. The interval between adjacent pulses is 371 ps, corresponding to a repetition rate of 2.695 GHz. The locked pulses are formed with clear pedestal, with slightly fluctuated amplitude. The output pulse amplitude is measured, and a peak power is recorded at 9.9 mW.

The pulse width is also measured and a value of 37 ps FWHM is obtained. However, as this value is taken from the trace of pulse on the oscilloscope, the rise time of the oscilloscope is included in the measured value. Since the rise time of the oscilloscope is 20 ps, which is comparable to the measured value, it cannot be ignored. Besides that the rise time of the photodiode also contributes to the total measured value. Therefore, the actual pulse width is much smaller than the measured value.

Figure 8.37b shows the optical spectrum of the output pulses. The FWHM bandwidth is 0.137 nm with central wavelength at 1552.926 nm. The time bandwidth product (TBP) was determined as follows:

$$TBP = T_{FWHM} \times BW$$

$$TBP = 37 \text{ ps} \times 17.125 \text{ THz} = 0.634$$

(8.106)

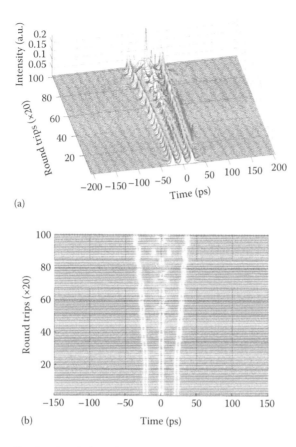

(a)

(b)

**FIGURE 8.31** Simulated evolution of triple-soliton bound state over 2000 roundtrips at the low phase modulation index $m=0.1\pi$: (a) 3D view and (b) contour plane view.

The large value obtained here again confirms that the measured value of pulse width is mainly contributed by the rise times of oscilloscope and photodiode. The accurate pulse width should be obtained from using an optical autocorrelator.

Alternatively, one can take advantage of the characteristic of the AM mode-locked fiber laser to approximate the pulse width. The pulses generated from AM MLFLs are nearly transform limited [15–18]. This means that the TBP takes the value of 0.44, assuming Gaussian shape pulse. Therefore, the pulse width can be estimated from

$$T_{FWHM} = \frac{TBP}{BW}$$

$$T_{FWHM} = \frac{0.44}{17.125} \text{ THz} = 26 \text{ ps}$$

(8.107)

It can be seen from Figure 8.37b that the pulse train spectrum has sidelobes with a separation of 0.022 nm, which corresponds to a longitudinal mode separation of 2.695 GHz. This spectrum profile is typical for the mode-locked laser and usually referred as the mode-locked structure spectrum [11,19,20].

The aforementioned spectrum profile can be explained by taking the Fourier transform of the periodic Gaussian pulse train:

$$p(t) = e^{-t^2/2T_0^2} * \text{III}\left(\frac{t}{T}\right) \qquad (8.108)$$

where
     $T_0$ is the width of the Gaussian pulse
     $T$ is the pulse repetition period
     $\text{III}(t/T)$ is the comb function given by the following:

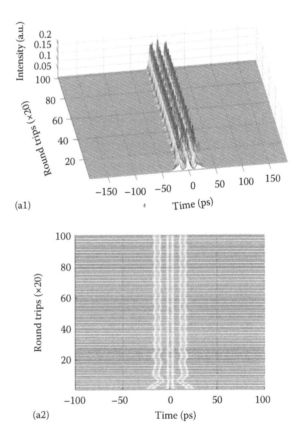

(a1)

(a2)

**FIGURE 8.32** Simulated evolutions over 2000 roundtrips in the fiber loop of (a1) triple-soliton bound state at $m = 1\pi$ and (a2) contour plane view of (a1).        *(Continued)*

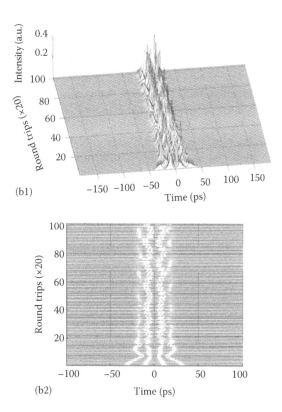

(b1)

(b2)

**FIGURE 8.32 (Continued)** Simulated evolutions over 2000 roundtrips in the fiber loop of (b1) quadruple-soliton bound state at $m = 0.7\pi$; (b2) contour plane views of (b1).

$$\mathrm{III}\left(\frac{t}{T}\right) = \sum_{n=-\infty}^{\infty} \delta(t - nT) \qquad (8.109)$$

$$P(\omega) = F(p(t)) = T_0\sqrt{2\pi}\,e^{-T_0^2\omega^2/2} \sum_{n=-\infty}^{\infty} \delta(\omega - n\Omega) \qquad (8.110)$$

where

$\Omega = 2\pi/T$

$P(\omega)$ has structure of a train of Dirac function pulses separated by the repetition frequency with a Gaussian envelope.

The laser wavelength is tunable over the whole C-band by tuning the central wavelength of the thin film bandpass filter inserted in the loop.

The repetition rate of the laser pulses can be increased by increasing the modulation frequency. Figure 8.38 shows the output pulses with the repetition rate of 4 GHz and its optical spectrum.

### 8.8.2.3 Measurement of the Fundamental Frequency

In active harmonic mode-locked fiber laser (HMLFL), the pulses are locked into one of the harmonic of the fundamental frequency $f_R$ by tuning the modulation frequency $f_m$ so that $f_m = kf_R$. In practice, the laser is mode-locked if $f_m$ is within the locking range around the harmonic frequency, that is, $f_m = kf_R \pm \Delta f$. If $f_m$ falls outside the locking range, the laser becomes unlocked, that is, no pulse trains are observed. If $f_m$ is adjusted until reaching the next harmonic frequency $(k+1)f_R$, the HMLFL is locked again into the next longitudinal resonant mode of the ring. By measuring those locked modulation frequencies, one can determine the fundamental frequency of the ring. The fundamental frequency is determined by

$$\frac{4,018,635,400 - 3,981,902,400}{10} \, \text{Hz} \leq f_R \leq \frac{4,018,636,000 - 3,981,901,400}{10} \, \text{Hz}$$

$$3,673,300 \, \text{Hz} \leq f_R \leq 3,673,460 \, \text{Hz}$$

$$\text{or } f_R = 3,673,380 \pm 80 \, \text{Hz}$$

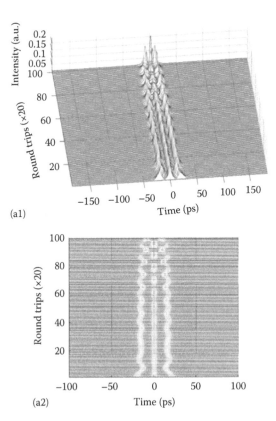

(a1)

(a2)

**FIGURE 8.33** Simulated dynamics of triple-soliton bound state at unoptimized power: (a1) $P_{sat} = 9$ dBm, $m = 0.7\pi$, (a2) contour plane views.                    *(Continued)*

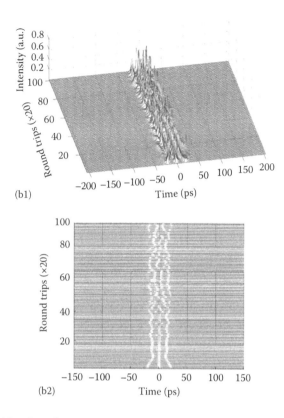

(b1)

(b2)

**FIGURE 8.33 (Continued)** Simulated dynamics of triple-soliton bound state at unoptimized power: (b1) $P_{sat} = 13$ dBm, $m = 0.6\pi$, (b2) contour plane views.

### 8.8.2.4 Effect of the Modulation Frequency

Increasing modulation frequency not only generates higher repetition rate mode-locked pulses but also improves other parameters of the pulses such as bandwidth and pulse width.

The solid curve shown in Figure 8.39 indicates the relationship between the pulse width of the generated pulse train and modulation frequency. The pulse width does not change so much when the modulation frequency increases above 2 GHz. This is because the pulse width in this region is small comparable to the rise time of the oscilloscope. The actual values should be smaller than the measured values. This is confirmed when examining the equivalent bandwidth of the pulse trains.

Figure 8.40 shows the pulse train bandwidth versus the modulation frequency. The modulation frequency increases when the bandwidth increases. Unlike the behavior of pulse width, the pulse train bandwidth continues to increase even when the modulation frequency is above 2 GHz. This indicates that the actual pulse width should be smaller than its measured value. The pulse width is calculated and plotted as shown by the dash curve in Figure 8.39.

### 8.8.2.5    Effect of the Modulation Depth/Index

The same setup shown in Figure 8.35 is used to study the effect of modulation depth on the performance of the mode-locked pulse trains. The settings of the cavity are as following: (1) the total pump power is 274 mW and (2) the modulator is biased at the quadrature point. The modulation depth is varied by changing the signal amplitude (from the signal synthesizer) applied to the MZIM. The pulse width and bandwidth of the output pulse trains are measured for different values of modulation depth.

Figure 8.41 shows the relationship between pulse width and modulation amplitude/power. As the modulation power increases, the pulse width slightly decreases. Therefore, the pulse can be shortened by increasing the modulation power (or the modulation depth). However, the pulse shortening effect of modulation depth is not as strong as that of modulation frequency. In addition, the modulation depth is limited to a maximum value of 1.

The relationship between pulse width and the modulation depth is verified by examining the effect of modulation power on the bandwidth. When the modulation power increases, the bandwidth decreases, as shown in Figure 8.42. This is consistent with the decrease of pulse width shown in Figure 8.41.

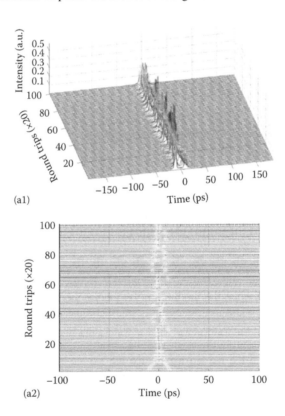

**FIGURE 8.34**  Time-domain dynamics of dual- and triple-soliton bound states, respectively, within phase pulses (a1) 3D view and (a2) plane contour view.          (*Continued*)

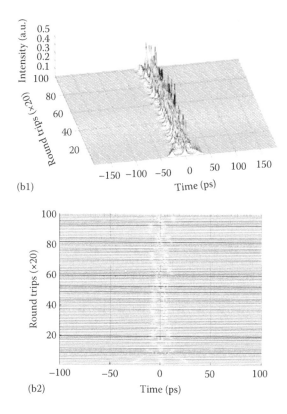

**FIGURE 8.34** (*Continued*) Time-domain dynamics of dual- and triple-soliton bound states, respectively, within phase pulses (b1) 3D view and (b2) plane contour view.

It is found that there is a threshold power of the modulation for locking. When the modulation power is reduced to −4 dBm corresponding to a modulation depth of 0.08, mode-locking does not occur and pulse train cannot be obtained.

### 8.8.2.6 Effect of Fiber Ring Length

The same setup as in Figure 8.35 has been used to study the effect of the fiber ring length on the mode-locked pulse trains. The DSF fiber has been extended from 18 to 118 m. A total pump power of 274 mW is used. The pulse width and bandwidth have been measured and compared with those described in Section 8.8.2.2. Figures 8.43 and 8.44 show the mode-locked pulse temporal train and its optical spectrum for the 18 m DSF ring laser and the 118 m DSF ring laser, respectively.

The pulse trains are very much similar for both cases. Their pulse widths (FWHM) are nearly equal, 33 and 29 ps. However, the pulse optical spectra are quite different. The longer ring laser has a bandwidth of 3 nm, larger than that of the shorter one, 0.184 nm. The broadening of the bandwidth of the pulse train of the 110 m ring length indicates that the pulse width is shorter than what is observed

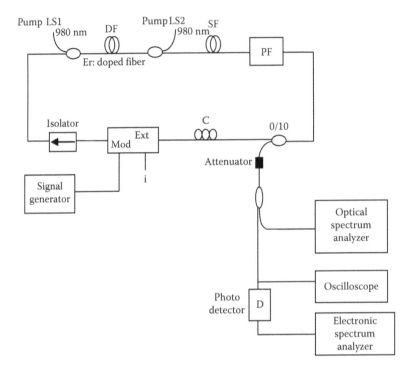

**FIGURE 8.35** Active harmonic mode-locked fiber laser experiment setup. PF = optical passband filter; D = photodetector; Mod = modulation port of the optical modulator; i = biasing port; ext = external modulator; C = controller for polarization; Pu = pump laser; SF = single mode fiber; DF = doped fiber or Erbium doped fiber for amplification.

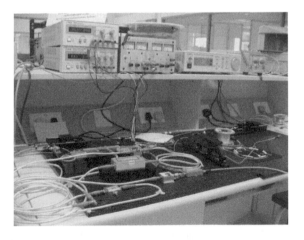

**FIGURE 8.36** Active harmonic mode-locked fiber laser.

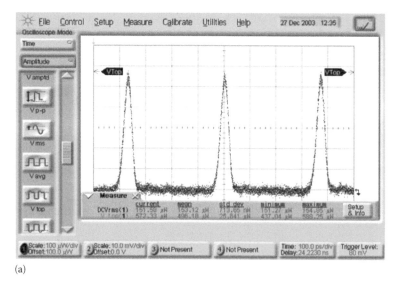

(a)

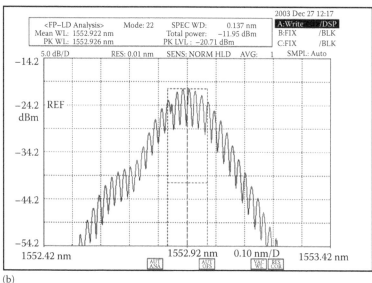

(b)

**FIGURE 8.37** Harmonic mode-locked pulse train with a repetition rate of 2.657 GHz (a) and its optical spectrum (b).

by the sampling oscilloscope. The discrepancy here is again due to the effect of the photodiode and oscilloscope's rise time on the observed pulse.

The shortening of the pulse in the longer length ring can be due to the self-phase modulation (SPM) effects or the nonlinear induced phase in the fiber. The high intensity pulse traveling in the fiber suffers from the nonlinear effect, which increases the peak of the pulse, shortens the pulse width, and hence the bandwidth broadening.

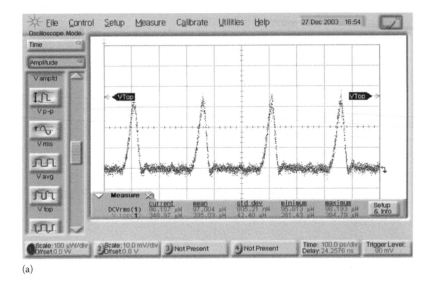

(a)

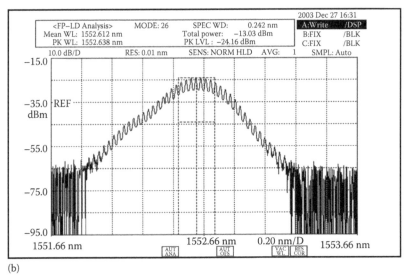

(b)

**FIGURE 8.38** Harmonic mode-locked pulse train with repetition rate of 4.000 GHz: (a) temporal profile, (b) optical spectrum.

To verify that the pulse shortening is due to the nonlinear effect, the two different length rings are studied under lower pump power. It is found that when the pump power is reduced to 50 mW, the pulse width and bandwidth of the laser does not change regardless of the fiber length.

One can estimate the effective nonlinear lengths of the fiber for the pump power of 274 and 50 mW. The peak pulse powers are 0.37 and 0.027 W, respectively. The nonlinear lengths can be calculated as follows [24]:

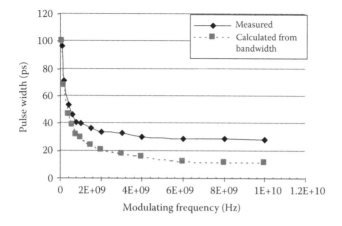

**FIGURE 8.39** Pulse width of the HMLFL for various modulation frequencies.

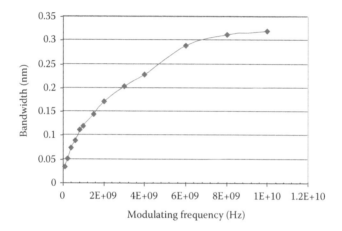

**FIGURE 8.40** Bandwidth of the HMLFL for various modulation frequencies.

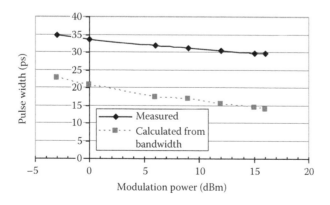

**FIGURE 8.41** Pulsewidth of the HMLFL for various modulation powers.

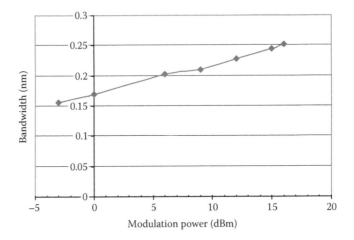

**FIGURE 8.42**  Bandwidth of the HMLFL for various modulation powers.

$$L_{NL}(P_p = 274 \text{ mW}) = \frac{1}{\gamma P_0} = \frac{1}{2 \times 10^{-3} \times 0.37} = 1351 \text{ (m)} \tag{8.111}$$

$$L_{NL}(P_p = 27 \text{ mW}) = \frac{1}{\gamma P_0} = \frac{1}{2 \times 10^{-3} \times 0.027} = 18,518 \text{ (m)} \tag{8.112}$$

It can be seen that the effective nonlinear length of the 118 m long fiber ring is comparable to that of the physical ring length when pumped at high power. Therefore, the nonlinear effect plays a significant role in this case. When the laser is pumped at lower power, the effective nonlinear length is much longer than the laser ring length; hence, the SPM effect is minute and cannot be observed. However, one can predict the existence of the SPM effect if the fiber length is increased to be comparable with the nonlinear length.

It is also noticed that the supermode noise was reduced in the case of longer fiber ring. The pulses are more temporally stable and less fluctuating. To verify this, the RF spectra of the lasers are recorded and compared. It can be seen from Figure 8.45 that the longer length laser has a lower noise floor than the shorter one.

### 8.8.2.7  Effect of Pump Power

The laser is configured as in Figure 8.35 except that the fiber length is 118 m. Different power levels are pumped into the EDFA to study the laser performance. Figure 8.46 shows the variation of the laser pulse widths under various pump powers. The pulse widths seem to be slightly affected by the change of pump power. However, the effect of the pump power is actually stronger, since the pulse widths measured here are not the true pulse widths. They are larger than their true values due to the rise time of the equipment as discussed earlier. The effect of pump power on the pulse width can be inferred from the relationship between pump power and output pulse bandwidth.

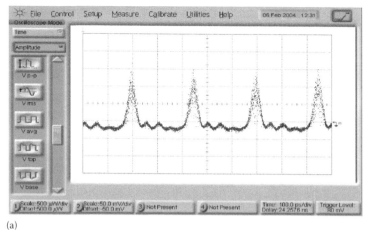

(a)

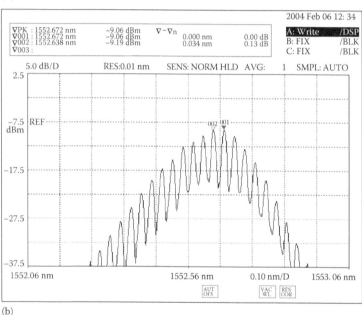

(b)

**FIGURE 8.43** Locked pulses (a) and its optical spectrum (b) for the 18 m DSF ring laser.

The dependence of the output pulse bandwidth is illustrated in Figure 8.47. As the pump power increases, the bandwidth increases. Since the bandwidth is inversely proportional to the pulse width, one can infer that the pulse width decreases when the pump power increases. The cause of the decrease in pulse width is attributed to the nonlinear effect. The higher the peak power, the stronger is the nonlinear effect, and hence the shorter pulse width. It is also noticed that the pulse shortening effect of the pump power is only observed when the fiber length is long compared to the effective nonlinear length.

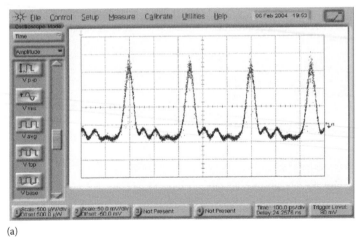

(a)

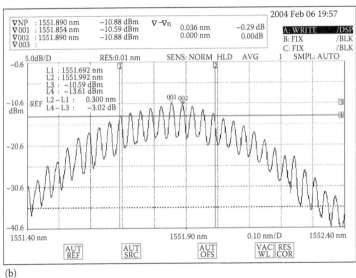

(b)

**FIGURE 8.44**  Locked pulses (a) and its optical spectrum (b) for the 118 m DSF ring laser.

### 8.8.2.7.1  Remarks

We have presented the propagation of solitons through optical fibers through analytical and ISM and numerical approaches. The Lax pairs have enabled the development of eigenvalue solutions for insight observation of the behavior of solitons and interactions of soliton pairs and triple solutions when their relative phase and amplitudes are changed. These dynamic soliton behaviors are then confirmed with experimental demonstration of the formation of solitons and their evolutions in a MLFL.

The time-dependent BPM, the SSFM, is employed to simulate the propagation of the soliton circulating in an MLFL. Our algorithm has been verified by the analytical approach using the ISM as described in Section 8.6. As both of these approaches

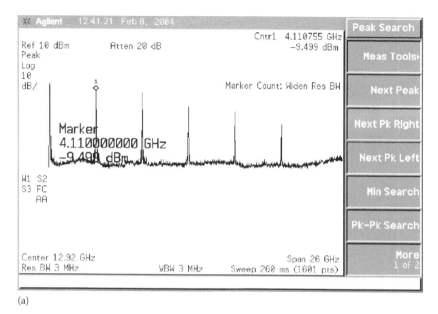

(a)

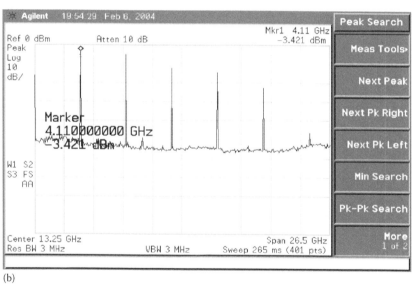

(b)

**FIGURE 8.45**   RF spectrum of HML laser with 18 m DSF (a) and 118 m DSF (b).

produce consistent results, we further conduct our investigation in interaction of two and three solitons by using the numerical approach.

As we can conclude, there are three effective methods to overcome the undesirable effect of soliton collapse in our optical soliton transmission systems: the method of setting the optimum pulse separation, relative amplitude, and relative phase. Pulse separation method could be used if we are not too concerned about the maximum

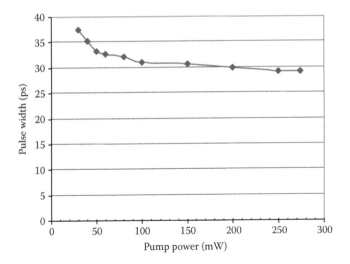

**FIGURE 8.46**   Pulsewidth of the HMLFL for various pump powers.

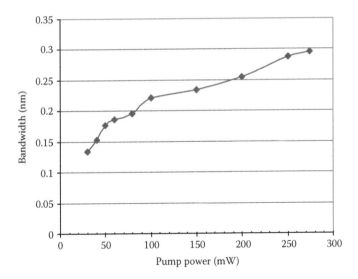

**FIGURE 8.47**   Change of bandwidth of the HMLFL versus pump power level.

achievable bit rate. Relative phase method is only useful at 0° and 180°. However, one can use the repulsive nature of the solitons by setting the relative phase to certain value in order to compensate for the attractive nature of the solitons. However, this method is difficult to implement in practice. Therefore, relative amplitude method would be the best in overcoming this problem. Only a slight difference in the amplitude of the solitons would prevent the solitons from collapsing.

Thus, we have to implement a device at the transmitting terminal to encode each binary digit with alternate value of amplitudes. This can be carried out by using an external modulator to modulate each pulse with different amplitude alternately: (1) The simulation assumes a zero-loss fibers. It would be very useful to take into account the losses due to propagation in the optical fiber, this could be implemented using distributed Raman amplification; (2) Also, simulation model assumes that the parameters can be controlled exactly. It would be more accurate if the simulation also considers jittering effects in the phase, amplitudes as well as the initial separation of the solitons; (3) Finally, the simulation should also consider the effects of the EDFA scheme to boost signals as well as Raman amplification.

This chapter has also experimentally demonstrated and simulated stable multi-soliton bound states, which have been generated in an active MLFL under phase matching via optical phase modulation. It is believed that this stable existence of multisoliton bound states is effectively supported by the phase modulation in an anomalous-dispersion fiber loop. Simulation results have confirmed the existence of multisoliton bound states in the FM MLFL. Created bound states can be easily harmonic mode locked to generate periodically multisoliton bound sequence at high repetition rate in this type of fiber laser that is much more prominent than those by passive types.

The pulse train of ultrashort pulse width at high repetition rate, up to 10 GHz and higher, generated from an HMLFL has been demonstrated. The characteristics of the pulse train such as pulse width and bandwidth have been studied by varying the settings of the cavity. It is found that as the modulation frequency increases, the pulse becomes shorter and its corresponding bandwidth increases. The increase of the modulation depth also makes the pulse shorter, but its effect on the pulse is not as strong as modulation frequency. High pump power is also found to help shorten the pulse width in the long cavity laser.

### 8.8.3 Transmission of Generated Multi-Bound Solitons

Temporal solitons are attractive for long haul fiber transmission systems due to their preservation of shape during propagation in nonlinear dispersive medium. In theory, the soliton character only remains unchanged in ideal transmission medium without any perturbation such as loss and noise. Therefore, the understanding of soliton propagation characteristics in real fiber systems is of significance to the design of an optical communication link. Moreover, there is a difference between propagation of solitons in a ring such as a MLFL and a real fiber transmission link. It must be noted that the length of the mode-locked fiber ring is much shorter than the dispersion length of the fiber, $L_D$, and the pulse sequence is thus operating in the near field region in which the chirp of the lightwave occurs significantly, especially at the edges of the pulses, while the distance of each span in the fiber link is longer than the dispersion length. Therefore, distinct transmission conditions are required to ensure that solitons can be recovered at the end of the link.

Although a number of reports on bound soliton states in MLFLs have been published, to date, there is no report on the propagation dynamics of multi-bound solitons in optical fiber. The difficulty of the generation of stable bound soliton sequence in passive MLFL may have prevented the investigation of their propagation and related dynamics. As described in earlier chapters, the actively FM

MLFL offers significant advantage in generation of ultrastable multi-bound soliton sequence at modulation frequency that would be important for propagation in optical fiber. Dynamics of multi-bound solitons is an important issue for investigation. The obtained results are significant for tapping potential applications of bound soliton lasers.

More comprehensive transmission of groups of solitons is later presented in Chapter 9.

### 8.8.3.1  Soliton Propagation in Optical Fibers

*8.8.3.1.1  Loss Management*

Because the dispersion or GVD value is unchanged along in optical fibers, the balance between GVD and SPM in soliton transmission would be only achieved if the soliton pulse was not attenuated during propagation. For a lossy fiber in the real transmission system, a reduction of the peak power over distance leads to a soliton broadening. The broadening of soliton can be understood by the reduction of SPM effect resulting from the reduced peak power that makes the impact of dispersion effect stronger during propagation. When the fiber loss is assumed as a weak perturbation of NLSE, change in soliton parameters under the influence of the fiber loss has been investigated by using variation method [25]. Variations of soliton amplitude and phase along the fiber can be given by

$$\eta(\xi) = \exp(-\Gamma\xi) \quad \text{and} \quad \phi(\xi) = \phi(0) + \frac{1 - \exp(-2\Gamma\xi)}{4\Gamma} \tag{8.113}$$

where
$\Gamma = \alpha L_D$
$\xi = z/L_D$
$\alpha$ is the loss coefficient of the fiber

An exponential decrease in soliton amplitude weakens the SPM effect; hence, soliton is also broadened in the same manner:

$$T(z) = T_0 \exp(\Gamma\xi) = T_0 \exp(\alpha z) \tag{8.114}$$

where
$T_0$ is the initial width of unperturbed soliton
$T$ is the width of soliton at distance $z$

It is to be noted that a linear increase in pulse width occurs in linear propagation scheme [24]. When the peak power is considerably attenuated at long distance, the SPM effect can be negligible and solitons behave like nonsolitary pulses. Therefore, the exponential dependence of soliton width over the distance is only valid at the distance where $\alpha z$ is less than 1.

To overcome the broadening problem due to the fiber loss effect, a periodic amplification of soliton is required to recover the soliton energy. A lumped amplification scheme can be used in fiber link for this purpose [26]. Solitons that propagate in this scheme are called the path-average solitons [27]. Similar to the mode-locked fiber ring system, the adjustment of soliton in the fiber following the amplifier can lead to a shedding a part of soliton energy as dispersive waves, which are accumulated to a considerable level over a large number of amplifiers. To keep variation of the soliton parameters negligible, two conditions to operate in the average-soliton regime must be satisfied [24,27]:

1. The amplifier spacing $L_A$ must be much smaller than the dispersion length $L_D$ ($L_A \ll L_D$). This condition is required to keep radiation of dispersive waves negligibly small, because the energy of dispersive waves is inversely proportional to the dispersion length $L_D$.
2. The input peak power of soliton must be larger than that of fundamental soliton by a factor

$$\frac{P_s}{P_0} = \frac{G \ln G}{G-1} \tag{8.115}$$

where

$P_s$, $P_0$ are the peak powers of the path-average soliton and the unperturbed soliton respectively
$G$ is the amplifier gain factor

This condition is to make certain that the average peak power over the $L_A$ is sufficient to balance the GVD effect (see Appendix E for definition and expressions).

The soliton energy varies periodically in each span due to the fiber loss; hence, other parameters of soliton such as the width and the phase also change accordingly. However, the soliton still remains unchanged at ultralong distance if the previous conditions are satisfied.

But it is impractical to satisfy the first condition when the existing transmission fibers with $L_D$ of about 10–20 km are used for the long-haul transmission system [28,29]. If the $L_A$ is short, there are some disadvantages such as the high cost and the large accumulated ASE noise resulting from a large number of optical amplifiers. To facilitate the first condition, a distributed amplification scheme such as Raman amplification can be employed [30,31]. It is understandable that the variation of peak power is considerably reduced, because the amplification process takes place along the transmission fiber. And this scheme allows a transmission with $L_A \gg L_D$. However, an unstable soliton transmission can occur due to the resonance of the dispersive waves and solitons when $L_A/L_D \geq 4\pi$ that need to be carefully considered in the design of soliton transmission system [32].

### 8.8.3.1.2  Dispersion Management

The problem of soliton broadening in lossy fiber can be also overcome by using dispersion-decrease-fiber (DDF) whose dispersion decreases exponentially corresponding to the reduction of the peak power. From the variation of soliton amplitude in Equation 8.113, the variation of the GVD in DDF follows a function as follows:

$$|\beta_2(z)| = |\beta_2(0)| \exp(-\alpha z) \tag{8.116}$$

Thus, the reduction of the GVD is proportional to the reduction of the SPM effect; hence, solitons can remain unchanged during propagation in the DDF. However, the use of DDF in practical transmission systems is not feasible because of the availability of the DDF as well as the complexity of the system design. Furthermore, the performance would be degraded, because the average dispersion along the link is large. Although this scheme is disadvantageous to the system, it is commonly applied for pulse compression based on soliton effect [33,34].

Therefore, another option is use of dispersion management that uses alternating positive and negative GVD fibers. This arrangement is commonly employed in high-speed transmission systems today, because it offers a relatively improved performance [35,36]. Hence, using periodic dispersion map along the fiber link has attracted extensive attention for its ability in soliton transmission. Both theoretical and experimental researches have shown the advantages of dispersion-managed (DM) solitons. Owing to alternating the sign of the fiber dispersion in one map period, average GVD of whole link can be kept in small value, while the GVD of each section is large enough to suppress the impairments such as four-wave mixing and third-order dispersion. Two important parameters of the map are the average GVD of the whole link $\bar{\beta}_2$ and the map strength $S_m$ that are defined as follows [24]:

$$\bar{\beta}_2 = \frac{\beta_{2n}l_n + \beta_{2a}l_a}{l_n + l_a}, \quad S_m = \frac{\beta_{2n}l_n - \beta_{2a}l_a}{\tau_{FWHM}^2} \quad (8.117)$$

where

    $\beta_{2n}$, $\beta_{2a}$ are the GVD parameters in the normal and anomalous sections of lengths
       $l_n$ and $l_a$ respectively
    $\tau_{FWHM}$ is the FWHM

It has been shown that the shape of DM solitons is closer to a Gaussian pulse rather than a "*sech*" shape and their parameters such as the width, peak power, and chirp vary considerably in each map period. Therefore, depending on the map configuration, input parameters of DM solitons should be carefully chosen to ensure that the pulse can recover its state after each map period [37]. In addition, the peak power enhancement of DM soliton compared to the constant-GVD soliton allows an improvement of performance in terms of signal-to-noise ratio during suppression of timing jitter by reducing the average GVD. Interestingly, it has been confirmed that DM solitons exist not only in anomalous average dispersion ($\bar{\beta}_2 < 0$) but also in normal average dispersion scheme ($\bar{\beta}_2 > 0$) [38]. However, the existence of DM soliton in the map with $\bar{\beta}_2 > 0$ is obtained only when the strength of map $S_m$ is greater than a critical value $S_{cr}$, which is approximately 4 [40]. Hence, it is not surprising to observe the existence of solitons in the MLFL with normal average dispersion.

Interaction between solitons is an important issue in practice. Because properties of DM solitons are different from the average GVD solitons, the evolution and dynamics of DM solitons have become more attractive for both theoretical and experimental studies. Result of theoretical study has shown that a strong interaction prevents the DM solitons in a positive average GVD map from practical high-speed transmission applications, although they can exist [40]. This also indicates a complexity in dynamics of DM solitons in propagation; beside a single soliton solution, some complex solutions of DM soliton can exist in specific dispersion maps. In one study on a high-order DM soliton, it has showed a stable evolution of antisymmetric (antiphase) soliton in the fiber at ultralong distance [41]. Recently, the dispersion management has been also proposed to support a transmission of bisoliton that consists of a couple of DM solitons with zero-(in-phase) or $\pi$-(antiphase) phase difference [42]. By selecting appropriate parameters of DM map and input pulse, a stable propagation of in-phase bisoliton for single channel and antiphase bisoliton for multi-channel has been confirmed through numerical simulation [42]. In other studies, a structure called soliton molecules consisting of one dark and two bright solitons, which is similar to an asymmetrical soliton, has been numerically and experimentally demonstrated [43–46].

We can see that the soliton complexes in earlier studies have a structure similar to pairs of bound solitons generated from the MLFLs. Stable propagation of the soliton complexes can offer new coding schemes for soliton transmission systems. In conventional coding scheme, data bits "1" or "0" is represented by the presence or absence of a soliton pulse in each time slot. In one proposal, using the soliton complex such as bisoliton for new coding scheme provides code states with residual bit for error prevention [42]. Therefore, it is necessary to have a new optical source that can generate the complex soliton like bisoliton in new coding schemes. With the ability of multi-bound solitons generation, the actively FM mode-locked fiber laser offers more states for coding scheme that allows an improvement of performance and transmission of more than one data bit in each time slot. Hence, propagation characteristics of multi-bound solitons generated from the actively FM MLFL are significant for potential applications of multi-bound soliton lasers in telecommunication systems.

### 8.8.3.2    Transmission of Multi-Bound Solitons: Experiments

#### 8.8.3.2.1    Experimental Setup

To investigate the propagation dynamics of multi-bound solitons, various multi-bound solitons with ultrahigh stability from dual- to quintuple-states are generated by carefully optimizing the locking conditions of the fiber ring. These states are then propagating through standard single mode optical fibers (SSMF) in order to investigate their propagation dynamics as shown in Figure 8.48. The length of the fiber is varied to prove the interaction between the soliton pulses. The estimated FWHM of individual pulse of original bound soliton pairs (BSP), triple-bound solitons (TBS), and quadruple-bound solitons (QBS) are 7.9, 6.9, and 6.0 ps, respectively. The time separation between two adjacent pulses is about three times that of the FWHM pulse width. The repetition rate of these multi-bound soliton sequences is about 1 GHz.

In this setup, a booster EDFA is used before the transmission section to specify the power of multi-bound solitons launched into the SSMF. Through adjustment of the launched power, the multi-bound solitons can propagate under various transmission conditions from linear to nonlinear schemes. Two rolls of SSMF with lengths of 1 and 50 km are used for the investigation. After propagation, the outputs such as time trace and spectrum are monitored by the oscilloscope and the optical spectrum analyzer (OSA), respectively.

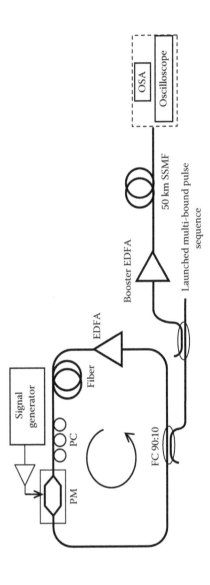

**FIGURE 8.48** Experimental setup for propagation of multi-bound solitons in standard single mode fiber. PM, phase modulator; EDFA, Er:doped fiber amplifier; SSMF, standard single mode fiber; PC, polarization controller; OSA, optical spectrum analyzer.

### 8.8.3.2.2   Results and Discussion

It is obvious but worth mentioning that the binding property of the solitons circulating in the fiber ring, is different under the case when they are propagating through the optical fiber. In the propagation, these individual solitons would be interacting with each other and naturally, they are no longer supported by the periodic phase modulation while circulating in the ring laser. When propagating through a dispersive fiber, the optical carrier under the envelope of generated multi-bound solitons is influenced by the chirping effects, hence the overlapping between the soliton pulses. When the accumulated phase difference of adjacent pulses is $\pi$, the solitons repel each other [7,47,49]. In another word, the solitons in their multi-bound states acquire the down-chirping effect when propagating in SSMF due to its GVD-induced phase shift. Hence, within the group of multi-bound solitons, the front-end soliton would travel with a higher positive frequency shift, and hence a higher group velocity than those at back-end of the multi-bound group that is influenced by a lower negative frequency shift. Therefore, the time separation between adjacent pulses varies with the propagation distance. The variation of the time separation between pulses depends on their relative frequency difference. Figure 8.49 shows the waveforms and their corresponding spectra of triple-bound solitons after propagating over 50 km fiber with different launched powers (Figures 8.50 and 8.51).

In all cases, the time separation significantly increases compared to the initial state due to the repulsion. In an ideal condition of propagation, direct interaction is the only factor that influences the temporal interval between bound solitons and the pulse shape of solitons in bound states. However, propagation in a real optical fiber is considerably influenced by perturbations such as loss of fiber and initial launched powers ($P_l$) of solitons. The loss of fiber always leads to the broadening of bound solitons. Figure 8.52a and b shows the dependence of pulse width and temporal separation of various multi-bound solitons on the launched power as a parameter. We observe similar dynamics in the propagation of dual, quadruple, and quintuple groups of solitons. In general, an increase in launched power leads to a enhanced shortening of the pulse width or a broadening of bound solitons at lower rate, and hence a reduction of the pulse temporal separation due to the enhancement of nonlinear self-phase-modulation phase shift. Unlike the nearly linear variation of the pulse width, variation curves of the temporal separation show nonlinear launched power dependence. As observed in Figure 8.52b, two distinct areas of the curves can be observed separately at the points $A$, $B$, and $C$. The intersection points $A$, $B$, $C$ of the two tangential curves correspond to the average soliton power ($P_{sol}$) of the bound state. On the other hand, the points $A$, $B$, and $C$ correspond to the minimum powers to launch the bound pulses for soliton propagation. For the transmission over SSMF, it is observed that with the experimental parameters of the multi-bound solitons, the estimated average soliton powers for BSP, TBS, and QBS are 11, 13.5, and 15.2 dBm, respectively.

At $P_l$ lower than $P_{sol}$, and in addition to the fiber attenuation, the power of multi-bound soliton is not sufficient to balance the GVD effect. As a result, the pulse widths are rapidly broadened by self-alignment of the multi-bound solitons due to

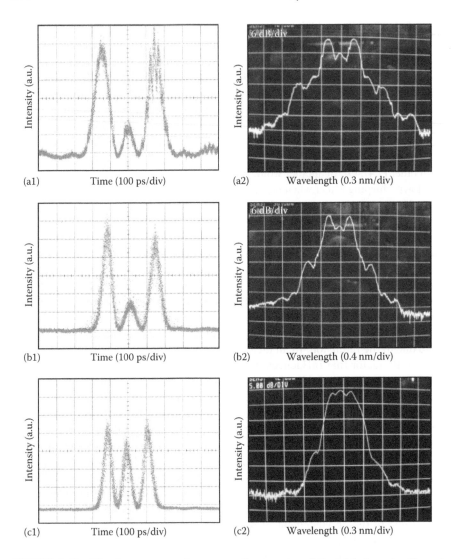

**FIGURE 8.49** The time traces and corresponding spectra of the triple-bound soliton at launching powers of: (a1 and a2) 4.5 dBm, (b1 and b2) 10.5 dBm, and (c1 and c2) 14.5 dBm respectively.

the perturbations accompanied by partly shading their energy in the form of dispersive waves [24]. The broadening of such dispersive wave is accumulated along the transmission fiber and oscillations are formed around the multi-bound solitons as shown by the ripple of the tail of the soliton group in Figure 8.49a1. Furthermore, the rate of broadening of the pulses is faster than that of the temporal separation at low launched powers, leading to the enhancement of the overlapping between pulses. Therefore, the pulse envelope is consequently modulated by the interference

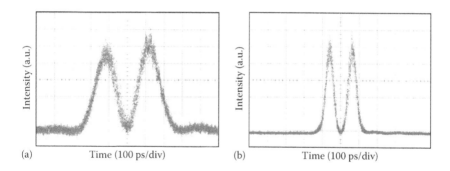

**FIGURE 8.50** The time traces of the dual-bound soliton at launching powers of (a) 4.5 dBm and (b) 17 dBm, respectively, after propagating through 50 km SSMF.

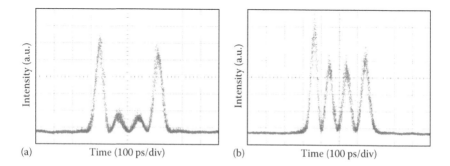

**FIGURE 8.51** The time traces of the quadruple-bound soliton at launching powers of (a) 5 dBm and (b) 16.5 dBm, respectively, after propagating through 50 km SSMF.

of waves with the phase modulation effect due to the GVD that is the same as the fractional temporal Talbot effect [49–51]. Because of the parabolic symmetry of the anomalous GVD-induced phase shift profile around multi-bound solitons with a relative phase difference of $\pi$, the energy of the inner pulses is shifted to the outer pulses, hence a decrease of the amplitude of inner pulses. The frequency components of the central parts of the soliton pulses propagates at higher relative velocity than those at the edge that result in the energy transfer from the central parts to the outer section of the soliton. This pattern also reconfirms the relative phase difference of $\pi$ in the bound state.

At $P_l < P_{sol}$, the multi-bound states operate in linear transmission mode where the pulse are the solitary waves. In contrast, at higher $P_l$, the SPM phase shift is increased to balance the GVD effect. The pulse width becomes narrower and the ripple of the pulse envelope is lower with higher level of launched power. For the dual-bound soliton, there is no difference in peak power of two pulses in bound state. For the bound states with the number of solitons greater than two, however, there is high difference between the inner and the outer pulses because of the energy transfer at $P_l < P_{sol}$. As the soliton content in pulses is enhanced at the average soliton power, there is a jump of peak power ratio between inner and outer pulses as shown in Figure 8.52c.

### 8.8.3.3 Dynamics of Multi-Bound Solitons in Transmission

To verify the experimental results as well as to identify the evolution of multi-bound solitons propagation in the fiber, we model the multi-bound solitons as [51]

$$u_{bs} = \sum_{i=1}^{N} u_i(z,t) \qquad (8.118)$$

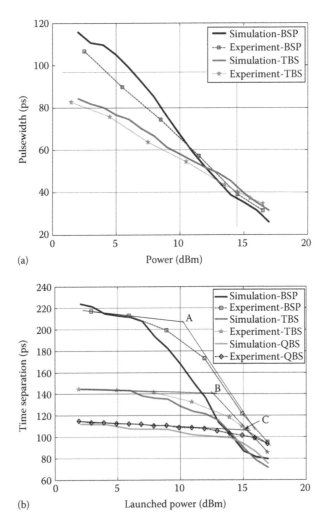

(a)

(b)

**FIGURE 8.52** The launching power dependent variation of (a) pulse width, (b) time separation.                   *(Continued)*

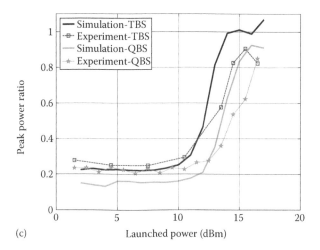

(c)                     Launched power (dBm)

**FIGURE 8.52 (*Continued*)** The launching power dependent variation of (c) peak power ratio of different multisoliton bound states after 50 km propagation distance.

with $N$ as number of solitons in bound state, and

$$u_i = A_i \text{sech}\{[t - iq_0]\} \exp(j\theta_i) \tag{8.119}$$

where

$A_i$ and $q_0$ are amplitude and time separation of solitons, respectively

the phase difference $\Delta\theta = \theta_{i+1} - \theta_i = \pi$

The propagation of multi-bound solitons in SSMF is governed by the NLSE with the input parameters as those obtained in the experiment. Shown together with experimental results in Figure 8.52a through c is the simulation evolution of the parameters (solid curves) over 50 km propagation of various orders of multisoliton bound states. The simulated evolution of the triple-bound soliton is shown in Figure 8.53 over 50 km SSMF, while Figure 8.54 shows the pulse shape and corresponding spectrum of the outputs at various launched power levels. The simulated results generally agree well with those obtained in the experiment (see Figures 8.49 and 8.52 for comparison). The small residual chirp caused small difference between the experimental and numerical results.

Dynamics of multi-bound solitons consisting of DBS, TBS, and QBS during propagation is shown in Figures 8.55 through 8.57. In each figure, respectively, the evolution of multi-bound solitons parameters, such as the pulse width, the temporal separation between solitons, the ratio between the pulse width and the pulse separation, and the peak power of the solitons, is simulated along the transmission fiber with different launched powers. The difference between lower and higher launching powers also obviously exhibits in simulated results. When the launched powers is far from the soliton power $P_{sol}$, there is oscillation or rapid variation of parameters at

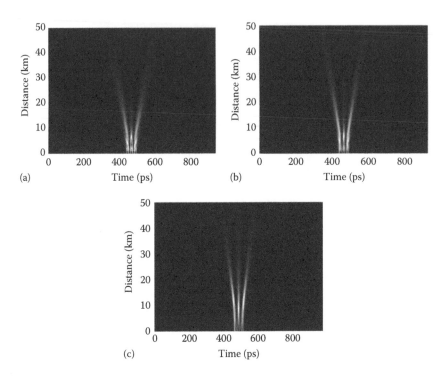

**FIGURE 8.53** Numerically simulated evolutions of the triple-bound solitons over 50 km SSMF propagation at $P_l$ of (a) 5 dBm, (b) 10 dBm, and (c) 14 dBm.

initial propagation distance due to the adjustment of multi-bound solitons to perturbations of propagation conditions as mentioned earlier. The pulses are compressed at $P_l$s higher than $P_{sol}$, while they are rapidly broadened at $P_l$s lower than $P_{sol}$. The slow variation of parameters occurs at $P_l$ close to $P_{sol}$. However, the time separation of solitons remains unchanged in the propagation distance of one soliton period. We have validated this prediction in our experiment.

Another important property of multi-bound solitons is the phase difference between pulses that can be determined by the shape of spectrum. We have monitored the optical spectrum of the multi-bound solitons at both the launched end and the output of the fiber length. The modulated spectrum of multi-bound solitons is symmetrical with carrier suppression due to a relative phase relationship of $\pi$ between adjacent solitons. After 50 km propagation at low $P_l$, both experimental and simulated results show that the spectrum of multi-bound solitons is nearly same as that at the launched end (see Figures 8.49d and 8.54d). In linear-like scheme where the SPM effect is negligible, the GVD only modulates the spectral phase, which modifies the temporal profile, but does not modify the spectrum of bound state. The modulation of spectrum is modified with increase in $P_l$. At sufficient high $P_l$, the nonlinear phase shift–induced chirp is increased at the edges of pulses. Although the nonlinear phase shift reduces the GVD effect, the phase transition between adjacent pulses is

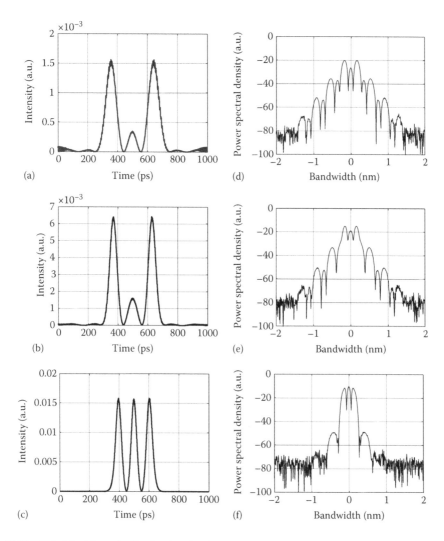

**FIGURE 8.54** Numerically simulated triple-soliton bound state pulse shape at the output (left) and its corresponding optical spectrum (right - same row) over 50 km SSMF propagation with the optical average power of 5 dBm (a and b), 10 dBm (c and d), and 14 dBm (e and f).

changed due to direct impact of the nonlinear phase shift. Hence, the small humps in the spectra of multi-bound states are strengthened and they may be comparable to main lobes due to enhancement of the far interaction between pulses as shown in Figures 8.49f and 8.54f. Figure 8.58 shows the variation of the relative phase difference between adjacent pulses in various multi-bound soliton states over 50 km propagation. The simulated results also indicate that the relative phase difference varies differently between two adjacent pulses in propagation. Although the phase difference of $\pi$ between two central pulses in even-soliton bound states such as DBS

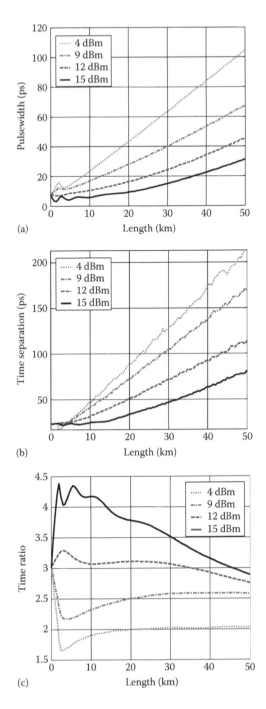

**FIGURE 8.55** Evolution of numerically simulated dual-soliton bound state after 50 km propagation distance of standard SMF-28 fiber: (a) pulse width, (b) temporal separation, (c) ratio between pulse width and separation.                                    (*Continued*)

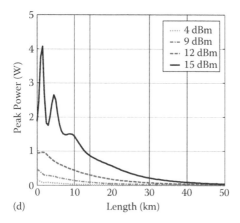

(d)

**FIGURE 8.55 (Continued)** Evolution of numerically simulated dual-soliton bound state after 50 km propagation distance of standard SMF-28 fiber: (d) peak power.

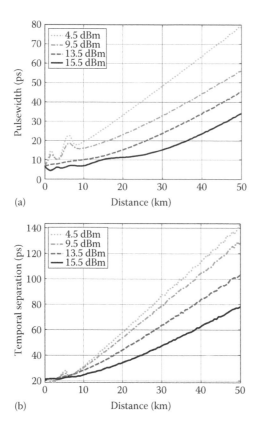

(a)

(b)

**FIGURE 8.56** Evolution of numerically simulated triple-soliton bound state after 50 km propagation distance of standard SMF-28 fiber: (a) pulse width, (b) temporal separation. *(Continued)*

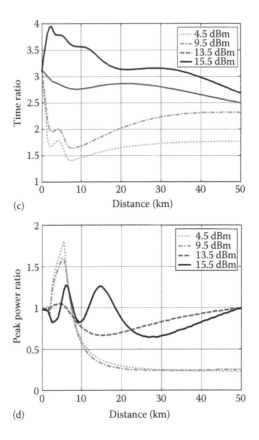

**FIGURE 8.56 (*Continued*)** Evolution of numerically simulated triple-soliton bound state after 50 km propagation distance of standard SMF-28 fiber: (c) ratio between pulse width and separation, and (d) peak power ratio between inner pulse and outer pulse.

and QBS can remain unchanged, it varies in general along propagation distance. At low $P_l$ in all cases, the phase difference is varied to zero or $\pi/2$ value, then recovered to $\pi$ at the output of the 50 km fiber. When the SPM phase shift becomes significant, the phase difference varies in the same manner in the first 40 km propagation, and then it tends to $\pi/2$ that modifies the corresponding spectrum.

### 8.8.3.4 Remarks

Characteristics of multi-bound soliton in propagation over 50 km SSM fiber have been investigated in this chapter. Depending on the launched power level, the variation of multi-bound states parameters can divide into two propagation schemes: linear and soliton schemes. At low launched power in linear propagation scheme, the modulated spectrum of multi-bound states remains unchanged due to the preservation of the phase difference of $\pi$, yet the temporal shape is modified that can be used to reconfirm the phase difference. At high launched power in soliton propagation scheme, the modulated spectrum of multi-bound states is modified due to the

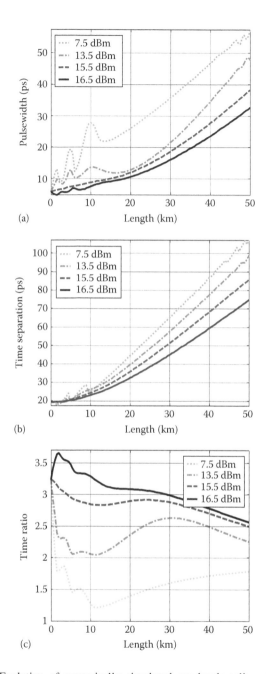

**FIGURE 8.57** Evolution of numerically simulated quadruple-soliton bound state after 50 km propagation distance of standard SMF-28 fiber: (a) pulse width, (b) temporal separation, (c) ratio between pulse width and separation. *(Continued)*

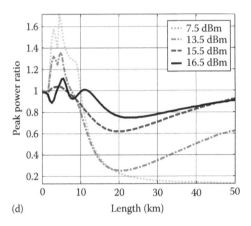

(d)

**FIGURE 8.57 (*Continued*)** Evolution of numerically simulated quadruple-soliton bound state after 50 km propagation distance of standard SMF-28 fiber: (d) peak power ratio between inner pulse and outer pulse.

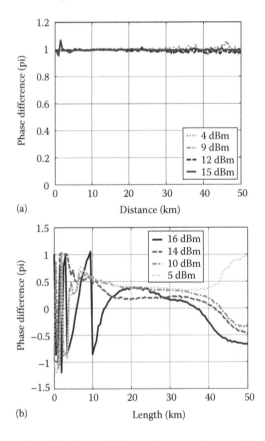

(a)

(b)

**FIGURE 8.58** Evolution of the phase difference between adjacent pulses over 50 km propagation in various bound states: (a) dual-bound soliton, (b) triple-bound soliton. (*Continued*)

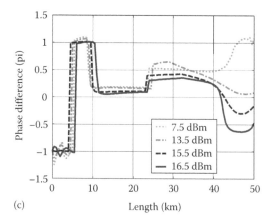

(c)

FIGURE 8.58 (*Continued*)   Evolution of the phase difference between adjacent pulses over 50 km propagation in various bound states: (c) quadruple-bound soliton.

enhancement of SPM effect; however, the pulses in bound state behave similar to the single soliton transmission in perturbed conditions.

## REFERENCES

1. N. Zabusky and M. D. Kruskal, Interaction of "solitons" in a collisionless plasma and the recurrence of initial states, *Phys. Rev. Lett.*, 15, 240, 1965.
2. G. P. Agrawal, *Nonlinear Fiber Optics*, Academic Press, London, U.K., 1992.
3. V. E. Zaharov and A. B. Shabat, Exact theory of two-dimensional self-focusing and one-dimensional self-modulation of wave in nonlinear media, *Sov. Phys. JETP*, 34(1), 62–69, 1972.
4. R. K. Dodd, J. C. Eilbeck, J. D. Gibbon, and H. C. Morris, *Solitons and Nonlinear Wave Equations*, Academic Press, London, U.K., 1982.
5. J. Satsuma and N. Yajima, Initial value problems of one-dimensional self-modulation of nonlinear waves in dispersive media, *Suppl. Prog. Theor. Phys.*, 55, 284–306, 1974.
6. C. Desem and P. L. Chu, Reducing soliton interaction in single-mode optical fibers, *IEE Proc.*, 134(Pt J, 3), 145–151, 1987.
7. J. P. Gordon, Interaction forces among solitons in optical fibers, *Opt. Lett.*, 8(10), 596–598, 1983.
8. D. Anderson and M. Lisak, Bandwidth limits due to mutual pulse interaction in optical soliton communication systems, *Opt. Lett.*, 10(3), 174–176, 1986.
9. A. Bondeson, M. Lisak, and D. Anderson, Soliton perturbations: A variational principle for soliton parameters, *Phys. Scr.*, 20, 479–485, 1979.
10. V. I. Karpman and V. V. Solov'ev, A perturbational approach to the two-soliton systems, *Physica*, 3D, 487–502, 1981.
11. A. P. Agrawal, *Fiber-Optic Communication Systems*, John Wiley, New York, 1992.
12. A. Hasegawa, *Optical Solitons in Fibers*, Spinger-Verlag, Berlin, Germany, 1989.
13. B. A. Malomed, Bound solitons in coupled nonlinear Schrodinger equation, *J. Phys. Rev. A*, 45, R8321–R8323 (1991).
14. B. A. Malomed, Bound solitons in the nonlinear Schrodinger–Ginzburg–Landau equation, *J. Phys. Rev. A*, 44, 6954–6957 (1991).

15. D. Y. Tang, B. Zhao, D. Y. Shen, and C. Lu, Bound-soliton fiber laser, *J. Phys. Rev. A*, 66, 033806 (2002).

16. Y. D. Gong, D. Y. Tang, P. Shum, C. Lu, T. H. Cheng, W. S. Man, and H. Y. Tam, Mechanism of bound soliton pulse formation in a passively mode locked fiber ring laser, *Opt. Eng.*, 41(11), 2778–2782 (2002).

17. P. Grelu, F. Belhache, and F. Gutty, Relative phase locking of pulses in a passively mode-locked fiber laser, *J. Opt. Soc. Am. B*, 20, 863–870 (2003).

18. L. M. Zhao, D. Y. Tang, T. H. Cheng, H. Y. Tam, and C. Lu, Bound states of dispersion-managed solitons in a fiber laser at near zero dispersion, *Appl. Opt.*, 46, 4768–4773 (2007).

19. W. W. Hsiang, C. Y. Lin, and Y. Lai, Stable new bound soliton pairs in a 10 GHz hybrid frequency modulation mode locked Er-fiber laser, *Opt. Lett.*, 31, 1627–1629 (2006).

20. C. R. Doerr, H. A. Hauss, E. P. Ippen, M. Shirasaki, and K. Tamura, Additive-pulse limiting, *Opt. Lett.*, 19, 31–33 (1994).

21. R. Davey, N. Langford, and A. Ferguson, Interacting solitons in erbium fiber laser, *Electron. Lett.*, 27, 1257–1259 (1991).

22. D. Krylov, L. Leng, K. Bergman, J. C. Bronski, and J. N. Kutz, Observation of the breakup of a prechirped N-soliton in an optical fiber, *Opt. Lett.*, 24, 1191–1193 (1999).

23. J. E. Prilepsky, S. A. Derevyanko, and S. K. Turitsyn, Conversion of a chirped Gaussian pulse to a soliton or a bound multisoliton state in quasi-lossless and lossy optical fiber spans, *J. Opt. Soc. Am. B*, 24, 1254–1261 (2007).

24. G. P. Agrawal, *Nonlinear Fiber Optics*, 3rd edn., Academic Press, San Diego, CA, 2001.

25. K. J. Blow and N. J. Doran, The asymptotic dispersion of soliton pulses in lossy fibers, *Opt. Commun.*, 52, 367–370, 1985.

26. M. Nakazawa, Y. Kimura, and K. Suzuki, Soliton amplification and transmission with Er:doped fiber repeater pumped by GaInAsP diode, *Electron. Lett.*, 25, 199–200, 1989.

27. A. Hasegawa and Y. Kodama, Guiding-center soliton, *Phys. Rev. Lett.*, 66, 161, 1991.

28. Y. Kodama and A. Hasegawa, Amplification and reshaping of optical solitons in glass fiber?—II, *Opt. Lett.*, 7, 339–341, 1982.

29. Y. Kodama and A. Hasegawa, Amplification and reshaping of optical solitons in glass fiber amplifiers with random gain, *Opt. Lett.*, 8, 342–344, 1983.

30. A. Hasegawa, Amplification and reshaping of optical solitons in a glass fiber—IV: Use of the stimulated Raman process, *Opt. Lett.*, 8, 650–652, 1983.

31. L. F. Mollenauer and K. Smith, Demonstration of soliton transmission over more than 4000 km in fiber with loss periodically compensated by Raman gain, *Opt. Lett.*, 13, 675–677, 1988.

32. L. Mollenauer, J. P. Gordon, and M. N. Islam, Soliton propagation in long fibers with periodically compensated loss, *IEEE J. Quant. Electron.*, 22, 157–173, 1986.

33. M. Nakazawa, Tb/s OTDM technology, in *27th European Conference on Optical Communication, 2001. ECOC '01*, Amsterdam, the Netherlands, vol. 2, pp. 184–187, 2001.

34. L. Ju Han, T. Kogure, and D. J. Richardson, Wavelength tunable 10-GHz 3-ps pulse source using a dispersion decreasing fiber-based nonlinear optical loop mirror, *IEEE J. Select. Top. Quant. Electron.*, 10, 181–185, 2004.

35. A. Sano and Y. Miyamoto, Performance evaluation of prechirped RZ and CS-RZ formats in high-speed transmission systems with dispersion management, *J. Lightwave Technol.*, 19, 1864–1871, 2001.

36. J. Fatome, C. Fortier, and S. Pitois, Practical design rules for single-channel ultra high-speed dense dispersion management telecommunication systems, *Opt. Commun.*, 282, 1427–1434, 2009.

37. E. Poutrina and G. P. Agrawal, Design rules for dispersion-managed soliton systems, *Opt. Commun.*, 206, 193–200, 2002.
38. J. H. B. Nijhof, N. J. Doran, W. Forysiak, and F. M. Knox, Stable soliton-like propagation in dispersion managed systems with net anomalous, zero and normal dispersion, *Electron. Lett.*, 33, 1726–1727, 1997.
39. J. H. B. Nijhof, N. J. Doran, W. Forysiak, and A. Berntson, Energy enhancement of dispersion-managed solitons and WDM, *Electron. Lett.*, 34, 481–482, 1998.
40. T. Inoue, H. Sugahara, A. Maruta, and Y. Kodama, Interactions between dispersion-managed solitons in optical-time-division-multiplexed systems, *Electron. Commun. Japan (Part II: Electron.)*, 84, 24–29, 2001.
41. C. Paré and P. A. Bélanger, Antisymmetric soliton in a dispersion-managed system, *Opt. Commun.*, 168, 103–109, 1999.
42. A. Maruta, T. Inoue, Y. Nonaka, and Y. Yoshika, Bisoliton propagating in dispersion-managed system and its application to high-speed and long-haul optical transmission, *IEEE J. Select. Top. Quant. Electron.*, 8, 640–650, 2002.
43. M. Stratmann, M. Bohm, and F. Mitschke, Dark solitons are stable in dispersion maps of either sign of path-average dispersion, in Summaries of papers presented at the *Quantum Electronics and Laser Science Conference, 2002 (QELS '02). Technical Digest*, Anaheim, CA, p. 226, 2002.
44. M. Stratmann and F. Mitschke, Bound states between dark and bright solitons in dispersion maps, in Summaries of papers presented at the *Quantum Electronics and Laser Science Conference, 2002 (QELS '02). Technical Digest*, Anaheim, CA, pp. 226–227, 2002.
45. M. Stratmann, T. Pagel, and F. Mitschke, Experimental observation of temporal soliton molecules, *Phys. Rev. Lett.*, 95, 143902, 2005.
46. I. Gabitov, R. Indik, P. Lushnikov, L. Mollenauer, and M. Shkarayev, Twin families of bisolitons in dispersion-managed systems, *Opt. Lett.*, 32, 605–607, 2007.
47. Y. Kodama and K. Nozaki, Soliton interaction in optical fibers, *Opt. Lett.*, 12, 1038–1040, 1987.
48. F. M. Mitschke and L. F. Mollenauer, Experimental observation of interaction forces between solitons in optical fibers, *Opt. Lett.*, 12, 355–357, 1987.
49. J. Azana and M. A. Muriel, Temporal self-imaging effects: Theory and application for multiplying pulse repetition rates, *IEEE J. Select. Top. Quant. Electron.*, 7, 728–744, 2001.
50. L. N. Binh, K. K. Pang, T. Vo, and T. Huynh, Temporal imaging and optical repetition multiplication via quadratic phase modulation, in *2007 6th International Conference on Information, Communications & Signal Processing*, Singapore, pp. 1–5, 2007.
51. N. D Nguyen and L. N. Binh, Generation of bound-solitons in actively phase modulation mode-locked fiber ring resonators, *Opt. Commun.*, 281, 2012–2022, 2008.

# 9 Concluding Remarks

In fiber and guided wave optics, third-order nonlinear effects have offered not only the challenges of transmission penalty and distortion effects in fiber systems such as mode-locked fiber ring lasers, but also opportunities to explore their potential in new mechanisms. Therefore, research issues about nonlinearity in some guided-wave media such as the actively FM mode-locked ring fiber for multi-pulse operation and the nonlinear waveguide for signal processing have been addressed for the control and formation of multi-bound solitons, in this book.

The important characteristics of FM mode-locked fiber lasers have been both experimentally and numerically investigated. In addition, significant applications in signal processing and transmission based on soliton bounding process in guided wave optical lines have been described.

We have demonstrated the existence of multi-bound solitons having more than two mutually binding solitons in an actively FM mode-locked fiber laser. By properly setting the parameters of the cavity, various multi-soliton bound states from dual to sextuple orders have been experimentally generated at the modulation frequency of 1.0 GHz applied to the optical phase modulator in the lasing cavity. In association with experimental results, numerical simulations have been conducted to show that multi-bound solitons with their relative phase difference of $\pi$ are stable solutions of the FM mode-locked fiber laser under some specific conditions of the cavity. When the cavity parameters deviate from the optimum levels, individual multi-bound solitons with the phases of their lightwave carriers can no longer hold with $\pi$ difference. Thus they would reach quasi-stable states with a periodic variation of the generated pulse sequence or unstable state with strong fluctuations in amplitude and temporal position.

We have also described both experimental and simulation techniques for generating higher-order bound states, from double to sextuple, with available optical power. The generated multi-bound states become more sensitive to environmental fluctuations and the changes in cavity settings.

Distinctively from passive mode-locking, the EO phase modulator incorporated in the ring laser as a mode locker, has played the essential role in the formation as well as the dynamic behavior of multi-bound solitons through effective interactions between adjacent pulses and the chirping of lightwave carriers under its envelope. The influence of the phase modulator on multi-soliton bound states can happen under two conditions: the artificial filtering response of the cavity and the chirping caused by optical phase modulation. The inherent birefringence in the $Ti–LiNbO_3$ waveguide of the EO phase modulators is identified through the formation of an artificial Lyot-based filtering response. This response not only relates to the discrete wavelength tunable function but also limits the pulse shortening in the cavity, which is very important in multi-bound soliton formation.

Therefore, the multi-bound solitons have been obviously obtained only in the cavity using the lumped-type phase modulator PM-315P, while it is difficult to generate multi-bound solitons in the cavity using the traveling-wave phase modulator Mach-40-27. In another aspect, the chirp rate induced by phase modulation from the EO phase modulator significantly affects important parameters of the multi-bound solitons such as temporal separation and optical power level for stabilization. At higher chirp rates, the temporal separation is reduced and the optical power level required for stability is increased. It has been also demonstrated that at the same optical power in the cavity, switching from a higher-order to a lower-order bound state can occur when the modulation index is sufficiently increased.

Moreover the bound solitons are split into lower-order bound states with controllable separation by the distortion of the phase modulation profile. The distortion is caused by the enhancement of higher-order harmonics in the driving signal, which is useful not only for rational harmonic mode-locking but also for soliton control inside the cavity.

The transmission characteristics of multi-bound solitons have been experimentally and numerically investigated in 50 km standard single mode fiber and also by inverse scattering method. Depending on the launched power of the multi-bound soliton, the obtained results showed two distinct soliton parameters after transmission. Multi-bound solitons have been under the influence of a strong repulsive interaction that is varied along the propagation distance. The changes in soliton parameters depend on the launching powers that can be lower or higher than the power needed to form a single soliton. Although soliton parameters such as the width and the time separation vary strongly due to the perturbation of loss and dispersion, the phase difference of $\pi$ still remains almost unchanged at low launching power even after a 50 km transmission. In contrast, the variation of the soliton parameters is reduced at sufficiently high optical launched powers, yet the phase difference deviates from the value of $\pi$.

The dynamic states of the multi-bound solitons have been also analyzed by triple correlation and its Fourier transforms, called bispectrum techniques. The bispectrum analysis showed distinguishable representation of various multi-bound soliton states that can be used to estimate their stability. In an effort to find an alternative way to estimate the triple correlation in the optical domain, the four wave mixing (FWM) parametric process or multi-bound solitons was experimented. The simulation results have shown that the limitations of this approach in practical applications.

Having demonstrated the existence of stable multi-bound solitons in the actively FM mode-locked fiber laser, it is expected to explore new interesting behaviors of the multi-bound soliton as well as to specify its potential applications in communication systems. As observed in passively mode-locked fiber lasers [1], the vector solitons also may exist in an actively mode-locked cavity. Furthermore, the polarization-dependent phase modulation of the $LiNbO_3$ modulators can affect the interaction and the coupling between two distinct polarization components that produce new dynamic states. In order to enhance controllability of the multi-bound solitons in the cavity beside active phase modulation, a change in the guided-wave medium can be applied. It has been shown that the gap solitons propagating in the fiber

Bragg grating (FBG) have a local interaction that is different from that in conventional fibers [2,3]. Hence, the insertion of an FBG in the cavity can allow a variation of propagation velocity of the pulses that influences an effective interaction in the bound state. In addition, photonic crystal fibers (PCFs) have recently emerged as a new class of optical fibers that offer interesting features such as high nonlinearity with optimized dispersion properties [4]. Therefore, PCFs can be another option for the nonlinear guided-wave medium in the generation of multi-bound solitons.

Although the triple correlation based on the FWM process shows the complexity of this technique and the limited performance due to the noise of other mixing processes, the FWM parametric effect will still play a key role in a variety of signal processing functions in future photonic networks. Moreover, distinct representation of the triple correlation and the bispectrum for a signal or process can be significantly applied for monitoring the signal in optical transmission systems. For ultrahigh-speed long-haul transmission systems, impairments such as dispersion and nonlinear phase noises can result in a non-Gaussian distribution of the distortions of the signal that can be easily identified by the bispectrum. Owing to the insensitivity of the bispectrum to a Gaussian random process, it has been demonstrated that the bispectrum is useful to reconstruct the signal corrupted by noise [5]. In this approach, we have numerically shown the potential of the signal recovery for Gaussian white noise corrupted BPSK system by using bispectrum estimation in the digital domain instead of the optical domain [6]. With the rapid development in high-speed analog-to-digital converters (ADCs) for optical receivers, it is becoming feasible to expand the ability of bispectrum estimation in the digital domain to practical transmission systems.

Currently, integrated photonics based on Si on insulator [7–11] have demonstrated their superiority in the integration of optical circuits and analog and digital electronic circuits. This new technology will open new horizons for nonlinear photonics. Multi-bound soliton generation techniques described in this book would be possible in such technological approaches. The main obstacles of Si photonics are the hybrid integration of III–V semiconductor laser sources and optical amplifiers so as to enable the implementation of ring lasers based on rib optical waveguides. However, the multi-bound soliton generator is very compact, a few μm square, in Si photonics, as shown in Figure 9.1. This would lead to highly stable and controllable multi-bound

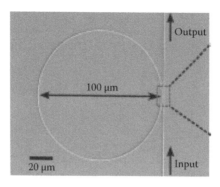

**FIGURE 9.1** Si photonic micro-ring resonator.

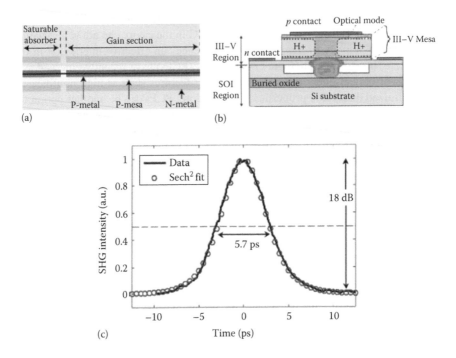

**FIGURE 9.2** Saturable mode-locked soliton fiber laser based on hybrid Si photonic technology: (a) structure (b) waveguide cross section and guided mode field distribution, (c) generated soliton pulse shape. *Legend*: SOI = SI on insulator.

soliton sequences for telecommunication systems or headers for ultrahigh-speed burst optical transmission. An example of a saturable mode-locked soliton generator and the pulse shape is shown in Figure 9.2a through c.

## REFERENCES

1. D. Y. Tang, H. Zhang, L. M. Zhao, and X. Wu, Observation of high-order polarization-locked vector solitons in a fiber laser, *Phys. Rev. Lett.*, 101, 153904, 2008.
2. A. Rosenthal and M. Horowitz, Bragg-soliton formation and pulse compression in a one-dimensional periodic structure, *Phys. Rev. E*, 74, 066611, 2006.
3. Y. P. Shapira and M. Horowitz, Optical AND gate based on soliton interaction in a fiber Bragg grating, *Opt. Lett.*, 32, 1211–1213, 2007.
4. P. Russell, Photonic crystal fibers, *Science*, 299, 358–362, 2003.
5. H. Bartelt, A. W. Lohmann, and B. Wirnitzer, Phase and amplitude recovery from bispectra, *Appl. Opt.*, 23, 3121–3129, 1984.
6. L. N. Binh and N. D. Nguyen, Nonlinear photonic pre-processing bi-spectrum optical receivers for long haul optically amplified transmission systems, in *2010 IEEE International Conference on Communication Systems (ICCS)*, Singapore, 2010, pp. 630–634.
7. R. Soref. and B. Bennett, Electrooptical effects in silicon, *IEEE J. Quant. Electron.*, 23(1), 123–129, 1987.

8. M. J. R. Heck, J. F. Bauters, M. L. Davenport, J. K. Doylend, S. Jain, G. Kurczveil, S. Srinivasan, Y. Tang, and J. E. Bowers, Hybrid silicon photonic integrated circuit technology, *IEEE J. Sel. Top. Quant. Electron.*, 9(4), 1–17, 2013.

9. S. Srinivasan, A. Arrighi, M. J. R. Heck, J. Hutchinson, E. Norberg, G. Fish, and J. E. Bowers, Harmonically mode-locked hybrid silicon laser with intra-cavity filter to suppress supermode noise, *IEEE J. Sel. Top. Quant. Electron.*, 20(4), 8–15, July–August 2014.

10. B. Jalali, and S. Fathpour, Silicon photonics, *IEEE J. Lightwave Technol.*, 24(12), 4600–4615, 2006.

11. L. Vivien and L. Pavesi, Eds., *Handbook of Silicon Photonics*, Series in Optics and Optoelectronics, CRC Press, Boca Raton, FL, April 26, 2013.

# Appendix A: Generic Mathematical Aspects of Nonlinear Dynamics

Nonlinear dynamics have attracted significant research interests and have been applied extensively in the generation of solitonic sequences and multi-bound solitons. This appendix gives some mathematical details for the generic nonlinear dynamic equations and simplified solutions whereby one can generate these nonlinear phenomena in practice.

It is appreciable that nonlinear dynamics can happen due to the nonlinear variations of parameters of physical systems or subsystems in which partial feedback of the system's outputs reaches an undetermined state or multiple states simultaneously at a particular instant. For example, in a system consisting of energy storage elements (e.g., inductors and capacitors or an optical recirculating resonant loop), the rate of charging and discharging of electrons or inducing currents through or across these elements or the optical energy or fields recirculating in the loop determine the states of dynamics and stability of the systems, so when both energy storage elements compete for the charges available in the system, chaotic states or bistability may occur depending on the rates of storage or release of energy. Note that for optical fields the feedback is negative only when the phase of the feedback field is equal to pi of that of the incident field. Hence, one must know exactly the phase of the optical fields in the optical systems.

The rates of competition of limited resources from different subsystems of a generic system would lead to chaotic states. This could happen in natural systems as well as in engineering systems, in particular nonlinear photonic ones. If there are more than one storage element or subsystems in energy storage systems, competition for energy would occur, and when they try to extract from or deposit their energy into each other, chaotic and/or bifurcation phenomena would occur. Indeed this energy competition is usually called coupled oscillating system. The system can thus behave in a nonlinear manner, hence the term nonlinear systems.

The most commonly known energy storage systems are those containing capacitors and inductors as charge and discharge competing elements, and photon storages with positive feedback into the resonance subsystem such as fiber lasers. Thus, higher-order differential equations can be employed to represent the dynamics of the evolution of the amplitude of the current or voltages or field amplitudes of the lightwaves in such systems.

Thus this appendix selects to treat the nonlinear systems in terms of fundamental principles and associated phenomena as well as their applications in signal processing in contemporary optical systems for communications and/or laser systems with a touch of mathematical representation of nonlinear equations that provide some insight into the nonlinear dynamics at different phases.

The basic mathematical dynamics of nonlinear systems are thus introduced in this appendix.

## A.1  INTRODUCTORY REMARKS

The rates of competition of limited resources from different subsystems of a generic system would lead into chaotic states. This could happen in natural systems as well as in engineering systems, in particular in nonlinear photonic ones. In energy storage systems, if there are more than one storage element or subsystems, competition for energy would occur and when they try to extract from or deposit their energy into each other, chaotic and/or bifurcation phenomena would occur. Indeed this energy competition is usually called coupled oscillating system. The system can thus behave in a nonlinear manner, hence the term nonlinear systems.

The most commonly known energy storage systems are those containing capacitors and inductors as a charge and discharge competing elements, and photon storages with positive feedback into the resonance subsystem such as fiber lasers. Thus higher-order differential equations can be employed to represent the dynamics of the evolution of the amplitude of the current or voltages or field amplitudes of the lightwaves in such systems.

We have emphasized that nonlinear dynamics in nature have been considered for the description of intricate patterns formed from simple shapes and the repeated application of dynamic procedures by introducing physics and mathematics. The fundamental rules underlying the shapes and the structures have led to the dynamic study of the nonlinear systems. Especially, modern nonlinear dynamics have focused on analytic solutions of the dynamic equations, and the nonlinearity is determined based on the principles of linear superposition under some approximations such as perturbation techniques. However, fundamental theoretical problems that arise in physics and mathematics, as well as in engineering systems, can be solved with the assistance of digital computing techniques.

Solitons and chaos are briefly introduced, and their potential applications to engineering are introduced and described in the field of nonlinear optics in this book. In the chapters, soliton and soliton lasers are treated in detail, especially their generation, dynamic behavior, and transmission over guided optical media.

Surprisingly, simple nonlinear systems are found to have chaotic solutions, which remain within a bounded region. In other words, the nonlinearity has been positively considered, and the result has been applied to the analysis and the design in engineering and technology. Thus, nonlinear dynamics in nature have played a key role in physics, mathematics, and engineering and would be fundamental tools for many branches of future research. This appendix gives a number of mathematical relationships of the principal functions in nonlinear system and states of the systems in stable chaotic, bifurcation, attractors, and repellers in such systems.

## A.2    NONLINEAR SYSTEMS: PHASE SPACES AND DYNAMICAL STATES

Nonlinear dynamical systems and chaotic phenomena are used widely in physics and engineering. This section describes some mathematical tools for the analysis of dynamical systems, especially the main pathways to chaos.

### A.2.1    PHASE SPACE

Consider a dynamical system represented by a differential equation subject to some initial conditions, for example, a well-known damped oscillation of a mass–spring system as a mechanical system is governed by the differential equation and initial conditions at $t_0$ given as

$$\frac{d^2x}{dt^2} = f(x, x') = -\frac{k}{m}x - \frac{c}{v}\frac{dx}{dt}$$

$$x(t_0) = x_0 \quad \text{and} \quad \frac{dx}{dt}\bigg|_{t=t_0} = v_0$$

(A.1)

where
   $m$ is the mass of the system
   $k$ is the stiffness of the spring
   $c$ is the damping coefficient

Equation A.1 is *autonomous* because its right-hand side (RHS) does not depend on the time variable. The prime indicates the order of differentiation. In general, one can deduce an $n$th order *autonomous* ordinary differential equation (ODE) to a system of $n$ first-order DEs, and then applying the fourth-order Runge–Kutta method to obtain the final solutions of the system.

Let a sinusoidal driving force $F$ excite on the system, then the DE (Equation A.1) can be rewritten as

$$\frac{d^2x}{dt^2} = g(t, x, x') = -\frac{k}{m}x - \frac{c}{v}\frac{dx}{dt} + F\cos\omega t$$

(A.2)

The RHS of Equation A.2 depends explicitly on time. Equation A.2 is *nonautonomous* and can be reduced to a system of $(n+1)$ *first-order DE* by replacing the variable $t$ by another dummy variable $z$. For example, Equation A.1 can be solved numerically by transforming it into two first-order DEs given by

$$\frac{dx}{dt} = v; \quad x(t_0) = x_0$$

$$\frac{dv}{dt} = -\frac{k}{m}x - \frac{c}{v}v; \quad v(t_0) = v_0$$

(A.3)

where

$x(t)$ and $v(t)$ represent the displacement and velocity, respectively

the prime indicates differentiation

In order to examine the behavior of the mass–spring system, a variation of the velocity versus displacement, $v(x)$, is used instead of the time-dependent displacement, $x(t)$. The $x-v$ coordinate system is termed as the phase space, or a two dimensional (2-D) phase state, of the system. Each coordinate represents a variable of the system.

The system can thus be represented by an $n$th-order autonomous differential equation leading to an $n$-dimensional phase space. The solutions of Equation A.3 are subject to a set of initial conditions under which $x(t)$ and $v(t)$ are determined by a point $P$ in the phase space. If time $t$ varies, then the locus of $P$ follows a trajectory in the phase space, namely, *the phase space trajectory* or *phase curve*. Furthermore, a set of trajectories can form a *phase portrait* illustrating the dynamical behavior of the system.

In general, there are three kinds of trajectories in the phase space: fixed points, closed trajectories, and nonclosed trajectories. Figure A.1 depicts a trajectory of the damped mass–spring system in the phase space operating under the following parameters: $m=1$ kg, $k=1.1$ N/m, $c=0.1$ kg/s, $x_0=1$, and $v_0=0$, $t=[0, 20\pi]$. The energy system dissipates its energy in air so the trajectory is a spiral curve toward the origin $O(0, 0)$; the point $(0, 0)$ is thus a stable attractor. Figure A.2 shows the phase portrait of a periodic mass–spring system in the phase space under different initial conditions given by $m=1$ kg, $k=1.1$ N/m, $c=0$, $x_0=[1, 0.5, 0.3, 0.1]$ and $v_0=0$, and $t=[0, 20\pi]$.

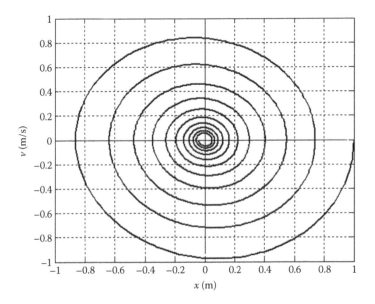

**FIGURE A.1**    Trajectory of a damped mass–spring in phase space with an attractor located at $(0, 0)$.

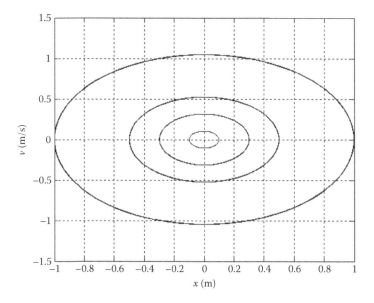

**FIGURE A.2**   Phase portrait of harmonic mass–spring in phase space.

Thus one can deduce a nonintersecting theorem as: "Two distinct state space trajectories can neither intersect (in a finite period of time), nor can a single trajectory cross itself at a phase space at a later time."

## A.2.2  CRITICAL POINTS

The critical point or fixed point in the phase space is the point that corresponds to the equilibrium of the dynamical system. The fixed point can also be the equilibrium point, which is an important position because a trajectory starts at a fixed point and then it stays at this point and the characteristic of the fixed point shows the behavior of trajectories on both its sides.

There are three kinds of fixed points that are the attractors, the repellers, and the saddle points. Attractors or nodes or sinks are stable fixed points that attract neighboring trajectories. Repellers or sources are unstable fixed points that repel neighboring trajectories. Saddle points are semi-stable fixed points that attract neighboring trajectories on one side and repel those on the other side. Fixed points can be identified without much difficulty. Let us consider the fixed points in one dimensional (1-D) phase space, which is just a line; we can then extend to those in 2-D phase space.

### A.2.2.1  Fixed Points in 1-D Phase Space

The dynamic equation in 1-D phase space can be written as

$$\frac{dx}{dt} = f(x) \tag{A.4}$$

where the dash indicates the differentiation of the position $x$ with respect to time. The fixed points of Equation A.4 are the locations in the phase space with their value $x^*$ satisfying the condition

$$\frac{dx^*}{dt} = f(x^*) = 0 \tag{A.5}$$

Thus to find the fixed points, one can solve the equation $f(x)=0$ and the solutions $x^*$ are the positions of the fixed points. Let us consider the point $x^r$ on the right side and the nearby fixed point $x^*$. If $f(x^r)>0$ so that $dx/dt>0$, the trajectory starts at $x^r$ and moves away from the fixed point; if $f(x^r)<0$ so that $dx/dt<0$, the trajectory starts at $x^r$ and moves toward the fixed point and vice versa for the point $x^l$ on the left side and in the neighborhood of the fixed point. If the trajectories on both sides of the fixed point move toward the fixed point, then it can now be called the *attractor*. On the other hand, if the trajectories on two sides of the fixed point move away from the fixed point, it is then called the *repeller*. If the trajectory on one side of the fixed point moves toward the fixed point and vice versa for the trajectory on the other side, the fixed point is classified as the *saddle point*.

Alternatively, we can use an eigenvalue, $\lambda$, *which is* a characteristic value, to distinguish different types of fixed points, which can be obtained by setting

$$\lambda = \frac{df(x)}{dx}\bigg|_{x^*} \tag{A.6}$$

We can then approximate this by using the Taylor series to distinguish the kind of fixed point. Let $\zeta = x - x^*$ and keep the first two terms of Taylor series expansion [1] in the neighborhood of $x^*$; we then have

$$f(\zeta + x^*) = f(x^*) + \zeta f'(x^*) + \Theta(\zeta^2) \tag{A.7}$$

On the RHS of this equation, the term $f(x^*)=0$ since $x^*$ is defined as the fixed point. So

$$\zeta' = f'(x^*)\zeta \tag{A.8}$$

The solution of Equation A.8 is

$$\zeta(t) = Ae^{\lambda t} \tag{A.9}$$

where $\zeta$ is the quantity that measures the distance between the trajectory and the fixed point. The exponential coefficient $\lambda$ influences the trajectory as follows:

(1) If $\lambda < 0$, the trajectory moves toward the fixed point exponentially so the fixed point is the attractor; (2) if $\lambda > 0$, the trajectory moves away from the fixed point exponentially so the fixed point is the repeller; and (3) if $\lambda = 0$, the fixed point may be the saddle point or the attractor or the repeller. In this case, we must compute the second derivative of $f(x)$. If $f''(x)$ has the same sign on both sides of the fixed point, it is the saddle point; if $f''(x) > 0$ on the left side of the fixed point and $f'(x) < 0$ on the right side of the fixed point, it is the attractor, otherwise the fixed point is the repeller.

As an example of the attractor and the repeller, consider the following equation:

$$\frac{dx}{dt} = f(x) = x^2 - 1 \tag{A.10}$$

$x^* = 1$ and $x^* = -1$ are the fixed points since $f(x^*) = 0$; the eigenvalue at $x^* = 1$ is $\lambda = 2 > 0$ so the fixed point at $x = 1$ is the repeller and its eigenvalue is located at $x^* = -1$ with $\lambda = -2 < 0$, and hence the fixed point at $x = -1$ is the attractor (Figure A.3). The saddle point of the equation

$$\frac{dx}{dt} = f(x) = x^2 \tag{A.11}$$

is depicted in Figure A.4 in which the saddle point is located at the origin $O(0,0)$.

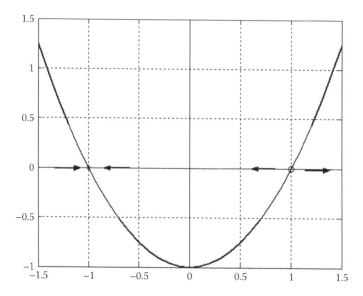

**FIGURE A.3**   Attractor at $(-1, 0)$. Repeller located at $(1, 0)$.

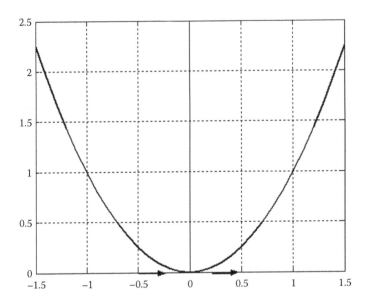

**FIGURE A.4**   Saddle point located at $O(0, 0)$.

## A.2.2.2   Fixed Points in 2-D Phase Space

We can now extend the fixed points in one-dimensional phase space to two-dimensional (2-D) phase space by considering the system of equations

$$\frac{dx}{dt} = f(x, y)$$

$$\frac{dy}{dt} = g(x, y)$$

(A.12)

The phase space is $(x - y)$ plane, and the behavior of the system is represented by the trajectories in the phase space. Let $(x_c, y_c)$ be the coordinate of the fixed points, then $(x_c, y_c)$ satisfies the following equations:

$$f(x_c, y_c) = 0$$

$$g(x_c, y_c) = 0$$

(A.13)

The first step is to find the positions of fixed points by solving the system of Equation A.13. The types of fixed points can be found by detecting the characteristic values of the fixed points. The functions $f$ and $g$ depend on the two variables, thus the characteristic values depend on their partial derivatives. Similar to the 1-D case, we are concerned with only the characteristic values in the $x$ and $y$ directions by setting

$$\lambda_x = \left.\frac{\delta f}{\delta x}\right|_{(x_c, y_c)}$$

$$\lambda_y = \left.\frac{\delta g}{\delta y}\right|_{(x_c, y_c)}$$

(A.14)

Thus we can extend the nature of the eigenvalues in 1-D to classify the types of fixed points in 2-D phase space. Consider the two distinct cases described next.

### A.2.2.2.1  $\lambda_x$ and $\lambda_y$ as Real Numbers

1. $\lambda_x < 0$ and $\lambda_y > 0$: Fixed points as attractors.
   Considering the equations

$$\frac{dx}{dt} = -x$$

$$\frac{dy}{dt} = -2y$$

(A.15)

whose solutions can be found as

$$x = C_1 e^{-t} \quad \text{and} \quad y = C_2 e^{-2t}$$

(A.16)

we can eliminate the variable $t$ in both $x$ and $y$ to give

$$y = Kx^2, \quad K = \frac{C_2}{C_1^2}$$

(A.17)

The fixed point is at the origin $O(0, 0)$ and the $\lambda_x = -1 < 0$ and $\lambda_y = -2 < 0$, so $O(0, 0)$ is the attractor. We can check this property by inspecting the trajectories. The phase space trajectories are a set of parabolas. For $K = 0$ the trajectory is the $x$-axis, when $t \to 0$, $x \to 0$, and $y \to 0$. So all trajectories move toward the origin—the attractor—as indicated in Figure A.5.

2. $\lambda_x > 0$ and $\lambda_y > 0$: Fixed points as repellers
   Consider the following equation:

$$\frac{dx}{dt} = x$$

$$\frac{dy}{dt} = 2y$$

(A.18)

The fixed point is also at the origin $O(0, 0)$ and $\lambda_x = 1 > 0$ and $\lambda_y = 2 > 0$, so $O(0, 0)$ is the repeller. The phase space trajectories are the same as the trajectories shown in Figure A.5, but moving away from the origin, the *repeller* (Figure A.6).

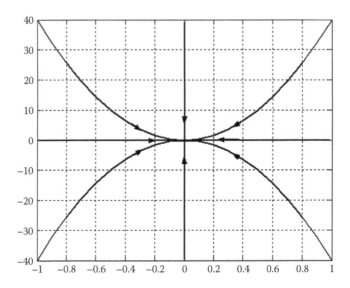

**FIGURE A.5**    Attractor in 2-D phase space.

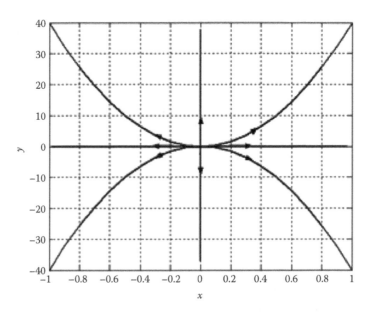

**FIGURE A.6**    Repeller type trajectory in 2-D phase space $(x, y)$.

3. $\lambda_x > 0$ and $\lambda_y < 0$: Fixed points as saddle points

When $\lambda_x > 0$ and $\lambda_y < 0$, we have the opposite situation and the fixed point now becomes the saddle point. For an example, consider the following set of equations:

$$\frac{dx}{dt} = x$$

$$\frac{dy}{dt} = -y$$

(A.19)

Eliminating the variable $t$ in the solution of Equation A.19, we have

$$y = \frac{K}{x}$$

(A.20)

The phase space trajectories shown in Figure A.7 are a set of hyperbolae. When $K=0$ the trajectory is $x$-axis, when $t \rightarrow 0$, $x \rightarrow \infty$ and $y \rightarrow 0$, so the origin is identified as the saddle point.

*A.2.2.2.2   $\lambda_x$ and $\lambda_y$ as Complex Conjugates*

Suppose

$$\lambda_{x,y} = \alpha + i\beta$$

(A.21)

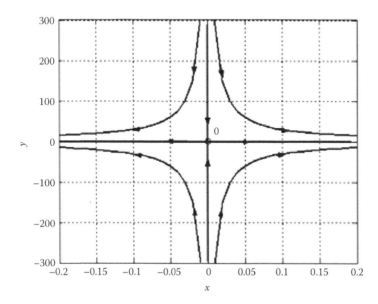

**FIGURE A.7**   Saddle point type trajectory in 2-D phase space.

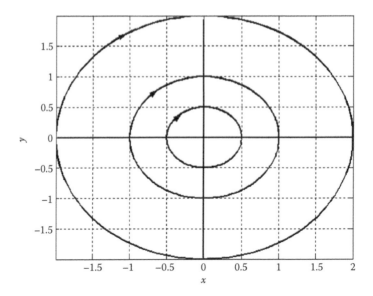

**FIGURE A.8** Central attractor.

Then the solutions of Equation A.19 can take possible solutions of

$$x(t) = e^{\alpha t} \cos \beta t \quad \text{and} \quad y(t) = e^{\alpha t} \sin \beta t \tag{A.22}$$

- *If $\alpha = 0$:* The solution $x(t) = \cos\beta t$ and $y(t) = \sin\beta t$, the trajectories in the phase space are circles and the fixed point at the origin is the central attractor (Figure A.8).
- *If $\alpha \neq 0$ and $\beta \neq 0$:* The fixed point is at the origin and is called the focal point; the solutions are given in Equation A.21 and the phase space trajectories oscillate with an angular frequency $\beta$. Its amplitudes can vary exponentially, the reason why the trajectories move spirally to the origin (attractor) ($\alpha < 0$) or out of the origin (repeller) ($\alpha > 0$) (Figure A.9).

### A.2.2.3 Limit Cycles

In 2-D phase space or higher dimensions, consider a new fixed point, which is a limit cycle defined as an isolated closed loop trajectory in phase space. The neighboring trajectories are opened loops moving spirally in or out of the limit cycle when $t \to \infty$ or $t \to -\infty$, respectively. If the neighboring trajectories on both sides move toward the limit cycle, it becomes a *stable limit cycle*; otherwise the limit cycle is an *unstable limit cycle*. If the neighboring trajectories at either side of the limit cycle move toward the limit cycle and the others move away, then the limit cycle is a *semi-stable limit cycle*. Limit cycles only appear in nonlinear dynamical systems. There are also closed loops in linear systems but they are not isolated, so these closed loops are not limit cycles, as shown in Figure A.2.

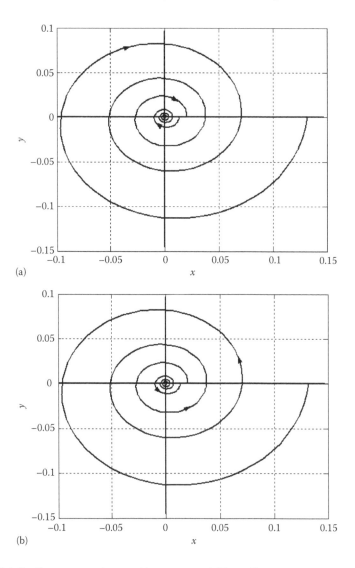

**FIGURE A.9**  Focus type trajectory: (a) attractor and (b) repeller.

Figure A.10 represents the stable limit cycle in phase space $(x_1 - x_2)$, where $x_1$ and $x_2$ are the solutions of the equations

$$\frac{dx_1}{dt} = \mu x_1 + x_2 - x_1 \left( x_1^2 + x_2^2 \right)$$

$$\frac{dx_2}{dt} = -x_1 + \mu x_2 - x_1 \left( x_1^2 + x_2^2 \right)$$

(A.23)

where $\mu$ is a parameter and the system has a stable limit cycle for $\mu > 0$, as in Figure A.10.

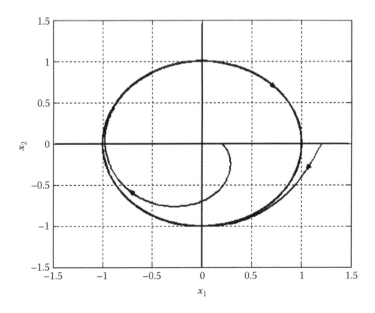

**FIGURE A.10**   Limit cycle.

This section can thus be concluded by stating the *Poincaré–Bendixson* theorem*: "Supposing that there is a bounded region that contains no fixed point, if a trajectory is confined in this region then only either of the following two possible cases can be true: (1) the trajectory is a closed orbit as $t \to \infty$ (2) the trajectory spirals toward a close orbit as $t \to \infty$."

This theorem implies that if a trajectory is confined in a closed region that does not contain any fixed point, then at last this trajectory spirals to a closed orbit. For a higher-order dimensional system ($n \geq 3$) the trajectories may wander forever in a bounded region. The trajectories are attracted to a complex geometrical object, the strange attractor. This is the characteristic of a chaotic phenomenon. So the chaotic trajectories cannot appear in 2-D phase spaces.

## A.3   BIFURCATION

As described earlier, a dynamical system can be represented by a differential equation. If the equation depends on only one or more parameters, the fixed points of the system can be altered accordingly. These qualitative variations of the

---

* The Poincaré–Bendixson theorem is a statement about the long-term behavior of orbits of continuous dynamical systems on the plane. The original theorem is as follows: "Given a differentiable real dynamical system defined on an open subset of the plane, then every non-empty compact ω-limit set of an orbit, which contains only finitely many fixed points, is either (1) a fixed point, (2) a periodic orbit, or (3) a connected set composed of a finite number of fixed points together with homoclinic and heteroclinic orbits connecting these."

dynamical system are termed as bifurcation. In this section, we briefly present here the bifurcation in one and 2-D phase spaces. Further, we restrict it to the most common types, such as the pitchfork, the saddle-node, the transcritical, and the Hopf bifurcations [1–3].

## A.3.1  PITCHFORK BIFURCATION

This bifurcation occurs most commonly in symmetrical systems. As an example of pitchfork bifurcation, consider the following equation;

$$\frac{dx}{dt} = f(x,\mu) = \mu x - x^3 \tag{A.24}$$

where $\mu$ is a real parameter that can take a value that is positive, negative, or zero. This equation is invariant when $x$ is replaced by $-x$. To study the bifurcation of this system, we have to find the fixed points and their characteristics.

- The positions of fixed points can be determined by

$$\frac{dx}{dt} = f(x,\mu) = \mu x - x^3 = 0 \tag{A.25}$$

So there are three fixed points at $x=0$, $x = \sqrt{\mu}$, and $x = -\sqrt{\mu}$.
- The characteristics of the fixed points are governed by

$$\lambda = \frac{df}{dx} = \mu - 3x^2 \tag{A.26}$$

When $\lambda < 0$, the fixed point is stable, the *attractor*. However, if $\lambda > 0$, then the fixed point is unstable, the *repeller*.
- If $\mu < 0$, there is only one fixed point at the origin ($x_1 = 0$) and it is stable ($\lambda = \mu < 0$).
- If $\mu > 0$, there are three cases: (1) the fixed point at $x_1 = 0$ is unstable ($\lambda = \mu > 0$), (2) the fixed point at $x_2 = \sqrt{\mu}$ is stable ($\lambda = -2\mu < 0$), and (3) the fixed point at $x_3 = -\sqrt{\mu}$ is stable ($\lambda = -2\mu < 0$).

We can thus solve the system of Equations A.25 and A.26 to obtain $x=0$ and $\mu=0$. The origin ($x=0$, $\mu=0$) is the *bifurcation* point.

The diagram of the pitchfork bifurcation is shown in Figure A.11, the solid lines indicate the positions of stable fixed points and the dashed line indicates the positions of unstable fixed points. This case is the supercritical pitchfork bifurcation because the bifurcation branches are stable. The other case is called the subcritical pitchfork bifurcation when the bifurcation branches are unstable.

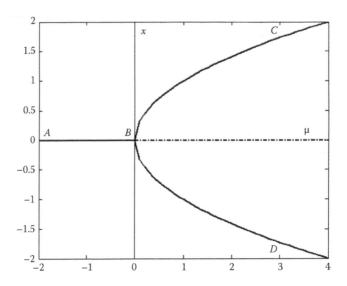

**FIGURE A.11**   Pitchfork bifurcation diagram. *Solid lines:* stable state equilibrium. *Dashed line:* unstable state equilibrium.

### A.3.2   SADDLE-NODE BIFURCATION

This bifurcation is the basic mechanism. In this case, the fixed points are created or destroyed as one or more parameters of the system change. As an example of the saddle-node bifurcation, consider

$$\frac{dx}{dt} = f(x,\mu) = \mu + x^2 \tag{A.27}$$

where $\mu$ is a parameter that can take values positive, negative, or zero. The fixed point and its characteristic are determined by

$$\frac{dx}{dt} = f(x,\mu) = \mu + x^2 = 0 \tag{A.28}$$

and

$$\lambda = \frac{df}{dx} = 2x \tag{A.29}$$

- If $\mu < 0$, the system has two fixed points, the stable fixed point $A$ at $x_1 = -\sqrt{-\mu}$ ($\lambda < 0$) and the unstable fixed point $B$ at $x_2 = \sqrt{-\mu}$ ($\lambda > 0$) (Figure A.12a).

- If $\mu \to 0$, the parabola moves up and the two fixed points $A$ and $B$ move toward each other. When $\mu=0$, the two fixed points $A$ and $B$ combine and become a half-stable fixed point at the origin $O$ ($x=0$) (Figure A.12b).
- If $\mu>0$, the system does not have any fixed point (Figure A.12c).

The system of Equations A.28 and A.29 can be solved to give solutions $x=0$ and $\mu=0$. So the bifurcation point is located at the origin ($x=0$, $\mu=0$).

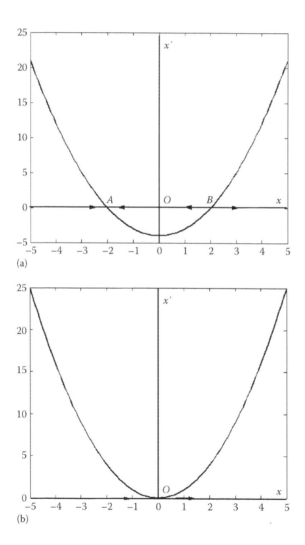

(a)

(b)

**FIGURE A.12**   Saddle-node bifurcation process in phase space: (a) $\mu<0$, (b) $\mu=0$.
(*Continued*)

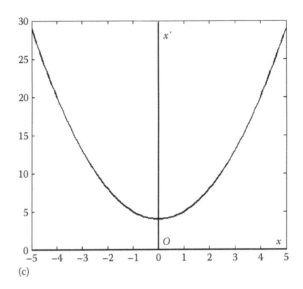

(c)

**FIGURE A.12 (*Continued*)** Saddle-node bifurcation process in phase space: (c) $\mu > 0$.

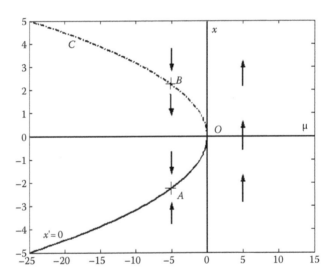

**FIGURE A.13** Saddle-node bifurcation diagram.

We can now depict the curves of $dx/dt = 0$ in the space $(\mu, x)$; the curves $(C)$ of function $\mu = -x^2$ represent equilibrium positions of $x$ on $\mu$ (Figure A.13). This is called the saddle-node bifurcation diagram. This is the common way to illustrate the saddle-node bifurcation, which can also be called a fold bifurcation or a turning-point bifurcation because the points $x = 0$ and $\mu = 0$ are turning points.

## A.3.3  TRANSCRITICAL BIFURCATION

In this bifurcation, the fixed point of the system always exists but its characteristic changes as one or more parameters change. As an example of the transcritical bifurcation, consider the following equation:

$$\frac{dx}{dt} = f(x,\mu) = x(\mu - x) \tag{A.30}$$

where $\mu$ is a parameter that can take values positive, negative, or zero. The fixed point and its characteristic can be determined by

$$\frac{dx}{dt} = x(\mu - x) = 0 \tag{A.31}$$

and

$$\lambda = \frac{df}{dx} = \mu - 2x \tag{A.32}$$

- If $\mu < 0$, the system has two fixed points, the stable fixed point at $x_1 = 0$ ($\lambda < 0$) and the unstable fixed point $x_2 = -\mu$ ($\lambda > 0$) (Figure A.14a).
- If $\mu = 0$, the system has a half-stable fixed point at the origin $O$ ($x = 0$) ($\lambda = 0$) (Figure A.14b).
- If $\mu > 0$, then the system has solutions of two fixed points, the unstable fixed point at $x_1 = 0$ ($\lambda > 0$), and the stable fixed point $x_2 = \mu$ ($\lambda < 0$) (Figure A.14c).

Combining three cases shown in Figure A.14a through c we see that the fixed point located at the origin ($x = 0$) always exists, but its stability changes as the parameter $\mu$ changes. So this is the case of transcritical bifurcation.

The solution of the system of Equations A.31 and A.32 gives the bifurcation point at the origin ($x = 0$ and $\mu = 0$). Thus we also use the space ($\mu$, $x$) to represent the transcritical bifurcation diagram as shown in Figure A.15.

## A.3.4  HOPF BIFURCATION

Hopf bifurcation occurs when a limit cycle is taken from a fixed point as one or more parameters of the system vary, and it takes place only in the phase space of a higher-order dimension ($\geq 2$). In this section we present the Hopf bifurcation in 2-D phase space. As an illustration of the Hopf bifurcation, consider the following set of equations:

$$\frac{dx_1}{dt} = \mu x_1 + x_2 - x_1\left(x_1^2 + x_2^2\right)$$

$$\frac{dx_2}{dt} = -x_1 + \mu x_2 - x_1\left(x_1^2 + x_2^2\right) \tag{A.33}$$

where $\mu$ is a real parameter. This system is simpler when transformed into polar coordinates $(r, \theta)$ by setting $x_1 = r \cos \theta$ and $x_2 = r \sin \theta$ with $r^2 = x_1^2 + x_2^2$ as given in Equation A.33

$$r' = r(\mu - r^2)$$

$$\theta' = -1$$

(A.34)

Equation A.34 shows that there are two fixed points at $r_1 = 0$, $r_2 = \sqrt{\mu}$ because the solution $r_2 = -\sqrt{\mu}$ can be eliminated by setting $r > 0$ and the eigenvalues $\lambda = \mu \pm i$.

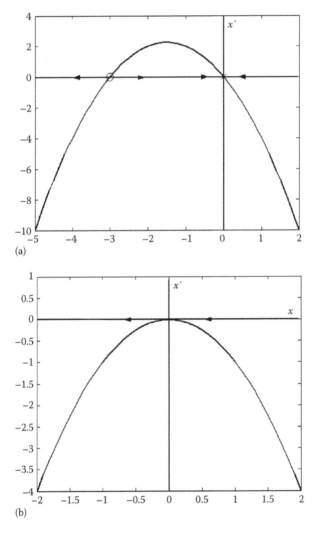

(a)

(b)

**FIGURE A.14** Transcritical bifurcation process in the phase space: (a) $\mu < 0$, (b) $\mu = 0$.

(*Continued*)

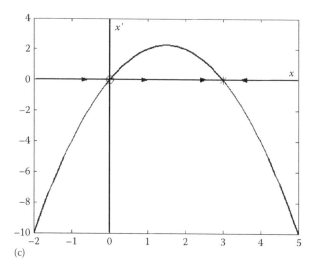

(c)

**FIGURE A.14 (*Continued*)**   Transcritical bifurcation process in the phase space: (c) $\mu > 0$.

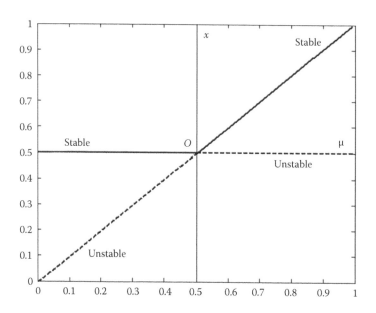

**FIGURE A.15**   Transcritical bifurcation diagram.

If $\mu < 0$, there is only a stable fixed point at the origin that is the focus; the trajectories spiral the origin (Figure A.16a). If $\mu = 0$, the origin, that is, a center, is also a stable fixed point, the trajectories also spiral the origin slowly (Figure A.16b). Then if $\mu > 0$, the fixed point at the origin ($r = 0$) is unstable and the fixed point at $r = \sqrt{\mu}$ is stable, the trajectories become a stable limit cycle (Figure A.16c).

We can see that when $\mu<0$, the origin is stable and when $\mu>0$ the origin is unstable and a stable limit cycle appears with $r=\sqrt{\mu}$. So the Hopf bifurcation occurs at $\mu=0$ and it is called the supercritical Hopf bifurcation because the limit cycle is stable. The diagram of the Hopf bifurcation is depicted in Figure A.17. We also have the subcritical Hopf bifurcation when the limit cycle is unstable. Hopf bifurcation is also called as the oscillatory bifurcation.

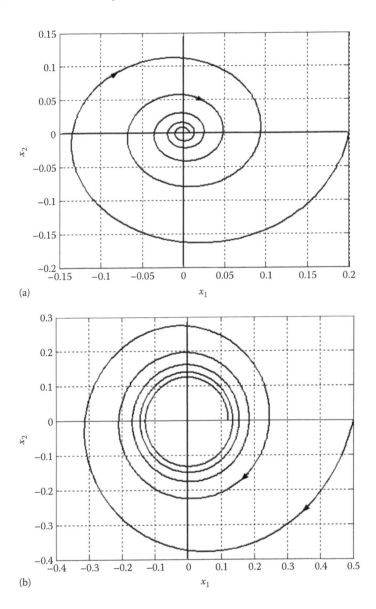

**FIGURE A.16**  The birth of the limit cycle of Hopf bifurcation: (a) $\mu=-0.1<0$, (b) $\mu=0$.

*(Continued)*

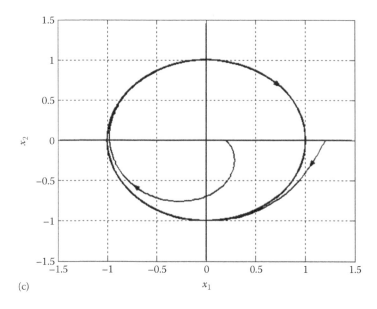

(c)

**FIGURE A.16 (Continued)**    The birth of the limit cycle of Hopf bifurcation: (c) $\mu = 0.3 > 0$.

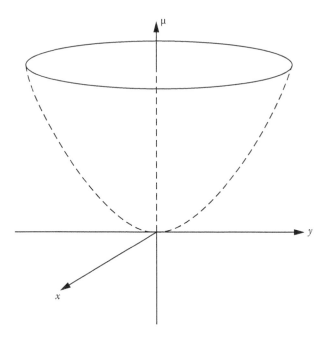

**FIGURE A.17**    Hopf bifurcation diagram.

Bifurcations play an important role, because in some dynamical systems, when the control parameter increases for a long time, bifurcations may appear several times. This results in the wandering unpredictability of the system's trajectories in the phase space. This is the chaotic phenomenon.

## A.4   CHAOS

### A.4.1   DEFINITION

There is no exact definition of chaos but the commonly accepted definition [3] is as follows: "Chaos is aperiodic long term behavior in a deterministic system which is strongly dependent on its initial conditions."

According to this definition, a chaotic phenomenon only appears when the system evolves for a long time and during which there are no intersections of different aperiodic trajectories produced by the system, which is represented by a DE associated with some certain deterministic initial conditions. The system is sensitive to its initial conditions, meaning that there are two neighboring trajectories beginning at the distance $d_0$ and the time-dependent distance is given by $d(t) = d_0 e^{\lambda t}(\lambda > 0)$, where $\lambda$ is known as the Lyapunov exponent, so the two neighboring trajectories can move away with respect to each other exponentially.

Lorenz in 1963 [3] conducted the first experiment on chaos. His model is a simplified model of the fundamental Navier–Stokes equation of the dynamics of fluids given by

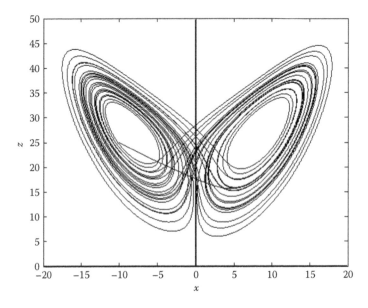

**FIGURE A.18**   Variation of $z$ with respect to $x$ as a strange attractor.

$$\frac{dx}{dt} = p(y - x)$$

$$\frac{dy}{dt} = rx - y - xz \tag{A.35}$$

$$\frac{dz}{dt} = xy - bz$$

where
  $p$ is the Prandtl number
  $r$ is the Rayleigh number
  $b > 0$ is an unknown parameter

Equation A.35 can be solved numerically to find that the system was sensitive to its initial conditions, that the system trajectories can neither settle at any point nor repeat again but always follow a spiral, that and no intersection of trajectories would occur. He then termed this image as the Lorenz attractor (Figure A.18) ($p = 10$, $r = 28$, $b = 8/3$). This is a chaotic behavior and termed as the attractor of a chaotic system, which is also called a strange attractor, a chaotic attractor, or a fractal attractor, as shown in Figure A.18.

## A.4.2 ROUTES TO CHAOS

The trajectories of nonlinear systems may move from a regular to a chaotic route pattern when a control parameter changes. This is the transitional route to chaos. There are several types of routes to chaos [2]. In this section, we review only some of the main routes. Electrical circuits are illustrated in this section so as to illustrate the concepts of energy storage and dissipating as well as nonlinear resistor with two slopes $I$–$V$ characteristics. Analogous to the electrical circuits, we can see that an optical ring can act as the energy storage element, that the loss of energy in the ring is equivalent to the resistance, and that the coupling of the optical coupler can be seen as adjustable resistance. An optical saturable absorber is equivalent to a nonlinear resistor.

### A.4.2.1 Period-Doubling

As described in the previous sections, the limit cycle originates from the bifurcation that relates to a fixed point. As a control parameter changes, the limit cycle would become unstable and there is an onset of periodic doubling limit cycles. When the control parameter continues to change, the period-doubling becomes unstable and a period-four cycle starts to appear. The process will continue to infinite-period cycle, and then trajectories follow the chaotic behavior.

As an illustration of period-doubling, consider Chua's circuit, a simple nonlinear circuit that leads to chaotic behavior.

Figure A.19a is Chua's circuit and Figure A.19b is the driving-point characteristic of the nonlinear resistor $N_R$ in Chua's circuit. Chua's equations are given as [4]

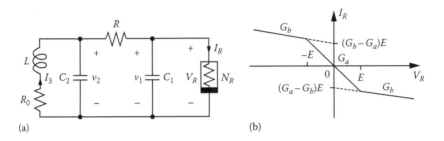

**FIGURE A.19**   (a) Chua's circuit [4]. (b) The driving-point characteristic of the nonlinear resistor $N_R$ in Chua's circuit [4]. Note the three linear sections creating nonlinearity.

$$\frac{dI_3}{dt} = -\frac{R_0}{L}I_3 - \frac{1}{L}V_2$$

$$\frac{dV_2}{dt} = \frac{1}{C_2}I_3 - \frac{G}{C_2}(V_2 - V_1)$$

$$\frac{dV_1}{dt} = \frac{G}{C_1}(V_2 - V_1) - \frac{1}{C_1}f(V_1)$$

(A.36)

$$= \begin{cases} \dfrac{G}{C_1}V_2 - \dfrac{G'_b}{C_1}V_1 - \left(\dfrac{G_b - G_a}{C_1}\right)E & \text{if } V_1 < -E \\[2mm] \dfrac{G}{C_1}V_2 - \dfrac{G'_a}{C_1}V & \text{if } -E \le V_1 \le E \\[2mm] \dfrac{G}{C_1}V_2 - \dfrac{G'_b}{C_1}V_1 - \left(\dfrac{G_a - G_b}{C_1}\right)E & \text{if } V_1 > E \end{cases}$$

in which the conductance of the elements is given as $G = 1/R$, $G'_a = G + G_a$ and $G'_b = G + G_b$.

Using Kennedy's values [4], $L = 18$ mH, $R_0 = 12.5\ \Omega$, $C_2 = 100$ nF, $G_a = -757.576$ $\mu$S, $G_b = -409.09\ \mu$S, $E = 1$ V, and $C_1 = 10$ nF. By changing the control parameter $G$, we have the phase portraits in Figure A.20. Figure A.20a shows a single-period limit cycle, $G = 530\ \mu$S. Figure A.20b shows a period-doubling limit cycle with $G = 537\ \mu$S and Figure A.20c is a quadrupling period limit cycle, $G = 539\ \mu$S. Figure A.20d depicts a spiral Chua's chaotic attractor with $G = 541\ \mu$S. Figure A.20 thus shows the sequence of dynamics leading to the doubling period and, consequently, the route to chaos.

## A.4.2.2   Quasi-Periodicity

The motion of a nonlinear dynamic system that associates with two frequencies is called the quasi-periodic motion and it appears when the ratio of two frequencies is not a ratio of integers. So the motion does not repeat itself exactly but is not chaotic.

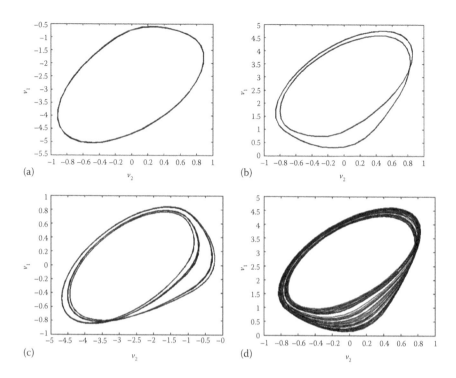

**FIGURE A.20**   Phase portraits: (a) $T$ period solution, $G=530$ μS, (b) $2T$ period solution, $G=537$ μS, (c) $4T$ period solution, $G=539$ μS, and (d) chaotic solution, $G=541$ μS.

The quasi-periodic route to chaos is connected to the Hopf bifurcation. In Section A.3, we considered the Hopf bifurcation that associates with the onset of the limit cycle, a periodic motion, from a fixed point. In some nonlinear dynamical systems, the first and then the second Hopf bifurcation appears when the control parameter changes. If the ratio of frequencies of the second and the first motion is not a rational ratio, the motion of the system is quasi-periodic. If the control parameter continues to change, a chaotic motion would appear. This route is called the Ruelle–Takens route.

Figure A.21 represented a quasi-periodic signal that is the voltage of the second capacitor in Chua's circuit using an inductor gyrator (see Figure A.23).

### A.4.2.3   Intermittency

The intermittency is the motion that is nearly periodic with irregular bursts appearing from time to time. The phenomenon appears pseudo-randomly and is not purely random because the system is represented by a deterministic equation. There can be two approaches to the intermittency: (1) the motion of system changes between the periodic motion and the chaotic motion and (2) the motion of system changes between the periodic motion and the quasi-periodic motion.

In some nonlinear dynamical systems, when the control parameters of the system change further and the irregular bursts appear frequently, the intermittency

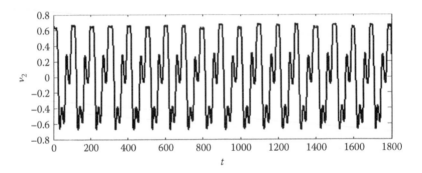

**FIGURE A.21**   Quasi-periodic signal.

behaviors become chaotic behaviors. We will look into the intermittency motion in detail in the next section.

### A.4.3   CHAOTIC NONLINEAR CIRCUIT

The Chua's circuit using the inductor gyrator was investigated by Muthuswamy et al. [5]. The schematic of the inductor gyrator is given in Figure A.22 and its impedance is given by

$$Z_{in} = R_L + j\omega R_L R_g C \tag{A.37}$$

In 1992, Murali and Lakshmanan [6] examined the effect of the external periodic excitation on Chua's circuit. We can now consider a Chua's circuit excited with an external driving sinusoidal excitation. This circuit uses an inductor gyrator (Figure A.22) instead of an inductor because the inductor is too cumbersome for the integrated circuit.

Figure A.23 shows the schematic of Chua's circuit using an inductor gyrator with a sinusoidal excitation. The state equations of the Chua's circuit are given as

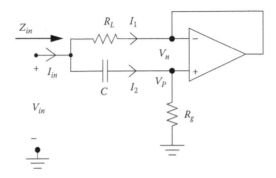

**FIGURE A.22**   Equivalent circuit of inductor gyrator [5] employing resistor, capacitor, and an op amp with shunt feedback.

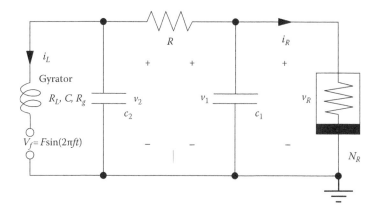

**FIGURE A.23**   Chua's circuit using inductor gyrator and sinusoidal excitation.

$$C_1 \frac{dv_1}{dt} = \frac{v_2 - v_1}{R} - i_R$$

$$C_2 \frac{dv_2}{dt} = \frac{v_2 - v_1}{R} - i_L - \frac{v_2 - i_L R_L - V_f}{R_g} \qquad \text{(A.38)}$$

$$C \frac{di_L}{dt} = \frac{v_2 - i_L R_L - V_f}{R_L R_g}$$

where $i_R = g(v_R) = g(v_1)$ is a piecewise-linear function given by

$$g(v_R) = G_b v_R + \frac{1}{2}(G_a - G_b)\left(\left|v_R + E\right| - \left|v_R - E\right|\right) \qquad \text{(A.39)}$$

$G_a$, $G_b$, and $E$ are shown on the curve depicted in Figure A.19b, $V_f = F \cdot \sin(2\pi ft - F)$, and $f$ is the frequency of the external sinusoidal excitable source connected to the induction gyrator.

From the schematic of Figures A.22 and A.23, one can design an electronic circuit. A signal generator excites a sinusoidal voltage to the circuit and an oscilloscope is used to capture the dynamic nonlinear behavior of the output point of the circuit versus the input excitation source.

### A.4.3.1   Simulation Results

Using Equations A.38 and A.39, the following values are set: $R_L = 10\ \Omega$, $R_g = 100\ \text{k}\Omega$, $C = 16\ \text{nF}$, $C_1 = 9.8\ \text{nF}$, $C_2 = 100\ \text{nF}$, $G_a = -0.756\ \text{mS}$, $G_b = -0.409\ \text{mS}$, $E = 1.08\ \text{V}$, and $R = 1770\ \Omega$. The amplitude of the excitation voltage $F = 275\ \text{mV}$ and its frequency can be employed as the control parameter. The following cases are observed:

- *Case 1*: $V_f = 0$, this leads to a self-excited attractor in the Chua's circuit incorporating the inductor gyrator; typical Chua's attractors can be obtained as shown in Figure A.24.

- *Case 2*: *F* = 275 mV and *f varies from 25 to 500 Hz*. We also get the typical double scroll of Chua's attractor, which looks like a combination of two single attractors. The motion behavior is more complex (Figure A.25).
- *Case 3*: *F* = 275 mV and *f varies from 516 to 1545 Hz*. We get a cascade of period adding when decreasing the frequency of sinusoidal excitation (Figure A.26).

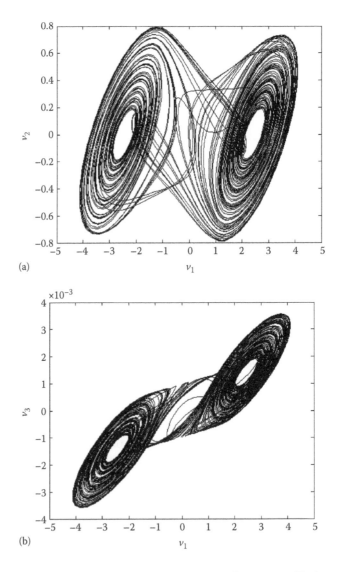

(a)

(b)

**FIGURE A.24**   Typical Chua's attractor: (a) ($v_1$_*vs*_$v_2$) phase space, (b) ($v_1$_*vs*_$v_3$) phase space. *(Continued)*

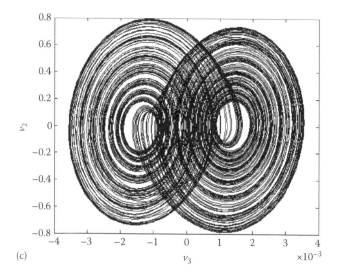

(c)

**FIGURE A.24 (*Continued*)**   Typical Chua's attractor: (c) ($v_3$_vs_$v_2$) phase space (in this figure $v_3 \equiv i_L$).

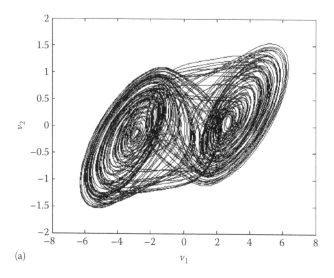

(a)

**FIGURE A.25**   Attractors at $F = 275$ mV: (a) $f = 27$ Hz.                                    (*Continued*)

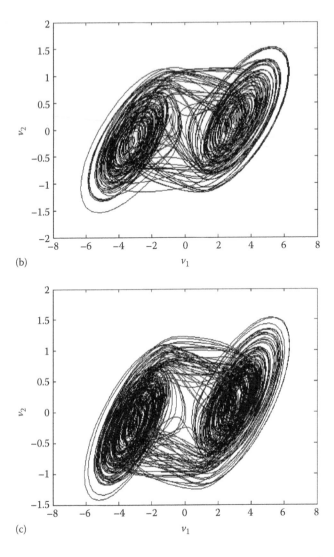

(b)

(c)

**FIGURE A.25 (Continued)**    Attractors at $F=275$ mV: (b) $f=275$ Hz, and (c) $f=400$ Hz.

- *Case 4*: Keeping $F=275$ mV and *f increasing from 1800 Hz*. We can observe some different phenomenon. At $f=1929$ Hz, the dynamical system is represented by a point-attractor (Figure A.27a) and after that the system will become chaotic. At $f=1967$ Hz, there exists two single scrolls attractors and some limit cycles (Figure A.27b). At $f=3333$ Hz, two single scroll attractors extend their sizes and come together to be a double scroll attractor with limit cycles on the outside. (Figure A.27c). At $f=4287$ Hz, there is a typical double scroll attractor (Figure A.27d).

## A.4.3.2  Experimental Results

We can set $R=1720\ \Omega$ so that the circuit behaves chaotically by itself without any sinusoidal excitation or self-excitation in order to study the dynamical behavior of the circuit when it is excited by an external sinusoidal force.

- *Bifurcation diagram*: We change the value of the driving source amplitude $F$ from 25 to 400 mV with the step of 25 mV; at every amplitude, we decrease the driving source frequency $f$ from 9 to 10 kHz with step varying

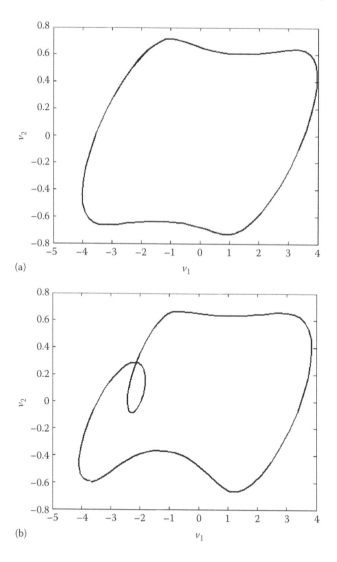

(a)

(b)

**FIGURE A.26**  Period adding at $F=275$ mV: (a) Period-1, $f=1545$ Hz, (b) Period-2, $f=1017$ Hz.                                                                    (*Continued*)

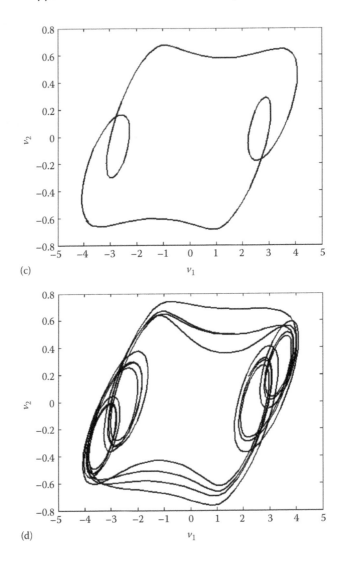

(c)

(d)

**FIGURE A.26 (*Continued*)** Period adding at $F = 275$ mV: (c) Period-3, $f = 798$ Hz, and (d) Period-4, $f = 624$ Hz.

from 5 to 20 Hz. We monitored $(v_1 - v_2)$ ($v_1$ and $v_2$ are voltages dropped across the capacitor $C_1$ and $C_2$ respectively), and the phase space on the oscilloscope depicts the bifurcation diagram in the ($F–f$) plane. Figure A.28 is the bifurcation diagram where $f$ changes from 200 to 4000 Hz, the numbers denote the periods (adding-periods) of attractors and the shaded regions denote chaos.

In the region of the driving source frequencies greater than 4000 Hz, the bifurcation only changes from a single scroll to double scroll attractors and vice versa as illustrated in Figure A.29.

- *Period-doubling*: A cascade of period-doubling bifurcations appear in the middle frequencies from 3000 to 4000 Hz (Figure A.28). At $F = 275$ mV, typically one period appears at the driving source frequency $f = 3582$ Hz (Figure A.30a). We then decrease $f$ to 3333 Hz to obtain a double scroll attractor (Figure A.30b). At $f = 3040$ Hz, a Hopf bifurcation appears with an outer limit cycle (Figure A.30c). However, we do not observe any period-doubling, period quadrupling, or period octupling trajectories in this

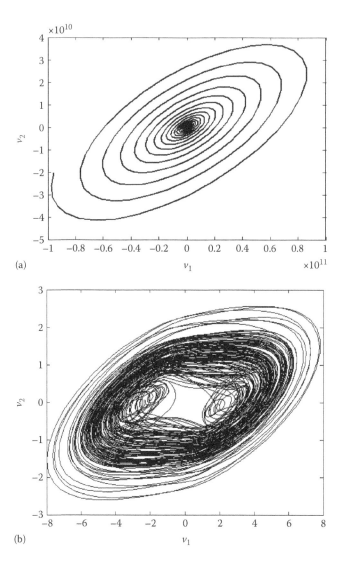

(a)

(b)

**FIGURE A.27**    Attractors at $F = 275$ mV: (a) $f = 1929$ Hz, (b) $f = 1967$ Hz.      (*Continued*)

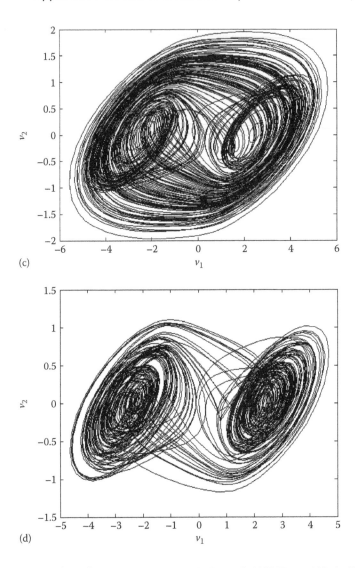

**FIGURE A.27 (*Continued*)**   Attractors at $F = 275$ mV: (c) $f = 3333$ Hz, and (d) $f = 4287$ Hz.

cascade structure. This may due to the fact that the state of the system has reached the limit of the bandwidth of the oscilloscope.

- *Period adding*: In Figure A.28, the region of excitable frequencies of less than 1200 Hz is called the region of periodic windows. In this region, chaotic behaviors and the period-doubling appear sequentially. Figure A.31 represents a cascade of period-doubling at the excitable amplitude $F = 275$ mV and the order of period increases when the excitable frequency decreases.

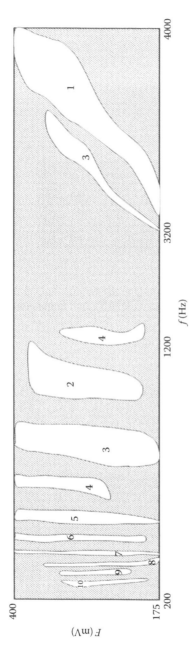

**FIGURE A.28**   Bifurcation diagram in $(F–f)$ plane. The numbers are the periods and the shaded regions are chaos.

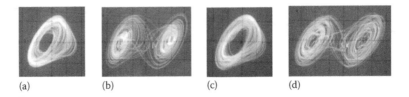

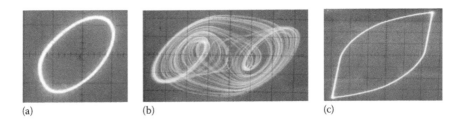

**FIGURE A.29** Relation of single scroll attractors and double scroll attractors in $(v_1 - v_2)$ phase space: (a) $f = 8.29$ kHz, (b) $f = 6.8$ kHz, (c) $f = 6$ kHz, and (d) $f = 4.4$ kHz. Horizontal axis: 1 V/div and vertical axis: 0.5 V/div.

**FIGURE A.30** Period-doubling at $F = 275$ mV. (b) Two-scroll attractor, $f = 3333$ Hz. (a) and (b) $x$-axis: 1 V/div and $y$-axis: 0.5 V/div. (c) Limit cycle, $f = 3040$ Hz. Horizontal axis: 5 V/div and vertical axis: 2 V/div.

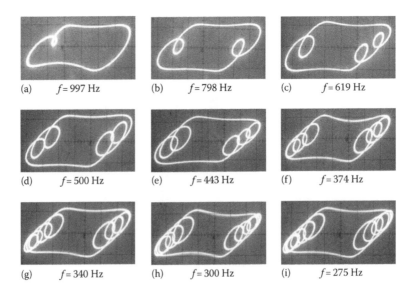

**FIGURE A.31** A cascade of period-doubling in $(v_1 - v_2)$ phase space at $F = 275$ mV. (a)–(i) From period-1 to period-10. Horizontal axis: 1 V/div and vertical axis: 0.5 V/div.

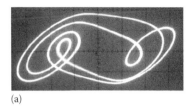 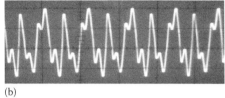

(a)  (b)

**FIGURE A.32**  Quasi-periodicity at $F = 100$ mV and $f = 2954$ Hz. (a) Trajectories in $(v_1 - v_2)$ phase space. Horizontal axis: 1 V/div, and vertical axis: 0.5 V/div. (b) $v_1$ time-domain voltage waveform.

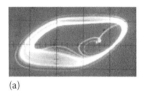 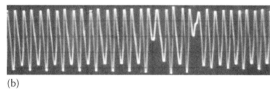

(a)  (b)

**FIGURE A.33**  Intermittency at $F = 275$ mV and $f = 1731$ Hz: (a) trajectories in $(v_1 - v_2)$ phase space. Horizontal axis: 1 V/div and vertical axis: 0.5 V/div. (b) $v_1$ voltage with intermittent burst.

- *Quasi-periodicity*: When the system has two different frequencies that associate with each other, the motion of the system is called quasi-periodic motion and it repeats itself inexactly. In our experiment, the quasi-periodic behaviors appear at the excitable frequencies less than 3000 Hz according to the excitable amplitude. Figure A.32 depicts the trajectories in phase space $(v_1 - v_2)$ and its signal $v_1$. When the excitable frequency is changed slowly, the quasi-periodic motion changes to a chaotic motion.

- *Intermittency*: In this experiment, the intermittent behaviors appear in the excitable region with a frequency greater than 1000 Hz. Figure A.33 depicts the trajectories in phase space $(v_1 - v_2)$ and its voltage $v_1$ to illustrate the intermittent behavior. We realized that the periodic oscillations are interrupted by the intermittent bursts. Figure A.34 depicts the intermittent behavior near the boundary crisis region.

(a)  (b)

**FIGURE A.34**  Intermittency near the boundary crisis region at $F = 400$ mV and $f = 1475$ Hz: (a) trajectories in $(v_1 - v_2)$ phase space. Horizontal axis: 1 V/div and vertical axis: 0.5 V/div. (b) $v_1$ voltage with intermittent bursts strongly.

## A.5 CONCLUDING REMARKS

In this appendix, we have presented briefly the main principles of nonlinear dynamical systems. The descriptions of fixed points, bifurcation, and chaos are explained. Other main routes to chaotic states of nonlinear systems are obtained and derived from a mathematical point of view and then through experimental demonstration. In the last section, Chua's nonlinear circuit is illustrated experimentally using an inductor gyrator in association with appropriate resistors and capacitors excited with the external sinusoidal source so that charging and discharging can happen simultaneously so that chaotic behavior can be observed in both simulation and experiment. These chaotic behaviors and nonlinear dynamical evolution is illustrated in this book's chapters in several fiber optic lasers of different positive feedback paths in the optical domain.

Bifurcation and chaotic dynamics of fiber laser systems are illustrated in Chapters 6 and 8 for active and passive mode-locked lasers in which the envelope of the lightwaves behaves in nonlinear motions illustrated in the mathematical representations described in this appendix.

## REFERENCES

1. R. C. Hilborn, *Chaos and Nonlinear Dynamics: An Introduction for Scientists and Engineers*, Oxford University Press Inc., New York, 1994.
2. T. Kapitaniak, *Chaos for Engineers: Theory, Applications, and Control*, 2nd edn., Springer, Berlin, Germany, 1998.
3. S. H. Strogatz, *Nonlinear Dynamics and Chaos: With Applications to Physics, Biology, Chemistry, and Engineering*, Wesview Press, Boulder, CO, 2000.
4. M. P. Kennedy, ABC (Adventures in Bifurcations and Chaos): A program for studying chaos, *J. Franklin Inst.*, 331B(6), 631–658, 1994.
5. B. Muthuswamy, T. Blain, and K. Sundqvist, A synthetic inductor implementation of Chua's circuit, Technical Report No. UCB/EECS-2009-20, 2009. http://www.eecs.berkeley.edu/Pubs/TechRpts/2009/EECS-2009-20.html (accessed June 2011).
6. K. Murali and M. Lakshmanan, Effect of sinusoidal excitation on the Chua's circuit, *IEEE Trans. Circuits Syst. I: Fundam. Theory Appl.*, 39(4), 264–270, 1992.

# Appendix B: Derivation of the Nonlinear Schrodinger Equation (NLSE)

## B.1  WAVE EQUATION IN NONLINEAR OPTICS

In order to derive the wave equation for the propagation of light in a nonlinear optical medium, we begin with Maxwell's equations for dielectric medium written as follows:

$$\nabla \times \vec{E} = -\frac{\partial \vec{B}}{\partial t} \tag{B.1}$$

$$\nabla \times \vec{H} = \frac{\partial \vec{D}}{\partial t} \tag{B.2}$$

$$\nabla \cdot \vec{D} = 0 \tag{B.3}$$

$$\nabla \cdot \vec{B} = 0 \tag{B.4}$$

where $\vec{E}, \vec{H}$ are the electric and magnetic fields, respectively. The electric and magnetic flux densities $\vec{D}$ and $\vec{B}$, respectively, are related to the electric and magnetic fields via

$$\vec{D} = \varepsilon_0 \vec{E} + \vec{P} \tag{B.5}$$

$$\vec{B} = \mu_0 \vec{H} \tag{B.6}$$

where
   $\varepsilon_0$ is the permittivity of free space
   $\mu_0$ is the permeability of free space
   $\vec{P}$ is the induced electric polarization

By taking curl of Equation B.1 and using Equation B.4 we obtain

$$\nabla \times \nabla \times \vec{E} = -\mu_0 \nabla \times \frac{\partial \vec{H}}{\partial t} \tag{B.7}$$

Substituting Equation B.2 into Equation B.7 yields the generic wave equation

$$\nabla \times \nabla \times \vec{E} = -\frac{1}{c^2}\frac{\partial^2 \vec{E}}{\partial t^2} - \mu_0 \frac{\partial^2 \vec{P}}{\partial t^2} \tag{B.8}$$

where $c$ is the speed of light in vacuum and is given by $c = 1/\sqrt{\varepsilon_0 \mu_0}$. The induced polarization consists of linear and nonlinear components as

$$\vec{P} = \vec{P}_L + \vec{P}_{NL} \tag{B.9}$$

which are defined as

$$\vec{P}_L(\vec{r},t) = \varepsilon_0 \int\limits_{-\infty}^{\infty} \chi^{(1)}(t-t') \cdot \vec{E}(\vec{r},t')dt' \tag{B.10}$$

$$\vec{P}_{NL}(\vec{r},t) = \varepsilon_0 \iint\limits_{-\infty}^{\infty} \int \chi^{(3)}(t-t_1,t-t_2,t-t_3) \times \vec{E}(\vec{r},t_1)\vec{E}(\vec{r},t_2)\vec{E}(\vec{r},t_3) \tag{B.11}$$

where $\chi^{(1)}$ and $\chi^{(3)}$ are the first- and third-order susceptibility tensors.

## B.2   GENERALIZED NONLINEAR SCHRODINGER EQUATION

The starting point for the derivation of the nonlinear Schrodinger equation (NLSE) is the wave equation (Equation B.8). In order to cover a larger number of nonlinear effects, the general form of the nonlinear polarization in Equation B.11 must be used, and an approximation of $\chi^{(3)}$ is given by

$$\chi^{(3)}(t-t_1,t-t_2,t-t_3) = \chi^{(3)}R(t-t_1)\delta(t-t_2)\delta(t-t_3) \tag{B.12}$$

where $R(t)$ is the nonlinear response function normalized in a manner similar to the delta function, that is, $\int\limits_{-\infty}^{\infty} R(t)dt = 1$. By introducing Equation B.12 into Equation B.11 with a slowly varying approximation, the nonlinear polarization in scalar form is given by

$$P_{NL}(\vec{r},t) = \frac{3}{4}\varepsilon_0 \chi^{(3)}_{xxxx}E(\vec{r},t) \int\limits_{-\infty}^{\infty} R(t-t_1)\left|E(\vec{r},t)\right|^2 dt_1 \tag{B.13}$$

The assumptions in Section B.1 are also applied for simplification of the NLSE derivation. It will be clearer to describe numerous effects in the frequency domain by using the following Fourier transforms:

$$E(\vec{r},\omega) = \int\limits_{-\infty}^{\infty} E(\vec{r},t)\exp\left[j(\omega-\omega_0)t\right]dt = FT\{E(\vec{r},t)\} \tag{B.14}$$

$$E(\vec{r},t) = \frac{1}{2\pi}\int_{-\infty}^{\infty} E(\vec{r},\omega)\exp\left[-j(\omega-\omega_0)t\right]d\omega = FT^{-1}\{E(\vec{r},\omega)\} \qquad \text{(B.15)}$$

In the frequency domain, the convolution operation becomes a simple multiplication and the time derivatives can be replaced by: $\partial/\partial t \rightarrow -j\omega$ and $\partial^2/\partial t^2 \rightarrow -\omega^2$. Hence, a modified Helmholtz equation can be derived from Equation B.8 by using Equations B.10 and B.13 and the Fourier transform

$$\nabla^2 E(\vec{r},\omega) + \varepsilon_L(\omega)\frac{\omega^2}{c^2}E(\vec{r},\omega) = -\mu_0\omega_0^2\left(1+\frac{\omega}{\omega_0}\right)P_{NL}(\vec{r},\omega) \qquad \text{(B.16)}$$

where $P_{NL}(\omega)$ are Fourier transforms of $P_{NL}$ in the time domain. Equation B.16 can be solved by using the method of separation of variables. The slowly varying part of the electric field $E(\vec{r},\omega)$ is approximated by

$$E(\vec{r},\omega) = F(x,y)A(z,\omega)\exp(j\beta_0 z) \qquad \text{(B.17)}$$

where
  $A(z, \omega)$ is the slowly varying function of $z$
  $\beta_0$ is the wave number
  $F(x, y)$ is the function of the transverse field distribution, which is assumed to be independent of $\omega$

Then substituting Equation B.17 into Equation B.16, the Helmholtz equation is split into two equations:

$$\frac{\partial^2 F}{\partial x^2} + \frac{\partial^2 F}{\partial y^2} + \left[\varepsilon_L(\omega)\frac{\omega^2}{c^2} - \beta^2\right]F = 0 \qquad \text{(B.18)}$$

$$F(x,y)\left[2j\beta_0\frac{\partial A(z,\omega)}{\partial z} + (\beta^2 - \beta_0^2)A(z,\omega)\right] = \mu_0\omega_0^2\left(1+\frac{\omega}{\omega_0}\right)P_{NL}(\vec{r},\omega) \qquad \text{(B.19)}$$

Equation B.18 is an eigenvalue equation that needs to be solved for the wave number $\beta$ and the fiber modes. In Equation B.18, $\varepsilon_L(\omega)$ can be approximated by $\varepsilon_L \approx n^2 + 2n\Delta n$, where $\Delta n$ is a small perturbation and can be determined from Equation B.11: $\Delta n = j\alpha c/(2\omega)$.

In the case of single mode, Equation B.19 can be solved using the first-order perturbation theory in which $\Delta n$ does not affect $F(x, y)$, but only the eigenvalues. Hence $\beta$ in Equation B.19 becomes $\beta(\omega) + \Delta\beta$, where $\Delta\beta$ accounts for the effect of the perturbation term (referred to $\Delta n$) to change the propagation constant for the fundamental mode. Using Equations B.17 and B.15, the electric field can be approximated by

$$E(\vec{r},t) = F(x,y)A(z,t)\exp(j\beta_0 z) \qquad \text{(B.20)}$$

where $A(z, t)$ is the slowly varying complex envelope propagating along $z$ in the optical fiber. From Equation B.19, after integrating over $x$ and $y$, the following equation in the frequency domain is obtained:

$$\frac{\partial A(z,\omega)}{\partial z} = j(\beta(\omega) - \beta_0)A(z,\omega) - \frac{\alpha(\omega)}{2}A(z,\omega) + j\gamma(\omega)\left(1 + \frac{\omega}{\omega_0}\right)$$

$$\times \int\int_{-\infty}^{\infty} R(\omega_1 - \omega_2)A(z,\omega_1)A^*(z,\omega_2)A(z,\omega - \omega_1 + \omega_2)d\omega_1 d\omega_2 \quad \text{(B.21)}$$

where $R(\omega)$ is the Fourier transform of $R(t)$, and the nonlinear coefficient $\gamma$, which has been introduced, is given by

$$\gamma = \frac{n_2\omega_0}{cA_{eff}} = \frac{2\pi n_2}{\lambda A_{eff}} \quad \text{(B.22)}$$

where $A_{eff}$ is the effective area of the optical fiber and is given by

$$A_{eff} = \frac{\left(\int_{-\infty}^{\infty}|F(x,y)|^2 dxdy\right)^2}{\int\int_{-\infty}^{\infty}|F(x,y)|^4 dxdy} \quad \text{(B.23)}$$

Equation B.21 is the NLSE that describes generally the pulse propagation in the frequency domain. It is useful to take into account the frequency dependence of the effects of the propagation constant $\beta$, the loss $\alpha$, and the nonlinear coefficient $\gamma$ by expanding them using Taylor series expansion as follows:

$$\beta(\omega) = \beta_0 + \beta_1(\omega - \omega_0) + \frac{1}{2}\beta_2(\omega - \omega_0)^2 + \frac{1}{6}\beta_2(\omega - \omega_0)^3 + \cdots$$

$$= \sum_{n=0}^{\infty} \frac{\beta_n}{n!}(\omega - \omega_0)^n \quad \text{(B.24)}$$

$$\alpha(\omega) = \sum_{n=0}^{\infty} \frac{\alpha_n}{n!}(\omega - \omega_0)^n \quad \text{(B.25)}$$

$$\gamma(\omega) = \sum_{n=0}^{\infty} \frac{\gamma_n}{n!}(\omega - \omega_0)^n \quad \text{(B.26)}$$

However, the pulse spectrum in most cases of practical interest is narrow enough such that $\gamma$ and $\alpha$ are constant over the pulse spectrum. Therefore, the NLSE in the time domain can be obtained by using the inverse Fourier transform

$$\frac{\partial A(z,t)}{\partial z} + \frac{\alpha}{2} A(z,t) - j \sum_{n=1}^{\infty} \frac{j^n \beta_n}{n!} \frac{\partial^n A(z,t)}{\partial t^n} = j\gamma \left( 1 + \frac{j}{\omega_0} \frac{\partial}{\partial t} \right)$$

$$\times A(z,t) \int_{-\infty}^{\infty} R(t') |A(z,t-t')|^2 \, dt' \tag{B.27}$$

Equation B.27 is the basic propagation equation, commonly known as the generalized nonlinear Schrodinger equation (NLSE), which is very useful for studying the evolution of the amplitude of the optical signal and the phase of the lightwave carrier under most effects of third-order nonlinearity in *optical waveguides* as well as *optical fibers*.

# Appendix C: Calculation Procedures of Triple Correlation and Bispectrum with Examples

## C.1 TRIPLE CORRELATION AND BISPECTRUM ESTIMATION

Definitions of triple correlation and bispectrum for continuous signal $x(t)$ are given in Chapter 6. However, the calculation of both triple correlation and bispectrum is normally achieved in the discrete domain. Thus the discrete triple correlation is estimated as follows:

$$C_3(\tau_1, \tau_2) = \sum_k x(kdt)x(kdt - \tau_1)x(kdt - \tau_2) \qquad \text{(C.1)}$$

where
$x(kdt)$ is the discrete version of $x(t)$
$k$ is integer number
$dt = 1/f_s$ is the sampling period, $f_s$ is the sampling frequency, the delay variables are also discretized as $\tau_i = mdt$, $m = 0, 1, 2, \ldots, N/2 - 1$.

Similarly, the discrete bispectrum is estimated by the discrete Fourier transform of $C_3$

$$B_i(f_1, f_2) = \sum_m \sum_n C_3(mdt, ndt)\exp(-2\pi j(f_1 mdt + f_2 ndt)) \qquad \text{(C.2)}$$

where
$m$, $n$ are integers and the frequency variables $f_i = Kdf$, $K = 0,1,2,\ldots,N/2-1$, the frequency resolution $df = 2/(Ndt)$
$N$ is the total number of samples in each computation window

In this book, the following steps are used to estimate the bispectrum:

- A discrete process or signal is divided into $M$ computation frames in which the number of samples $N$ in each frame chosen is 1024. The sampling time $dt$ is properly selected to ensure that whole significant frequency components of $x(kdt)$ are in the range from $-B$ to $B$, where $B = 1/(2dt)$.

- The triple correlation of each data frame is computed by using Equation C.1. The result obtained is an array with the size 512 × 512. Each value in the array is represented for the amplitude of the triple correlation.
- The bispectrum of each frame is then calculated by using discrete Fourier transform Equation C.2. Thus the bispectrum is also an array with the same size. Finally, the bispectrum is averaged over $M$ data frames via the following expression:

$$B(f_1, f_2) = \frac{1}{M} \sum_{i=1}^{M} B_i(f_1, f_2) \tag{C.3}$$

- Both the triple correlation and the bispectrum can be displayed in a 3D graph, as shown in Figure C.1, in which the magnitude is normalized in logarithmic scale. However, the contour representation is selected to display effectively the variation in the bispectrum structure.

## C.2   PROPERTIES OF BISPECTRUM

Important properties of the bispectrum are briefly summarized as follows [1]:

- The bispectrum is generally complex. It contains both magnitude and phase information, which is important for signal recovery as well as identifying nonlinear response and processes.
- The bispectrum has the lines of symmetry $f_1 = f_2$, $2f_1 = -f_2$, and $2f_2 = -f_1$ corresponding to the permutation of the frequencies $f_1, f_2$.
- The bispectrum of a stationary, zero-mean Gaussian process is zero. Thus a nonzero bispectrum indicates a non-Gaussian process.
- The bispectrum suppresses linear phase information or constant phase shift information.
- The bispectrum is flat for non-Gaussian white noise and is zero for Gaussian white noise.

## C.3   BISPECTRUM OF OPTICAL PULSE PROPAGATION

In this section, an example of propagation of optical pulses through optical fiber is characterized and analyzed by the triple correlation and the bispectrum. Figures C.2 and C.3 show, respectively, the triple correlations and the bispectra of the 6.25 ps Gaussian pulse propagating at different lengths of the optical fiber with the second-order GVD coefficient $\beta_2 = -21.6$ ps$^2$/km. Figures C.4 and C.5 show, respectively, the triple correlations and the bispectra of the 6.25 ps super-Gaussian pulse propagating at different lengths of the same fiber.

More importantly, the triple correlation can easily detect the asymmetrical distortion of the pulse that is impossible in autocorrelation estimation. Figures C.6 and C.7 show, respectively, the triple correlations and the bispectra of the super-Gaussian pulse propagating through the fiber with the third-order dispersion coefficient $\beta_3 = 0.133$ ps$^3$/km.

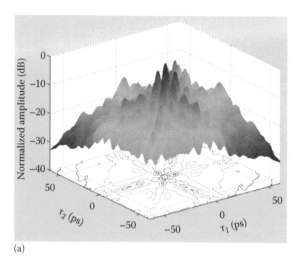

(a)

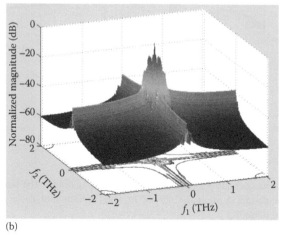

(b)

**FIGURE C.1** Graphical representations of (a) triple correlation and (b) magnitude bispectrum. (Bottom: corresponding contour representations.)

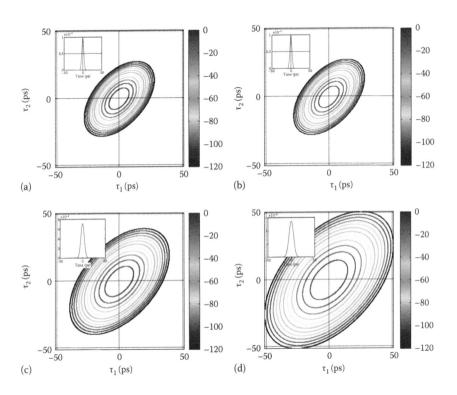

**FIGURE C.2** Triple correlations of Gaussian pulse propagating in the fiber with $\beta_2 = -21.6$ ps$^2$/km at different distances (a) $z = 0$, (b) $z = 50$ m, (c) $z = 650$ m, and (d) $z = 1$ km. (Insets: the waveforms in time domain.)

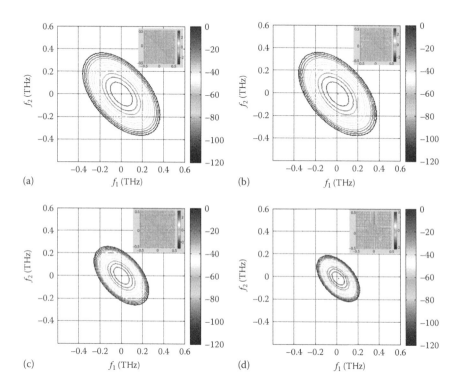

**FIGURE C.3**  Corresponding bispectra of Gaussian pulse propagating in the fiber at different distances (a) $z=0$, (b) $z=50$ m, (c) $z=650$ m, and (d) $z=1$ km. (Insets: the corresponding phase bispectra.)

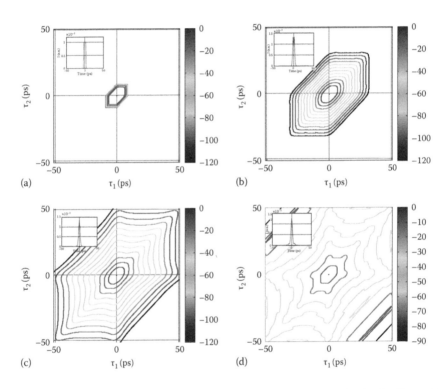

**FIGURE C.4** Triple correlations of a super-Gaussian pulse propagating in the fiber with $\beta_2 = -21.6$ ps$^2$/km at different distances (a) $z=0$, (b) $z=50$ m, (c) $z=100$ m, and (d) $z=300$ m. (Insets: the waveforms in time domain.)

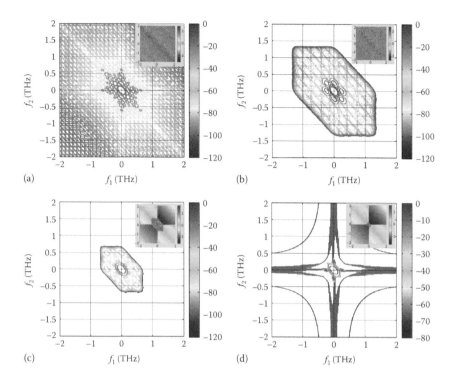

**FIGURE C.5** Corresponding bispectra of the super-Gaussian pulse propagating in the fiber at different distances (a) $z=0$, (b) $z=50$ m, (c) $z=100$ m, and (d) $z=300$ m. (Insets: the corresponding phase bispectra.)

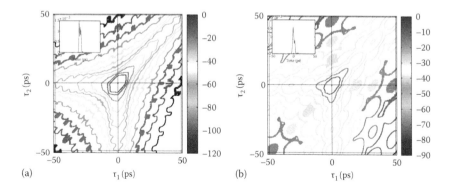

**FIGURE C.6** Triple correlations of a super-Gaussian pulse propagating in the fiber with $\beta_2=0$, $\beta_3=0.133$ ps$^3$/km at different distances (a) $z=100$ m and (b) $z=500$ m. (Insets: the waveforms in time domain.)

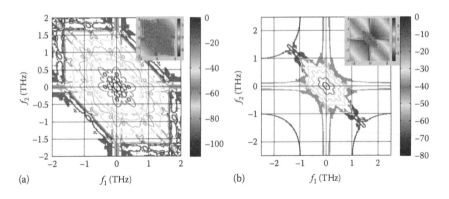

**FIGURE C.7** Corresponding bispectra of the super-Gaussian pulse propagating in the fiber at different distances (a) $z = 100$ m and (b) $z = 500$ m. (Insets: the corresponding phase bispectra.)

## REFERENCE

1. C. L. Nikias and M. R. Raghuveer, Bispectrum estimation: A digital signal processing framework, *Proc. IEEE*, 75, 869–891, 1987.

# Appendix D: Simulink Models

## D.1 MATLAB® AND SIMULINK® MODELING PLATFORMS

A modeling and simulation platform for optical fiber transmission systems has been developed using MATLAB® and Simulink®, an environment for simulation and model-based design [1]. Some of the advantages of the Simulink modeling platform are as follows:

- Subsystem blocks for a complicated transmission system can be set up from the basic blocks available in the toolboxes and block-sets of the Simulink [1]. It is noted that there is no block-sets for optical communication in Simulink. Therefore, the main functional blocks of the optical communication system in the Simulink platform have been developed for years. Details of operational principles as well as examples of the optical components and transmission systems can be found in [1].
- Signals can be easily monitored at any point along the simulation system by available scopes in Simulink block-sets.
- Numerical data can be stored for post-processing in MATLAB to estimate the performance of the system.

An example of an optical fiber transmission system is shown in Figure D.1. Depending on different problems or targets, various Simulink models are set up for investigation in this appendix.

## D.2 WAVELENGTH CONVERTER IN WDM SYSTEM

See Figures D.2 and D.3.

## D.3 NONLINEAR PHASE CONJUGATION FOR MID-LINK SPECTRAL INVERSION

See Figures D.4 through D.7.

## D.4 PULSE GENERATOR

See Figures D.8 and D.9.

## D.5 OTDM DEMULTIPLEXER

For the optical time division multiplexing (OTDM) system using the on-off keying (OOK) scheme, the Simulink models of the transmitter and receiver are similar to those shown in Figure D.6 but using an mode-locked fiber laser (MLFL) instead

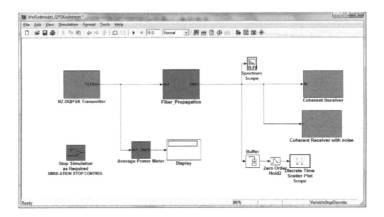

**FIGURE D.1**  An example of an optical fiber transmission system consisting of the main blocks: optical transmitter, fiber transmission link, and optical receiver.

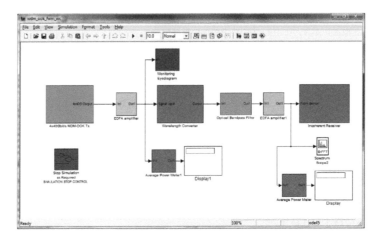

**FIGURE D.2**  The Simulink® model of optical parametric amplifier used as a wavelength converter in the WDM system.

of a carver (Figures D.10 and D.11). Figure D.12 shows the Simulink models of the transmitter and receiver in the OTDM system using the differential quadrature phase shift keying (DQPSK) modulation scheme (Figure D.13).

## D.6  TRIPLE CORRELATION

Figure D.14 shows the Simulink model for triple-correlation based on FWM in non-linear waveguide. The structural block consists of two variable delay lines to generate delayed versions of the original signal, as shown in Figure D.15, and frequency converters to convert the signal into three different waves before combining at the optical coupler to launch into the nonlinear waveguide (Figures D.16 through D.18).

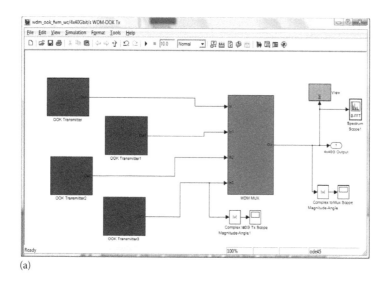

(a)

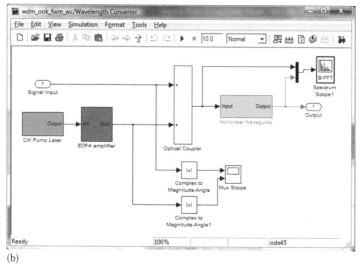

(b)

**FIGURE D.3** (a) Simulink model of the WDM transmitter consisting of four optical transmitters at different wavelengths and a wavelength multiplexer. (b) Simulink setup of the parametric amplifier using the model of nonlinear waveguide used for wavelength conversion in the WDM system.

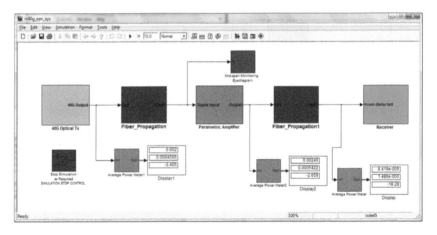

**FIGURE D.4**   Simulink setup of a long-haul 40 Gbps transmission system using nonlinear phase conjugator (NPC) for distortion compensation.

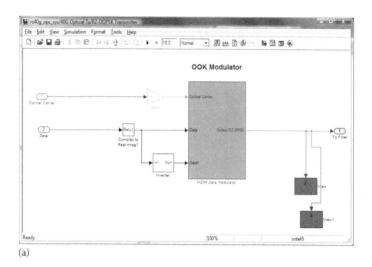

(a)

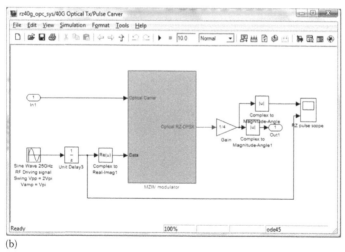

(b)

**FIGURE D.5** (a) Simulink model of an intensity optical modulator driven by data. (b) Simulink model of an optical pulse carver driven by a sinusoidal signal for RZ pulse generation.

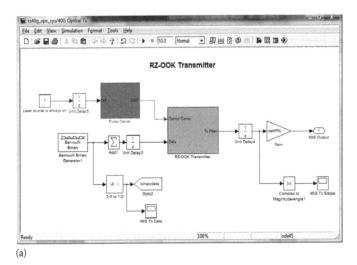

(a)

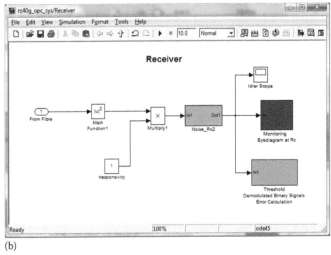

(b)

**FIGURE D.6**    (a) Simulink model of an optical transmitter for RZ-OOK modulation scheme. (b) Simulink model of an optical receiver for the OOK signal.

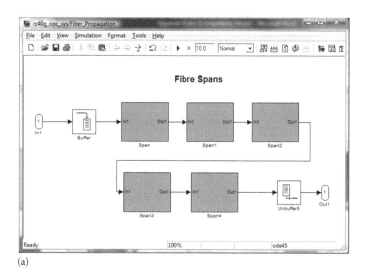

(a)

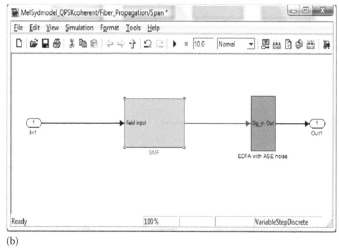

(b)

**FIGURE D.7** (a) Simulink model of each fiber transmission section consisting of five spans. (b) Simulink model of each span consisting of one SMF fiber and one EDFA for loss compensation.

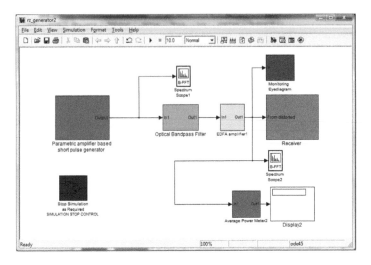

**FIGURE D.8**  Simulink setup of a short pulse generator at 40 GHz based on the parametric amplifier.

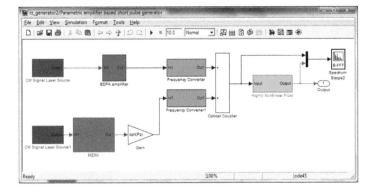

**FIGURE D.9**  Simulink blocks inside the 40 GHz short-pulse generator to demonstrate ultrahigh speed switching based on parametric amplification.

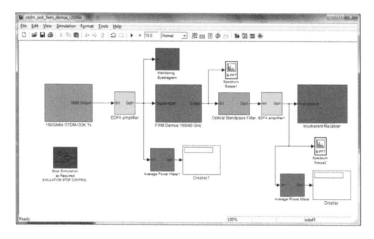

**FIGURE D.10** Simulink setup of the 160 Gbps OTDM system using the FWM process for demultiplexing.

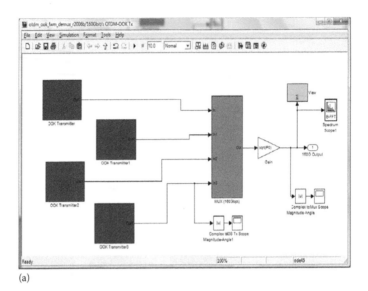

(a)

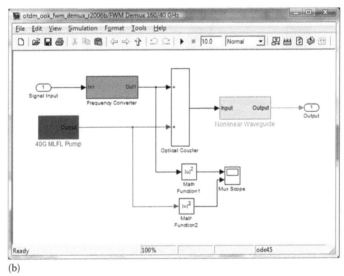

(b)

**FIGURE D.11**   Simulink model of (a) an OTDM transmitter consisting of four optical transmitters and an OTDM multiplexer and (b) the FWM-based demultiplexer.

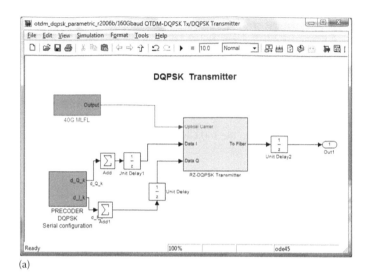

(a)

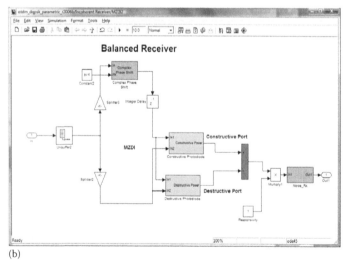

(b)

**FIGURE D.12** Simulink model of (a) an optical transmitter for DQPSK modulation scheme and (b) an optical balanced receiver for DQPSK signal.

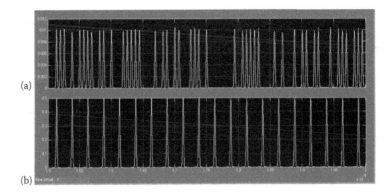

**FIGURE D.13**　Time traces of (a) the 160 Gbps OTDM signal and (b) the 40 Gbps demultiplexed signal.

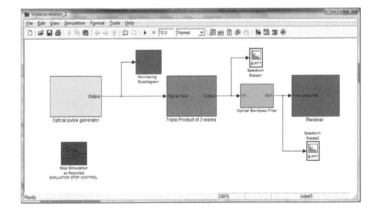

**FIGURE D.14**　Simulink setup for the investigation of the triple-correlation based on the FWM process.

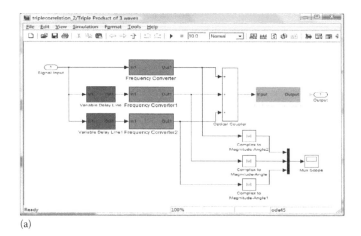

(a)

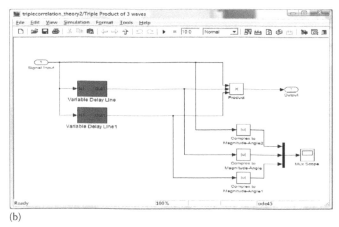

(b)

**FIGURE D.15** Simulink setup of (a) the FWM-based triple-product generation and (b) the theory-based triple-product generation.

**FIGURE D.16** The variation in time domain of the time delay variable (middle grey), the original signal (dark grey) and the delayed signal (light grey).

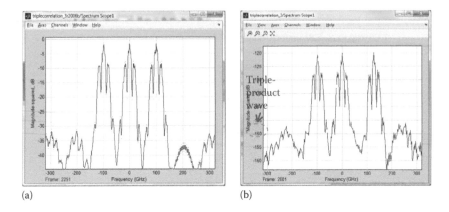

(a)                                    (b)

**FIGURE D.17** Spectrum with (a) equal and (b) unequal wavelength spacing at the output of the nonlinear waveguide.

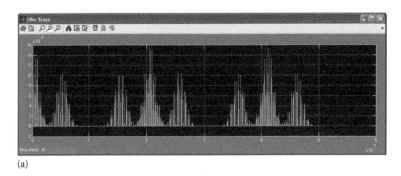

(a)

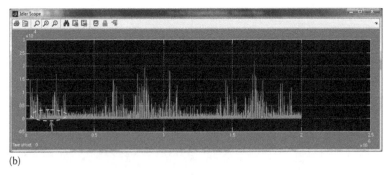

(b)

**FIGURE D.18** Generated triple-product waves in the time domain of the dual-pulse signal based on (a) theory and (b) FWM in the nonlinear waveguide.

## REFERENCE

1. L. N. Binh, *Optical Fiber Communications Systems: Theory and Practice with MATLAB® and Simulink® Models*, CRC Press, Boca Raton, FL, 2010.

# Appendix E: Optical Waveguides

## E.1 OVERVIEW

The propagation of multi-bound solitons in optical waveguides will require an understanding of the properties of the guided wave optical medium. This appendix gives an introduction to the waveguide geometrical properties and their effects on the modulated lightwaves when propagating through such medium.

Planar optical waveguides compose of a guiding region, a slab imbedded between a substrate, and a superstrate having identical or different refractive indices. The lightwaves are guided by their confinement with oscillation solution and penetration into the cladding region, the evanescent waves. The number of oscillating solutions that satisfy the boundary constraints is the number of modes that can be guided. The guiding of lightwaves in an optical fiber is similar to that of the planar waveguide except the lightwaves are guiding through a circular core embedded in a circular cladding layer.

Within the context of this book, optical fibers are treated as the circular optical waveguides that can support single mode with two polarized modes or few modes with different polarizations.

## E.2 OPTICAL FIBER: GENERAL PROPERTIES

### E.2.1 GEOMETRICAL STRUCTURES AND INDEX PROFILE

An optical fiber consists of two concentric dielectric cylinders. The inner cylinder, or core, has a refractive index of $n(r)$ and radius $a$. The outer cylinder, or cladding, has index $n_2$ with $n_2 < n(r)$ for all positions in the core region. The size of the core of about 4–9 μm and a cladding diameter of 125 μm are the typical values for silica-based single mode optical fiber. A schematic diagram of the structure of a circular optical fiber is shown in Figure E.1. Figure E.1a shows the core and cladding region of the circular fiber, while Figure E.1b and c show the figure of the etched cross sections of a multimode and single mode, respectively. The silica fibers are etched in a hydroperoxide solution so that the core region doped with impurity would be etched faster than that of pure silica; thus, the exposure of the core region as observed. Figure E.2 shows the index profile and the structure of circular fibers. The refractive index profile can be step or graded.

The refractive index $n(r)$ of a circular optical waveguide is usually changed with radius $r$ from the fiber axis ($r = 0$) and is expressed by

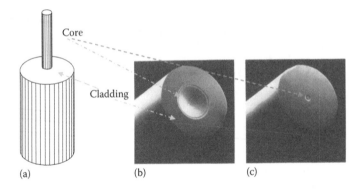

(a)                          (b)                          (c)

**FIGURE E.1** (a) Schematic diagram of the step-index fiber: coordinate system, structure. The refractive index of the core is uniform and slightly larger than that of the cladding. For silica glass the refractive index of the core is about 1.478 and that of the cladding about 1.47 at 1550 nm wavelength region. (b) Cross section of an etched fiber—multimode type—50 μm diameter. (c) Single mode optical fiber etched cross section.

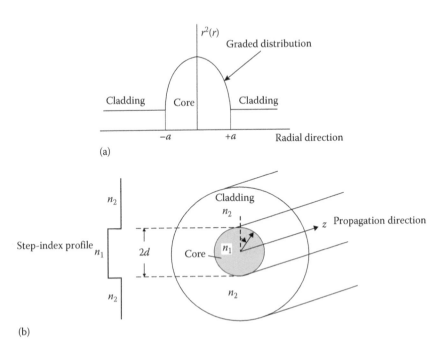

**FIGURE E.2** (a) Refractive index profile of a graded index profile. (b) Fiber cross section and step-index profile with a as the radius of fiber.

$$n^2(r) = n_2^2 + NA^2 s\left(\frac{r}{a}\right) \tag{E.1}$$

where

NA is the numerical aperture at the core axis

$s(r/a)$ represents the profile function that characterizes any profile shape ($s = 1$ at maximum) with a scaling parameter (usually the core radius)

For a step-index profile the refractive index remains constant in the core region, thus

$$s\left(\frac{r}{a}\right) = \begin{cases} 1 & r \leq a \\ 0 & r > a \end{cases} \xrightarrow{\text{hence ref\_index}} n^2(r) = \begin{cases} n_1^2 & r \leq a \\ n_2^2 & r > a \end{cases} \tag{E.2}$$

For graded index profile, we can consider two most common types of graded-index profiles: power law index and the Gaussian profile.

For power law profile, the core refractive index of optical fiber usually follows a graded profile. In this case the refractive index rises gradually from the value $n_1$ of the cladding glass to value $n_1$ of the fiber axis. Therefore, $s(r/a)$ can be expressed as

$$s\left(\frac{r}{a}\right) = \left\{ 1 - \left(\frac{r}{a}\right)^\alpha \right\} \quad \text{for } r \leq a \text{ and } = 0 \text{ for } r < a \tag{E.3}$$

where $\alpha$ is the power exponent. Thus the index profile distribution $n(r)$ can be expressed in the usual way by using Equations E.3 and E.2 and by substituting $NA^2 = n_1^2 - n_2^2$:

$$n^2(r) = \begin{cases} n_1^2 \left[ 1 - 2\Delta \left(\frac{r}{a}\right)^\alpha \right] & \text{for } r \leq a \\ n_2^2 & \text{for } r > a \end{cases} \tag{E.4}$$

where $\Delta = NA^2/n_1^2$ is the relative refractive difference with a small difference between that of the cladding and the core regions. The profile shape given in Equation E.4 offers three special distributions: (1) $\alpha = 1$: the profile function $s(r/a)$ is linear and the profile is called a triangular profile; (2) $\alpha = 2$: the profile is a quadratic function with respect to the radial distance and the profile is called the parabolic profile; and (3) $\alpha = \infty$: the profile is a step type.

For Gaussian profile, the refractive index changes gradually from the core centre to a great distance very far away from it and $s(r)$ can be expressed as

$$s\left(\frac{r}{a}\right) = e^{-(r/a)^2} \tag{E.5}$$

### E.2.2 FUNDAMENTAL MODE OF WEAKLY GUIDING FIBERS

The electric and magnetic fields $E(r, \phi, z)$ and $H(r, \phi, z)$ of the optical fibers in cylindrical coordinates can be found by solving Maxwell's equations. Only the lower order modes of ideal step-index fibers are important for digital optical communication systems. The fact is that for $\Delta < 1\%$, the optical waves are confined very weakly and are thus gently guided. Thus, the electric and magnetic fields $E$ and $H$ can take approximate solutions of the scalar wave equation in a cylindrical coordinate system $(x, \theta, \phi)$

$$\left[ \frac{\delta^2}{\delta r^2} + \frac{1}{r} \frac{\delta}{\delta r} + k^2 n_j^2 \right] \varphi(r) = \beta^2 \varphi(r) \tag{E.6}$$

where

$n_j = n_1, n_2$

$\varphi(r)$ is the spatial field distribution of the nearly transverse EM waves

$$E_x = \psi(r) e^{-i\beta z}$$

$$H_y = \left( \frac{\varepsilon}{\mu} \right)^{1/2}, \quad E_x = \frac{n_2}{Z_0} E_x \tag{E.7}$$

with $E_y, E_z, H_x, H_z$ being negligible, $\varepsilon = \varepsilon_0 n_2^2$ and $Z_0 = (\varepsilon_0 \mu_0)^{1/2}$ are vacuum impedance. We can assume that the waves are plane waves traveling down along the fiber tube. These plane waves are reflected between the dielectric interfaces, in other words, they are trapped and guided along the core of the optical fiber. Note that the electric and magnetic components are spatially orthogonal with each other. Thus, for a single mode, there are always two polarized components that are then the polarized modes of single mode fiber. It is further noted that Snell's law of reflection would not be applicable for single mode propagation, but Maxell's equations must be used. However we will see in the next section that the field distribution of single mode optical fibers follows closely to that of a Gaussian shape. Hence the solution of the wave equation (Equation E.6) can be assumed and the eigenvalue or the propagation constant of the guided wave can be found or optimized to achieve the best fiber structure. However, currently due to the potentials of digital signal processing, the uses of the modes of multimode fibers can be beneficial and so a few mode optical fibers are intensively investigated.

In the next section, we give a brief analysis of the wave equations subject to the boundary conditions so that the eigenvalue equation can be found and, hence, the propagation or wave number of the guided modes can be found and, in turn, the propagation delay of these group of lightwaves along the fiber transmission line. Thence we will revisit the single mode fiber with a Gaussian mode field profile to give an insight into the weakly guiding phenomena that is critical for the understanding of the guiding of lightwaves over very long distance with minimum loss and optimum dispersion, the group delay difference.

### E.2.2.1 Solutions of the Wave Equation for Step-Index Fiber

The field spatial function $\varphi(r)$ would have the form of Bessel functions given by Equation E.6:

$$\varphi(r) = \begin{cases} A\dfrac{J_0(ur/a)}{J_0(u)} & 0 < r < a\text{—core} \\[3mm] A\dfrac{K_0(vr/a)}{K_0(v)} & r > a\text{—cladding} \end{cases} \tag{E.8}$$

where $J_0$, $K_0$ are the Bessel functions of the first kind and modified of the second kind, respectively, and $u$, $v$ are defined as:

$$\frac{u^2}{a^2} = k^2 n_1^2 - \beta^2 \tag{E.9}$$

$$\frac{v^2}{a^2} = -k^2 n_2^2 + \beta^2 \tag{E.10}$$

Thus, following the Maxwell's equations relation, we can find that $E_z$ can take two possible solutions that are orthogonal:

$$E_z = -\frac{A}{kan_2} \begin{pmatrix} \sin\phi \\ \cos\phi \end{pmatrix} \begin{cases} \dfrac{uJ_1(ur/a)}{J_0(u)} & \text{for } 0 \le r < a \\[3mm] \dfrac{vK_1(vr/a)}{K_0(v)} & \text{for } r > a \end{cases} \tag{E.11}$$

The terms $u$ and $v$ must simultaneously satisfy two equations:

$$u^2 + v^2 = V^2 = ka\left(n_1^2 - n_2^2\right)^{1/2} = kan_2(2\Delta)^{1/2} \tag{E.12}$$

$$u\frac{J_1(u)}{J_0(u)} = v\frac{K_1(v)}{K_0(v)} \tag{E.13}$$

where Equation E.13 is obtained by applying the boundary conditions at the interface $r = a$ ($E_z$ is the tangential component and must be continuous at this dielectric interface). Equation E.13 is commonly known as the eigenvalue equation of the wave equation bounded by the continuity at the boundary of the two dielectric media; thence, the condition for guiding in the transverse plane such that the maximum or fastest propagation velocity in the axial direction. The solution of this equation would give specific discrete values of $\beta$ and the propagation constants of the guided lightwaves.

### E.2.2.2 Single and Few Mode Conditions

Over the years since the demonstration of the guiding in circular optical waveguides, the eigenvalue equation (Equation E.13) is employed to find the number of modes supported by the waveguide and their specific propagation constants. Thence the Gaussian mode spatial distribution can be approximated for the fundamental mode based on experimental measurement of the mode fields and the eigenvalue equation is no longer needed when single mode fiber is extensively used. However under current extensive research interests in the spatial multiplexing in DSP-based coherent optical communication systems, few mode fibers have attracted quite a lot of interests due to their potential supporting of many channels with their modes and their related filed polarizations. This section is thus extended to consider both the fundamental mode and higher order modes.

Equation E.12 shows that the longitudinal field is in the order of $u/(kan_2)$ with respect to the transverse component. In practice it is very common that $\Delta \ll 1$, by using Equation E.12, we observe that this longitudinal component is negligible compared with the transverse component. Thus the guided mode is transversely polarized. The fundamental mode is then usually denominated as $LP_{01}$ mode ($LP$ = Linearly Polarized) for which the field distribution is shown in Figure E.4a and b. The graphical representation of the eigenvalue equation (Equation E.13) calculated as the variation of $b = \beta/k$ as the normalized propagation constant and the $V$-parameter is shown in Figure E.5d. There are two possible polarized modes, the horizontal and vertical polarizations which are orthogonal to each other. These two polarized modes can be employed for transmission of different information channels. They are currently exploited, in the first two decades of the 21st century, in optical transmission systems employing polarization division multiplexed so as to offer the transmission bit rate to 100 Gbps and beyond.

Furthermore when the number of guided modes is higher than two polarized modes, they do form a set of modes over which information channels can be simultaneously carried and spatially demultiplexed at the receiving end so as to increase the transmission capacity as illustrated in Figures E.4a and b [1,2]. Such few mode fibers are employed in the most modern optical transmission system whose schematic is shown in Figure E.6. Mode multiplexer acts as mode spatial mixing and likewise the demultiplexer split the modes $LP_{11}$, $LP_{01}$ into individual mode and then converted to $LP_{01}$ mode field and thence injecting into SMF to feed into the coherent receiver. Obviously, there must be mode spatial de-multiplexing and then modulation and then multiplexing back into the transmission fiber for transmission. Similar structures would be available at the receiver to separate and detect the channels. Note the two possible polarizations of the mode $LP_{11}$. Note that there are four polarized modes of the $LP_{11}$ mode, only two polarized modes are shown in this diagram. The delay due to the propagation velocity, from the difference in the propagation constant, can be easily compensated in the DSP processing algorithm, similar to that due to the polarization mode dispersion (PMD). The main problems to resolve in this spatial mode multiplexing optical transmission system is the optical amplification for all modes so that long haul transmission can be achieved, that is the amplification in multimode fiber structure.

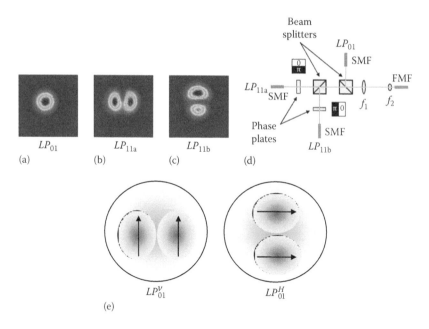

**FIGURE E.3** (a–c) Intensity profiles distributed in the core of first few order modes of a few-mode optical fiber employed for 5×65 Gbps optical transmission system [3] and optical system arrangement for spatially demux and mux of modal channels. (d) Optical setup for launching spatial mode into few mode fiber. (e) Horizontal (*H*) and vertical (*V*) polarized modes $LP_{01}^{V,H}$, polarization directions indicated by arrows.

The number of guided modes is determined by the number of intersecting points of the circle of radius *V* and the curves representing the eigenvalue solutions (Equation E.13). Thus for a single mode fiber the *V*-parameter must be less than 2.405 and for few mode fiber this value is higher, for example if $V = 2.8$ we have three intersecting points between the circle of radius *V* and three curves, then the number of modes would be $LP_{01}$, $LP_{11}$ and their corresponding alternative polarized modes as shown in Figures E.3 and then E.4b and c. For single mode there are two polarized modes whose polarizations can be vertical or horizontal. Thus a single mode fiber is not a monomode but supports two polarized modes! The main issues are also on the optical amplification gain for the transmission of modulated signals in such few mode fibers. This remains to be the principal obstacles.

We can illustrate the propagation of the fundamental mode and higher order modes as in Figure E.5a and b. The rays of these modes can be axially straight or skewed and twisted around the principal axis of the fiber. Thus there are different propagation times between these modes. This property can be employed to compensate for chromatic dispersion effect [4]. Figure E.5 shows a spectrum of the graphical solution of the modes of optical fibers. In Figure E.5d the regions of single operation and then a higher order, second order mode regions as determined by the value of the *V*-parameter are indicated. Naturally, due to manufacturing accuracy the mode regions would be variable from fibers to fibers.

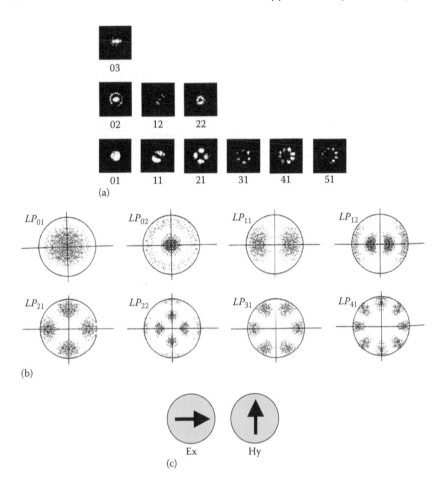

**FIGURE E.4** (a) Spectrum of guided modes in a FMF, numbers indicate order of modes. (b) Calculated intensity distribution of *LP* guided modes in a step-index optical fibers with $V=7$. (c) Electric and magnetic field distribution of an $LP_{01}$ mode polarized along Ox (*H*-mode) and Oy (*V*-mode) of the fundamental mode of a single mode fiber.

Figure E.6 shows a set up for ultra-high capacity transmission using a FMF with spatial division multiplexing (SDM) and demultiplexing devices. For example, a lightwave source can be launched into the *f* of wavelength 850 nm launched into a SSMF, which is single mode at 1,550 nm but becomes FMF at 850 nm supporting mode up to $LP_{41}$; thus, 16 partial mode can be launched into the SSMF in which the bit rate can reach 100 Gbps per spatial channel resulting into 1.6 Tbps for the transmission over the FMF. The transmission would be less than 10 km though. Another advantage of this SDM transmission is that at the receiver multiple DSP systems can be combined and processing under multiple inputs multiple outputs (MIMO) to optimize and reduce significantly the bit error rate (BER) (Figure E.7).

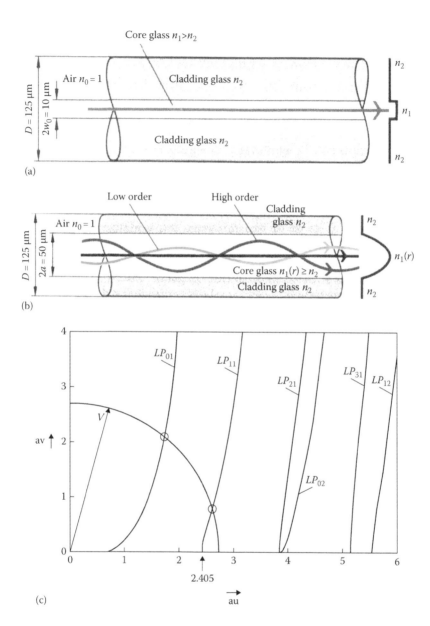

**FIGURE E.5**  (a) Guided modes as seen by "a ray" in the transverse plane of a circular optical fiber, "Ray" model of lightwave propagating in single mode fiber. (b) Ray model of propagation of different modes guided in a few/multi-mode graded-index fiber. (c) Graphical illustration of solutions for eigenvalues (propagation constant—wave number of optical fibers). *(Continued)*

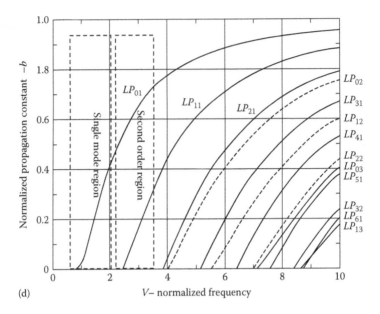

(d)

**FIGURE E.5 (*Continued*)**    (d) $b - V$ characteristics of guided fibers.

**FIGURE E.6**   Few mode fiber employed as a spatial multiplexing and demultiplexing in DSP-based coherent optical transmission systems operating at 100 Gbps and higher bit rate.

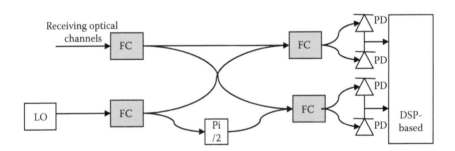

**FIGURE E.7**   pi/2 hybrid coupler for polarization demultiplexing and mixing with local oscillator in a coherent receiver of modern DSP-based optical receiver for detection of phase modulated schemes.

### E.2.2.3   Gaussian Approximation: Fundamental Mode Revisited

We note again that the $E$ and $H$ are approximate solutions of the scalar wave equation and the main properties of the fundamental mode of weakly guiding fibers that can be observed as follows:

- The propagation constant $\beta$ (in $z$-direction) of the fundamental mode must lie between the core and cladding wave numbers. This means the effective refractive index of the guided mode lies with the range of the cladding and core refractive indices.
- Accordingly, the fundamental mode must be nearly a transverse electromagnetic wave as described by Equation E.7.

$$\frac{2\pi n_2}{\lambda} < \beta < \frac{2\pi n_1}{\lambda} \qquad (E.14)$$

- The spatial dependence $\psi(r)$ is a solution of the scalar wave equation (Equation E.6).

The main objectives are to find a good approximation for the field $\psi(r)$ and the propagation constant $\beta$. This can be found though the eigenvalue equation and the Bessel's solutions as shown in the previous section. It is desirable we are able to approximate the field to a good accuracy to obtain simple expressions that should result in a clearer understanding of light transmission on single mode optical fiber without going through graphical or numerical methods. Furthermore, experimental measurements and numerical solution for step and power-law profiles show that $\psi(r)$ is approximately Gaussian in appearance. We thus approximate the field of the fundamental mode as

$$\varphi(r) \cong A e^{-(1/2)(r/r_0)^2} \qquad (E.15)$$

where $r_0$ is defined as the spot-size, i.e., at which the intensity equals to $e^{-1}$ of the maximum. Thus if the wave equation (Equation E.6) is multiplied by $r\psi(r)$ and using the identity

$$r\varphi \frac{\delta^2 \varphi}{\delta r^2} + \varphi \frac{\delta \varphi}{\delta r} = \frac{\delta}{\delta r}\left( r\varphi \frac{\delta \varphi}{\delta r} \right) - r\left( \frac{\delta \varphi}{\delta r} \right)^2 \qquad (E.16)$$

then, by integrating from 0 to infinitive and using $[r\psi(d\varphi/dr)]_0^\infty = 0$ we have

$$\beta^2 = \frac{\int_0^\infty \left[ -(\delta\varphi/\delta r)^2 + k^2 n^2(r)\varphi^2 \right] r\delta r}{\int_0^\infty r\varphi^2 \delta r} \qquad (E.17)$$

The procedure to find the spot-size is then followed by substituting $\psi(r)$ (Gaussian) in Equation E.15 into E.17 then differentiating and set $\delta^2\beta/\delta r$ evaluated at $r_0$ to zero, that is the propagation constant $\beta$ of the fundamental mode *must* give the largest value of $r_0$. Therefore, knowing $r_0$ *and* $\beta$ the fields $E_x$ and $H_y$ (Equation E.7) are fully specified.

### E.2.2.3.1    Step-Index Profile

Substituting the step-index profile given by Equation E.7 and $\psi(r)$ into Equation E.15 and then E.17 leads to an expression for $\beta$ in term of $r_0$ given by

$$V = NA \cdot k \cdot a = NA \frac{2\pi}{\lambda} a \tag{E.18}$$

The spotsize is thus evaluated by setting

$$\frac{\delta^2 \beta}{\delta r_0} = 0 \tag{E.19}$$

and $r_0$ is then found as

$$r_0^2 = \frac{a^2}{\ln V^2} \tag{E.20}$$

Substituting Equation E.20 into Equation E.18 we have

$$(a\beta)^2 = (akn_1)^2 - \ln V^2 - 1 \tag{E.21}$$

This expression is physically meaningful only when $V > 1$, that is when $r_0$ is positive which is naturally feasible.

### E.2.2.3.2    Gaussian Index Profile Fiber

Similarly, for the case of a Gaussian index profile, by following the procedures for step-index profile fiber we can obtain

$$(a\beta)^2 = (an_1k)^2 - \left(\frac{a}{r_0}\right)^2 + \frac{V^2}{\left((a/r_0)+1\right)} \tag{E.22}$$

and

$$r_0^2 = \frac{a^2}{V-1} \quad \text{by using} \quad \frac{\delta^2 \beta}{\delta^2 r_0} = 0 \tag{E.23}$$

That is maximizing the propagation constant of the guided waves. The propagation constant is at maximum when the "light ray" is very close to the horizontal direction. Substituting Equation E.23 into Equation E.22 we have

$$(a\beta)^2 = (akn_1)^2 - 2V + 1 \qquad\qquad (E.24)$$

thus Equations E.23 and E.24 are physically meaningful only when $V > 1 (r_0 > 0)$.

It is obvious from Equation E.23 that the spot-size of the optical fiber with a $V$-parameter of one is extremely large. It is very important that one must not design the optical fiber with a near unit value of the $V$-parameter; under this scenario all the optical field is distributed in the cladding region. In practice we observe that the spot-size is large but finite (observable). In fact if $V$ smaller than 1.5, the spot-size becomes large and in the next chapter this will be investigated in details.

## E.3   SIGNAL PROPAGATION IN OPTICAL FIBERS

Optical loss in optical fibers is one of the two main fundamental limiting factors as it reduces the average optical power reaching the receiver. The optical loss is the sum of three major components: intrinsic loss, micro-bending loss and splicing loss.

### E.3.1   Attenuation

#### E.3.1.1   Attenuation: Intrinsic or Material Absorption Losses

Intrinsic loss consists mainly of absorption loss due to OH impurities and Rayleigh scattering loss. The intrinsic is a function of $\lambda^{-4}$. Thus the longer the operating wavelength, the lower the loss. However, it also depends on the transparency of the optical materials that is used to form the optical fibers. For silica fiber the optical material loss is low over the wavelength range 0.8–1.8 μm. Over this wavelength range there are three optical windows in which optical communication are utilized. The first window over the central wavelength 810 nm is about 20.0 nm spectral window over the central wavelength. The second and third windows most commonly used in present optical communications are over 1,300 and 1,550 nm with a range of 80 nm and 40 nm, respectively. The intrinsic losses are about 0.3 and 0.15 dB/km at 1,310 and 1,550 nm regions respectively (Figure E.8).

This is a few hundred thousand times improvement over the original transmission of signal over 5.0 m with a loss of about 60 dB/km. Most communication fibers systems are operating at 1,300 nm due to their minimum dispersion at this range. For "power hungry" system optical or extra-long systems should operate at 1,550 nm. The absorption loss in silica glass is composed mainly of ultraviolet (UV) and infrared (IR) absorption tales of pure silica. The IR absorption tale of pure silica has been shown due to the vibration of the basic tetrahedron and thus strong resonances occurs around 8–13 μm with a loss about $10^{-10}$ dB/km. This loss is shown in Curve IR of Figure E.1. Overtones and combinations of these vibrations lead to various absorption peaks in the low wavelength range as shown by Curve UV. Various impurities that also lead to spurious absorption effects in the wavelength range of interest (1.2–1.6 μm) are transition metal ions and water in the form of OH ions. These sources of absorptions have been practically reduced in recent years. The Raleigh scattering loss, $L_R$, which is due to microscopic non-homogeneities of the material, shows a $\lambda^{-4}$ dependence and is given by

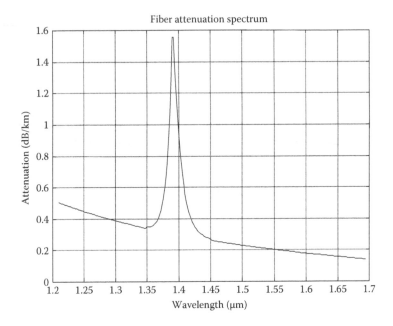

**FIGURE E.8**  Attenuation of optical signals as a function of wavelength. The minimum loss at wavelength: at $\lambda = 1.3$ μm about 0.3 dB/km and at $\lambda = 1.5$ μm loss of about 0.13 dB/km. For cabled fibers the attenuation factor at 1550 nm is 0.25 dB/km.

$$L_R = (0.75 + 4.5\Delta)\lambda^{-4} \text{ dB/km} \qquad (E.26)$$

where
   $\Delta$ is the relative index difference as defined above
   $\lambda$ is the wavelength in μm

Thus to minimize the loss $\Delta$ should be made as low as possible.

### E.3.1.2  Waveguide Losses

The losses due to waveguide structure arise from power leakage, bending, micro-bending of the fiber axis and defects and joints between fibers. The power leakage is significant only for depressed cladding fibers. When a fiber is bent, the plane wave fronts associated with the guided mode are pivoted at the center of the curvature and their longitudinal velocity along the fiber axis increases with the distance from the center of curvature. As the fiber is bent further over a critical curve, the phase velocity would exceed that of plane wave in the cladding and radiation occurs.

The bending loss $L_B$ for a radius $R$, the radius of curvature, is given by

$$L_B = -10\log_{10}(1 - 890)\frac{r_0^6}{\lambda^4 R^2} \quad \text{for silica} \qquad (E.27)$$

Microbending loss results from power coupling from the guided fundamental mode of the fiber to radiation modes. This coupling takes place when the fiber axis is bent randomly in a high spatial frequency. Such bending can occur while packing of the fiber during the cabling process. The microbending loss of a SM fiber is a function of the fundamental mode spot-size $r_0$. Fibers with large spot-size are extremely sensitive to microbending. It is therefore desirable to design the fiber to have as small a spot-size as possible to minimize bending loss. The microbending loss can be expressed by the relation

$$L_m = 2.15 \times 10^{-4} r_0^6 \lambda^{-4} L_{mm} \text{ dB/km} \tag{E.28}$$

where $L_{mm}$ is the microbending loss of a 50 µm core multimode fiber having a $NA$ of 0.2.

Ultimately, the fibers will have to be spliced together to form the final transmission link. With fiber cable that averages 0.4–0.6 dB/km, splice loss in excess of 0.2 dB/splice drastically reduces the non-repeated distance that can be achieved. It is therefore extremely important that the fiber be designed such that splicing loss be minimized.

### E.3.1.3  Attenuation Coefficient

Under a general conditions of power attenuation inside an optical fiber the attenuation coefficient of the optical power $P$ can be expressed as

$$\frac{dP}{dz} = -\alpha P \tag{E.29}$$

where $\alpha$ is the attenuation factor in linear scale. This attenuation coefficient can include all effects of power loss when signals are transmitted though the optical fibers. Considering optical signals with an average optical power entering at the input of the fiber length $L$ is $P_{in}$ and $P_{out}$ is the output optical power, then we have $P_{in}$ and $P_{out}$ are related to the attenuation coefficient $\alpha$ as

$$P_{out} = P_{in} e^{(-\alpha L)} \tag{E.30}$$

It is customary to express $\alpha$ in dB/km by using the relation

$$\alpha(\text{dB/km}) = -\frac{10}{L} \log_{10} \left( \frac{P_{out}}{P_{in}} \right) = 4.343\alpha \tag{E.31}$$

Standard optical fibers with a small $\Delta$ would exhibit a loss of about 0.2 dB/km, i.e., the purity of the silica is very high. Such purity of a bar of silica would allow us to see though a 1-km glass bar a person standing at the other end without distortion! The attenuation curve for silica glass is shown in Figure E.1.

### E.3.2   DISTORTION

Consider *a monochromatic* field given by

$$E_x = A\cos(\omega t - \beta z) \tag{E.32}$$

where
   $A$ is the wave amplitude
   $\omega$ is the radial frequency
   $\beta$ is the propagation constant along the $z$-direction

If setting $(\omega t - \beta z)$ constant, the wave phase velocity is given by

$$v_p = \frac{dz}{dt} = \frac{\omega}{\beta} \tag{E.33}$$

Now consider the propagating wave consists of two monochromatic fields of frequencies of frequencies $\omega + \delta\omega$; $\omega - \delta\omega$

$$E_{x1} = A\cos[(\omega + \delta\omega)t - (\beta + \delta\beta)z)] \tag{E.34}$$

$$E_{x2} = A\cos[(\omega - \delta\omega)t - (\beta - \delta\beta)z)] \tag{E.35}$$

The total field is then given by:

$$E_x = E_{x1} + E_{x2} = 2A\cos(\omega t - \beta z)\cos(\delta\omega t - \delta\beta z) \tag{E.36}$$

If $\omega \gg \delta\omega$ then $\cos(\omega t - \beta z)$ varies much faster than $\cos(\delta\omega t - \delta\beta z)$, hence by setting $(\delta\omega t - \delta\beta z)$ *invariant,* we can define the group velocity as

$$v_g = \frac{d\omega}{d\beta} \rightarrow v_g^{-1} = \frac{d\beta}{d\omega} \tag{E.37}$$

The group delay $t_g$ per unit length (setting $L$ at 1.0 km) is thus given as

$$t_g = \frac{L(of\ 1\ km)}{v_g} = \frac{d\beta}{d\omega} \tag{E.38}$$

The pulse spread $\Delta\tau$ per unit length due to group delay of light sources of spectral width $\sigma_\lambda$, i.e., the Full-Width-Half-Mark (FWHM) of the optical spectrum of the light source, is

$$\Delta\tau = \frac{dt_g}{d\lambda}\sigma_\lambda \tag{E.39}$$

The spread of the group delay due to the spread of source wavelength can be in ps/km. Thus the linewidth of the light source contributes significantly to the distortion of optical signal transmitted through the optical fiber due to the fact that the delay differences between the guided modes carried by the spectral components of the lightwaves. Hence the narrower the source linewidth, the less dispersed the optical pulses are. Typical linewidth of Fabry–Perot semiconductor lasers is about 1–2.0 nm while the DFB (Distributed Feedback) laser would exhibit a linewidth of 100 MHz (how many nm is this 100 MHz optical frequency equivalent to?). Later we will see that under the case that the source linewidth is very narrow such as the external cavity laser (ECL) than the components of the modulated sources, the bandwidth of the channel would play the principal role in the distortion.

Optical signal traveling along a fiber becomes increasingly distorted. This distortion is a consequence of *intermodal* delay effects and *intramodal* dispersion. Intermodal delay effects are significant in multimode optical fibers due to each mode having different value of group velocity at a specific frequency. While intermodal dispersion is pulse spreading that occurs within a single mode. It is the result of the group velocity being a function of the wavelength $\lambda$ and is therefore referred as chromatic dispersion.

Two main causes of intermodal dispersion are: (1) material dispersion which arises from the variation of the refractive index $n$ ($\lambda$) as a function of wavelengths. This causes a wavelength dependence of the group velocity of any given mode and (2) waveguide dispersion, which occurs because the mode propagation constant $\beta(\lambda)$ is a function of wavelength and core radius $a$ and the refractive index difference.

The group velocity associated with the fundamental mode is frequency dependent because of chromatic dispersion. As a result different spectral components of the light pulse travel at different group velocities, a phenomenon referred to as *the group-velocity-dispersion* (GVD), intra-modal dispersion or as material dispersion and waveguide dispersion.

### E.3.2.1   Material Dispersion

The refractive index of silica as a function of wavelength is shown in Figure E.9. The refractive index is plotted over the wavelength region of 1.0–2.0 μm which is the most important range for silica base optical communications systems as the loss is lowest at 1,300 and 1,550 nm windows.

The propagation constant $\beta$ of the fundamental mode guided in the optical fiber can be written as

$$\beta(\lambda) = \frac{2\pi n(\lambda)}{\lambda} \tag{E.40}$$

The group delay $t_{gm}$ per unit length then Equation E.40 can be obtained as

$$t_{gm} = \frac{d\beta}{d\omega} \tag{E.41}$$

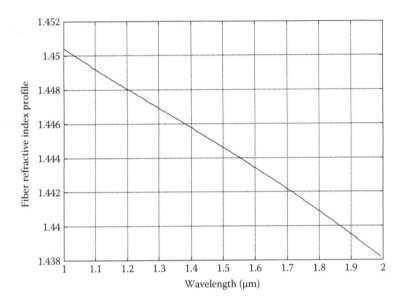

**FIGURE E.9** Variation in the refractive index as a function of optical wavelength of silica.

where we can replace

$$dω = d\left(\frac{2\pi c}{\lambda}\right) = -\frac{2\pi c}{\lambda^2} d\lambda \qquad (E.42)$$

Thus, leading to

$$t_{gm} = -\frac{\lambda^2}{2\pi c}\frac{d\beta}{d\lambda} \qquad (E.43)$$

Substituting Equation E.40 to Equation E.43 we have

$$t_{gm} = \frac{1}{c}\left[n(\lambda) - \frac{\lambda dn(\lambda)}{d\lambda}\right] \qquad (E.44)$$

Thus the pulse dispersion per unit length $\Delta\tau_m/\Delta\lambda$ due to material (using Equation E.44) for a source having RMS spectral width $\sigma_\lambda$ of

$$\Delta\tau_m = -\frac{\lambda}{c}\frac{d^2n}{d\lambda^2}\sigma_\lambda \qquad (E.45)$$

if setting $\Delta\tau_m = M(\lambda)\sigma_\lambda$, thence

$$M(\lambda) = -\frac{\lambda}{c}\frac{d^2 n}{d\lambda^2} \qquad (E.46)$$

$M(\lambda)$ is assigned as *the material dispersion factor or "material dispersion parameter,"* its unit is commonly expressed in ps/(nm km). Thus if the refractive index can be expressed as a function of the optical wavelength then the material dispersion can be calculated. In fact in practice optical material engineers have to characterize all optical properties of new materials.

The refractive index $n(\lambda)$ can usually be expressed in Sellmeier's dispersion formula as

$$n^2(\lambda) = 1 + \sum_k \frac{G_k \lambda^2}{\left(\lambda^2 - \lambda_k^2\right)} \qquad (E.47)$$

where

$G_k$ are Sellmeier's constants

$k$ is integer and normally taken a range of $k = 1-3$

In late-1970s, several silica-based glass materials have been manufactured and their properties are measured. The refractive indices are usually expressed using Sellmeier's coefficients. These coefficients for several optical fiber materials are given in Table E.1.

By using curve fitting, the refractive index of pure silica $n(\lambda)$ can be expressed as:

$$n(\lambda) = c_1 + c_2 \lambda^2 + c_3 \lambda^{-2} \qquad (E.48)$$

where

$c_1 = 1.45084$

$c_2 = -0.00343 \ \mu m^{-2}$

$c_3 = 0.00292 \ \mu m^2$

---

**TABLE E.1**

**Sellmeier's Coefficients for Several Optical Fiber Silica-Based Materials with Germanium Doped in the Core Region**

| Sellmeier's Constants | Germanium Concentration, C (mol%) | | | |
| --- | --- | --- | --- | --- |
| | 0 (Pure Silica) | 3.1 | 5.8 | 7.9 |
| $G_1$ | 0.6961663 | 0.7028554 | 0.7088876 | 0.7136824 |
| $G_2$ | 0.4079426 | 0.4146307 | 0.4206803 | 0.4254807 |
| $G_3$ | 0.8974794 | 0.8974540 | 0.8956551 | 0.8964226 |
| $\lambda_1$ | 0.0684043 | 0.0727723 | 0.0609053 | 0.0617167 |
| $\lambda_2$ | 0.1162414 | 0.1143085 | 0.1254514 | 0.1270814 |
| $\lambda_3$ | 9.896161 | 9.896161 | 9.896162 | 9.896161 |

Thus from Table E.1 and either Equation E.48, we can use Equation E.46 to determine the material dispersion factor for certain wavelength range. Figure E.10 illustrates the time-domain signal and its corresponding spectrum when the carrier modulated by this signal. Figure E.11 illustrates the effects on the complex signal using vector phasor diagram representing the complex envelope. Figure E.12 then illustrate the temporal signal by the magnitude of the signal complex envelope when not sinusoidal or the envelope subject to nonlinear distortions.

For the doped core of the optical fiber the Sellmeier's expression Equation E.47 can approximated by using curve fitting technique to approximate it to the form in (Equation E.48). The material dispersion factor $M(\lambda)$ becomes zero at wavelengths around 1,350 nm and about −10 ps/(nm km) at 1,550 nm. However the attenuation at 1,350 nm is about 0.4 dB/km compared with 0.2 dB/km at 1,550 nm, as shown in Table E.1. Plots of the material dispersion factors and their total due to material and waveguide group velocity as a function of wavelength are shown in Figure E.13.

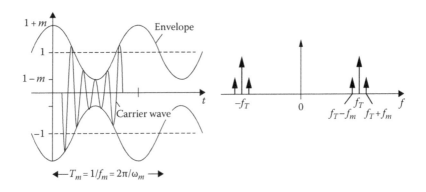

**FIGURE E.10**   Time signal and spectrum.

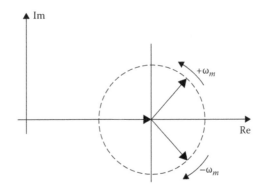

**FIGURE E.11**   Vector phasor diagram of the complex envelope.

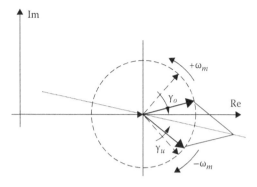

**FIGURE E.12** Magnitude of complex envelope when not sinusoidal, the envelope subject to nonlinear distortions.

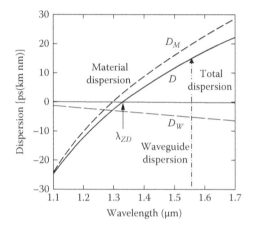

**FIGURE E.13** Chromatic dispersion factor of SSMF. Dotted curves representing the material dispersion factor as a function of optical wavelength for silica base optical fiber with a zero dispersion wavelength at 1,290 nm. This curve is generated as an example for standard single mode optical fibers which are currently installed throughout the world. The total dispersion is around +17 ps/(nm km) at 1,550 nm and almost zero at 1,310 nm. Can we estimate the waveguide dispersion curve for the standard single mode (SM) optical fiber at around 1,300 and 1,550 nm windows?

## E.3.2.2 Waveguide Dispersion

The effect of waveguide dispersion can be approximated by assuming that the refractive index of the material is independent of wavelength. Let us now consider the group delay, i.e., the time required for a mode to travel along a fiber of length $L$. This kind of dispersion depends strongly on $\Delta$ and $V$ parameters. To make the results of fiber parameters, we define a *normalized propagation constant* $b$ as

$$b = \frac{(\beta^2/k^2) - n_2^2}{n_1^2 - n_2^2} \tag{E.49}$$

for small $\Delta$. We note that the $\beta/k$ is in fact the "*effective*" *refractive index* of the guided optical mode propagating along the optical fiber, that is the guided waves traveling the axial direction of the fiber "sees" it as a medium with a refractive index of an equivalent "effective" index.

In case that that fiber is a weakly guiding waveguide with the effective refractive index takes a value of significantly close to that of the core or cladding index, Equation E.49 can then be approximated by

$$b \cong \frac{\beta/k - n_2}{n_1 - n_2} \tag{E.50}$$

solving Equation E.50 for $\beta$, we have

$$\beta = n_2 k(b\Delta + 1)$$

the group delay for waveguide dispersion is then given by (per unit length)

$$t_{wg} = \frac{d\beta}{d\omega} = \frac{1}{c}\frac{d\beta}{dk} \tag{E.51}$$

$$t_{wg} = \frac{1}{c}\left[n_1 + n_2\Delta\frac{d(bk)}{dk}\right] = \frac{1}{c}\left[n_1 + n_2\Delta\frac{d(bk)}{dk}\right] = \frac{1}{c}\left[n_1 + n_2\Delta\frac{d(bV)}{dV}\right] \tag{E.52}$$

Equation E.52 can be obtained from Equation E.51 by using the expression of $V$. Thus the pulse spreading $\Delta\tau_\omega$ due to the waveguide dispersion per unit length by a source having an optical bandwidth (or linewidth $\sigma_\lambda$) is given by

$$\Delta\tau_\omega = \frac{dt_{gw}}{d\lambda}\sigma_\lambda = -\frac{n_2\Delta}{c\lambda}V\frac{d^2(Vb)}{dV^2}\sigma_\lambda \tag{E.53}$$

and similar to the definition of the material dispersion factor, the *waveguide dispersion factor or "waveguide dispersion parameter"* can be defined as:

$$D(\lambda) = -\frac{n_2(\lambda)\Delta}{c\lambda}V\frac{d^2(Vb)}{dV^2} \tag{E.54}$$

This waveguide factor can take unit of ps/(nm km). In the range of $0.9 < \lambda/\lambda_c < 2.6$, the factor $V(d^2(Vb)/dV^2)$ can be approximated (to <5% error) by

$$V\frac{d^2(Vb)}{dV^2} \cong 0.080 + 0.549(2.834 - V)^2 \tag{E.55}$$

or alternatively using the definition of cut-off wavelength and the expression of the $V$-parameters we obtain

$$V \frac{d^2(Vb)}{dV^2} \cong 0.080 + 3.175\left(1.178 - \frac{\lambda_c}{\lambda}\right)^2 \tag{E.56}$$

It is not so difficult to prove the equivalent equation of Equation E.51 through Equation E.56. Further the sign assignment of the material and waveguide dispersion factors must be the same. Otherwise a negative and positive of these dispersion factors would create confusion. Can you explain what would happen to the pulse if it is transmitted through an optical fiber having a total negative dispersion factor?

Thus from Equations E.56 and E.55 we can calculate the waveguide dispersion factor and hence the pulse dispersion factor for a particular source spectral width $\sigma_\lambda$. It is noted that the dispersion considered in this chapter is for step-index fiber only. For grade index fiber, ESI parameters must be found and the chromatic dispersion can then be calculated. Figure E.13 shows a design of single mode optical fibers with the total dispersion factor contributed by material and waveguide effects.

### E.3.2.2.1  Alternative Expression for Waveguide Dispersion Parameter

Alternatively the waveguide dispersion parameter can be expressed as function of the propagation constant $\beta$ by using $\omega = 2\pi c/\lambda_g$ and Equation E.56, the waveguide dispersion factor can be written as

$$D(\lambda) = -\frac{2\pi c}{\lambda^2}\beta_2 = -\frac{2\pi c}{\lambda^2}\frac{d\beta^2}{d\omega^2} \tag{E.57}$$

$\beta_2$ is defined also as the GVD or group delay dispersion factor. Thus the waveguide dispersion factor is directly related to the second order derivative of the propagation constant with respect to the optical radial frequency. The cladding material is pure silica is shown in Figure E.13, the curves of the material dispersion factor, waveguide dispersion factor and total dispersion for a single mode optical fiber with non-uniform refractive index profile in the core.

### E.3.2.2.2  Higher Order Dispersion

We observe that the bandwidth-length product of the optical fiber can be extended to infinitive if the system is operating at the wavelength at which the total dispersion factor is zero. However the dispersive effects do not disappear completely at this zero-dispersion wavelength. Optical pulses still experience broadening because of higher order dispersion effects. It is easily imagined that the total dispersion factor cannot be made zero "flatten" over the optical spectrum. This is higher order dispersion which govern by the slope of the total dispersion curve, called the dispersion slope $S = d[D(\lambda) + M(\lambda)]/d\lambda$; $S(\lambda)$ can thus be expressed as:

$$S(\lambda) = \left(\frac{2\pi c}{\lambda^2}\right)^2 \frac{d^3\beta}{d\lambda^3} + \left(\frac{4\pi c}{\lambda^3}\right)\frac{d^2\beta}{d\lambda^2} \tag{E.58}$$

$S(\lambda)$ is also known as the differential-dispersion parameter.

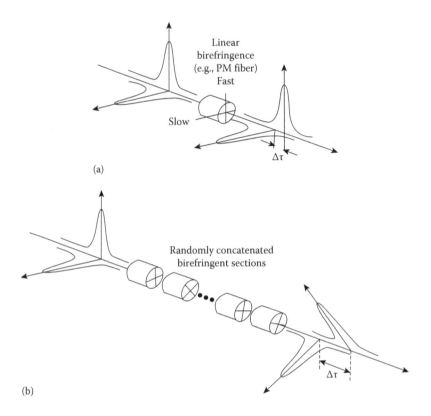

**FIGURE E.14** Conceptual model of PMD (a) simple birefringence device and (b) randomly concatenated birefringence.

### E.3.3 POLARIZATION MODE DISPERSION

The delay between two PSPs is normally negligibly small at bit rate less than 10 Gbps. However, at high bit rate and in ultra long-haul transmission, PMD severely degrades the system performance [5–8] (Figure E.14). The differential delay between the polarized modes create the PMD effect can be illustrated in Figure E.15. The instantaneous value of DGD ($\Delta\tau$) varies along the fiber and follows a Maxwellian distribution [9,10].

## E.4 TRANSFER FUNCTION OF SINGLE MODE FIBERS

### E.4.1 LINEAR TRANSFER FUNCTION

The treatment of the propagation of modulated lightwaves through single mode fiber in the linear and nonlinear regimes has been well documented [11–16]. For completeness of the transfer function of single mode optical fibers, in this section we restrict our study the frequency transfer function and impulse responses of the fiber to the linear region of the media. Furthermore, the delay term in the NLSE can be ignored, as

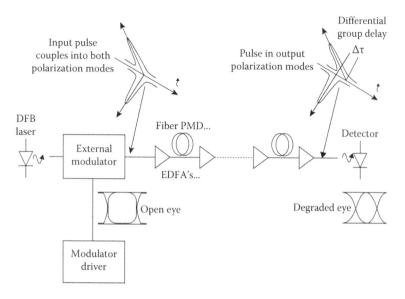

**FIGURE E.15** Effect of PMD in a digital optical communication system, degradation of the received eye diagram.

it has no bearing on the size and shape of the pulses. From NLSE we can thus model the fiber simply as a quadratic phase function. This is derived from the fact that the nonlinear term of NLSE can be removed and Taylors series approximation around the operating frequency (central wavelength) can be obtained and a frequency and impulse responses of the single mode fiber. The input-output relationship of the pulse can therefore be depicted. Equation E.59 expresses the time-domain impulse response $h(t)$ and the frequency domain transfer function $H(\omega)$ as a Fourier transform pair:

$$h(t) = \sqrt{\frac{1}{j4\pi\beta_2}} \exp\left(\frac{jt^2}{4\beta_2}\right) \leftrightarrow H(\omega) = \exp(-j\beta_2\omega^2) \qquad (E.59)$$

where $\beta_2$ is well known as the group velocity dispersion (GVD) parameter. The input function $f(t)$ is typically a rectangular pulse sequence and $\beta_2$ is proportional to the length of the fiber. The output function $g(t)$ is the dispersed waveform of the pulse sequence. The propagation transfer function in Equation E.59 is an exact analogy of diffraction in optical systems (see Item 1, Table 2.1, p. 14, Papoulis [17]). Thus, the quadratic phase function also describes the diffraction mechanism in one-dimensional optical systems, where distance $x$ is analogous to time $t$. The establishment of this analogy affords us to borrow many of the imageries and analytical results that have been developed in the diffraction theory. Thus, we may express the step response $s(t)$ of the system $H(\omega)$ in terms of Fresnel cosine and sine integrals as follows:

$$s(t) = \int_0^t \sqrt{\frac{1}{j4\pi\beta_2}} \exp\left(\frac{jt^2}{4\beta_2}\right) dt = \sqrt{\frac{1}{j4\pi\beta_2}}\left[C\left(\sqrt{1/4\beta_2 t}\right) + jS\left(\sqrt{1/4\beta_2 t}\right)\right] \quad (E.60)$$

with

$$C(t) = \int\limits_0^t \cos\left(\frac{\pi}{2}\tau^2\right)d\tau$$

$$S(t) = \int\limits_0^t \sin\left(\frac{\pi}{2}\tau^2\right)d\tau$$

(E.61)

where $C(t)$ and $S(t)$ are the Fresnel cosine and sine integrals.

Using this analogy, one may argue that it is always possible to restore the original pattern $f(x)$ by refocusing the blurry image $g(x)$ (e.g., image formation, item 5, Table 2.1, Ref. [15]). In the electrical analogy, it implies that it is possible to compensate the quadratic phase media perfectly. This is not surprising. The quadratic phase function $H(\omega)$ in Equation E.59 is an all-pass transfer function, thus it is always possible to find an inverse function to recover $f(t)$. One can express this differently in information theory terminology that the quadratic phase channel has a theoretical bandwidth of infinity; hence its information capacity is infinite. Shannon's channel capacity theorem states that there is no limit on the reliable rate of transmission through the quadratic phase channel.

Figure E.16 shows the pulse and impulse responses of the fiber. It is noted that only the envelope of the pulse is shown and the phase of the lightwave carrier is included as the complex values of the amplitudes. As observed the chirp of the carrier is significant at the edges of the pulse. At the center of the pulse, the chirp is almost negligible at some limited fiber length, thus the frequency of the carrier remains nearly the same as at its original starting value. One could obtain the impulse response quite easily but in this work we believe that the pulse response is much more relevant in the investigation of the uncertainty in the pulse sequence detection. Rather the impulse response is much more important in the process of equalization.

The uncertainty of the detection depends on the modulation formats and detection process. The modulation can be implemented by manipulation of the amplitude, the phase or the frequency of the carrier or both amplitude and phase or multi-sub-carriers such as the orthogonal frequency division multiplexing (OFDM) [16]. The amplitude detection would be mostly affected by the ripples of the amplitudes of the edges of the pulse. The phase of the carrier is mostly affected near the edge due to the chirp effects. However if differential phase detection is used then the phase change at the transition instant is the most important and the opening of the detected eye diagram. For frequency modulation the uncertainty in the detection is not very critical provided that the chirping does not enter into the region of the neighborhood of the center of the pulse in which the frequency of the carrier remains almost constant.

The picture changes completely if the detector/decoder is allowed only a finite time window to decode each symbol. In the convolution coding scheme for example, it is the decoder's constraint length that manifests due to the finite time window. In adaptive equalization scheme, it is the number of equalizer coefficients that determines the decoder window length. Since the transmitted symbols have already been

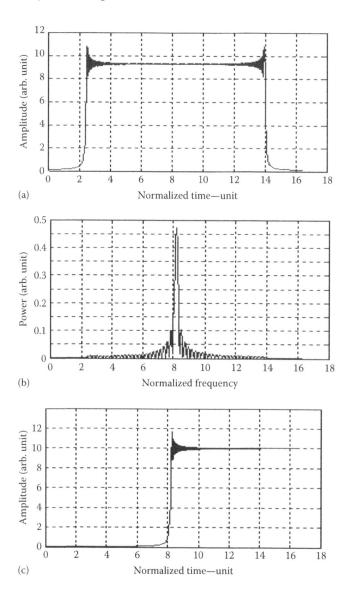

(a)

(b)

(c)

**FIGURE E.16** Rectangular pulse transmission of through a SMF (a) pulse response, (b) frequency spectrum, and (c) step response of the quadratic-phase transmittance function. Note Horizontal scale in normalized unit of time.

broadened by the quadratic phase channel, if they are next gated by a finite time window, the information received could be severely reduced. The longer the fiber, the more the broadening of the pulses is widened, hence the more uncertain it becomes in the decoding. It is the interaction of the pulse broadening on one hand, and the restrictive detection time window on the other, that gives rise to the finite channel capacity.

It is observed that the chirp occurs mainly near the edge of the pulses when it is in the near field region, about a few kms for standard single mode fibers. In this near field distance the accumulation of nonlinear effects is still very weak and thus this chirp effects dominate the behavior of the single mode fiber. The nonlinear Volterra transfer function presented in the next section would thus have minimum influence. This point is important for understanding the behavior of lightwaves circulating in shot length fiber devices in which the both linear and nonlinear effects are to be balanced such as active mode locked soliton and multi-bound soliton lasers [18,19]. In the far field the output of the fiber is a Gaussian like for the square pulse launched at the input. In this region the nonlinear effects would dominate over the linear dispersion effect as they have been accumulated over a long distance.

The linear time variant system such as the single mode fiber would have a transfer function of

$$H(f) = |H(f)| e^{-j\alpha(f)} \tag{E.62}$$

where $\alpha = \pi^2 \beta_2 L = (-\pi D L \lambda^2 / 2c)$ is proportional to the length $L$ and the dispersion factor $D(\lambda)$ (ps/(nm km)). The phase of the frequency transfer response is a quadratic function of the frequency thus the group delay would follow a linear relationship with respect to the frequency as observed in Figure E.17. The frequency response in amplitude term is infinite and is a constant while the phase response is a quadratic function with respect to the frequency of the base band signals. The carrier is chirped accordingly as observed in Figures E.18 and E.19. The chirping effect is very significant near the edge of the rectangular pulse and almost nil at the center of the pulse, in the near field region of less than 1 km of standard single mode fiber. In the far field region the pulse becomes Gaussian like. Thus the response of the fiber in the linear region can be seen as shown in Figure E.20 for a Gaussian pulse input to the fiber. The output pulse is also Gaussian by taking the Fourier transform of the input pulse and multiplied by the fiber transfer function. Thence an inverse Fourier would indicate the output pulse shape follows a Gaussian profile.

This leads to a rule of thumb for consideration of the scaling of the bit rate and transmission distance as "*Given that a modulated lightwave of a bit rate B can be transmitted over a maximum distance L of single mode optical fiber with a BER of error-free level, than if the bit rate is halved then the transmission distance can be increase by four times.*" For example for 10 Gbps amplitude shift keying modulation format signals can be transmitted over 80 km of standard single mode optical fiber then at 40 Gbps only 5 km can be transmitted for a bit error rate (BER) of $10^{-9}$.

Figure E.17 shows a typical frequency response in magnitude phase and the low ps property. Ideally one could see that the amplitude response of the fiber is constant throughout all frequency if no attenuation or constant attenuation throughout all frequency range. Only the phase of the lightwaves is altered, that is the chirping of the carrier. It is this chirping of the carrier that would then limit the response frequency range. In the case of phase modulation this chirp would rotate the constellation of signals modulated by the QAM s given in Chapter 1. The digital signal processing could then be used to determine the exact dispersion and hence the

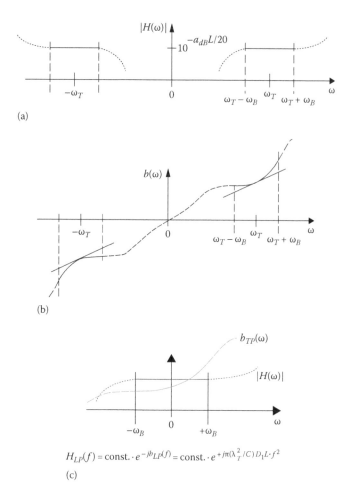

$$H_{LP}(f) = \text{const.} \cdot e^{-jb_{LP}(f)} = \text{const.} \cdot e^{+j\pi(\lambda_T^2/C)D_1 L \cdot f^2}$$

(c)

**FIGURE E.17**  Frequency response of a single mode optical fiber (a) magnitude, (b) phase response in bandpass regime, and (c) baseband equivalence.

rotation of the constellation by an angle such that it is recovered back to its original position. The chirp of the carrier can be seen in Figure E.18 in which the chirp is much less at the center but heavily chirped near the edge of the pulse. The step responses shown in Figures E.19 and E.20 indicate the damping oscillation of the step pulse which is due to the chirp of the carrier as we also observe from the calculated impulse and step responses given in Figure E.16. Figure E.19 shows the Gaussian-like impulse response when the transmission distance is large. This is the typical pulse shape in long haul nondispersion compensating transmission. These dispersive pulse sequences are then coherently detected and processed by the digital signal processors. The number of samples must be long enough to cover the dispersive sequence and the number of taps of the finite impulse response (FIR) filter must be high enough to ensure the whole dispersive pulse is covered and fallen within the

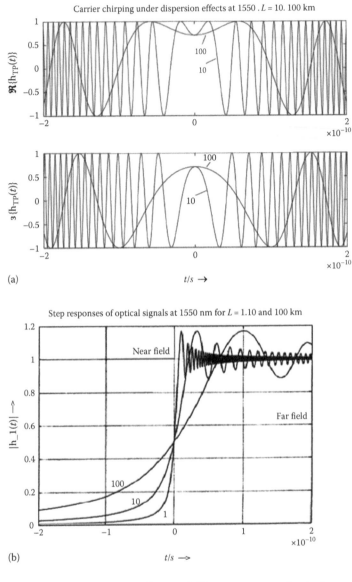

**FIGURE E.18** (a) Carrier chirping effects and (b) step response of a single mode optical fiber of $L = 1$, 10 and 100 km.

filter length. So the longer the transmission length the higher the number of taps of the FIR is. This is the first step in the DSP-based optical receiver, by using the constant modulus amplitude (CMA) algorithm to compensate for the dispersion effects before compensating and recovering the phase constellation of the modulated and transmitted channels. It is noted that the chirp of the pulse envelope is much higher when the pulse in the near field region than that in the far field as observed in the

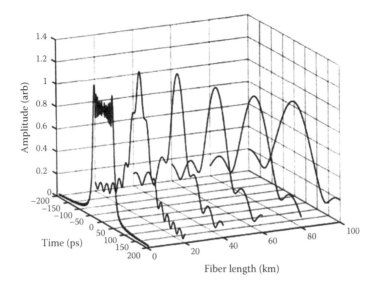

**FIGURE E.19** Pulse response from near field (~<2 km) to far field (>80 km).

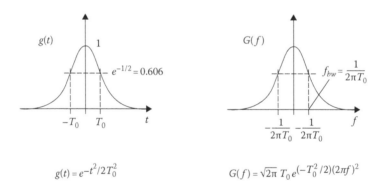

$$g(t) = e^{-t^2/2T_0^2}$$

$$G(f) = \sqrt{2\pi}\, T_0\, e^{(-T_0^2/2)(2\pi f)^2}$$

**FIGURE E.20** Fiber response to Gaussian pulse. Gaussian shape → Gaussian shape after propagation under chromatic dispersion.

step responses given in Figures E.18a and E.19. Thus if a Gaussian pulse is launched and propagating in the single mode fiber we would obtain Gaussian pulse shape at the output (see Figure E.19) provided that the fiber length is sufficient long, typically more than 20 km which is commonly met in real transmission system.

### E.4.2 NONLINEAR FIBER TRANSFER FUNCTION

The weakness of most of the recursive methods in solving the NLSE is that they do not provide much useful information to help the characterization of nonlinear effects. The Volterra series model provides an elegant way of describing a system's nonlinearities, and enables the designers to see clearly where and how the nonlinearity affects the system performance. Although Refs. [20,21] have given an outline of the kernels of the transfer function using the Volterra series, it is necessary for clarity and physical representation of these functions, brief derivations are given here on the nonlinear transfer functions of an optical fiber operating under nonlinear conditions.

The Volterra series transfer function of a particular optical channel can be obtained in the frequency-domain as a relationship between the input spectrum $X(\omega)$ and the output spectrum $Y(\omega)$, as

$$Y(\omega) = \sum_{n=1}^{\infty} \int_{-\infty}^{\infty} \cdots \int_{-\infty}^{\infty} H_n(\omega_1,\ldots,\omega_{n-1},\omega-\omega_1-\cdots-\omega_{n-1})$$

$$\times X(\omega_1)\cdots X(\omega_{n-1})X(\omega-\omega_1-\cdots-\omega_{n-1})d\omega_1\cdots d\omega_{n-1} \tag{E.63}$$

where $H_n(\omega_1, \ldots, \omega_n)$ is the $n$th-order frequency domain Volterra kernel including all signal frequencies of orders 1 to $n$. The wave propagation inside a single-mode fiber can be governed by a simplified version of the NLSE given above in this chapter with only the SPM effect included as

$$\frac{\partial A}{\partial z} = -\frac{\alpha_0}{2}A - \beta_1\frac{\partial A}{\partial t} - j\frac{\beta_2}{2}\frac{\partial^2 A}{\partial t^2} - \frac{\beta_3}{6}\frac{\partial^3 A}{\partial t^3} + j\gamma|A|^2 A \tag{E.64}$$

where $A = A(t,z)$. The proposed solution of the NLSE can be written with respect to the VSTF model of up to fifth order as

$$A(\omega,z) = H_1(\omega,z)A(\omega) + \int_{-\infty}^{\infty}\int_{-\infty}^{\infty} H_3(\omega_1,\omega_2,\omega-\omega_1+\omega_2,z)$$

$$\times A(\omega_1)A^*(\omega_2)A(\omega-\omega_1+\omega_2)d\omega_1 d\omega_2$$

$$+ \int_{-\infty}^{\infty}\int_{-\infty}^{\infty}\int_{-\infty}^{\infty}\int_{-\infty}^{\infty} H_5(\omega_1,\omega_2,\omega_3,\omega_4,\omega-\omega_1+\omega_2-\omega_3+\omega_4,z)$$

$$\times A(\omega_1)A^*(\omega_2)A(\omega_3)A^*(\omega_4)$$

$$\times A(\omega-\omega_1+\omega_2-\omega_3+\omega_4)d\omega_1 d\omega_2 d\omega_3 d\omega_4 \tag{E.65}$$

where $A(\omega) = A(\omega,0)$, that is the amplitude envelop of the optical pulses at the input of the fiber. Taking the Fourier transform of Equation E.3 and assuming $A(t,z)$ is of sinusoidal form we have

$$\frac{\partial A(\omega,z)}{\partial z} = G_1(\omega)A(\omega,z)\int_{-\infty}^{\infty}\int_{-\infty}^{\infty} G_3(\omega_1,\omega_2,\omega-\omega_1+\omega_2)A(\omega_1,z)A^*(\omega_2,z)$$

$$\times A(\omega-\omega_1+\omega_2,z)d\omega_1 d\omega_2 \tag{E.66}$$

where

$$G_1(\omega) = -(\alpha_0/2) + j\beta_1\omega + j(\beta_2/2)\omega^2 - j(\beta_3/6)\omega_3$$
$$G_3(\omega_1,\omega_2,\omega_3) = j\gamma$$

$\omega$ is taking the values over the signal bandwidth and beyond in overlapping the signal spectrum of other optically modulated carriers while $\omega_1\cdots\omega_3$ are all also taking values over similar range as that of $\omega$. For general expression the limit of integration is indicted over the entire range to infinitive.

Substituting Equation E.65 with Equation E.66 and equating both sides, the kernels can be obtained after some algebraic manipulations

$$\frac{\partial}{\partial z}\left[ H_1(\omega,z)A(\omega) + \int_{-\infty}^{\infty}\int_{-\infty i}^{\infty} H_3(\omega_1,\omega_2,\omega-\omega_1+\omega_2,z)\, A(\omega_1)A^*(\omega_2) \right.$$

$$\times A(\omega-\omega_1+\omega_2)d\omega_1 d\omega_2$$

$$+ \int_{-\infty}^{\infty}\int_{-\infty}^{\infty}\int_{-\infty}^{\infty}\int_{-\infty}^{\infty} H_5(\omega_1,\omega_2,\omega_3,\omega_4,\omega-\omega_1+\omega_2-\omega_3+\omega_4,z)$$

$$\left. \times A(\omega_1)A^*(\omega_2)A(\omega_3)A^*(\omega_4)A(\omega-\omega_1+\omega_2-\omega_3+\omega_4)d\omega_1 d\omega_2 d\omega_3 d\omega_4 \right]$$

$$= G_1(\omega)\left[ H_1(\omega,z)A(\omega)i + \int_{-\infty}^{\infty}\int_{-\infty i}^{\infty} H_3(\omega_1,\omega_2,\omega-\omega_1+\omega_2,z) \right.$$

$$\times A(\omega_1)A^*(\omega_2)A(\omega-\omega_1+\omega_2)d\omega_1 d\omega_2$$

$$\left. + \int_{-\infty}^{\infty}\int_{-\infty}^{\infty}\int_{-\infty}^{\infty}\int_{-\infty}^{\infty} H_5(\omega_1,\omega_2,\omega_3,\omega_4,\omega-\omega_1+\omega_2-\omega_3+\omega_4,z)\, A(\omega_1)A^*(\omega_2)A(\omega_3)A^*(\omega_4) \right.$$

$$\left. \times A(\omega-\omega_1+\omega_2-\omega_3+\omega_4)d\omega_1 d\omega_2 d\omega_3 d\omega_4 \right] + \int_{-\infty}^{\infty}\int_{-\infty}^{\infty} G_3(\omega_1,\omega_2,\omega-\omega_1+\omega_2)$$

$$\times \left[ H_1(\omega_1,z)A(\omega_1) + \int_{-\infty}^{\infty}\int_{-\infty i}^{\infty} H_3(\omega_{11},\omega_{12},\omega_1-\omega_{11}+\omega_{12},z) \right.$$

$$\times A(\omega_{11})A^*(\omega_{12})A(\omega_1-\omega_{11}+\omega_{12})d\omega_{11}d\omega_{12}$$

$$+ \int_{-\infty}^{\infty}\int_{-\infty}^{\infty}\int_{-\infty}^{\infty}\int_{-\infty}^{\infty} H_5(\omega_{11},\omega_{12},\omega_{13},\omega_{14},\omega_1-\omega_{11}+\omega_{12}-\omega_{13}+\omega_{14},z)$$

$$\left. \times A(\omega_{11})A^*(\omega_{12})A(\omega_{13})A^*(\omega_{14})\times A(\omega_1-\omega_{11}+\omega_{12}-\omega_{13}+\omega_{14})d\omega_{11}d\omega_{12}d\omega_{13}d\omega_{14} \right]$$

$$\times \left[ H_1(\omega_1,z)A(\omega_1) + \int_{-\infty}^{\infty}\int_{-\infty i}^{\infty} H_3(\omega_{11},\omega_{12},\omega_1-\omega_{11}+\omega_{12},z)\times A(\omega_{11})A^*(\omega_{12}) \right.$$

$$\times A(\omega_1 - \omega_{11} + \omega_{12}) d\omega_{11} d\omega_{12} + \int_{-\infty}^{\infty} \int_{-\infty}^{\infty} \int_{-\infty}^{\infty} \int_{-\infty}^{\infty} H_5(\omega_{21}, \omega_{22}, \omega_{23}, \omega_{24},$$

$$\omega_2 - \omega_{21} + \omega_{22} - \omega_{23} + \omega_{24}, z)$$

$$\times A(\omega_{21}) A^*(\omega_{22}) A(\omega_{23}) A^*(\omega_{24}) \times A^*(\omega_2 - \omega_{21} + \omega_{22} - \omega_{23} + \omega_{24}) d\omega_{21} d\omega_{22} d\omega_{23} d\omega_{24} \Big]$$

$$\times \Big[ H_1(\omega - \omega_1 + \omega_2, z) A(\omega - \omega_1 + \omega_2) + \int_{-\infty}^{\infty} \int_{-\infty}^{\infty} H_3(\omega_{31}, \omega_{32}, \omega - \omega_1 + \omega_2 - \omega_{31} + \omega_{32}, z)$$

$$\times A(\omega_{31}) A^*(\omega_{32}) A(\omega - \omega_1 + \omega_2 - \omega_{31} + \omega_{32}) d\omega_{31} d\omega_{32}$$

$$+ \int_{-\infty}^{\infty} \int_{-\infty}^{\infty} \int_{-\infty}^{\infty} \int_{-\infty}^{\infty} H_5\big(\omega_{31}, \omega_{32}, \omega_{33}, \omega_{34}, \omega - \omega_1 + \omega_2 - \omega_{31} + \omega_{32} - \omega_{33} + \omega_{34},$$

$$z \times A(\omega_{31}) A^*(\omega_{32}) A(\omega_{33}) A^*(\omega_{34})\big)$$

$$\times A(\omega - \omega_1 + \omega_2 - \omega_{31} + \omega_{32} - \omega_{33} + \omega_{34}) \times d\omega_{31} d\omega_{32} d\omega_{33} d\omega_{34} \Big]\Big] \tag{E.67}$$

Equating the first order terms on both sides we obtain

$$\frac{\partial}{\partial z} H_1(\omega, z) = G_1(\omega) H_1(\omega, z) \tag{E.68}$$

Thus the solution for the first order transfer function Equation E.68 is then given by

$$H_1(\omega, z) = e^{G_1(\omega) z} = e^{(-(\alpha_0/2) + j\beta_1 \omega + j(\beta_2/2)\omega^2 - j(\beta_3/6)\omega^3) z} \tag{E.69}$$

This is in fact the linear transfer function of a single mode optical fiber with the dispersion factors $\beta_2$ and $\beta_3$ as already shown in the previous section.

Similarly for the third order terms we have

$$\frac{\partial}{\partial z} \int_{-\infty}^{\infty} \int_{-\infty}^{\infty} H_3(\omega_1, \omega_2, \omega - \omega_1 + \omega_2, z) \times A(\omega_1) A^*(\omega_2) A(\omega - \omega_1 + \omega_2) d\omega_1 d\omega_2$$

$$= \int_{-\infty}^{\infty} \int_{-\infty}^{\infty} G_3(\omega_1, \omega_2, \omega - \omega_1 + \omega_2) H_1(\omega_1, z) A(\omega_1) H_2^*(\omega_2, z)$$

$$\times A(\omega_2) H_1(\omega - \omega_1 + \omega_2) A(\omega - \omega_1 + \omega_2) d\omega_1 d\omega_2 \tag{E.70}$$

Now letting $\omega_3 = \omega - \omega_1 + \omega_2$ then it follows

$$\frac{\partial H_3(\omega_1, \omega_2, \omega_3, z)}{\partial z} = G_1(\omega_1 - \omega_2 + \omega_3) H_3(\omega_1, \omega_2, \omega_3, z)$$

$$+ G_3(\omega_1, \omega_2, \omega_3) H_1(\omega_1, z) H_1^*(\omega_2, z) H_1(\omega_3, z) \tag{E.71}$$

The third kernel transfer function can be obtained as

$$H_3(\omega_1,\omega_2,\omega_3,z) = G_3(\omega_1,\omega_2,\omega_3) \times \frac{e^{(G_1(\omega_1)+G_1^*(\omega_2)+G_1(\omega_3))z} - e^{G_1(\omega_1-\omega_2+\omega_3)z}}{G_1(\omega_1)+G_1^*(\omega_2)+G_1(\omega_3)-G_1(\omega_1-\omega_2+\omega_3)}$$

(E.72)

The fifth order kernel can similarly be obtained as

$$H_5(\omega_1,\omega_2,\omega_3,\omega_4,\omega_5,z)$$

$$= \frac{H_1(\omega_1,z)H_1^*(\omega_2,z)H_1(\omega_3,z)H_1^*(\omega_4,z)H_1(\omega_5,z)}{G_1(\omega_1)+G_1^*(\omega_2)+G_1(\omega_3)+G_1^*(\omega_4)+G_1(\omega_5)}$$
$$\frac{-H_1(\omega_1-\omega_2+\omega_3-\omega_4+\omega_5,z)}{-G_1(\omega_1-\omega_2+\omega_3-\omega_4+\omega_5)}$$

$$\times \left[ \frac{G_3(\omega_1,\omega_2,\omega_3-\omega_4+\omega_5)G_3(\omega_3,\omega_4,\omega_5)}{G_1(\omega_3)+G_1^*(\omega_4)+G_1(\omega_5)-G_1(\omega_3-\omega_4+\omega_5)} \right.$$

$$+ \frac{G_3(\omega_1,\omega_2-\omega_3+\omega_4,\omega_5)G_3^*(\omega_2,\omega_3,\omega_4)}{G_1^*(\omega_2)+G_1(\omega_3)+G_1^*(\omega_4)-G_1^*(\omega_2-\omega_3+\omega_4)}$$

$$\left. + \frac{G_3(\omega_1-\omega_2+\omega_3,\omega_4,\omega_5)G_3(\omega_1,\omega_2,\omega_3)}{G_1(\omega_1)+G_1^*(\omega_2)+G_1(\omega_3)-G_1(\omega_1-\omega_2+\omega_3)} \right]$$

$$- \frac{G_3(\omega_1,\omega_2,\omega_3-\omega_4+\omega_5)G_3(\omega_3,\omega_4,\omega_5)}{G_1(\omega_3)+G_1^*(\omega_4)+G_1(\omega_5)-G_1(\omega_3-\omega_4+\omega_5)}$$

$$\times \frac{H_1(\omega_1,z)H_1^*(\omega_2,z)H_1(\omega_1-\omega_2+\omega_3,z)-H_1(\omega_1-\omega_2+\omega_3-\omega_4+\omega_5,z)}{G_1(\omega_1)+G_1^*(\omega_2)+G_1(\omega_3-\omega_4+\omega_5)-G_1(\omega_1-\omega_2+\omega_3-\omega_4+\omega_5)}$$

$$- \frac{G_3(\omega_1,\omega_2-\omega_3+\omega_4,\omega_5)G_3^*(\omega_2,\omega_3,\omega_4)}{G_1^*(\omega_2)+G_1(\omega_3)+G_1^*(\omega_4)-G_1^*(\omega_2-\omega_3+\omega_4)}$$

$$\times \frac{H_1(\omega_1,z)H_1^*(\omega_2-\omega_3+\omega_4,z)H_1(\omega_5,z)-H_1(\omega_1-\omega_2+\omega_3-\omega_4+\omega_5,z)}{G_1(\omega_1)+G_1^*(\omega_2-\omega_3+\omega_4)+G_1(\omega_5)-G_1(\omega_1-\omega_2+\omega_3-\omega_4+\omega_5)}$$

$$- \frac{G_3(\omega_1-\omega_2+\omega_3,\omega_4,\omega_5)G_3(\omega_1,\omega_2,\omega_3)}{G_1(\omega_1)+G_1^*(\omega_2)+G_1(\omega_3)-G_1(\omega_1-\omega_2+\omega_3)}$$

$$\times \frac{H_1(\omega_1-\omega_2+\omega_3,z)H_1^*(\omega_4,z)H_1(\omega_5,z)-H_1(\omega_1-\omega_2+\omega_3-\omega_4+\omega_5,z)}{G_1(\omega_1-\omega_2+\omega_3)+G_1^*(\omega_4)+G_1(\omega_5)-G_1(\omega_1-\omega_2+\omega_3-\omega_4+\omega_5)}$$

(E.73)

Higher order terms can be derived with ease if higher accuracy is required. However in practice such higher order would not exceed the fifth rank. We can understand that for a length of a uniform optical fiber the 1st to $n$th order frequency spectrum transfer can be evaluated indicating the linear to nonlinear effects of the optical signals transmitting through it. Indeed the third and fifth order kernel transfer functions based on the Volterra series indicate the optical filed amplitude of the frequency components which contribute to the distortion of the propagated pulses. An inverse of these higher order functions would give the signal distortion in the time domain. Thus the VSTFs allow us to conduct distortion analysis of optical pulses and hence an evaluation of the bit-error-rate of optical fiber communications systems.

The superiority of such Volterra transfer function expressions allow us to evaluate each effect individually, especially the nonlinear effects so that we can design and manage the optical communications systems under linear or nonlinear operations. Currently this linear-nonlinear boundary of operations is critical for system implementation, especially for optical systems operating at 40 Gbps where linear operation and carrier suppressed return-to-zero format is employed. As a norm in series expansion the series need converged to a final solution. It is this convergence that would allow us to evaluate the limit of nonlinearity in a system.

## E.5   FIBER NONLINEARITY

This section re-visits these effects and their influence on the propagation of optical signals over long length of fibers. The nonlinearity and linear effects in optical fibers can be classified as shown in Figure E.21.

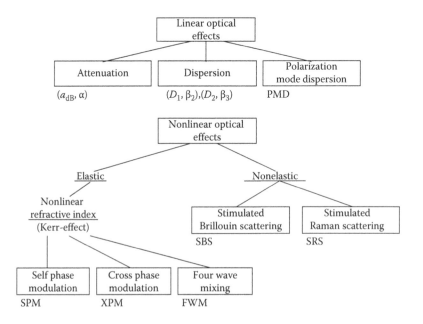

**FIGURE E.21**   Linear and nonlinear fiber properties in single mode optical fibers.

The elastic nonlinear effects are due to the change of refractive index of the core region by the intensity of the lightwaves, hence the corresponding change of their phases and interference effects or to these different propagation time of the components of the modulated spectra. The fiber RI is dependent on both operating wavelengths and lightwave intensity. This intensity-dependent phenomenon is known as the Kerr effect and is the cause of fiber nonlinear effects.

On the other hand the nonelastic nonlinear effects are due to the electronic vibration of the atoms of the impurities doped in the fiber core. The electronic vibrations can then transfer the energy of the lightwaves to another components whose frequency would be some distance away, for example 100 nm away or a few THz by Raman scattering or only 100 MHz amplitude modulation by Brillouin scattering.

### E.5.1   SPM, XPM EFFECTS

The power dependence of RI is expressed as:

$$n' = n + \bar{n}_2 \left( \frac{P}{A_{eff}} \right) \tag{E.74}$$

where
   $P$ is the average optical power of the guided mode
   $\bar{n}_2$ is the fiber nonlinear coefficient
   $A_{eff}$ is the effective area of the fiber

Fiber nonlinear effects include intra-channel SPM, inter-channel XPM, FWM, Stimulated Raman Scattering (SRS) and Stimulated Brillouin Scattering (SBS). SRS and SBS are not the main degrading factors as their effects are only getting noticeably large with very high optical power. On the other hand, FWM degrades severely the performance of an optical system with the generation of ghost pulses only if the phases of optical signals are matched with each other. However, with high local dispersions such as in SSMF, effects of FWM become negligible. In terms of XPM, its effects can be considered to be negligible in a DWDM system in the following scenarios (1) highly locally dispersive system and (2) large channel spacing. However, XPM should be taken into account for optical transmission systems deploying NZ-DSF fiber where local dispersion values are small. Thus, SPM is usually the dominant nonlinear effect for systems employing transmission fiber with high local dispersions, e.g., SSMF and DCF. The effect of SPM is normally coupled with the nonlinear phase shift $\phi_{NL}$ defined as

$$\phi_{NL} = \int_0^L \gamma P(z)dz = \gamma L_{eff} P$$

$$\gamma = \frac{\omega_c \bar{n}_2}{(A_{eff} c)} \tag{E.75}$$

$$L_{eff} = \frac{(1 - e^{-\alpha L})}{\alpha}$$

where

    $\omega_c$ is the lightwave carrier

    $L_{eff}$ is the effective transmission length

    $\alpha$ is the fiber attenuation factor which normally has a value of 0.17–0.2 dB/km in the 1,550 nm spectral window

The temporal variation of the nonlinear phase $\phi_{NL}$ results in the generation of new spectral components far apart from the lightwave carrier $\omega_c$, indicating the broadening of the signal spectrum. This spectral broadening $\delta\omega$ can be obtained from the time dependence of the nonlinear phase shift as follows:

$$\delta\omega = -\frac{\partial \phi_{NL}}{\partial T} = -\gamma \frac{\partial P}{\partial T} L_{eff} \tag{E.76}$$

Equation E.76 indicates that $\delta\omega$ is proportional to the time derivative of the average signal power $P$. Additionally, the generation of new spectral components occur mainly at the rising and falling edges of optical pulses, i.e., the amount of generated chirps is substantially larger for an increased steepness of the pulse edges.

The wave propagation equation can be represented as

$$\frac{\partial A(z,t)}{\partial z} + \frac{\alpha}{2} A(z,t) + \beta_1 \frac{\partial A(z,t)}{\partial t} + \frac{j}{2}\beta_2 \frac{\partial^2 A(z,t)}{\partial t^2} - \frac{1}{6}\beta_3 \frac{\partial^3 A(z,t)}{\partial t^3}$$

$$= -j\gamma |A(z,t)|^2 A(z,t) - \frac{1}{\omega_0} \frac{\delta}{\delta t}\left(|A|^2 A\right) - T_R A \frac{\delta\left(|A|^2\right)}{\delta t} \tag{E.77}$$

in which we have ignored the pure delay factor involving $\beta_1$. The last term in the RHS represents the Raman scattering effects.

## E.5.2 SPM AND MODULATION INSTABILITY

Nonlinear effects such as the Kerr effect where the refractive index of the fiber medium strongly depends on the intensity of the optical signal can severely impair the transmitted signals. In a non-soliton system the Kerr effect broadens the signal optical spectrum through self-phase modulation (SPM), this broadened spectrum is mediated by fiber dispersion and causes a performance degradation. In addition, four wave mixing (FWM) between the signal and the amplified spontaneous emission (ASE) noise generated from the inline optical amplifiers, has been reported to cause performance degradations, in transmission systems employing in-line optical amplifiers. This later effect is commonly referred to as modulation instability. Thus the instability is resulted from the conversion of the phase noises to intensity noises. The gain spectrum of the modulation instability is shown in Figure E.22 [22].

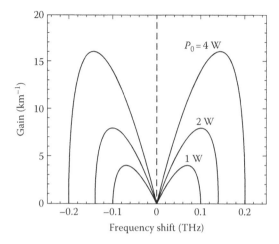

**FIGURE E.22**   Spectrum of the optical gain due to modulation instability at three different average power level in an optical fiber with $\beta_2 = 20$ ps²/km and $\gamma = 2$ W/km.

### E.5.3   SPM AND INTRA-CHANNEL NONLINEAR EFFECTS

Under considerations only the SPM of all the nonlinear effects on the optical signals transmitting through dispersive transmission link we can drop the cross coupling terms but $\Delta\Omega$, the nonlinear effect is thus contributed by additional intra- channel effects with $\omega_1, \omega_2$ take the values with the spectra of the optical signal and not cross over the spectra of other adjacent channels. With the substituting of the fundamental order transfer function we arrive at,

$$H_3(\omega_1, \omega_2, \omega)_{\text{SPM,inter}} = -\frac{j\gamma_L}{4\pi^2}(e^{-j\omega^2 N_s \beta_2 L_s/2})(1 - e^{-(\alpha + j\beta_2 \Delta\Omega)L_s})$$

$$\times x\left(L_s\sqrt{\alpha^2 + \beta_2^2\Delta\Omega^2}\, e^{j\tan^{-1}(\alpha/(\beta_2\Delta\Omega))}\right)\sum_{k=0}^{N_s-1} e^{-jk\beta_2 L_s\Delta\Omega} \quad \text{(E.78)}$$

The nonlinear distortion noises contributed to the signals when operating under the two regimes of large and negligible dispersion are given in Ref. [13], depending on the dispersion factor of the fiber spans. The nonlinear transfer function $H_3$ indicates the power penalty due to nonlinear distortion, and can be approximated as [23]

$$H_3^i(\omega_1, \omega_2, \omega) \approx \begin{cases} j\gamma_L e^{-[\alpha/2 - j\beta_2(\omega_1 - \omega_2)(\omega - \omega_2)]} \\[2mm] \times\left[L_s^{\text{eff}} - j\beta_2(\omega_1 - \omega_2)(\omega - \omega_2)\right]\left(\dfrac{L_s - L_s^{\text{eff}}}{\alpha} - L_s L_s^{\text{eff}}\right) \end{cases} \quad \text{(E.79)}$$

Thus if the ASE noise of the inline optical amplifier is weak compared with signal power then we can obtain the nonlinear distortion noises for *highly dispersive* fiber spans (e.g. G.652 standard single mode fiber SSMF) as

$$K_{N,\beta}(\omega_0) = N_s \left[ Q(\omega_0) + 2\left(\frac{\gamma_L}{2\pi}\right)^2 \frac{\Omega^2}{\alpha^2} \frac{P^3}{\Delta\omega_c^3} \partial\left(\frac{\omega_0}{\Delta\omega_c}, \frac{\beta\Omega^2}{\alpha}\right) \right]$$ (E.80)

and for *mildly dispersive* fiber spans (e.g., G.655 fiber spans):

$$K_{N,\beta \ll 20}(\omega_0) \approx N_s \left[ Q(\omega_0) + 2\left(\frac{\gamma_L}{2\pi}\right)^2 \frac{\Omega^2}{\alpha^2} \frac{P^3}{\Delta\omega_c^3} \partial\left(\frac{\omega_0}{\Delta\omega_c}, 0\right) \right]$$ (E.81)

with

$$\partial(x,\xi) = \int_{-\infty}^{\infty} dx_1 \int_{-\infty}^{\infty} dx_2$$

$$\times \left( \frac{\left(1 + e^{-\alpha L}\right) - 2e^{-\alpha L} \cos\left[\alpha L\xi(x - x_1)(x_1 - x_2)\right]}{1 + \xi^2 (x - x_1)^2 (x_1 - x_2)^2} \times \eta(x_1)\eta(x_2)\eta(x - x_1 + x_2) \right)$$

(E.82)

and

$$\eta(x) = \begin{cases} 1 & \text{for } x = \left[\frac{-1}{2}, \frac{1}{2}\right] \\ 0 & \text{else\_where} \end{cases}$$ (E.83)

The nonlinear power penalty thus consists of the linear optical amplifier noises; the second is the SPM noises from the input signal and nonlinear interference between the input and the optical amplifier noises which may be ignored when the ASE is weak. Equations E.80 and E.81 show the variation of the penalty hence channel capacity of dispersive fibers, of transmission systems operating under the influence of nonlinear effects with optically amplified multi-span transmission line whose fiber dispersion parameter varying from 0 to −20 ps²/km. Under the scenario of non-dispersive fiber the spectral efficiency has been limited to about 3–4 and 9–6 bps/Hz with 4 and 32 spans, respectively, for a dispersion factor of −20 ps²/km with 100 DWDM channels of 50 GHz spacing between the channels with the optical spectral noise density of 1 μW/GHz. The fiber length of each span is 80 km.

By the definition, the nonlinear threshold is determined at 1 dB penalty deviation level from the linear OSNR, the contribution of the nonlinear noise term, from Equation E.80, we can obtain the maximal launched power at which there is an onset of the degradation of the channel capacity as

$$\max\_P = \sqrt{0.1 \frac{\omega_c^3}{2N_s \left(\gamma_L/2\pi\right)^2 \left(\Omega^2/\alpha^2\right)}} \qquad (E.84)$$

An example of the estimation of the maximum level of power per channel to be launched to the fiber before reaching the nonlinear threshold 1 dB penalty level: for an overall 100 channels of 150 GHz spacing $\Omega_T \approx 200$nm then $P_{th} \approx 58\mu W/GHz$. Thus for 25 GHz bandwidth, we have the threshold power level at $P_{th\beta low} = 0.15$mW per channel. For highly dispersive and 8 wavelength channels we have $P_{th\beta high} \rightarrow 7-10P_{th\beta low}$ or the threshold level may reach 1.5 mW/channel. The estimations given here, as an example, are consistent with the analytical expression obtained in Equation E.84. Thus this shows clearly that (1) Dispersive multi-span long distance transmission under coherent ideal receiver would lead to better channel capacity than low dispersive transmission line. (2) If a combination of low and high dispersive fiber spans is used then we expect that, from our analytical Volterra approach, the penalty would reach the same level of threshold power so that a 1 dB penalty on the OSNR is reached. Note that this approach relies on the average level of optical power of the lightwave modulated sequence. This may not be easy to estimate if simulation model is employed. (3) However under simulation, the estimation of average power cannot be done without costing extremely high time, thus commonly the instantaneous power is estimated at the sampled time interval of a symbol. This sampled amplitude and hence the instantaneous power can be deduced. Thence the nonlinear phase is estimated and superimposed on the sampled complex envelope for further propagation along the fiber length. The sequence high-low dispersive spans would offer slightly better performance than low-high combination. This can be due to the fact that for low dispersive fiber the output optical pulse would be higher in amplitude which is to be launched into the high dispersive fibers, thus this would suffers higher nonlinear effects due to the fact that the instantaneous power launched into the fiber would be different even the average power would be the same for both cases.

The argument in step (3) can be further strengthened by representing a fiber span by the Volterra series Transfer Functions (VSTF) as shown in Figure E.23. Any swapping of the sequence of low and high dispersion fiber spans would offer the same power penalty due to nonlinear phase distortion except the accumulated noises contributed from the ASE noises of the in-line optical amplifiers of all spans. Thus we could see that the noise figure (NF) of both configurations can be approximated the same. This is contrasting to the simulated results reported in [24]. We believe that the difference in the power penalty in different order of arrangement of low and high dispersion fiber spans reported in [15] is due numerical error as possibly the split step Fourier method (SSFM) was employed and the instantaneous amplitude of the complex envelope was commonly used. This does not indicate the total average signal power of all channels. Therefore we can conclude that the simulated nonlinear threshold power level would be suffering additional artificial OSNR penalty due to the instantaneous power of the sampled complex amplitude of the propagating amplitude.

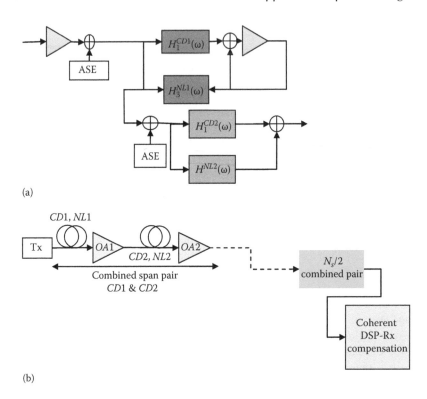

(a)

(b)

**FIGURE E.23**  (a) System of concatenation of fiber spans consisting of pair of different CDs and NL, model for simulation. (b) Optically amplified $N_s$-span fiber link without DCF an DSP-based optical receiver with compensation of nonlinear effects by back propagation or Volterra series transfer function conducted by digital signal processing. This model represents realtime processing in practical optical transmission systems.

Reference [13] reported the variation of the channel capacity against the input power/channel with dispersion as a parameter $-2 \rightarrow -20\text{ps}^2/\text{km}$ with a noise power spectral density of 10 μW/GHz over 4 spans and that for 4 and 32 spans of dispersive fibers of 0 and $-20\text{ps}^2/\text{km}$ with a channel spacing of 50 GHz and 100 Channels, the noise spectral density of 10 μW/GHz. The deviation of the capacity is observed at the onset of the power per channel of 0.1, 2 and 5 mW. Further observations can be made here. The noises responses indicate that the nonlinear frequency transfer function of a highly dispersive fiber link is related directly with the fundamental linear transfer function of the fiber link. When the transmission is highly dispersive the linear transfer function acts as a low pass filter and thus all the energy concentrates in the passband of this filter which may be lower than that of the signal at the transmitting end. This may thus lower the nonlinear effects as given in Equation E.78. While for lower dispersive fiber this transfer function would represent a low pass filter with 3 dB roll off frequency much higher than that of a dispersive fiber. For example the G.655 would have a dispersion-factor of about three times lower than that of the G.652 fiber. This wideband low pass filter will allow the nonlinear effects of intra-channels and inter-channel interactions. The dispersive accumulation term $\sum_{k=0}^{N_s-1} e^{-jk\beta_2 L_s \Delta\Omega}$ dominates when the number of spans is high.

In the simulation results given in Ref. [25], the Volterra series transfer functions were applied for dispersive fiber spans. It is expected from Equation E.78, that the arrangement of alternating position between G.655 and G.652 would not exert any penalty. The simulation reported in Ref. [21] indicates 1.5 dB difference at $10 \times 2$ spans (SSMF + TWC) and no difference at $20 \times 2$ spans. The contribution by the ASE noises of the optical amplifiers at the end of each span would influence the phase noises and hence the effects on the error vector magnitude (EVM) of the sampled signal detected constellation.

From the transfer functions including both linear and nonlinear kernels of the dispersive fibers we could see that if the noises are the same then the nonlinear effects would not be different regardless whether high or low dispersive fiber spans placed at the front or back. However if the nonlinear noises are accounted for and especially the intra-channel effects we could see that if less dispersive fibers are placed in the front then higher noises are expected and thus lower nonlinear threshold (at which 1 dB penalty is reached on the OSNR). This is in opposite to the simulation results presented in Ref. [16]. However these accumulated noises are much smaller than the average signal power. Under simulation, depending on the numerical approach to solve the NLSE, the estimation of signal power at the sampled instant is normally obtained from the sampled amplitude at this instant and thus different with the average launched power into the fiber span. This creates discrepancies in the order of the high or low dispersion fiber spans. Thus there are possibilities that the peak above average amplitude of the very dispersive pulse sequence at some instants along the propagation path at which there is a superposition of several pulses. This amplitude may reach a level much higher than the nonlinear threshold and thus create different distortion penalty due to nonlinear effects.

Volterra transfer function (VSTF) offers better accuracy and covers a number of SPM and parametric scattering but suffers costs of computing resources due to two-dimensional FFT for the SPM and XPM. This model should be employed when such extra nonlinear phase noises are required such as in the case of superchannel transmission.

### E.5.4   NONLINEAR PHASE NOISES IN CASCADED MULTI-SPAN OPTICAL LINK

Gordon and Mollenauer [26] showed that when optical amplifiers are used to compensate for fiber loss, the interaction of amplifier noise and the Kerr effect causes phase noise, even in systems using constant-intensity modulation. This nonlinear phase noise, often called the Gordon–Mollenauer effect or, more precisely, SPM-induced nonlinear phase noise, or simply nonlinear phase noise (NLPN), corrupts the received phase and limits transmission distance in systems using $M$-ary QAM. The NLPN in turn would create random variation in the intensity, thus a transfer or conversion of the NLPN into intensity noise, or modulation intensity.

Under the cascade of optically amplified spans to form a multi-span long haul link without using dispersion compensating fiber (DCF) has emerged as the most modern optical link structure with coherent detection and digital signal processing at the receivers. Ho and Kahn [27] have studied and derived the variances and co-variances of DWDM optical transmission systems. For an electric field $E_0$ of the optical waves launched at the input of the first span, the field at the input of the $k$th span would be the launched field superimposed by the noises accumulated over $k$ spans as $E_k = E_0 + n_1 + n_2 + \cdots + n_{k-1}$, then the variance $\sigma_{\phi NL}$ of the nonlinear phase shift is given as

$$\sigma_{NLPN}(\alpha_{NL}) = (\gamma L_{eff})^2 \left[ \sigma_{NL}^2 (N-1) + (\alpha_{NL} - 1)^2 f\left(N\sigma_1^2\right) - 2(\alpha_{NL} - 1) \sum_{k=1}^{N} f\left(N\sigma_1^2\right) \right]$$

(E.85)

where
  $\alpha$ is the scaling factor
  $f(N\sigma^2)$ is the expected value of the optical electric field between two consecutive spans
  $\sigma$ is the variance of the field under superposition with the noises
  $N$ is the total number of optically amplified spans > the optimal factor can be found by differentiating Equation E.85 with respect to this factor to give:

$$\alpha_{NL} \approx -\gamma L_{eff} \frac{N+1}{2}$$

(E.86)

At high OSNR $\gg 1$, this variance can be found to be

$$\sigma_{NLPN}^2 \approx \frac{4}{3} N^3 \left(\gamma L_{eff} \sigma |E_0|^2\right)^2$$

(E.87)

With $\sigma \equiv \sigma_{|E_0 + n_1 + n_2 + \cdots + n_k|}$ as the variance of the field superimposed by noises after $k$th span of $k$ cascade spans. The expected value of the nonlinear phase shift can be approximated as

$$\langle \phi_{NL} \rangle \simeq N \gamma L_{eff} |E_0|^2$$

(E.88)

Then the NLPN variance of $N$ cascaded spans can then be given as

$$\sigma_{NLPN}^2 \approx \frac{4}{3} N^3 \left(\gamma L_{eff} \sigma |E_0|^2\right) = \frac{4}{3} N \frac{\langle \phi_{NL} \rangle^2}{OSNR_L^2}$$

(E.89)

where $OSNR_L$ is the optical signal to noise ration in linear scale and the mean phase noise is given by Equation E.88. Thus we can see that the nonlinear phase rotation due to SPM in $N$-cascade-span link is the total phase rotation accumulated over the spans.

The variances of the residual NLPN is also given as

$$\sigma_{NLPN,res}^2 \approx \frac{1}{6} \frac{\langle \phi_{NL} \rangle^2}{OSNR_L^2} \tag{E.90}$$

The NLPN power variance is also proportional to the square of the accumulated phase rotation. This allows a compensation algorithm for nonlinear impairments by rotating the phase of the digital sampled of the in phase and quadrature phase components at the end of each span. This is indeed a linear operation based on the derived and observed rotation of the constellation due to SPM. These linear phase rotation simplifies the numerical and hence the computing resources of the DSP. The phase to intensity conversion, the instability problem will create some degradation of the OSNR due to this increase of the noise intensity over the 0.4 nm band commonly measured of the noises in practice. Thus we expect a logarithmic reduction of the OSNR with respect to the number of cascaded optically amplified fiber spans to the SPM-induced and modulation instability.

## E.6  NUMERICAL SOLUTION: SPLIT-STEP FOURIER METHOD

In practice with the extremely high speed operation of the transmission systems, it is very costly to simulate by experiment, especially when the fiber transmission line is very long, e.g., few thousands of kilometers. Then it is preferred to conduct computer simulations to guide the experimental set up. In such simulation the propagation of modulated lightwave channels play a crucial role so as to achieve transmission performance of the systems closed to practical ones. The main challenge in the simulation of the propagation of lightwaves channel employing the nonlinear Schrodinger equation (NLSE) which can be derived from Maxwell equations is whether the signal presented in the time domain be propagating though the fiber and its equivalent in the frequency domain when the nonlinearity is effective. The propagation techniques for such modulated channels are described in the subsection of this part.

### E.6.1  SYMMETRICAL SPLIT-STEP FOURIER METHOD

The evolution of slow varying complex envelopes $A(z, t)$ of optical pulses along a single-mode optical fiber is governed by Nonlinear Schrodinger Equation (NLSE):

$$\frac{\partial A(z,t)}{\partial z} + \frac{\alpha}{2} A(z,t) + \beta_1 \frac{\partial A(z,t)}{\partial t} + \frac{j}{2}\beta_2 \frac{\partial^2 A(z,t)}{\partial t^2} - \frac{1}{6}\beta_3 \frac{\partial^3 A(z,t)}{\partial t^3} = -j\gamma |A(z,t)|^2 A(z,t)$$

$$\tag{E.91}$$

where
   $z$ is the spatial longitudinal coordinate, $\alpha$ accounts for fiber attenuation
   $\beta_1$ indicates DGD
   $\beta_2$ and $\beta_3$ represent second and third order dispersion factors of fiber CD
   $\gamma$ is the nonlinear coefficient as also defined above

In a single channel transmission, Equation E.91 includes the following effects: fiber attenuation, fiber CD and PMD, dispersion slope and SPM nonlinearity. Fluctuation of optical intensity caused by Gordon–Mollenauer effect is also included in this equation. We can observe that the term involve $\beta_2$ and $\beta_3$ relates to the phase evolution of optical carriers under the pulse envelop. Respectively the term $\beta_1$ relates to the delay of the pulse when propagating through a length of the fiber. So if the observer is situation on the top of the pulse envelop then this delay term can be eliminated.

The solution of NLSE and hence the modeling of pulse propagation along a single mode optical fiber is solved numerically by using Split Step Fourier Method (SSFM) so as to facilitate the solution of nonlinear equation when nonlinearity is involved. In SSFM, fiber length is divided into a large number of small segments $\delta z$. In practice, fiber dispersion and nonlinearity are mutually interactive at any distance along the fiber. However, these mutual effects are small within $\delta z$ and thus effects of fiber dispersion and fiber nonlinearity over $\delta z$ are assumed to be statistically independent with each other. As a result, SSFM can separately define two operators: (1) the linear operator that involves fiber attenuation and fiber dispersion effects and (2) the nonlinearity operator that takes into account fiber nonlinearities. These linear and nonlinear operators are formulated as follows:

$$\hat{D} = -\frac{j\beta_2}{2}\frac{\partial^2}{\partial T^2} + \frac{\beta_3}{6}\frac{\partial^3}{\partial T^3} - \frac{\alpha}{2} \tag{E.92}$$

$$\hat{N} = j\gamma |A|^2$$

where
$$j = \sqrt{-1}$$
$A$ replaces $A(z, t)$ for simpler notation
$T = t - z/v_g)$ is the reference time frame moving at the group velocity, meaning that the observer is situated on top of the pulse envelop

Equation E.92 can be rewritten in a shorter form, given by:

$$\frac{\partial A}{\partial z} = (\hat{D} + \hat{N})A \tag{E.93}$$

and the complex amplitudes of optical pulses propagating from $z$ to $z + \delta z$ are calculated using the following approximation

$$A(z + h, T) \approx \exp(h\hat{D})\exp(h\hat{N})A(z, T) \tag{E.94}$$

Equation E.94 is accurate to the second order of the step size $\delta z$. The accuracy of SSFM can be improved by including the effect of fiber nonlinearity in the middle of the segment rather than at the segment boundary. This modified SSFM is known as the symmetric SSFM as illustrated in Figure E.24.

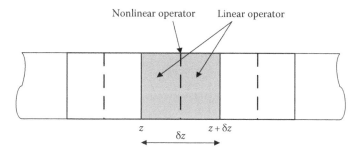

**FIGURE E.24** Schematic illustration of symmetric SSFM.

Equation E.94 can now be modified as:

$$A(z+\delta z,T) \approx \exp\left(\frac{\delta z}{2}\hat{D}\right)\exp\left(\int_{z}^{z+\delta z}\hat{N}(z')dz'\right)\exp\left(\frac{\delta z}{2}\hat{D}\right)A(z,T) \qquad \text{(E.95)}$$

This method is accurate to the third order of the step size $\delta z$. In symmetric SSFM, the optical pulse propagates along a fiber segment $\delta z$ in two stages. Firstly, the optical pulse propagates through the linear operator that has a step of $\delta z/2$ in which the fiber attenuation and dispersion effects are taken into account. A FFT is used here in the propagation step so that the output of this half-size step is in the frequency domain. Note that the carrier is removed here and the phase of the carrier is represented by the complex part, the phase evolution. Hence the term complex amplitude is coined.

Then, the fiber nonlinearity is superimposed to the frequency domain pulse spectrum at the middle of the segment. After that, the pulse propagates through the second half of the linear operator using the inverse FFT to recover the pulse envelop in the time domain.

The process continues repetitively over consecutive segments of size $\delta z$ until the end of the fiber length. It should be again noted that the linear operator is computed in the time domain while the nonlinear operator is calculated in the frequency domain.

### E.6.1.1 Modeling of Polarization Mode Dispersion

As described above, PMD is resulted from the delay difference between the propagation of each polarized mode of the linearly polarized modes $LP_{01}^{H}$ and $LP_{01}^{V}$ of the horizontal and vertical directions respectively, as illustrated in Figure E.3. The parameter DGD, differential group delay, determines the he first-order PMD which can be implemented by modeling the optical fiber as two separate paths representing the propagation of two polarization states (PSPs). The symmetrical SSFM can be implemented in each step on each polarized transmission path and then their outputs are superimposed to form the output optical field of the propagated signals. The transfer function to represent the first-order PMD is given by:

$$H(f) = H^{+}(f) + H^{-}(f) \qquad \text{(E.96)}$$

where

$$H^+(f) = \sqrt{k}\exp\left[j2\pi f\left(-\frac{\Delta\tau}{2}\right)\right] \tag{E.97}$$

and

$$H^-(f) = \sqrt{k}\exp\left[j2\pi f\left(-\frac{\Delta\tau}{2}\right)\right] \tag{E.98}$$

where
   $k$ is the power splitting ratio; $k=0.5$ when a 3 dB or 50:50 optical coupler/splitter
      is used
   $\Delta\tau$ is the instantaneous DGD value which the average value of a statistical distri-
      bution that follows a Maxwell distribution [28,29]

This randomness is due to the random variations of the core geometry, the fiber
stress and hence anisotropy due to the drawing process, variation of temperatures
etc. in installed fibers.

### E.6.1.2  Optimization of Symmetrical SSFM

#### E.6.1.2.1  Optimization of Computational Time

A huge amount of time can be spent for the symmetric SSFM via the uses of FFT and
IFFT operations, in particular when fiber nonlinear effects are involved. In practice,
when optical pulses propagate towards the end of a fiber span, the pulse intensity
has been greatly attenuated due to the fiber attenuation. As a result, fiber nonlinear
effects are getting negligible for the rest of that fiber span and hence, the transmission is operating in a linear domain in this range. In this research, a technique to
configure symmetric SSFM is proposed in order to reduce the computational time.
If the peak power of an optical pulse is lower than the nonlinear threshold of the
transmission fiber, for example around −4 dBm, symmetrical SSFM is switched to
a linear mode operation. This linear mode involves only fiber dispersions and fiber
attenuation and its low-pass equivalent transfer function for the optical fiber is:

$$H(\varpi) = \exp\left\{-j\left[\left(\frac{1}{2}\right)\beta_2\varpi^2 + \left(\frac{1}{6}\right)\beta_3\varpi^3\right]\right\} \tag{E.99}$$

If $\beta_3$ is not considered in this fiber transfer function, which is normally the case due
to its negligible effects on 40 Gbps and lower bit rate transmission systems, the above
transfer function has a parabolic phase profile [28,29].

#### E.6.1.2.2  Mitigation of Windowing Effect and Waveform Discontinuity

In symmetric SSFM, mathematical operations of FFT and IFFT play very significant
roles. However, due to a finite window length required for FFT and IFFT operations,

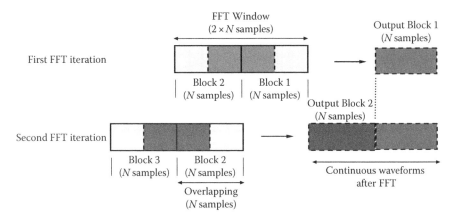

**FIGURE E.25** Proposed technique for mitigating windowing effect and waveform discontinuity caused by FFT/IFFT operations.

these operations normally introduce overshooting at two boundary regions of the FFT window, commonly known as the windowing effect of FFT. In addition, since the FFT operation is a block-based process, there exists the issue of waveform discontinuity, i.e., the right-most sample of the current output block does not start at the same position of the left-most sample of the previous output block. The windowing effect and the waveform discontinuity problems are resolved with the following technique, see also Figure E.25.

The actual window length for FFT/IFFT operations consists of two blocks of samples, hence $2N$ sample length. The output, however, is a truncated version with the length of one block ($N$ samples) and output samples are taken in the middle of the two input blocks. The next FFT window overlaps the previous one by one block of $N$ samples.

## REFERENCES

1. A. Safaai-Jazi and J. C. McKeeman, Synthesis of intensity patterns in few-mode optical fibers, *IEEE J. Lightwave Technol.*, 9(9), 1047, September 1991.
2. M. Salsi et al., Transmission at 2 × 100 Gb/s, over two modes of 40 km-long prototype few-mode fiber, using LCOS based mode multiplexer and demultiplexer, in *Optical Fiber Conference (OFC 2011)*, and *National Fiber Optic Engineers Conference (NFOEC)*, Los Angeles, CA, March 6, 2011, Postdeadline Session B (PDPB).
3. S. Randel, R. Ryf, A. Sierra, P. J. Winzer, A. H. Gnauck, C. A. Bolle, R.-J. Essiambre, D. W. Peckham, A. McCurdy, and R. Lingle, Jr., 6 × 56-Gb/s mode-division multiplexed transmission over 33-km few-mode fiber enabled by 6 × 6 MIMO equalization, *Opt. Exp.*, 19(17), 16697, August 2011.
4. C. D. Poole, J. M. Wiesenfeld, D. J. DiGiovanni, and A. M. Vengsarkar, Optical fiber-based dispersion compensation using higher order modes near cutoff, *IEEE J. Lightwave Technol.*, 12(10), 1746, October 1994.
5. J. P. Gordon and H. Kogelnik, PMD fundamentals: Polarization mode dispersion in optical fibers, *PNAS*, 97(9), 4541–4550, April 2000.

6. S. Ten and M. Edwards. An introduction to the fundamentals of PMD in fibers, White paper, Corning Inc., July 2006.

7. A. Galtarossa and L. Palmieri, Relationship between pulse broadening due to polarisation mode dispersion and differential group delay in long singlemode fiber, *Electron. Lett.*, 34(5), 492–493, March 1998.

8. J. M. Fini and H. A. Haus, Accumulation of polarization-mode dispersion in cascades of compensated optical fibers, *IEEE Photon. Technol. Lett.*, 13(2), 124–126, February 2001.

9. A. Carena, V. Curri, R. Gaudino, P. Poggiolini, and S. Benedetto, A time-domain optical transmission system simulation package accounting for nonlinear and polarization-related effects in fiber, *IEEE J. Select. Areas Commun.*, 15(4), 751–765, 1997.

10. S. A. Jacobs, J. J. Refi, and R. E. Fangmann, Statistical estimation of PMD coefficients for system design, *Electron. Lett.*, 33(7), 619–621, March 1997.

11. G. P. Agrawal, *Fiber Optic Communication Systems*, Academic Press, New York, 2002.

12. A. F. Elrefaie, R. E. Wagner, D. A. Atlas, and D. G. Daut, Chromatic dispersion limitations in coherent lightwave transmission systems, *IEEE J. Lightwave Technol.*, 6(6), 704–709, 1998.

13. J. Tang, The channel capacity of a multispan DWDM system employing dispersive nonlinear optical fibers and an ideal coherent optical receiver, *IEEE J. Lightwave Technol.*, 20(7), 1095–1101, July 2002.

14. B. Xu and M. Brandt-Pearce, Comparison of FWM- and XPM-induced crosstalk using the Volterra series transfer function method, *IEEE J. Lightwave Technol.*, 21(1), 40–54, 2003.

15. J. Tang, The Shannon channel capacity of dispersion-free nonlinear optical fiber transmission, *IEEE J. Lightwave Technol.*, 19(8), 1104–1109, 2001.

16. J. Tang, A comparison study of the shannon channel capacity of various nonlinear optical fibers, *IEEE J. Lightwave Technol.*, 24(5), 2070–2075, 2006.

17. A. Papoulis, *Systems and Transforms with Applications in Optics*, Krieger Pub. Co., Malabar, FL, June 1981.

18. L. N. Binh, *Digital Optical Communications*, CRC Press, Boca Raton, FL, 2009.

19. L. N. Binh and N. Nguyen, Generation of high-order multi-bound-solitons and propagation in optical fibers, *Opt. Commun.*, 282, 2394–2406, 2009.

20. K. V. Peddanarappagari and M. Brandt-Pearce, Volterra series transfer function of single-mode fibers, *J. Lightwave Technol.*, 15(12), 2232–2241, December 1997.

21. L. N. Binh, L. Liu, and L. C. Li, Volterra series transfer function in optical transmission and nonlinear compensation, Chapter 10, in L. N. Binh and D. V. Liet, *Nonlinear Optical Systems*, CRC Press, Taylor & Francis Group, Boca Raton, FL, January 2012.

22. G. P. Agrawal, *Nonlinear Fiber Optics*, 3rd edn., Academic Press, San Diego, CA, 2001.

23. K. V. Peddanarappagari and M. Brandt-Pearce, Volterra series approach for optimizing fiber-optic communications systems designs, *IEEE J. Lightwave Technol.*, 16(11), 2046–2055, 1998.

24. J. Pina, C. Xia, A. G. Strieger, and D. V. D. Borne, Nonlinear tolerance of polarization-multiplexed QPSK transmission over fiber links, in *ECOC 2011*, Geneva, Switzerland, 2011.

25. L. N. Binh, Linear and nonlinear transfer functions of single mode fiber for optical transmission systems, *JOSA A*, 26(7), 1564–1575, 2009.

26. J. P. Gordon and L. F. Mollenauer, Phase noise in photonic communications systems using linear amplifiers, *Opt. Lett.*, 15, 1351–1353, December 1990.

27. K.-P. Ho and J. M. Kahn, Electronic compensation technique to mitigate nonlinear phase noise, *IEEE J. Lightwave Technol.*, 22(3), 779, March 2004.
28. A. F. Elrefaie and R. E. Wagner, Chromatic dispersion limitations for FSK and DPSK systems with direct detection receivers, *IEEE Photon. Technol. Lett.*, 3(1), 71–73, 1991.
29. A. F. Elrefaie, R. E. Wagner, D. A. Atlas, and A. D. Daut, Chromatic dispersion limitation in coherent lightwave systems, *IEEE J. Lightwave Technol.*, 6(5), 704–710, 1988.

# Index